NUMERICAL MATHEMATICS AND SCIENTIFIC COMPUTATION

NUMERICAL MATHEMATICS AND SCIENTIFIC COMPUTATION

---

*P. Dierckx: *Curve and surface fittings with splines*
*H. Wilkinson: *The algebraic eigenvalue problem*
*I. Duff, A. Erisman, and J. Reid: *Direct methods for sparse matrices*
*M. J. Baines: *Moving finite elements*
*J. D. Pryce: *Numerical solution of Sturm–Liouville problems*
K. Burrage: *Parallel and sequential methods for ordinary differential equations*
Y. Censor and S. A. Zenios: *Parallel optimization: theory, algorithms and applications*
M. Ainsworth, J. Levesley, M. Marletta and W. Light: *Wavelets, multilevel methods and elliptic PDEs*
W. Freeden, T. Gervens, and M. Schreiner: *Constructive approximation on the sphere: with applications to geomathematics*

Monographs marked with an asterisk (*) appeared in the series 'Monographs in Numerical Analysis' which has been folded into, and is continued by, the current series.

Constructive Approximation on the Sphere

With Applications to Geomathematics

W. FREEDEN

Geomathematics Group, University of Kaiserslautern

T. GERVENS

University of Osnabrück (FH)

M. SCHREINER

Geomathematics Group, University of Kaiserslautern

CLARENDON PRESS · OXFORD

1998

Oxford University Press, Great Clarendon Street, Oxford OX2 6DP

Oxford New York
Athens Auckland Bangkok Bogota Bombay
Buenos Aires Calcutta Cape Town Dar es Salaam
Delhi Florence Hong Kong Istanbul Karachi
Kuala Lumpur Madras Madrid Melbourne
Mexico City Nairobi Paris Singapore
Taipei Tokyo Toronto Warsaw
and associated companies in
Berlin Ibadan

Oxford is a trade mark of Oxford University Press

Published in the United States by
Oxford University Press Inc., New York

A catalogue record for this book is available from the British Library

Library of Congress Cataloging in Publication Data
(Data available)

ISBN 0 19 853682 8

Typeset by the authors
Printed in Great Britain by
Biddles Ltd., Guildford and King's Lynn

PREFACE

This book provides the necessary foundation for students interested in any of the diverse areas of constructive approximation in sphere-oriented geomathematics. It is designed as a graduate-level textbook and assumes some basic undergraduate training in linear algebra and (functional) analysis, plus some basic knowledge of numerical analysis. But the primary objective of this book is to help geophysicists and geoengineers to understand future aspects of approximation by spherical harmonics and their modern counterparts (such as splines, wavelets, etc.). In the opinion of the authors, there has been a wide gap between the level of the textbooks available to geoscientists and the work published in research journals. No previous work has bridged this gap, no unified setup of the material is to be found. We hope that our treatise fulfils this objective. For those who wish to gain additional knowledge of the topics covered in the text, bibliographical notes are given at the end of each chapter. It should be mentioned that all our constructive approximation methods have been studied not only in theoretical nature and practical applicability, but also in the computational sense. Although all the algorithms in the book have been tested successfully on georelevant data sets, we have not attempted to put together a software package.

The book mainly derives from the first author's experiences of teaching several courses at the Universities of Aachen and Kaiserslautern, and our joint research during recent years. U. Windheuser, Geomathematics Group Kaiserslautern, has contributed by his PhD thesis to the continuous variant of spherical wavelet theory. The vectorial and tensorial theory of spherical harmonics and the techniques for converting vectorial or tensorial problems to scalar quantities are strongly influenced by the PhD theses of the second and third authors. The roots of this book are the pioneering work of G. Herglotz and C. Müller on the scalar theory of spherical harmonics. In learning this subject, we have benefited from conversations and correspondence with many colleagues, to whom we are very grateful. In particular, we would like to mention the members of the 'Laboratory of Technomathematics, Kaiserslautern', and the 'International Association of Geodesy' who made many valuable suggestions and comments.

Many people generously devoted considerable time to reading early versions of the manuscript and making many corrections. We particularly wish to thank S. Beth, J. Brudowsky, V. Michel, and F. Schneider. Our special thanks go to Mrs C. Korb for her excellent typing job.

Kaiserslautern W.F. and M.S.
Osnabrück T.G.

March 1997

CONTENTS

LIST OF SYMBOLS

In the following definitions the first digit refers to the the chapter in which the notation occurs and the second to the section within the chapter.

$\mathbb{N}_0$	set of non-negative integers:	1.1
$\mathbb{N}$	set of positive integers:	1.1
$\mathbb{Z}$	set of integers:	1.1
$\mathbb{R}$	set of real numbers:	1.1
$\mathbb{C}$	set of complex numbers:	1.1
$\mathbb{R}^3$	three-dimensional Euclidean space:	1.1
x, y, z	elements of $\mathbb{R}^3$:	1.1
$x \cdot y$	scalar product of vectors:	1.1
$x \wedge y$	vector product:	1.1
$x \otimes y$	dyadic product:	1.1
$\lvert x\rvert$	Euclidean norm of x:	1.1
$\varepsilon^i,\ i = 1, 2, 3$	canonical orthonormal basis in $\mathbb{R}^3$:	1.2
Ω	unit sphere in $\mathbb{R}^3$:	1.2
ξ, η, ζ	elements of Ω:	1.2
ν	unit normal field on Σ directed exterior to Σ:	1.2
$\mathbf{i}$	unit matrix:	1.2
$\mathbf{i}_{\tan}$	surface identity tensor:	1.2
$\mathbf{j}_{\tan}$	surface rotation tensor:	1.2
∇	gradient:	1.2
Δ	Laplace operator:	1.2
∇^*	surface gradient:	1.2
Δ^*	Beltrami operator:	1.2
L^*	surface curl gradient:	1.2
F, G	scalar-valued functions:	1.2
f, g	vector-valued funtions:	1.2
$\mathbf{f}, \mathbf{g}$	tensor-valued functions:	1.2
$\mathcal{C}^{(k)}, \mathcal{L}^p$	classes of scalar-valued functions:	1.2
$c^{(k)}, l^p$	classes of vector-valued functions:	1.2
$\mu(F, \delta)$	modulus of continuity:	1.2
$\mathcal{X}$	space $\mathcal{C}$ or $\mathcal{L}^p$:	1.2
$(\cdot, \cdot)_{\mathcal{X}}$	inner product of space $\mathcal{X}$:	1.2
$\lVert \cdot \rVert_{\mathcal{X}}$	norm of space $\mathcal{X}$:	1.2
$\overline{\mathcal{V}}$	completion of $\mathcal{V}$:	1.2

Symbol	Meaning	Section
$\delta_{i,j}$	Kronecker symbol:	2.1
$d(\mathcal{V})$	dimension of space $\mathcal{V}$:	2.1
H_n, K_n, L_n, U_n	homogeneous polynomial of order n:	2.1
$Hom_n, Hom_n(\mathbb{R}^3)$	space of homogeneous polynomials of order n:	2.1
$K_{\mathcal{V}}(\cdot,\cdot)$	reproducing kernel in $\mathcal{V}$:	2.1
$\mathbf{matr}_{\eta_1,\ldots,\eta_N}(U_1,\ldots,U_N)$	$\{U_1,\ldots,U_N\}$-matrix taken at $\eta_1,\ldots,\eta_N$:	2.1
$\mathcal{V}_1 \oplus \mathcal{V}_2$	direct sum of spaces:	2.2
$\mathcal{V}^{\perp}$	orthogonal complement:	2.2
$Pol_{0,\ldots,m}(\mathbb{R}^3)$	space of polynomials in $\mathbb{R}^3$ of degree $\leq m$:	2.2
P_n	Legendre polynomial:	2.3
μ_n	normalization factor:	2.3
$Harm_n, Harm_n(\mathbb{R}^3)$	space of homogeneous harmonic polynomials of order n:	3.1
Y_n	scalar spherical harmonic of degree n:	3.1
$Y_{n,j}, j = 1,\ldots,2n+1$	orthonormal system of scalar spherical harmonics of order n:	3.1
$\mathbf{t}$	orthogonal matrix:	3.3
$\mathbf{t}^T$	transpose of the matrix $\mathbf{t}$:	3.3
$\det \mathbf{t}$	determinant of the matrix $\mathbf{t}$:	3.3
$O(3)$	group of real orthogonal matrices:	3.3
$SO(3)$	group of real orthogonal matrices with $\det \mathbf{t} = 1$:	3.3
span(...)	linear span:	3.3
X_{2n+1}	fundamental system relative to $Harm_n(\Omega)$:	3.3
$\{x_1,\ldots,x_{d(\mathcal{V})}\}$	fundamental system relative to $\mathcal{V}$:	3.3
$Lip(\Omega)$	space of Lipschitz-continuous functions on Ω:	3.4
FT	Fourier transform:	3.4
LT	Legendre transform:	3.6
$G * F$	convolution of G and F:	3.6
T_h	translation operator:	3.6
$(\Delta^*)^\wedge(n) = -n(n+1)$	eigenvalues of the Beltrami operator:	3.5
$\partial_n, \partial_{0,\ldots,m}$	(iterated) Beltrami operators:	4.1
$Harm_{0,\ldots,m}$	space of homogeneous harmonic polynomials of orders $\leq m$:	4.3
$\{A_n\}$	sequence of real numbers:	5.1
$\mathcal{H}, \mathcal{H}(\{A_n\};\Omega)$	Sobolev space:	5.1

INTRODUCTION

During recent decades technological progress has completely changed the observational methods in all fields of geosciences and geoengineering: the trend is to achieve immediate results, thus reducing time and costs. Modern high speed computers and satellite based techniques are entering more and more disciplines such as cartography, climate investigation, geodesy, geology, geophysics, meteorology, navigation, oceanography, and many others. The increasing observational accuracy demands adequate mathematical tools; mathematics concerned with geoscientific problems, i.e. *geomathematics*, is becoming more and more indispensable. Geomathematics offers appropriate means of assimilating, assessing and reducing to comprehensible form the readily increasing flow of data from geochemical, geodetic, geological, geophysical and satellite sources and providing an objective basis for scientific interpretation, classification, testing of concepts and solution of problems. Undoubtly, the stage is set for geomathematics to play a major role in all earth sciences. Today's geoengineering believes in the usefulness and effectivity of modern mathematics, and there is a growing interest in geomathematical publications.

Unfortunately we are confronted with the historical paradox that mathematics, the oldest of the sciences, is, when applied to studies of the earth, one of the youngest geosciences. Thus, even for sphere-oriented ideas and concepts of constructive approximation, there is no topical comprehensive treatment of geomathematical methods and structures: either monographs on special functions of mathematical physics are available or constructive approximation theory is dealt within the framework of Euclidean spaces. For this reason the authors felt that there was a definite need for a textbook on spherical approximation theory in (geo)mathematics (including an adapted theory of spherical harmonics).

Our approach, of course, has to be selective, and greater weight has to be given to some topics than to others. Thus some basic principles are imposed on the whole text: whenever reasonable arguments can be found, properties of spherical harmonics will be derived within the concept of homogeneous harmonic polynomials. In this way, operations such as integration and differentiation can be carried out directly within the usual context of cartesian coordinates. In particular, differential equations can be attacked without transformation into spherical coordinates. Moreover, most of the difficulties originating from the topological structure of the sphere, which is such that there is no coordinate system free of singularities, are under control. This is also substantial for the vectorial and tensorial theory. It avoids sources of errors in models of geophysical processes on the sphere. We also make use of the rotational invariance on the sphere from which a double simplification of a mathematical problem is attained. On the one hand, the number of the independent (spatial) variables is reduced, resulting in a lower dimensional problem, and, on the other hand, the dependent variables for vectorial and tensorial problems can be expressed simply as scalar quantities. For example, elliptic partial differential equations are then reduced to a system of uncoupled differential equations. Finally, throughout the book, another strong principle is the axial symmetry (radial dependence) of all trial functions occurring as well as the spherical harmonics in our constructive approximation theory. By taking single trial functions of axial symmetry with local or global support and varying the axes, approximation techniques in pseudodifferential equations can be constructed so as to preserve the invariance of the corresponding pseudodifferential operator (with respect to the Riemannian geometry of the sphere). This is fundamental for many important problems, particularly in physical geodesy and geophysics.

Our purpose is to present a textbook that allows the reader to concentrate on special topics (such as spherical harmonics, splines, and wavelets). The schedule of topics is characterized by a twofold subdivision into three parts. On the one hand, the contents of the book are split into scalar, vectorial, and tensorial parts. On the other hand, major emphasis is laid on the theory of spherical harmonics, approximation by splines, and approximation by wavelets. Constructive approximation by splines and wavelets is mainly presented for the scalar case: only if extensions of scalar results to the vectorial or tensorial case are not straightforward are they discussed in more detail.

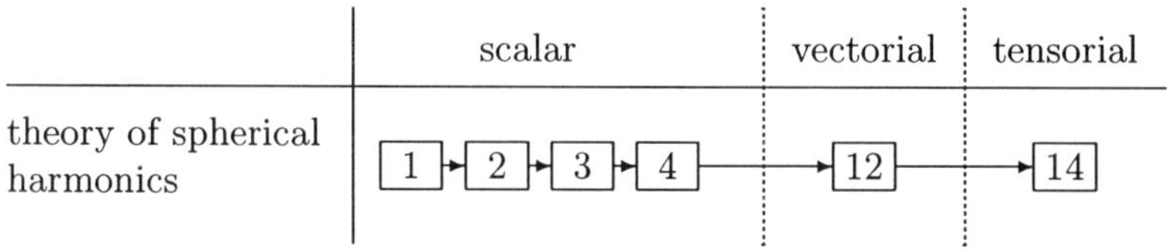

The ‘spherical harmonic path’ gives an overview of the theory of spherical harmonics. The investigations indicate several possibilities for introducing spherical harmonics. Even in the vectorial or tensorial case, most of the important properties can be outlined by discussing homogeneous harmonic polynomials in cartesian coordinates. Another approach is indicated by the representation of the special orthogonal group $O(3)$ within the space $\mathcal{L}^2(\Omega)$ of square-integrable functions on the unit sphere Ω. It turns out that spherical harmonics of given order constitute an irreducible invariant subspace of $\mathcal{L}^2(\Omega)$. A third approach is emphasized by establishing the spherical harmonics as eigenfunctions of invariant pseudodifferential operators (such as the Beltrami operator). Essential results of the spherical harmonic theory are the addition theorem and the Funk–Hecke formula. The decomposition theorems provide us with the appropriate equipment to convert vectorial or tensorial problems into scalar ones. The traditional approach to approximate a function on the sphere is to use the expansion properties of spherical harmonics. Indeed, 'Fourier analysis' on the sphere by means of spherical harmonics is an established subject in the case of geomathematics. Both 'spherical Fourier transform' and the 'Fourier series' (orthogonal expansion) in terms of spherical harmonics have significant physical interpretations due to the fact that spherical harmonics ideally localize in 'frequency space'. In addition, the computational aspects of spherical harmonic expansions are especially attractive, mainly because of the polynomial structure and the orthogonality properties of the series. In many instances, therefore, some incoming information of a function is given as a discrete table of numbers (i.e. Fourier coefficients) rather than its original form (remember the tables of the earth's gravitational or magnetic field). Although Fourier series techniques are very powerful, they also have some deficiencies. Working with spherical harmonics is in many respects computationally complex. In order to compensate for convergence problems there is a growing need for fast summation of spherical harmonic expansions. Spherical harmonics are globally supported; that is, localization in the 'space domain' does not occur. Thus, local changes of a function will affect the whole table of Fourier coefficients, which is a feature we would like to avoid in future geomathematical applications.

It is, however, only relatively recently that radial basis function techniques such as splines and wavelets, which play a fundamental role in modern numerical analysis in Euclidean spaces, have been applied to spherical problems. As a matter of fact, when one looks at the wide range of applications for which the globally supported spherical harmonics might be invoked, it is clear that radial basis functions on the sphere have much to offer. Of particular interest in constructive approximation are radial basis functions localizing sufficiently well in the 'space domain' as well as the 'frequency domain'. The class of radial basis functions of this type is large (e.g. Abel–Poisson kernel, Gauß–Weierstraß kernel, axisymmetric 'finite elements'). The uncertainty principle is a restriction which tells us that sharp localization in space and frequency is mutually exclusive.

The essential features of the 'spline path' are organized as follows. Spline functions are known to be canonical generalizations of 'spherical polynomials', i.e. spherical harmonics, having desirable characteristics as interpolating and best approximating functions. When we talk of interpo-

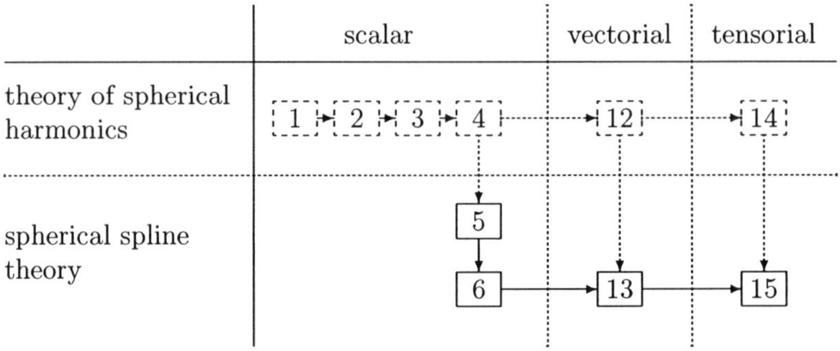

lation or smoothing by spherical splines we mean well-posed problems of minimizing a (semi-)norm of a suitable Sobolev space. According to the choice of the norm, (rotation-invariant) splines of different types are listed: for example, 'polynomial splines' related to (iterated) Beltrami derivatives, splines related to (invariant) pseudodifferential operators, 'multiquadric–splines', localizing (singular integral kernel) splines of Abel–Poisson and Gauß–Weierstraß type, and locally supported splines (axisymmetric 'finite-elements'). It is remarkable that our spherical interpolatory processes by splines open new perspectives for solving boundary value problems to elliptic equations (e.g. of potential theory and elasticity) even for georelevant non-spherical boundaries. This is of basic interest in the determination of the earth's external gravitational field or the deformation of the earth's internal elastic field using the actual surface of the earth.

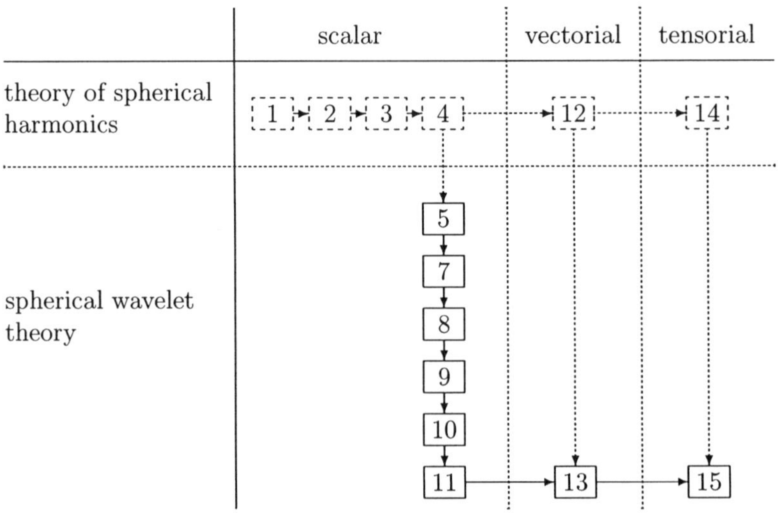

The 'wavelet path' is the youngest one. It essentially includes three variants: Gabor transform, Toeplitz transform, and wavelet transform. The basic tool is the theory of singular integrals on the sphere. Of great interest for numerical purposes are singular integrals corresponding to the Abel–Poisson kernel (which is closely related to Laplace's equation), the Gauß–Weierstraß kernel (which is intimately connected with the diffusion (heat) equation), and 'finite elements' (which are understood to be of piecewise polynomial nature, compact support and rotational symmetry around a prescribed axis). Analogous to spherical Fourier analysis there are two important mathematical entities in all wavelet variants: the 'integral transform' and the 'series expansion'. However, our series expansion is generally non-orthogonal. While the constituting elements of the Gabor transform consists of 'rotations' and 'modulations' of a 'window function', Toeplitz transform and wavelet transform are generated from the 'generating wavelet' by 'rotations' and 'dilations'. The Toeplitz transform displays information of a function at various levels (multilevel method) by virtue of Toeplitz summation of singular integrals with localizing axisymmetric kernels. Although the construction of the multilevel method is intended to accelerate the convergence, two drawbacks must be mentioned: the positioning of hierarchical point distributions on the sphere, which are comparably structured as the dyadic positions on the real line, and the need for an additional (somehow artificial) Toeplitz summation process. The latter drawback can be overcome by the spherical wavelet transform. There are at least two ways to introduce wavelets: one is through the continuous wavelet transform, another is through the discrete wavelet transform. Both methods involve representing general functions on the sphere in terms of simple, fixed building blocks at different scales and positions. In the continuous wavelet transform the dilation and rotation parameter vary continuously, while the discrete wavelet transform provides approximation in form of a sequence of nested subspaces of $\mathcal{L}^2(\Omega)$. In other words, the wavelet transform breaks up a complicated function into many simple pieces and studies each of the pieces separately with zooming-in capability by virtue of a multiresolution procedure. As in the well known theory of one variable, the coefficients of the wavelet expansion of any square-integrable function are precisely the values of the integral wavelet transform, evaluated at (binary) scale levels. Since there are efficient real-time algorithms for obtaining the coefficient sequences (based on a pyramid scheme) and existing spherical harmonic models may be combined in canonical manner with wavelet approximation, we are convinced that the list of future spherical wavelet applications in geomathematics will be long.

1

NOTATION

For the convenience of the reader we provide in this introductory chapter a summary of mathematical notation that will be used throughout the scalar part of the book. We begin with the customary set-theoretical preliminaries and some basic settings in Euclidean spaces.

1.1 Basic Settings

The symbols $\cap, \cup$, and $\subset$ denote the intersection, union, and inclusion of sets, respectively; $\emptyset$ is the empty set, and $x \in X$ reads: x belongs to X. If X is a set, then $\{x \in X \mid P(x)\}$ is the subset of X consisting of all those elements which have the property P. The cartesian product $A \times B$ of the sets A and B consists of all pairs (a, b) such that $a \in A$ and $b \in B$. A finite set may be described by enumerating its elements.

If X and Y are sets, then $F : X \to Y$ means F is a 'map (function) from X to Y' and $X \ni x \mapsto F(x) \in Y$ means '$F(x) \in Y$ is the value of F at $x \in X$'. The composition of maps $F : X \to Y$ and $G : Y \to Z$ is the map $G \circ F : X \to Z, x \mapsto (G \circ F)(x) = G(F(x))$. For every set X, there is the identity map $I_X : X \to X$ such that $I_X(x) = x$ for every $x \in X$. One often writes I instead of I_X whenever this does not lead to misunderstandings.

The letters $\mathbb{N}$, $\mathbb{N}_0$, $\mathbb{Z}$, $\mathbb{R}$, and $\mathbb{C}$ denote the set of positive integers, non-negative integers, integers, real numbers, and complex numbers, respectively. We write $x, y,$... to represent the elements of the three-dimensional (real) Euclidean space $\mathbb{R}^3$. In components we have

$$x = (x_1, x_2, x_3)^T, \qquad y = (y_1, y_2, y_3)^T.$$

The inner product, norm, vector product, and tensor product in $\mathbb{R}^3$ are defined, as usual, by

$$\begin{aligned} x \cdot y &= x^T y &&= x_1 y_1 + x_2 y_2 + x_3 y_3 , \\ x^2 &= x \cdot x &&= x_1^2 + x_2^2 + x_3^2 , \\ |x| &= \sqrt{x^2} &&= \sqrt{x_1^2 + x_2^2 + x_3^2} , \end{aligned}$$

$$x \wedge y \;=\; (x_2y_3 - x_3y_2,\ x_3y_1 - x_1y_3,\ x_1y_2 - x_2y_1)^T ,$$

and

$$x \otimes y \;=\; \begin{pmatrix} x_1y_1 & x_2y_1 & x_3y_1 \\ x_1y_2 & x_2y_2 & x_3y_2 \\ x_1y_3 & x_2y_3 & x_3y_3 \end{pmatrix} .$$

Let $\alpha = (\alpha_1, \alpha_2, \alpha_3)^T$ be a triple of non-negative integers $\alpha_1, \alpha_2, \alpha_3$. We set $\alpha! = \alpha_1!\ \alpha_2!\ \alpha_3!$ and $[\alpha] = \alpha_1 + \alpha_2 + \alpha_3$. We say $\alpha = (\alpha_1, \alpha_2, \alpha_3)^T$ is a multiindex of degree n if $[\alpha] = n$. As usual we let

$$x^\alpha \;= x_1^{\alpha_1} x_2^{\alpha_2} x_3^{\alpha_3} \tag{1.1.1}$$

and

$$(\nabla_x)^\alpha \;= \left(\frac{\partial}{\partial x_1}\right)^{\alpha_1} \left(\frac{\partial}{\partial x_2}\right)^{\alpha_2} \left(\frac{\partial}{\partial x_3}\right)^{\alpha_3} \;=\; \frac{\partial^{[\alpha]}}{(\partial x_1)^{\alpha_1} (\partial x_2)^{\alpha_2} (\partial x_3)^{\alpha_3}} . \tag{1.1.2}$$

The binomial theorem gives

$$(x \cdot y)^n = (x_1y_1 + x_2y_2 + x_3y_3)^n \;=\; \sum_{[\alpha]=n} \frac{n!}{\alpha!}\, x^\alpha y^\alpha . \tag{1.1.3}$$

Furthermore, we have

$$(\nabla_x)^\alpha x^\beta = \begin{cases} 0 & \text{for } \alpha \neq \beta \text{ and } [\alpha] = [\beta] \\ \alpha! & \text{for } \alpha = \beta \end{cases} . \tag{1.1.4}$$

1.2 Spherical Nomenclature

Each $x \in \mathbb{R}^3$, $x = (x_1, x_2, x_3)^T$, $|x| \neq 0$, admits a representation of the form $x = r\xi$, $r = |x|$, $\xi = (\xi_1, \xi_2, \xi_3)^T$, where $\xi \in \mathbb{R}^3$, $|\xi| = 1$, is the uniquely determined directional (unit) vector of x. The unit sphere in $\mathbb{R}^3$ will be denoted by Ω:

$$\Omega = \{\xi \in \mathbb{R}^3 \mid \mid \xi \mid = 1\}. \tag{1.2.1}$$

We set Ω_{int} for the 'inner space' of Ω, while Ω_{ext} denotes the 'outer space' of Ω. More explicitly,

$$\Omega_{\text{int}} = \{x \in \mathbb{R}^3 \mid \mid x \mid < 1\}, \qquad \Omega_{\text{ext}} = \{x \in \mathbb{R}^3 \mid \mid x \mid > 1\}. \tag{1.2.2}$$

It is well known that the total surface $\|\Omega\|$ of Ω is equal to 4π: $\|\Omega\| = \int_\Omega d\omega = 4\pi$ ($d\omega$ is the surface element).

Let $\varepsilon^1, \varepsilon^2, \varepsilon^3$ form the (canonical) orthonormal basis in $\mathbb{R}^3$:

$$\varepsilon^1 = \begin{pmatrix} 1 \\ 0 \\ 0 \end{pmatrix}, \qquad \varepsilon^2 = \begin{pmatrix} 0 \\ 1 \\ 0 \end{pmatrix}, \qquad \varepsilon^3 = \begin{pmatrix} 0 \\ 0 \\ 1 \end{pmatrix}. \tag{1.2.3}$$

We may represent the points $\xi \in \Omega$ as follows:

$$\begin{aligned} \xi &= t\varepsilon^3 + \sqrt{1-t^2}(\cos\varphi\varepsilon^1 + \sin\varphi\varepsilon^2), \\ & \quad -1 \le t \le 1 \ , \ 0 \le \varphi < 2\pi \ , \ t = \cos\vartheta \ , \end{aligned} \tag{1.2.4}$$

(ϑ: latitude, φ: longitude, t: polar distance):

$$\xi = (\sin\vartheta \ \cos\varphi, \ \sin\vartheta \ \sin\varphi, \ \cos\vartheta)^T \ . \tag{1.2.5}$$

By a scalar, vector, or tensor function (synonymously: field) on Ω we mean a map that assigns to each point $\xi \in \Omega$, respectively, a scalar, vector, or tensor. Unless otherwise specified, all fields are assumed to be real-valued. It will be of advantage to use the following general scheme of notations:

capital letters F, G	: scalar functions,
lower-case letters f, g	: vector fields,
boldface lower-case letters $\mathbf{f}, \mathbf{g}$	: tensor fields of second rank,
boldface capital letters $\mathbf{F}, \mathbf{G}$	: tensor fields of fourth rank.

The class of (scalar) functions $F : \Omega \to \mathbb{R}$ which are measurable and for which

$$\|F\|_{\mathcal{L}^p(\Omega)} = \left(\int_\Omega | \ F(\xi) \ |^p \ d\omega(\xi) \right)^{1/p} < \infty, \quad 1 \le p < \infty, \tag{1.2.6}$$

is known as $\mathcal{L}^p(\Omega)$. Clearly, $\mathcal{L}^p(\Omega) \subset \mathcal{L}^q(\Omega)$ for $1 \le q \le p$. A function $F : \Omega \to \mathbb{R}$ possessing k continuous derivatives on the unit sphere Ω is said to be of class $\mathcal{C}^{(k)}$ $(0 \le k \le \infty)$. $\mathcal{C}(\Omega)$ $(= \mathcal{C}^{(0)}(\Omega))$ is the class of continuous scalar valued functions on Ω. $\mathcal{C}(\Omega)$ is a complete normed space endowed with

$$\|F\|_{\mathcal{C}(\Omega)} = \sup_{\xi \in \Omega} | \ F(\xi) \ | \ . \tag{1.2.7}$$

By $\mu(F;\delta)$ we denote the modulus of continuity of the function $F \in \mathcal{C}(\Omega)$

$$\mu(F;\delta) = \max_{\xi,\zeta \in \Omega; \ 1-\xi\cdot\zeta \le \delta} | \ F(\xi) - F(\zeta) \ |, \quad 0 < \delta < 2. \tag{1.2.8}$$

A function $F : \Omega \to \mathbb{R}$ is said to be *Lipschitz-continuous* if there exists a (Lipschitz) constant $C_F > 0$ such that the inequality

$$|F(\xi) - F(\eta)| \le C_F \ |\xi - \eta| = \sqrt{2} \ C_F \ \sqrt{1 - \xi \cdot \eta} \tag{1.2.9}$$

holds for all $(\xi, \eta) \in \Omega \times \Omega$. The class of all Lipschitz-continuous functions on Ω is denoted by $Lip(\Omega)$. Clearly, $\mathcal{C}^{(1)}(\Omega) \subset Lip(\Omega)$.

In connection with $(\cdot,\cdot)_{\mathcal{L}^2(\Omega)}$, $\mathcal{C}(\Omega)$ is a pre-Hilbert space. For each $F \in \mathcal{C}(\Omega)$ we have the *norm estimate*

$$\|F\|_{\mathcal{L}^2(\Omega)} \leq \sqrt{4\pi}\|F\|_{\mathcal{C}(\Omega)}. \tag{1.2.10}$$

$\mathcal{L}^2(\Omega)$ is a Hilbert space with respect to the inner product $(\cdot,\cdot)_{\mathcal{L}^2(\Omega)}$. $\mathcal{L}^2(\Omega)$ is the completion of $\mathcal{C}(\Omega)$ with respect to the norm $\|\cdot\|_{\mathcal{L}^2(\Omega)}$, i.e.

$$\mathcal{L}^2(\Omega) = \overline{\mathcal{C}(\Omega)}^{\|\cdot\|_{\mathcal{L}^2(\Omega)}}.$$

We let $\mathcal{X}(\Omega)$ stand either for the space $\mathcal{C}(\Omega)$ or $\mathcal{L}^p(\Omega)$ (with corresponding $\|\cdot\|_{\mathcal{X}(\Omega)}$). In what follows we are mainly interested, however, in results for the Banach space $(\mathcal{C}(\Omega), \|\cdot\|_{\mathcal{C}(\Omega)})$ and the Hilbert space $(\mathcal{L}^2(\Omega), \|\cdot\|_{\mathcal{L}^2(\Omega)})$, respectively.

Any function of the form $G_\xi : \Omega \to \mathbb{R}, \eta \mapsto G_\xi(\eta) = G(\xi\cdot\eta)$, $\eta \in \Omega$, is called a *ξ-zonal function* on Ω. Zonal functions are constant on the sets $M(\xi;h) = \{\eta \in \Omega | \xi\cdot\eta = h\}$, $h \in [-1,+1]$. The set of all ξ-zonal functions is isomorphic to the set of functions $G : [-1,+1] \to \mathbb{R}$. This gives rise to interpret $\mathcal{C}[-1,+1]$ and $\mathcal{L}^p[-1,+1]$ with norms defined correspondingly as subspaces of $\mathcal{C}(\Omega)$ and $\mathcal{L}^p(\Omega)$. We let $\mathcal{X}[-1,+1]$ stand either for the space $\mathcal{C}[-1,+1]$ or $\mathcal{L}^p[-1,+1]$ (with corresponding norm $\|\cdot\|_{\mathcal{X}[-1,+1]}$).

In particular, we have

$$\|G\|_{\mathcal{X}[-1,+1]} = \|G(\varepsilon^3\cdot)\|_{\mathcal{X}(\Omega)},$$

so that

$$\begin{aligned} \|G\|_{\mathcal{L}^p[-1,+1]} &= \|G(\varepsilon^3\cdot)\|_{\mathcal{L}^p(\Omega)} \\ &= \left(\int_\Omega |G(\eta\cdot\varepsilon^3)|^p \, d\omega(\eta)\right)^{1/p} \\ &= \left(2\pi \int_{-1}^{1} |G(t)|^p \, dt\right)^{1/p}. \end{aligned} \tag{1.2.11}$$

Next we give some preliminaries for the study of vector fields defined on the unit sphere Ω. Using the canonical orthonormal basis $\{\varepsilon^1,\varepsilon^2,\varepsilon^3\}$ of $\mathbb{R}^3$ we may write any vector field $f : \Omega \to \mathbb{R}^3$ in the form

$$f(\xi) = \sum_{i=1}^{3} F_i(\xi)\varepsilon^i, \quad \xi \in \Omega, \tag{1.2.12}$$

where the component functions F_i are given by $F_i(\xi) = f(\xi)\cdot\varepsilon^i, \xi \in \Omega$.

$l^2(\Omega)$ denotes the space consisting of all square-integrable vector fields on Ω. In connection with the inner product

$$(f,g)_{l^2(\Omega)} = \int_\Omega f(\xi)\cdot g(\xi)\,d\omega(\xi), \quad f,g\in l^2(\Omega), \tag{1.2.13}$$

$l^2(\Omega)$ is a Hilbert space. We say a vector field f is p times continuously differentiable at a point $\xi\in\Omega$ if its component functions are. The space $c^{(p)}(\Omega), 0\le p\le\infty$, consists of all p times continuously differentiable vector fields on Ω. For brevity, we write $c(\Omega)$ instead of $c^{(0)}(\Omega)$. The space $c(\Omega)$ is complete with respect to the norm

$$\|f\|_{c(\Omega)} = \sup_{\xi\in\Omega}|f(\xi)|, \quad f\in c(\Omega). \tag{1.2.14}$$

Furthermore $\overline{c(\Omega)} = l^2(\Omega)$, where the completion is understood with respect to the $l^2(\Omega)$ topology. Analogous to (1.2.10) we have for all $f\in c(\Omega)$ the *norm estimate*

$$\|f\|_{l^2(\Omega)} \le \sqrt{4\pi}\|f\|_{c(\Omega)}. \tag{1.2.15}$$

In order to separate vector fields into their tangential and normal parts we introduce the projection operators p_{nor} and p_{tan} by

$$p_{\text{nor}}f(\xi) = (f(\xi)\cdot\xi)\,\xi, \qquad \xi\in\Omega, \quad f\in c(\Omega), \tag{1.2.16}$$

$$p_{\text{tan}}f(\xi) = f(\xi)-p_{\text{nor}}f(\xi), \qquad \xi\in\Omega, \quad f\in c(\Omega). \tag{1.2.17}$$

We extend their definition in canonical way to vector fields in $l^2(\Omega)$. Hence,

$$l^2_{\text{nor}}(\Omega) = \{f\in l^2(\Omega)|f=p_{\text{nor}}f\}, \tag{1.2.18}$$

$$l^2_{\text{tan}}(\Omega) = \{f\in l^2(\Omega)|f=p_{\text{tan}}f\}. \tag{1.2.19}$$

We say $f\in l^2(\Omega)$ is *normal* if $f=p_{\text{nor}}f$ and *tangential* if $f=p_{\text{tan}}f$. Clearly, we have the orthogonal decomposition

$$l^2(\Omega) = l^2_{\text{nor}}(\Omega)\oplus l^2_{\text{tan}}(\Omega). \tag{1.2.20}$$

The spaces $c^{(p)}_{\text{nor}}(\Omega)$ and $c^{(p)}_{\text{tan}}(\Omega), 0\le p\le\infty$, are defined in the same fashion.

As usual, the *identity tensor* $\mathbf{i}$ is defined by

$$\mathbf{i} = \sum_{i=1}^{3}\varepsilon^i\otimes\varepsilon^i. \tag{1.2.21}$$

Its projection onto the tangential components at a point $\xi\in\Omega$ defines the *surface identity tensor field* $\mathbf{i}_{\text{tan}}$. To be specific,

$$\mathbf{i}_{\tan}(\xi) = \mathbf{i} - \xi \otimes \xi, \quad \xi \in \Omega. \tag{1.2.22}$$

Moreover, we define the *surface rotation tensor field* $\mathbf{j}_{\tan}$ by

$$\mathbf{j}_{\tan}(\xi) = \sum_{i=1}^{3} (\xi \wedge \varepsilon^i) \otimes \varepsilon^i, \quad \xi \in \Omega. \tag{1.2.23}$$

Obviously,

$$\begin{array}{rclcrcl} \mathbf{i}_{\tan}(\xi)\ \xi & = & 0, & \quad & \mathbf{j}_{\tan}(\xi)\ \xi & = & 0, \\ \mathbf{i}_{\tan}(\xi)\ u_\xi & = & u_\xi, & & \mathbf{j}_{\tan}(\xi)\ u_\xi & = & \xi \wedge u_\xi, \end{array}$$

if $u_\xi \in \mathbb{R}^3,\ u_\xi \cdot \xi = 0$.

The most important tools for investigating spherical vector fields (and tensor fields) are certain types of differential operators. The use of these operators enables us in the second part of the book to define vector spherical harmonics in a formulation that avoids completely the use of a special spherical coordinate system. Indeed, putting the fundamentals of our definitions on the usual orthonormal basis of $\mathbb{R}^3$ (as in the definition of scalar spherical harmonics via homogeneous harmonic polynomials in $\mathbb{R}^3$), provides us with a powerful concept which can be easily generalized to the tensorial case. However, since our approach is also based on the separation of vector fields into normal and tangential parts, a representation of differential operators in local coordinates is sometimes of advantage. For this reason, we develop and formulate several results also in terms of a local coordinate system. The local coordinate system we use here is the polar one defined by (1.2.24) which turns out to be very advantageous for discussing radially symmetric functions on the sphere:

$$\Phi : (0, \infty) \times (0, 2\pi) \times (-1, 1) \to \mathbb{R}^3$$

with

$$\Phi(r, \varphi, t) = \begin{pmatrix} r\sqrt{1-t^2}\ \cos\ \varphi \\ r\sqrt{1-t^2}\ \sin\ \varphi \\ rt \end{pmatrix}. \tag{1.2.24}$$

Setting $r = 1$ we get a local coordinate system on the unit sphere Ω. If no confusion is likely to arise we identify in the following an element $\xi \in \Omega$ with its coordinates (φ, t), that is to say,

$$\xi = \sqrt{1-t^2}(\cos\varphi\varepsilon^1 + \sin\varphi\varepsilon^2) + \ t\varepsilon^3.$$

In doing so we (locally) obtain the vector fields

$$\varepsilon^r(\varphi, t) \quad = \quad \left| \frac{\partial \Phi(r, \varphi, t)}{\partial r} \right|_{r=1} \left|^{-1} \frac{\partial \Phi(r, \varphi, t)}{\partial r} \right|_{r=1},$$

$$\begin{aligned}
\varepsilon^\varphi(\varphi,t) &= \left|\frac{\partial\Phi(1,\varphi,t)}{\partial\varphi}\right|^{-1}\frac{\partial\Phi(1,\varphi,t)}{\partial\varphi},\\
\varepsilon^t(\varphi,t) &= \left|\frac{\partial\Phi(1,\varphi,t)}{\partial t}\right|^{-1}\frac{\partial\Phi(1,\varphi,t)}{\partial t},
\end{aligned}$$

which define a moving orthonormal triad on the unit sphere Ω. Simple calculations yield

$$\varepsilon^r(\varphi,t) = \begin{pmatrix}\sqrt{1-t^2}\ \cos\varphi\\ \sqrt{1-t^2}\ \sin\varphi\\ t\end{pmatrix}, \tag{1.2.25}$$

$$\varepsilon^\varphi(\varphi,t) = \begin{pmatrix}-\sin\varphi\\ \cos\varphi\\ 0\end{pmatrix}, \tag{1.2.26}$$

$$\varepsilon^t(\varphi,t) = \begin{pmatrix}-t\cos\varphi\\ -t\sin\varphi\\ \sqrt{1-t^2}\end{pmatrix}. \tag{1.2.27}$$

Notice that

$$\varepsilon^r(\varphi,t)\wedge\varepsilon^\varphi(\varphi,t) = \varepsilon^t(\varphi,t).$$

On the other hand, we easily find

$$\begin{aligned}
\varepsilon^1 &= \sqrt{1-t^2}\cos\varphi\varepsilon^r(\varphi,t) - \sin\varphi\varepsilon^\varphi(\varphi,t) - t\cos\varphi\varepsilon^t(\varphi,t),\\
\varepsilon^2 &= \sqrt{1-t^2}\sin\varphi\varepsilon^r(\varphi,t) + \cos\varphi\varepsilon^\varphi(\varphi,t) - t\sin\varphi\varepsilon^t(\varphi,t), \quad (1.2.28)\\
\varepsilon^3 &= t\varepsilon^r(\varphi,t) + \sqrt{1-t^2}\varepsilon^t(\varphi,t).
\end{aligned}$$

Observing these facts the tensors $\mathbf{i}_{\tan}$ and $\mathbf{j}_{\tan}$ can be rewritten locally as follows:

$$\mathbf{i}_{\tan}(\varphi,t) = \varepsilon^\varphi(\varphi,t)\otimes\varepsilon^\varphi(\varphi,t) + \varepsilon^t(\varphi,t)\otimes\varepsilon^t(\varphi,t), \tag{1.2.29}$$

$$\mathbf{j}_{\tan}(\varphi,t) = \varepsilon^t(\varphi,t)\otimes\varepsilon^\varphi(\varphi,t) - \varepsilon^\varphi(\varphi,t)\otimes\varepsilon^t(\varphi,t). \tag{1.2.30}$$

Using the local coordinates, the *gradient* ∇ in $\mathbb{R}^3$ can be written out as follows:

$$\nabla = \varepsilon^r\frac{\partial}{\partial r} + \frac{1}{r}\nabla^*, \tag{1.2.31}$$

where the *surface gradient* on Ω has the representation

$$\nabla^* = \varepsilon^\varphi\frac{1}{\sqrt{1-t^2}}\frac{\partial}{\partial\varphi} + \varepsilon^t\sqrt{1-t^2}\frac{\partial}{\partial t}. \tag{1.2.32}$$

For $F \in \mathcal{C}^{(1)}(\Omega)$ we introduce the *surface curl gradient* L^* by

$$L^*_\xi F(\xi) = \xi \wedge \nabla^*_\xi F(\xi), \qquad \xi \in \Omega. \tag{1.2.33}$$

This operator will be of particular importance in the following considerations. Its local coordinate expression is

$$L^* = -\varepsilon^\varphi \sqrt{1-t^2}\frac{\partial}{\partial t} + \varepsilon^t \frac{1}{\sqrt{1-t^2}}\frac{\partial}{\partial \varphi}. \tag{1.2.34}$$

Moreover, the *Laplace operator* Δ in $\mathbb{R}^3$ given by

$$\Delta_x = \nabla_x \cdot \nabla_x = \left(\frac{\partial}{\partial x_1}\right)^2 + \left(\frac{\partial}{\partial x_2}\right)^2 + \left(\frac{\partial}{\partial x_3}\right)^2 \tag{1.2.35}$$

allows the representation

$$\Delta = \left(\frac{\partial}{\partial r}\right)^2 + \frac{2}{r}\frac{\partial}{\partial r} + \frac{1}{r^2}\Delta^*, \tag{1.2.36}$$

where Δ^* denotes the *Beltrami operator* of the unit sphere Ω. In local coordinates the operator Δ^* reads as follows:

$$\Delta^* = \frac{\partial}{\partial t}(1-t^2)\frac{\partial}{\partial t} + \frac{1}{1-t^2}\left(\frac{\partial}{\partial \varphi}\right)^2. \tag{1.2.37}$$

Let $\eta \in \Omega$ be fixed. Then it is not difficult to see from the above representation that

$$\nabla^*_\xi(\xi\cdot\eta) = \eta - (\xi\cdot\eta)\xi, \tag{1.2.38}$$
$$L^*_\xi(\xi\cdot\eta) = \xi \wedge \eta, \tag{1.2.39}$$

and

$$\Delta^*_\xi(\xi\cdot\eta) = -2(\xi\cdot\eta). \tag{1.2.40}$$

Furthermore, if $F \in \mathcal{C}^{(1)}[-1,+1]$, then

$$\nabla^*_\xi F(\xi\cdot\eta) = F'(\xi\cdot\eta)(\eta - (\xi\cdot\eta)\xi), \tag{1.2.41}$$
$$L^*_\xi F(\xi\cdot\eta) = F'(\xi\cdot\eta)(\xi\wedge\eta), \tag{1.2.42}$$

whereas for $F \in \mathcal{C}^{(2)}[-1,+1]$

$$\Delta^*_\xi F(\xi\cdot\eta) = -2(\xi\cdot\eta)F'(\xi\cdot\eta) + (1-(\xi\cdot\eta)^2)F''(\xi\cdot\eta). \tag{1.2.43}$$

The formal scalar product of the operators ∇^* and L^* with tangential vector fields yields the *surface divergence* $\nabla^*\cdot$ and the *surface curl* $L^*\cdot$.

More explicitly, let $f \in c^{(1)}_{\mathrm{tan}}(\Omega)$ possess the component functions F_i, that is to say $f(\xi) = \sum_{i=1}^{3} F_i(\xi)\varepsilon^i$. Then we let

$$\nabla^*_\xi \cdot f(\xi) = \sum_{i=1}^{3} (\nabla^*_\xi F_i(\xi)) \cdot \varepsilon^i, \tag{1.2.44}$$

$$L^*_\xi \cdot f(\xi) = \sum_{i=1}^{3} (L^*_\xi F_i(\xi)) \cdot \varepsilon^i. \tag{1.2.45}$$

It is not hard to verify that for $\varphi \in (0, 2\pi), t \in (-1, 1)$

$$\begin{aligned} \nabla^* \cdot \varepsilon^\varphi(\varphi, t) &= 0, & \nabla^* \cdot \varepsilon^t(\varphi, t) &= -\frac{t}{\sqrt{1-t^2}}, \\ L^* \cdot \varepsilon^\varphi(\varphi, t) &= \frac{t}{\sqrt{1-t^2}}, & L^* \cdot \varepsilon^t(\varphi, t) &= 0. \end{aligned}$$

These results can be used for the calculation of the surface divergence or surface curl of a tangential vector field $f \in c^{(1)}_{\mathrm{tan}}(\Omega)$, locally given by

$$f(\varphi, t) = F_\varphi(\varphi, t)\varepsilon^\varphi(\varphi, t) + F_t(\varphi, t)\varepsilon^t(\varphi, t).$$

Using the product rule we find

$$\begin{aligned} \nabla^* \cdot f(\varphi, t) &= (\nabla^* F_\varphi(\varphi, t)) \cdot \varepsilon^\varphi(\varphi, t) + F_\varphi(\varphi, t)(\nabla^* \cdot \varepsilon^\varphi(\varphi, t)) \\ &\quad + (\nabla^* F_t(\varphi, t)) \cdot \varepsilon^t(\varphi, t) + F_t(\varphi, t)(\nabla^* \cdot \varepsilon^t(\varphi, t)) \\ &= \frac{1}{\sqrt{1-t^2}} \frac{\partial}{\partial \varphi} F_\varphi(\varphi, t) + 0 \\ &\quad + \sqrt{1-t^2} \frac{\partial}{\partial t} F_t(\varphi, t) - \frac{t}{\sqrt{1-t^2}} F_t(\varphi, t). \end{aligned}$$

Similarly we get

$$\begin{aligned} L^* \cdot f(\varphi, t) &= -\sqrt{1-t^2} \frac{\partial}{\partial t} F_\varphi(\varphi, t) + \frac{1}{\sqrt{1-t^2}} \frac{\partial}{\partial \varphi} F_t(\varphi, t) \\ &\quad + \frac{t}{\sqrt{1-t^2}} F_\varphi(\varphi, t). \end{aligned}$$

If $F \in \mathcal{C}^{(1)}(\Omega)$, we clearly have

$$\nabla^*_\xi F(\xi) \cdot L^*_\xi F(\xi) = 0, \quad \xi \in \Omega. \tag{1.2.46}$$

Furthermore,

$$\nabla^* \cdot \nabla^* = L^* \cdot L^* = \Delta^*. \tag{1.2.47}$$

Let $\partial\Gamma$ be the (sufficiently smooth) boundary curve of $\Gamma \subset \Omega$. Moreover, denote by ν and τ unit surface vectors normal (outward of Γ) and tangential to $\partial\Gamma$, respectively. Let σ denote the arc length along $\partial\Gamma$. Then the *surface theorem of Gauß* reads

$$\int_\Gamma \nabla_\xi^* \cdot f(\xi)\, d\omega(\xi) = \int_{\partial\Gamma} \nu_\xi \cdot f(\xi)\, d\sigma(\xi), \tag{1.2.48}$$

while the *surface theorem of Stokes* takes the form

$$\int_\Gamma L_\xi^* \cdot f(\xi)\, d\omega(\xi) = \int_{\partial\Gamma} \tau_\xi \cdot f(\xi)\, d\sigma(\xi)$$

provided that $f : \Omega \to \mathbb{R}^3$ is continuously differentiable on $\bar{\Gamma} = \Gamma \cup \partial\Gamma$. Applying the Gauß formula towards $f = F\ \nabla^* G$ with suitable F, G we obtain *Green's first surface identity*

$$\int_\Gamma \nabla_\xi^* G(\xi) \cdot \nabla_\xi^* F(\xi)\, d\omega(\xi) + \int_\Gamma F(\xi) \Delta_\xi^* G(\xi)\, d\omega(\xi) = \int_{\partial\Gamma} F(\xi) \frac{\partial}{\partial \nu_\xi} G(\xi)\, d\sigma(\xi). \tag{1.2.49}$$

Interchanging F, G and subtracting (1.2.49) yields *Green's second surface identity*

$$\int_\Gamma \{F(\xi)\Delta_\xi^* G(\xi) - G(\xi)\Delta_\xi^* F(\xi)\}\, d\omega(\xi) = \int_{\partial\Gamma} \{F(\xi) \frac{\partial}{\partial \nu_\xi} G(\xi) - G(\xi) \frac{\partial}{\partial \nu_\xi} F(\xi)\}\, d\sigma(\xi). \tag{1.2.50}$$

There are immediate consequences of the above formulae due to the fact that the surface Ω is closed:

$$\int_\Omega f(\xi) \cdot \nabla_\xi^* F(\xi)\, d\omega(\xi) = -\int_\Omega F(\xi) \nabla_\xi^* \cdot f(\xi)\, d\omega(\xi), \tag{1.2.51}$$

$$\int_\Omega f(\xi) \cdot L_\xi^* F(\xi) d\omega(\xi) = -\int_\Omega F(\xi) L_\xi^* \cdot f(\xi)\, d\omega(\xi), \tag{1.2.52}$$

$$\int_\Omega \nabla_\xi^* F(\xi) \cdot \nabla_\xi^* G(\xi) d\omega(\xi) = -\int_\Omega F(\xi) \Delta_\xi^* G(\xi)\, d\omega(\xi) \tag{1.2.53}$$

$$= -\int_\Omega G(\xi) \Delta_\xi^* F(\xi)\, d\omega(\xi).$$

Furthermore,

$$\int_\Omega \nabla_\xi^* \cdot f(\xi)\, d\omega(\xi) = 0, \tag{1.2.54}$$

$$\int_\Omega L^*_\xi F(\xi) \cdot L^*_\xi G(\xi)\, d\omega(\xi) = -\int_\Omega F(\xi)\Delta^*_\xi G(\xi)\, d\omega(\xi), \tag{1.2.55}$$

$$\int_\Omega \nabla^*_\xi \cdot (f(\xi) \wedge \xi)\, d\omega(\xi) = 0, \tag{1.2.56}$$

provided that $F : \Omega \to \mathbb{R}$ (resp. $f : \Omega \to \mathbb{R}^3$) are sufficiently often continuously differentiable.

The strongest motivation for the definition of the operators ∇^* and L^* is the fact (see for example G.E. BACKUS (1966, 1967)) that any vector field $f \in c^{(1)}(\Omega)$ allows a decomposition of the form

$$f(\xi) = \xi F_1(\xi) + \nabla^*_\xi F_2(\xi) + L^*_\xi F_3(\xi), \quad \xi \in \Omega, \tag{1.2.57}$$

with certain functions $F_i : \Omega \to \mathbb{R}$. To be more specific, the tangent vector representation theorem asserts that if $p_{\tan} f$ is the tangent part of f, there exist unique scalar fields $F_2, F_3 \in \mathcal{C}^{(1)}(\Omega)$ such that

$$\int_\Omega F_2(\eta)\, d\omega(\eta) = \int_\Omega F_3(\eta)\, d\omega(\eta) = 0$$

and

$$p_{\tan} f(\xi) = \nabla^*_\xi F_2(\xi) + L^*_\xi F_3(\xi), \quad \xi \in \Omega.$$

This result will be applied in Chapter 12 to determine the explicit form of the functions F_2, F_3.

Part I

Scalars

2

POLYNOMIALS

In this chapter we are concerned with the theory of homogeneous harmonic polynomials in Euclidean space $\mathbb{R}^3$. The main result is the addition theorem. When restricted to the unit sphere this theorem will play later on a central role in the theory of spherical radial basis functions.

The outline of the chapter is as follows. In Section 2.1 we define an inner product $(\cdot,\cdot)_{Hom_n}$ on the space Hom_n of homogeneous polynomials of degree n in $\mathbb{R}^3$ by letting $(H_n, K_n)_{Hom_n} = H_n(\nabla)K_n$ for all $H_n, K_n \in Hom_n$, where $H_n(\nabla)$ is the differential operator formally defined by replacing x by ∇_x in the expression of $H_n(x)$. Section 2.2 deals with homogeneous harmonic polynomials. We develop the basic result that any homogeneous polynomial H_n of degree n may be decomposed into a sum of functions $|\cdot|^{2j}K_{n-2j}, j = 0,\dots,[n/2]$, where K_{n-2j} are homogeneous harmonic polynomials of degree $n-2j$. In Section 2.3 we prove the addition theorem for homogeneous harmonic polynomials by determination of the projection of homogeneous polynomials H_n to the space $Harm_n$ of homogeneous harmonic polynomials of degree n in $\mathbb{R}^3$. Moreover, we verify that, apart from a multiplicative constant μ_n, the usual $\mathcal{L}^2$-norm defined on $Harm_n$ equals the Hom_n-norm mentioned before.

2.1 Homogeneous Polynomials

Let Hom_n (more accurately: $Hom_n(\mathbb{R}^3)$) consist of all polynomials H_n in three variables which are homogeneous of degree n (i.e. $H_n(\lambda x) = \lambda^n H_n(x)$ for all $\lambda \in \mathbb{R}$ and all $x \in \mathbb{R}^3$). Thus, if $H_n \in Hom_n$, then there exist real numbers $C_\alpha = C_{\alpha_1\alpha_2\alpha_3}$ such that

$$\begin{aligned} H_n(x) = H_n(x_1,x_2,x_3) &= \sum_{[\alpha]=n} C_\alpha\, x^\alpha \\ &= \sum_{\alpha_1+\alpha_2+\alpha_3=n} C_{\alpha_1\alpha_2\alpha_3}\, x_1^{\alpha_1}x_2^{\alpha_2}x_3^{\alpha_3}. \end{aligned} \tag{2.1.1}$$

It is obvious that the set of monomials $x \mapsto x^\alpha$, $[\alpha] = n$, is a basis for the space Hom_n. The number of such monomials is precisely the number of ways a triple can be chosen so that we have $[\alpha] = n$, i.e. the number of ways selecting 2 elements out of a collection of $n+2$. This means that the dimension $d(Hom_n)$ of Hom_n is equal to

$$d(Hom_n) \; = \; \binom{n+2}{2} \; = \; \frac{(n+1)(n+2)}{2} \; . \tag{2.1.2}$$

Let $H_n(\nabla_x)$ be the differential operator associated to $H_n(x)$ (i.e. replace x^α formally by $(\nabla_x)^\alpha$ in the expression of $H_n(x)$):

$$H_n(\nabla_x) = \sum_{\alpha_1+\alpha_2+\alpha_3=n} C_{\alpha_1\alpha_2\alpha_3} \frac{\partial^{[\alpha]}}{\partial x_1^{\alpha_1} \partial x_2^{\alpha_2} \partial x_3^{\alpha_3}} = \sum_{[\alpha]=n} C_\alpha (\nabla_x)^\alpha \; . \tag{2.1.3}$$

If such an operator is applied to a homogeneous polynomial U_n of the same degree

$$U_n(x) \; = \; \sum_{[\beta]=n} D_\beta \; x^\beta \tag{2.1.4}$$

we obtain as result a real number:

$$\begin{aligned}
&(H_n(\nabla_x)) \; U_n(x) \\
&= \sum_{[\alpha]=n} \sum_{[\beta]=n} C_\alpha \; D_\beta \left(\frac{\partial}{\partial x_1}\right)^{\alpha_1} x_1^{\beta_1} \left(\frac{\partial}{\partial x_2}\right)^{\alpha_2} x_2^{\beta_2} \left(\frac{\partial}{\partial x_3}\right)^{\alpha_3} x_3^{\beta_3} \\
&= \sum_{[\alpha]=n} C_\alpha \; D_\alpha \; \alpha! \; .
\end{aligned} \tag{2.1.5}$$

Clearly, we find

$$\begin{aligned}
(H_n(\nabla_x)) \; U_n(x) \;&=\; (U_n(\nabla_x)) \; H_n(x), \\
(H_n(\nabla_x)) \; H_n(x) \;&\geq\; 0.
\end{aligned} \tag{2.1.6}$$

This enables us to introduce an inner product $(\cdot,\cdot)_{Hom_n}$ on the space Hom_n by letting

$$(H_n, U_n)_{Hom_n} = (H_n(\nabla_x)) \; U_n(x). \tag{2.1.7}$$

The space Hom_n equipped with the inner product $(\cdot,\cdot)_{Hom_n}$ is a finite-dimensional Hilbert space. The set of monomials

$$\{x \mapsto (\alpha!)^{-1/2} x^\alpha \mid [\alpha] = n\}$$

forms an orthonormal system in the space Hom_n. For each $H_n \in Hom_n$ we have in connection with (1.1.3)

$$H_n(x) \;\; = \;\; \sum_{[\alpha]=n} \frac{1}{\alpha!} \; (H_n(\nabla_y)) \; y^\alpha \; x^\alpha$$

$$\begin{aligned} &= (H_n(\nabla_y))\ \frac{1}{n!} \sum_{[\alpha]=n} \frac{n!}{\alpha!}\ x^\alpha y^\alpha \\ &= (H_n(\nabla_y))\ \frac{(x\cdot y)^n}{n!} \\ &= \frac{1}{n!}\ (x\cdot\nabla_y)^n\ H_n(y). \end{aligned} \tag{2.1.8}$$

In other words,

$$H_n(x) = (\frac{(x\cdot\ )^n}{n!}, H_n)_{Hom_n}. \tag{2.1.9}$$

Theorem 2.1.1. *Hom_n equipped with the inner product $(\cdot,\cdot)_{Hom_n}$ is a $d(Hom_n)$-dimensional Hilbert space with the reproducing kernel*

$$K_{Hom_n}(x,y) = \frac{(x\cdot y)^n}{n!}, \qquad x,y\in\mathbb{R}^3, \tag{2.1.10}$$

i.e.

(i) for every fixed y, the function $K_{Hom_n}(\cdot,y)$ belongs to Hom_n,

(ii) for any $H_n \in Hom_n$ and any point x the reproducing property

$$H_n(x) = (K_{Hom_n}(\cdot,x), H_n)_{Hom_n}$$

is valid.

Now, let $\{H_{n,j}\}_{j=1,\ldots,d(Hom_n)}$, $\{U_{n,j}\}_{j=1,\ldots,d(Hom_n)}$ be two orthonormal systems in the space Hom_n:

$$\begin{aligned} (H_{n,j}, H_{n,k})_{Hom_n} &= \delta_{jk}, \\ (U_{n,j}, U_{n,k})_{Hom_n} &= \delta_{jk}, \end{aligned} \tag{2.1.11}$$

where δ_{jk} is the usual Kronecker symbol. Then, for $j = 1,\ldots,d(Hom_n)$, we have

$$\begin{aligned} H_{n,j} &= \sum_{k=1}^{d(Hom_n)} (H_{n,j}, U_{n,k})_{Hom_n} U_{n,k}, \\ U_{n,j} &= \sum_{k=1}^{d(Hom_n)} (U_{n,j}, H_{n,k})_{Hom_n} H_{n,k}. \end{aligned} \tag{2.1.12}$$

Therefore it follows that

$$\sum_{j=1}^{d(Hom_n)} H_{n,j}(x)\ H_{n,j}(y) = \sum_{j=1}^{d(Hom_n)} U_{n,j}(x) U_{n,j}(y). \tag{2.1.13}$$

Hence, in particular for the orthonormal system of monomials, we obtain the following result.

Theorem 2.1.2. *Let* $\{H_{n,j}\}_{j=1,\ldots,d(Hom_n)}$ *be an orthonormal system in* Hom_n. *Then*

$$K_{Hom_n}(x,y) = \frac{(x\cdot y)^n}{n!} = \sum_{j=1}^{d(Hom_n)} H_{n,j}(x)H_{n,j}(y), \quad x,y \in \mathbb{R}^3. \quad (2.1.14)$$

$K_{Hom_n}(\cdot,\cdot)$ *is the only reproducing kernel in* Hom_n.

Suppose that there are given $d(Hom_n)$ points $x_1, ..., x_{d(Hom_n)} \in \mathbb{R}^3$ and $d(Hom_n)$–values $d_1, ...d_{d(Hom_n)} \in \mathbb{R}$. We are able to solve the Hom_n interpolation problem

$$\sum_{j=1}^{d(Hom_n)} b_j H_{n,j}(x_k) = d_k, \quad k = 1, ..., d(Hom_n), \quad (2.1.15)$$

if and only if the matrix

$$\mathbf{matr}_{\{x_1,\ldots,x_{d(Hom_n)}\}}(H_{n,1},\ldots,H_{n,d(Hom_n)}) \quad (2.1.16)$$

$$= \begin{pmatrix} H_{n,1}(x_1) & \ldots & H_{n,1}(x_{d(Hom_n)}) \\ \vdots & \ddots & \vdots \\ H_{n,d(Hom_n)}(x_1) & \ldots & H_{n,d(Hom_n)}(x_{d(Hom_n)}) \end{pmatrix}$$

is non-singular. A system of $d(Hom_n)$ points $x_1, ..., x_{d(Hom_n)}$ is called a *fundamental system relative to* Hom_n if the matrix (2.1.16) is non-singular. We guarantee the existence of a fundamental system relative to Hom_n (cf. C. MÜLLER (1966)).

Lemma 2.1.3. *There exists a system* $\{x_1,\ldots,x_{d(Hom_n)}\} \subset \mathbb{R}^3$ *such that (2.1.16) is non-singular.*

Proof. As orthonormal system, the functions $H_{n,1}, ..., H_{n,d(Hom_n)}$ are linearly independent. Hence, there exists a point x_1 for which

$$H_{n,1}(x_1) \neq 0. \quad (2.1.17)$$

Now, there must also be a point x_2 such that

$$\begin{vmatrix} H_{n,1}(x_1) & H_{n,1}(x_2) \\ H_{n,2}(x_1) & H_{n,2}(x_2) \end{vmatrix} \neq 0,$$

for else we would have a contradiction to the linear independence of $H_{n,1}, H_{n,2}$. In the same way the existence of a point x_3 can be deduced by the requirement

$$\begin{vmatrix} H_{n,1}(x_1) & H_{n,1}(x_2) & H_{n,1}(x_3) \\ H_{n,2}(x_1) & H_{n,2}(x_2) & H_{n,2}(x_3) \\ H_{n,3}(x_1) & H_{n,3}(x_2) & H_{n,3}(x_3) \end{vmatrix} \neq 0.$$

Finally, by induction, we obtain a system of points $x_1, ..., x_{d(Hom_n)}$ such that

$$\begin{vmatrix} H_{n,1}(x_1) & \cdots & H_{n,1}(x_{d(Hom_n)}) \\ \vdots & \ddots & \vdots \\ H_{n,d(Hom_n)}(x_1) & \cdots & H_{n,d(Hom_n)}(x_{d(Hom_n)}) \end{vmatrix} \neq 0, \qquad (2.1.18)$$

i.e. a fundamental system relative to Hom_n. □

To every $H_n \in Hom_n$, there exist real numbers $b_1, ..., b_{d(Hom_n)}$ such that

$$H_n = \sum_{k=1}^{d(Hom_n)} b_k \, H_{n,k}. \qquad (2.1.19)$$

Under the assumption that $\{x_1, ..., x_{d(Hom_n)}\}$ is a fundamental system relative to Hom_n, the linear equations

$$\sum_{j=1}^{d(Hom_n)} a_j \, H_{n,k}(x_j) = b_k, \quad k = 1, ..., d(Hom_n) \,, \qquad (2.1.20)$$

are uniquely solvable in the unknowns $a_1, ..., a_{d(Hom_n)}$. Thus we obtain

$$H_n = \sum_{k=1}^{d(Hom_n)} \sum_{j=1}^{d(Hom_n)} a_j \, H_{n,k}(x_j) \, H_{n,k}. \qquad (2.1.21)$$

Theorem 2.1.4. *Let $\{H_{n,j}\}_{j=1,\ldots,d(Hom_n)}$ be an orthonormal system in Hom_n. Assume that $\{x_k\}_{k=1,\ldots,d(Hom_n)}$ is a fundamental system relative to Hom_n. Then, each $H_n \in Hom_n$ is uniquely representable in the form*

$$H_n(x) = \sum_{j=1}^{d(Hom_n)} a_j \, K_{Hom_n}(x_j, x) = \sum_{j=1}^{d(Hom_n)} a_j \frac{(x_j \cdot x)^n}{n!}.$$

2.2 Homogeneous Harmonic Polynomials

Let $Harm_n$ (more accurately: $Harm_n(\mathbb{R}^3)$) be the class of all polynomials in Hom_n that are harmonic:

$$Harm_n = \{H_n \in Hom_n \mid \Delta_x H_n(x) = 0, \ x \in \mathbb{R}^3\} \,. \qquad (2.2.1)$$

For $n < 2$, of course, all homogeneous polynomials are harmonic.

Any homogeneous harmonic polynomial of degree n can be represented in the form

$$H_n(x) = H_n(x_1, x_2, x_3) = \sum_{j=0}^{n} x_3^j A_{n-j}(x_1, x_2), \tag{2.2.2}$$

where A_{n-j} is a homogeneous polynomial of degree $n-j$ in the variables x_1, x_2. Application of the Laplace operator gives

$$\begin{aligned} 0 = \Delta_x H_n(x) &= \sum_{j=0}^{n} \left(\left(\frac{\partial}{\partial x_1}\right)^2 + \left(\frac{\partial}{\partial x_2}\right)^2 + \left(\frac{\partial}{\partial x_3}\right)^2 \right) x_3^j A_{n-j}(x_1, x_2) \\ &= \sum_{j=0}^{n-2} x_3^j \left(\left(\frac{\partial}{\partial x_1}\right)^2 + \left(\frac{\partial}{\partial x_2}\right)^2 \right) A_{n-j}(x_1, x_2) \\ &\quad + \sum_{j=0}^{n-2} x_3^j (j+2)(j+1) A_{n-j-2}(x_1, x_2), \end{aligned}$$

where we have used the facts that

$$\begin{aligned} \left(\left(\frac{\partial}{\partial x_1}\right)^2 + \left(\frac{\partial}{\partial x_2}\right)^2 \right) A_0(x_1, x_2) &= 0, \\ \left(\left(\frac{\partial}{\partial x_1}\right)^2 + \left(\frac{\partial}{\partial x_2}\right)^2 \right) A_1(x_1, x_2) &= 0. \end{aligned} \tag{2.2.3}$$

Thus the functions $A_{n-j} : \mathbb{R}^2 \to \mathbb{R}$ satisfy the recursion relation

$$\left(\left(\frac{\partial}{\partial x_1}\right)^2 + \left(\frac{\partial}{\partial x_2}\right)^2 \right) A_{n-j}(x_1, x_2) + (j+2)(j+1) A_{n-j-2}(x_1, x_2) = 0 \tag{2.2.4}$$

for $j = 0, 1, ..., n-2$. Therefore, all polynomials A_{n-j} are determined if we know A_n and A_{n-1}.

Theorem 2.2.1. *Let A_n and A_{n-1} be homogeneous polynomials of degree n and $n-1$ in $\mathbb{R}^2$, respectively. For $j = 0, ..., n-2$ we set recursively*

$$A_{n-j-2}(x_1, x_2) = -\frac{1}{(j+1)(j+2)} \left(\left(\frac{\partial}{\partial x_1}\right)^2 + \left(\frac{\partial}{\partial x_2}\right)^2 \right) A_{n-j}(x_1, x_2). \tag{2.2.5}$$

Then $H_n : \mathbb{R}^3 \to \mathbb{R}$ given by

$$H_n(x_1, x_2, x_3) = \sum_{j=0}^{n} x_3^j A_{n-j}(x_1, x_2)$$

is a homogeneous harmonic polynomial of degree n in $\mathbb{R}^3$, i.e. $H_n \in Harm_n$. The number of linearly independent homogeneous harmonic polynomials is equal to the number of coefficients of A_n and A_{n-1}, i.e.

$$d(Harm_n) = n + n + 1 = 2n + 1.$$

Assume that $n \geq 2$. Let H_{n-2} be a homogeneous polynomial of degree $n-2$, i.e. $H_{n-2} \in Hom_{n-2}$. Then, for each homogeneous harmonic polynomial K_n, we have

$$(|\cdot|^2 H_{n-2}, K_n)_{Hom_n} = (H_{n-2}(\nabla_x))\Delta_x K_n(x) = 0 \ . \tag{2.2.6}$$

This means $|\cdot|^2 H_{n-2}$ is orthogonal to K_n in the sense of the inner product $(\cdot,\cdot)_{Hom_n}$. Conversely, suppose that $K_n \in Hom_n$ is orthogonal to all elements L_n of the form

$$L_n(x) = |x|^2 H_{n-2}(x) \ , \quad H_{n-2} \in Hom_{n-2}.$$

Then it follows that

$$0 = (|\cdot|^2 H_{n-2}, K_n)_{Hom_n} = (H_{n-2}(\nabla_x))\Delta_x K_n(x) = (H_{n-2}, \Delta K_n)_{Hom_{n-2}}$$

for all $H_{n-2} \in Hom_{n-2}$. This is true only if $\Delta K_n = 0$, i.e. K_n is a homogeneous harmonic polynomial.

Theorem 2.2.2 *(decomposition theorem of Hom_n). Hom_n, $n \geq 2$, is the orthogonal direct sum of $Harm_n$ and $Harm_n^{\perp}$, where $Harm_n^{\perp} = |\cdot|^2 Hom_{n-2}$ is the space of all L_n with $L_n(x) = |x|^2 H_{n-2}(x)$, $H_{n-2} \in Hom_{n-2}$, i.e. each homogeneous polynomial H_n of degree n can be decomposed uniquely in the form*

$$H_n(x) = K_n(x) + |x|^2 H_{n-2}(x) \ ,$$

where K_n is a homogeneous harmonic polynomial of degree n and H_{n-2} is a homogeneous polynomial of degree $n-2$.

Denote by Proj_{Harm_n} and $\mathrm{Proj}_{Harm_n^{\perp}}$ the projection operators in Hom_n onto $Harm_n$ and $Harm_n^{\perp}$, respectively. Then

$$H_n = \mathrm{Proj}_{Harm_n} H_n + \mathrm{Proj}_{Harm_n^{\perp}} H_n \ . \tag{2.2.7}$$

In other words,

$$K_n(x) = \mathrm{Proj}_{Harm_n} H_n(x), \tag{2.2.8}$$

$$|x|^2 H_{n-2}(x) \quad = \quad \mathrm{Proj}_{Harm_n^\perp} H_n(x)\ . \tag{2.2.9}$$

For all $H_n, U_n \in Hom_n$,

$$(\mathrm{Proj}_{Harm_n} H_n, U_n)_{Hom_n} \quad = \quad (H_n, \mathrm{Proj}_{Harm_n} U_n)_{Hom_n}\ . \tag{2.2.10}$$

Moreover, we have $\mathrm{Proj}_{Harm_n} H_n = \mathrm{Proj}_{Harm_n}\, K_n = K_n$. Observe that

$$\begin{aligned} d(Harm_n) &= d(Hom_n) - d(Harm_n^\perp) \\ &= d(Hom_n) - d(Hom_{n-2}) \\ &= \binom{n+2}{2} - \binom{n}{2} = 2n+1. \end{aligned} \tag{2.2.11}$$

If we apply Theorem 2.2.2 recursively to H_{n-2}, $H_{n-4}, \ldots$, we obtain the following result.

Theorem 2.2.3. *Each homogeneous polynomial of degree n can be decomposed uniquely in the form*

$$H_n(x) = \sum_{i=0}^{[n/2]} |x|^{2i} K_{n-2i}(x), \quad K_{n-2i} \in Harm_{n-2i}, \quad x \in \mathbb{R}^3, \tag{2.2.12}$$

where

$$\left[\frac{n}{2}\right] = \frac{1}{2}\left(n - \frac{1}{2}\left(1-(-1)^n\right)\right) .$$

In other words, Hom_n admits the direct sum decomposition

$$Hom_n(\mathbb{R}^3) = \bigoplus_{i=0}^{[n/2]} |\cdot|^{2i} Harm_{n-2i}(\mathbb{R}^3). \tag{2.2.13}$$

This result gives rise to the following corollary.

Corollary 2.2.4. *For $n = 0, 1, \ldots$*

$$Hom_n(\mathbb{R}^3)|\Omega = Hom_n(\Omega) = \bigoplus_{i=0}^{[n/2]} Harm_{n-2i}(\mathbb{R}^3)|\Omega.$$

Since the space $Pol_{0,\ldots,n}(\mathbb{R}^3)$ of polynomials in three variables of degree $\leq n$ can be written as direct sum decomposition of $Hom_n(\mathbb{R}^3)$ and $Hom_{n-1}(\mathbb{R}^3)$, when restricted to Ω, i.e.

$$Pol_{0,\ldots,n}(\mathbb{R}^3)|\Omega = (Hom_n(\mathbb{R}^3)|\Omega) \oplus (Hom_{n-1}(\mathbb{R}^3)|\Omega) \tag{2.2.14}$$

we finally obtain the following.

Corollary 2.2.5. *For $n = 0, 1, \ldots$*

$$Pol_{0,\ldots,n}(\mathbb{R}^3)|\Omega = \bigoplus_{i=0}^{n} Harm_i(\mathbb{R}^3)|\Omega.$$

In other words, the restriction to the unit sphere Ω of any polynomial of three variables is a sum of restrictions to Ω of homogeneous harmonic polynomials.

2.3 Addition Theorem

We are now interested in giving the explicit representation of the orthogonal projection $\mathrm{Proj}_{Harm_n} H_n$ of a given homogeneous polynomial H_n. For that purpose we need some preliminaries. By induction we are able to prove that for $i = 1, 2, 3$ and $|x| \neq 0$ (cf. E.W. HOBSON (1955))

$$\begin{aligned} &\left(\frac{\partial}{\partial x_i}\right)^n \frac{1}{|x|} \\ &= (-1)^n \frac{(2n)!}{n!2^n} \frac{1}{|x|^{2n+1}} \left(\sum_{s=0}^{[n/2]} (-1)^s \frac{n!(2n-2s)!}{(2n)!(n-s)!s!} |x|^{2s} \Delta^s \right) x_i^n. \end{aligned}$$

In other words, we find

$$\begin{aligned} &(\varepsilon^i \cdot \nabla_x)^n \frac{1}{|x|} \\ &= (-1)^n \frac{(2n)!}{n!2^n} \frac{1}{|x|^{2n+1}} \left(\sum_{s=0}^{[n/2]} (-1)^s \frac{n!(2n-2s)!}{(2n)!(n-s)!s!} |x|^{2s} \Delta^s \right) (\varepsilon^i \cdot x)^n \end{aligned}$$

($i = 1, 2, 3$). Since the differential operator Δ is invariant with respect to orthogonal transformations it is easy to see that

$$\begin{aligned} &(y \cdot \nabla_x)^n \frac{1}{|x|} \\ &= (-1)^n \frac{(2n)!}{n!2^n} \frac{1}{|x|^{2n+1}} \left(\sum_{s=0}^{[n/2]} (-1)^s \frac{n!(2n-2s)!}{(2n)!(n-s)!s!} |x|^{2s} \Delta^s \right) (y \cdot x)^n \end{aligned}$$

is valid for every $y \in \mathbb{R}^3$. Now, as we have seen in Section 2.1, each $H_n \in Hom_n$ may be represented in the form

$$H_n(x) = \sum_{j=1}^{d(Hom_n)} c_j (x_j \cdot x)^n, \quad x \in \mathbb{R}^3,$$

where c_j, $j = 1, \ldots, d(Hom_n)$, are suitable coefficients and $x_1, \ldots, x_{d(Hom_n)}$ is a fundamental system relative to Hom_n.

Consequently, we have the following result.

Theorem 2.3.1. *Let H_n be a homogeneous polynomial of degree n. Then, for each $x \in \mathbb{R}^3$, $|x| \neq 0$*

$$\begin{aligned}(H_n(\nabla_x))\frac{1}{|x|} \\ = \quad (-1)^n \frac{(2n)!}{n!2^n}\frac{1}{|x|^{2n+1}}\left(\sum_{s=0}^{[n/2]}(-1)^s\frac{n!(2n-2s)!}{(2n)!(n-s)!s!}|x|^{2s}\Delta^s\right) H_n(x).\end{aligned}$$

From the considerations given above it follows that

$$(H_n(\nabla_x))\frac{1}{|x|} = (K_n(\nabla_x))\frac{1}{|x|} + (H_{n-2}(\nabla_x))\Delta_x\frac{1}{|x|}, \quad |x| \neq 0. \quad (2.3.1)$$

Thus, in connection with

$$\begin{aligned}\Delta_x\frac{1}{|x|} &= 0, \quad |x| \neq 0, \\ \Delta_x K_n(x) &= 0, \quad x \in \mathbb{R}^3,\end{aligned} \quad (2.3.2)$$

we obtain for $|x| \neq 0$

$$(H_n(\nabla_x))\frac{1}{|x|} = (K_n(\nabla_x))\frac{1}{|x|} = (-1)^n\frac{(2n)!}{n!2^n}\frac{1}{|x|^{2n+1}}K_n(x). \quad (2.3.3)$$

Therefore, by comparison of Theorem 2.3.1 and (2.3.3), we get the following lemma.

Lemma 2.3.2. *Let H_n be a homogeneous polynomial of degree n. Then*

$$\mathrm{Proj}_{Harm_n} H_n(x) = \left(\sum_{s=0}^{[n/2]}(-1)^s\frac{n!(2n-2s)!}{(2n)!(n-s)!s!}|x|^{2s}\Delta^s\right) H_n(x). \quad (2.3.4)$$

Observing

$$\Delta_x(x\cdot y)^n = n(n-1)|y|^2(x\cdot y)^{n-2}, \quad y \in \mathbb{R}^3, \quad (2.3.5)$$

we obtain, in particular,

$$\begin{aligned}&\mathrm{Proj}_{Harm_n}\left(\frac{(x\cdot y)^n}{n!}\right) \\ &= \frac{1}{n!}\sum_{s=0}^{[n/2]}(-1)^s\frac{(2n-2s)!(n!)^2}{(n-2s)!(n-s)!s!(2n)!}|x|^{2s}|y|^{2s}(x\cdot y)^{n-2s}.\end{aligned} \tag{2.3.6}$$

Hence, after an easy calculation, we find by using $x = |x|\xi$, $y = |y|\eta$, $(\xi,\eta)\in\Omega^2$, the equation

$$\begin{aligned}&\mathrm{Proj}_{Harm_n}\left(\frac{(x\cdot y)^n}{n!}\right) \\ &= \frac{(2n+1)2^n\cdot n!}{(2n+1)!}(|x|\,|y|)^n\sum_{s=0}^{[n/2]}(-1)^s\frac{(2n-2s)!}{2^n(n-2s)!(n-s)!s!}(\xi\cdot\eta)^{n-2s}.\end{aligned} \tag{2.3.7}$$

Suppose that $\{H_{n,j}\}_{j=1,\ldots,d(Harm_n)}$ is an orthonormal system in $Harm_n$ with respect to $(\cdot,\cdot)_{Hom_n}$. Let $\{U_{n,j}\}_{j=1,\ldots,d(Hom_n)-d(Harm_n)}$ be an orthonormal system in $Harm_n^{\perp}$. Then the union of both systems

$$\{H_{n,j}\}_{j=1,\ldots,d(Harm_n)}\cup\{U_{n,j}\}_{j=1,\ldots,d(Hom_n)-d(Harm_n)} \tag{2.3.8}$$

forms an orthonormal system in Hom_n. Therefore it follows that

$$\begin{aligned}&\frac{(x\cdot y)^n}{n!} \\ &= \sum_{j=1}^{d(Harm_n)} H_{n,j}(x)\,H_{n,j}(y) + \sum_{j=1}^{d(Hom_n)-d(Harm_n)} U_{n,j}(x)\,U_{n,j}(y)\end{aligned} \tag{2.3.9}$$

for any pair $x,y\in\mathbb{R}^3$. Furthermore, in view of the definition of the projection operator Proj_{Harm_n}, we get

$$\begin{aligned}\mathrm{Proj}_{Harm_n}&\left(\sum_{j=1}^{d(Harm_n)} H_{n,j}(x)H_{n,j}(y) + \sum_{j=1}^{d(Hom_n)-d(Harm_n)} U_{n,j}(x)U_{n,j}(y)\right) \\ &= \sum_{j=1}^{d(Harm_n)} H_{n,j}(x)H_{n,j}(y).\end{aligned} \tag{2.3.10}$$

On the other hand, as we have shown above,

$$\mathrm{Proj}_{Harm_n}\left(\frac{(x\cdot y)^n}{n!}\right) \tag{2.3.11}$$

$$= \frac{(2n+1)2^n n!}{(2n+1)!}|x|^n|y|^n \sum_{s=0}^{[n/2]}(-1)^s \frac{(2n-2s)!}{2^n(n-2s)!(n-s)!s!}(\xi\cdot\eta)^{n-2s}.$$

By comparison of (2.3.10) and (2.3.11) we obtain the *addition theorem of homogeneous harmonic polynomials in* $\mathbb{R}^3$.

Theorem 2.3.3. *Let* $\{H_{n,j}\}_{j=1,\ldots,d(Harm_n)}$, $d(Harm_n) = 2n+1$, *be an orthonormal system in* $Harm_n$ *with respect to* $(\cdot,\cdot)_{Hom_n}$. *Then, for* $x, y \in \mathbb{R}^3$, $x = |x|\xi$, $y = |y|\eta$, *we have*

$$\sum_{j=1}^{2n+1} H_{n,j}(x)\, H_{n,j}(y) = \frac{2^n n!}{(2n)!}|x|^n|y|^n P_n(\xi\cdot\eta),$$

where we have used the abbreviation

$$P_n(t) = \sum_{s=0}^{[n/2]}(-1)^s \frac{(2n-2s)!}{2^n(n-2s)!(n-s)!s!}t^{n-2s}, \quad t\in[-1,1]. \qquad (2.3.12)$$

Next we discuss the important question of how, for any pair of elements $H_n \in Harm_n$, $K_n \in Harm_n$, the inner product $(\cdot,\cdot)_{Hom_n}$ defined by (2.1.7) is related to the (usually used) inner product $(\cdot,\cdot)_{\mathcal{L}^2(\Omega)}$.

Theorem 2.3.4. *For* $H_m \in Harm_m, K_n \in Harm_n$,

$$(H_m, K_n)_{\mathcal{L}^2(\Omega)} = \frac{\delta_{nm}}{\mu_n}(H_m(\nabla_x))K_n(x), \qquad (2.3.13)$$

where μ_n *is given by*

$$\mu_n = \frac{(2n+1)!}{4\pi 2^n n!} = \frac{1\cdot 3\cdot \ldots \cdot (2n+1)}{4\pi} = \frac{(2n+1)!!}{4\pi}. \qquad (2.3.14)$$

Proof. By virtue of the fundamental theorem of potential theory (see for example O.D. Kellogg (1929))

$$K_n(x) = \frac{1}{4\pi}\int_\Omega \left\{ \frac{1}{|x-y|}\frac{\partial}{\partial\nu_y}K_n(y) - K_n(y)\frac{\partial}{\partial\nu_y}\frac{1}{|x-y|}\right\} d\omega(y)$$

for all $x \in \mathbb{R}^3$ with $|x| < 1$, where $\partial/\partial\nu$ denotes the derivative in the direction of the outer normal to Ω. Therefore we find

$$(H_m(\nabla_x))K_n(x) = \frac{1}{4\pi}\int_\Omega \left\{ (H_m(\nabla_x))\frac{1}{|x-y|}\frac{\partial}{\partial\nu_y}K_n(y)\right.$$

$$- \quad K_n(y)\frac{\partial}{\partial \nu_y}(H_m(\nabla_x))\frac{1}{|x-y|}\Bigg\} \, d\omega(y). \tag{2.3.15}$$

For $x \neq y$ we get from (2.3.3)

$$(H_m(\nabla_x))\frac{1}{|x-y|} \;=\; (-1)^m \frac{(2m)!}{m!2^m}\frac{H_m(x-y)}{|x-y|^{2m+1}}. \tag{2.3.16}$$

Because H_m is homogeneous, this is equivalent to

$$(H_m(\nabla_x))\frac{1}{|x-y|} \;=\; \frac{(2m)!}{m!2^m}\frac{H_m(y-x)}{|x-y|^{2m+1}}. \tag{2.3.17}$$

Inserting (2.3.17) into (2.3.15) gives

$$\begin{aligned}(H_m(\nabla_x))K_n(x) \;=\; & \frac{(2m)!}{(m!)2^m}\frac{1}{4\pi}\int_\Omega \left\{ \frac{H_m(y-x)}{|x-y|^{2m+1}}\frac{\partial}{\partial \nu_y}K_n(y) \right. \\ & \left. -K_n(y)\frac{\partial}{\partial \nu_y}\frac{H_m(y-x)}{|x-y|^{2m+1}} \right\} d\omega(y). \end{aligned} \tag{2.3.18}$$

It is easy to see that for $m \neq n$

$$(H_m(\nabla_x))K_n(x)\,|_{x=0} \;=\; 0, \tag{2.3.19}$$

while for $m = n$

$$(H_m(\nabla_x))K_n(x)\,|_{x=0} \;=\; (H_m(\nabla_x))K_n(x) \;=\; (H_m, K_n)_{Hom_n}. \tag{2.3.20}$$

Therefore we obtain

$$\begin{aligned}& \frac{1}{4\pi}\int_\Omega \left\{ \frac{H_m(y)}{|y|^{2m+1}}\frac{\partial}{\partial \nu_y}K_n(y) - K_n(y)\frac{\partial}{\partial \nu_y}\frac{H_m(y)}{|y|^{2m+1}} \right\} d\omega(y) \\ & = \begin{cases} 0 & \text{for} \quad m \neq n \\ \left(\frac{2^m m!}{(2m)!}\right)(H_m, K_n)_{Hom_n} & \text{for} \quad m = n \end{cases}. \end{aligned} \tag{2.3.21}$$

Since the normal derivatives of K_n and H_m are equal to

$$\frac{\partial}{\partial r}K_n(r\xi)\,|_{r=1} \;=\; nK_n(\xi)\,, \qquad \frac{\partial}{\partial r}H_m(r\xi)\,|_{r=1} \;=\; mH_m(\xi), \tag{2.3.22}$$

respectively, it follows that

$$\begin{aligned}& \frac{1}{4\pi}\int_\Omega \left\{ \frac{H_m(y)}{|y|^{2m+1}}\frac{\partial}{\partial \nu_y}K_n(y) - K_n(y)\frac{\partial}{\partial \nu_y}\frac{H_m(y)}{|y|^{2m+1}} \right\} d\omega(y) \\ & = \frac{1}{4\pi}\int_\Omega \{nH_m(\xi)K_n(\xi) + (m+1)H_m(\xi)K_n(\xi)\}\, d\omega(\xi) \\ & = \frac{n+m+1}{4\pi}\int_\Omega H_m(\xi)K_n(\xi)\, d\omega(\xi). \end{aligned} \tag{2.3.23}$$

Thus, by combination of (2.3.21) and (2.3.23), we finally obtain the desired result stated in Theorem 2.3.4. □

In other words, to any orthonormal system $\{H_{n,j}\}_{j=1,\ldots,2n+1}$ in $Harm_n$ with respect to $(\cdot,\cdot)_{Hom_n}$ there corresponds the $\mathcal{L}^2(\Omega)$-orthonormal system $\{\sqrt{\mu_n}H_{n,j}\}_{j=1,\ldots,2n+1}$, and vice versa.

Finally, we mention the following *reformulation of the addition theorem.*

Theorem 2.3.5. *Let $\{H_{n,j}\}_{j=1,\ldots,2n+1}$ be an orthonormal system in $Harm_n$ with respect to $(\cdot,\cdot)_{Hom_n}$. Then, for $x, y \in \mathbb{R}^3$, we have*

$$\sum_{j=1}^{2n+1} \sqrt{\mu_n}H_{n,j}(x)\ \sqrt{\mu_n}H_{n,j}(y) = \frac{2n+1}{4\pi}|x|^n|y|^n P_n(\xi\cdot\eta),$$

where $x = |x|\xi,\ y = |y|\eta;\ (\xi,\eta) \in \Omega^2$.

Bibliographical Notes

The decomposition of Hom_n into a direct sum of $Harm_{n-2i}, i = 0,\ldots,[n/2]$ is well known in the theory of homogeneous polynomials (see for example E.M. STEIN, G. WEISS (1971)). The addition theorem as presented here can be found in W. FREEDEN (1979) for the three-dimensional case and W. FREEDEN, R. REUTER (1984) for the higher-dimensional case. The constructive character of the proof of the addition theorem can be exploited to generate (orthonormal systems of) homogeneous harmonic polynomials in integer arithmetic. A first variant of exact computation of homogeneous harmonic polynomials has been proposed by W. FREEDEN, R. REUTER (1984), while a more efficient second variant was presented by W. FREEDEN, R. REUTER (1990) (including the multidimensional case). The role of homogeneous harmonic polynomials as basis system in boundary-value problems of potential theory corresponding to not necessarily spherical (georelevant) boundaries (such as ellipsoid, spheroid, (regular) earth's surface etc.) is studied in W. FREEDEN (1980A, 1983, 1985A,B, 1987A,B,C).

3

SCALAR SPHERICAL HARMONICS

The restriction to the unit sphere Ω of any homogeneous harmonic polynomial of degree n is called a spherical harmonic of order n. It turns out that any spherical harmonic is an eigenfunction of the Beltrami operator. On the sphere, spherical harmonics are the analogues of the exponential functions for Fourier analysis. They were introduced in the 1780s to study gravitational theory (cf. P.S. DE LAPLACE (1785), A.M. LEGENDRE (1785)). Early books on the theory of potentials and spherical harmonics are due to E. HEINE (1878) and F. NEUMANN (1887). Indeed, spherical harmonics are essential for the analysis of any phenomena with spherical symmetry, e.g. celestial mechanics, terrestrial magnetism, earthquakes, solar corona etc. The use of spherical harmonics in all geosciences for the purpose of representing fields is a well established technique. A great incentive came from the fact that global geomagnetic data became available in the first half of the last century (cf. C.F. GAUSS (1839)). Nowadays reference models for (earth's) gravitational or magnetic field, for example, are widely given by tables of coefficients for the spherical harmonic expansions of their potentials. This chapter is devoted to the development of important properties of spherical harmonics. It will enable us, in particular, to show how $\mathcal{L}^2(\Omega)$ can be decomposed into a direct sum of finite-dimensional subspaces $Harm_n$, where $Harm_n$ is spanned by a spherical harmonic basis $\{Y_{n,1}, \ldots, Y_{n,2n+1}\}$ of order n. Explicit representations of the functions $Y_{n,j}$ are obtainable, for instance, from the structure of homogeneous harmonic polynomials in $\mathbb{R}^3$ or by separation of the Laplace differential equation in terms of polar coordinates. However, our purpose is not only to give a comprehensive description of the theory of spherical harmonics, but also to use its developments as basic tools in constructive approximation techniques over the sphere.

3.1 Definition

Firstly, we introduce spherical harmonics as restrictions of homogeneous harmonic polynomials to the unit sphere.

Definition 3.1.1. Let H_n be a homogeneous harmonic polynomial of degree n in $\mathbb{R}^3$, i.e. $H_n \in Harm_n(\mathbb{R}^3)$. The restriction $Y_n = H_n \mid \Omega$ is called a spherical harmonic of order n. The space of all spherical harmonics of order n, that is the set of all restrictions $Y_n = H_n \mid \Omega$, $H_n \in Harm_n(\mathbb{R}^3)$, is denoted by $Harm_n(\Omega)$. More explicitly, $Harm_n(\Omega) = Harm_n(\mathbb{R}^3) \mid \Omega$.

Remark. In what follows we simply write $Harm_n$ instead of $Harm_n(\mathbb{R}^3)$ (or $Harm_n(\Omega)$) if no confusion is likely to arise.

We know already that the linear space $Harm_n$ is of dimension $2n+1$, that is $d(Harm_n) = 2n+1$. From Theorem 2.3.4 it follows that spherical harmonics of different orders are orthogonal in the sense of the $\mathcal{L}^2$-inner product

$$(Y_n, Y_m)_{\mathcal{L}^2(\Omega)} = \int_\Omega Y_n(\xi) Y_m(\xi)\, d\omega(\xi) = 0, \quad n \neq m.$$

Using the standard method of separation we have $H_n(x) = r^n Y_n(\xi), x = r\xi, r = |x|, \xi \in \Omega$. Observing the identity

$$\frac{1}{r^2}\frac{d}{dr}\left(r^2 \frac{d}{dr}\right) r^n = n(n+1) r^{n-2}$$

we obtain

$$0 = \Delta_x H_n(x) = r^{n-2} n(n+1) Y_n(\xi) + r^{n-2} \Delta^*_\xi Y_n(\xi).$$

Thus we are able to formulate the following lemma.

Lemma 3.1.2. *Any spherical harmonic $Y_n, n = 0, 1, \ldots,$ is an infinitely often differentiable eigenfunction of the Beltrami operator corresponding to the eigenvalue $-n(n+1)$. More explicitly,*

$$(\Delta^*_\xi - (\Delta^*)^\wedge(n)) Y_n(\xi) = 0, \quad \xi \in \Omega, \quad Y_n \in Harm_n,$$

where the 'spherical symbol' $\{(\Delta^*)^\wedge(n)\}_{n=0,1,\ldots}$ of the Beltrami operator Δ^* is given by

$$(\Delta^*)^\wedge(n) = -n(n+1), \quad n = 0, 1, \ldots.$$

Remark. Throughout the book, for convenience, the capital letter Y followed by double indices, for example, $Y_{n,j}$, denotes a member of an orthonormal system $\{Y_{n,1}, \ldots, Y_{n,2n+1}\}$ with respect to $(\cdot,\cdot)_{\mathcal{L}^2(\Omega)}$.
In terms of spherical harmonics the *addition theorem* then allows the following reformulation.

Theorem 3.1.3. *Let $\{Y_{n,j}\}_{j=1,\ldots,2n+1}$ be an $\mathcal{L}^2(\Omega)$-orthonormal system in $Harm_n$. Then, for any pair $(\xi,\eta) \in \Omega^2$,*

$$\sum_{j=1}^{2n+1} Y_{n,j}(\xi)Y_{n,j}(\eta) = \frac{2n+1}{4\pi}P_n(\xi\cdot\eta).$$

Proof. Theorem 3.1.3 follows immediately from Theorem 2.3.5. □

The addition theorem is of particular importance, since it relates restrictions of homogeneous harmonics polynomials to the unit sphere to a univariate polynomial.

Because of $P_n(1) = 1$ we find

$$\sum_{j=1}^{2n+1} (Y_{n,j}(\xi))^2 = \frac{2n+1}{4\pi}, \quad \xi \in \Omega. \tag{3.1.1}$$

If we remember that every $Y_n \in Harm_n$ can be written in the form

$$Y_n = \sum_{j=1}^{2n+1} (Y_n, Y_{n,j})_{\mathcal{L}^2(\Omega)} Y_{n,j} \tag{3.1.2}$$

we immediately get the following lemma.

Lemma 3.1.4. *For every $Y_n \in Harm_n$*

$$\frac{2n+1}{4\pi}\int_\Omega Y_n(\eta)P_n(\xi\cdot\eta)\,d\omega(\eta) = Y_n(\xi), \quad \xi\in\Omega,$$

that is

$$(\xi,\eta)\mapsto K_{Harm_n}(\xi,\eta) = \frac{2n+1}{4\pi}P_n(\xi\cdot\eta), \quad (\xi,\eta)\in\Omega^2,$$

represents the (uniquely determined) reproducing kernel in $Harm_n$. Moreover,

$$(\Delta^*_\eta - (\Delta^*)^\wedge(n))K_{Harm_n}(\xi,\eta) = 0, \quad \eta\in\Omega,$$

holds for all $\xi \in \Omega$.

Observing that

$$\int_\Omega |Y_n(\xi)|^2\,d\omega(\xi) = \sum_{j=1}^{2n+1} (Y_n, Y_{n,j})^2_{\mathcal{L}^2(\Omega)} \tag{3.1.3}$$

we find in connection with (3.1.1) and (3.1.2)

$$|Y_n(\xi)|^2 \leq \sum_{j=1}^{2n+1} (Y_n, Y_{n,j})^2_{\mathcal{L}^2(\Omega)} \sum_{j=1}^{2n+1} (Y_{n,j}(\xi))^2$$

$$= \frac{2n+1}{4\pi} \sum_{j=1}^{2n+1} (Y_n, Y_{n,j})^2_{\mathcal{L}^2(\Omega)}.$$

This yields the following lemma.

Lemma 3.1.5. *For every* $Y_n \in Harm_n$

$$\sup_{\xi \in \Omega} |Y_n(\xi)| \leq \left(\frac{2n+1}{4\pi}\right)^{\frac{1}{2}} \|Y_n\|_{\mathcal{L}^2(\Omega)}.$$

In particular,

$$\sup_{\xi \in \Omega} |Y_{n,j}(\xi)| \leq \left(\frac{2n+1}{4\pi}\right)^{\frac{1}{2}}, \quad j = 1, \ldots, 2n+1. \tag{3.1.4}$$

3.2 Legendre Polynomials

The function $P_n : [-1, +1] \to \mathbb{R}$, $n = 0, 1, \ldots$, defined by (2.3.12) is called the *Legendre polynomial.* P_n is uniquely determined by the following properties:

(i) P_n is a polynomial of degree n on the interval $[-1, 1]$,

(ii) $\int_{-1}^{+1} P_n(t) P_m(t)\, dt = 0$ for $n \neq m$,

(iii) $P_n(1) = 1$.

This is easily seen from the usual process of orthogonalization. A graphical impression can be found in Figure 3.2.1. In particular, we have

$$P_0(t) = 1, \qquad P_1(t) = t, \qquad P_2(t) = \frac{1}{2}(3t^2 - 1).$$

Furthermore,

$$\int_{-1}^{+1} P_n(t) P_m(t)\, dt = \frac{2}{2n+1} \delta_{nm}. \tag{3.2.1}$$

Applying the Cauchy–Schwarz inequality to the addition theorem (Theorem 3.1.3) we obtain

$$\begin{aligned} \frac{2n+1}{4\pi} \left| P_n(\xi \cdot \eta) \right| &= \left| \sum_{j=1}^{2n+1} Y_{n,j}(\xi) Y_{n,j}(\eta) \right| \\ &\leq \sqrt{\sum_{j=1}^{2n+1} \left(Y_{n,j}(\xi)\right)^2} \sqrt{\sum_{j=1}^{2n+1} \left(Y_{n,j}(\eta)\right)^2} \end{aligned}$$

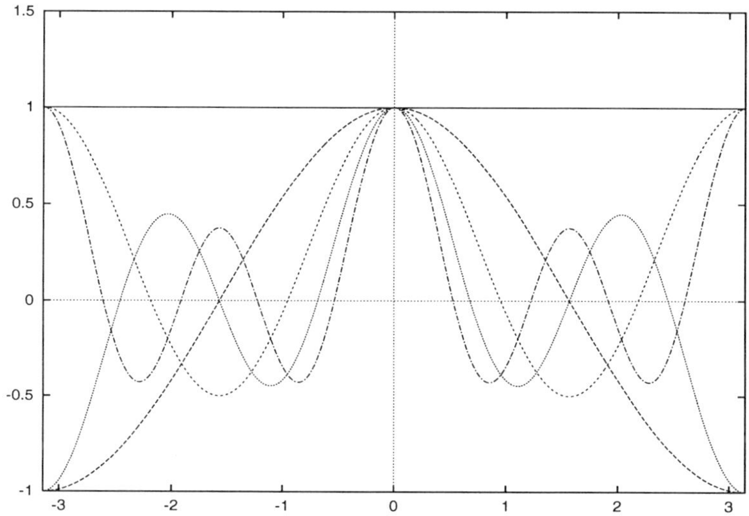

FIG. 3.2.1. Legendre polynomials $P_n(\cos\vartheta)$, $n = 0, \ldots, 4$

$$= \frac{2n+1}{4\pi} \mid P_n(1) \mid .$$

Therefore it follows that

$$|P_n(t)| \le |P_n(1)| = 1 \; , \; t \in [-1, +1]. \tag{3.2.2}$$

Moreover, the Legendre polynomial P_n satisfies the estimate (see for example C. MÜLLER (1952))

$$|P_n^{(k)}(t)| \le |P_n^{(k)}(1)|, \tag{3.2.3}$$

where $P_n^{(k)}(1) = O(n^{2k})$. In particular, we have

$$P_n'(1) = \frac{n(n+1)}{2}. \tag{3.2.4}$$

Furthermore, for $k = 2, 3, \ldots, n$,

$$P_n^{(k)}(1) = \left(\frac{1}{2}\right)^k \frac{1}{k!} \, n(n+1)\left((n(n+1) - 1 \cdot 2) \ldots (n(n+1) \quad k(k-1))\right).$$

For all $s, t \in [-1, +1]$ it follows that

$$| P_n(t) - P_n(s) | \leq P_n'(1)|t-s| = \frac{n(n+1)}{2}|t-s|. \tag{3.2.5}$$

Therefore we obtain

$$|P_n(\xi \cdot \zeta) - P_n(\eta \cdot \zeta)| \leq \frac{n(n+1)}{2}|\xi - \eta| \tag{3.2.6}$$

for any pair $(\xi, \eta) \in \Omega^2$ and every $\zeta \in \Omega$. A simple calculation gives

$$P_n(\xi \cdot \xi) - 2P_n(\xi \cdot \eta) + P_n(\eta \cdot \eta) = \frac{2n+1}{4\pi} \int_\Omega |P_n(\xi \cdot \zeta) - P_n(\eta \cdot \zeta)|^2 \, d\omega(\zeta). \tag{3.2.7}$$

Thus we find

$$|P_n(\xi \cdot \xi) - 2P_n(\xi \cdot \eta) + P_n(\eta \cdot \eta)| \leq (2n+1) \left(\frac{n(n+1)}{2}|\xi - \eta| \right)^2. \tag{3.2.8}$$

Summarizing our results we obtain the following.

Lemma 3.2.1. *For every triple* $(\xi, \eta, \zeta) \in \Omega^3$

$$\begin{aligned} |P_n(\xi \cdot \zeta) - P_n(\eta \cdot \zeta)| &\leq 2 \left(\frac{n(n+1)}{2}|\xi - \eta| \right)^\sigma, \\ |P_n(\xi \cdot \xi) - 2P_n(\xi \cdot \eta) + P_n(\eta \cdot \eta)| &\leq 4(2n+1) \left(\frac{n(n+1)}{2}|\xi - \eta| \right)^{2\sigma} \end{aligned}$$

provided that $\sigma \in [0, 1]$.

From Theorem 3.1.3 it follows that

$$\left((1-t^2) \left(\frac{d}{dt} \right)^2 - 2t\frac{d}{dt} + n(n+1) \right) P_n(t) = 0, \quad t \in [-1, +1]. \tag{3.2.9}$$

We therefore obtain the following lemma.

Lemma 3.2.2. *The Legendre polynomial* P_n *is the only infinitely often differentiable eigenfunction of the 'Legendre operator'*

$$L_t = (1-t^2)(d/dt)^2 - 2t(d/dt)$$

on $[-1, +1]$, *corresponding to the eigenvalues* $-n(n+1)$, $n = 0, 1, \ldots$, *and bounded on* $[-1, +1]$ *with* $P_n(1) = 1$.

From the binomial theorem it follows that

$$(t^2-1)^n = \sum_{s=0}^{n}(-1)^s \frac{n!}{(n-s)!s!} t^{2n-2s}, \quad n=0,1,\ldots. \tag{3.2.10}$$

For all $s \le [n/2]$ we find

$$\left(\frac{d}{dt}\right)^n t^{2n-2s} = \frac{(2n-2s)!}{(n-2s)!} t^{n-2s}, \tag{3.2.11}$$

while for $n \ge s > [n/2]$ we get

$$\left(\frac{d}{dt}\right)^n t^{2n-2s} = 0. \tag{3.2.12}$$

Therefore we see that

$$\left(\frac{d}{dt}\right)^n (t^2-1)^n = \sum_{s=0}^{[n/2]}(-1)^s \frac{n!}{(n-s)!s!} \frac{(2n-2s)!}{(n-2s)!} t^{n-2s}. \tag{3.2.13}$$

By comparison with the definition of the Legendre polynomial (2.3.12) we therefore obtain

$$P_n(t) = \frac{1}{2^n n!}\left(\frac{d}{dt}\right)^n (t^2-1)^n, \quad t \in [-1,+1]. \tag{3.2.14}$$

This result is known as *Rodriguez's formula.* Integrating by parts we obtain

$$\int_{-1}^{1} F(t)P_n(t)\,dt = \frac{1}{2^n n!}\int_{-1}^{1} F^{(n)}(t)(1-t^2)^n\,dt \tag{3.2.15}$$

for every $F \in \mathcal{C}^{(n)}[-1,+1]$. As an application of this formula we discuss the integrals

$$\int_{-1}^{1} P_k(t)(P_n'(t)-P_{n-2}'(t))\,dt = -\int_{-1}^{1}(P_n(t)-P_{n-2}(t))P_k'(t)\,dt. \tag{3.2.16}$$

$P_n' - P_{n-2}'$ is a polynomial of degree $n-1$ on $[-1,+1]$. Consequently the integral (3.2.16) vanishes for $k \ge n$. On the other hand, P_k' is a polynomial of degree $k-1$. This means that the integral vanishes also for $k < n-1$ so that (3.2.16) differs from zero only for $k = n-1$. Thus

$$P_n'(t) - P_{n-2}'(t) \;=\; c_n P_{n-1}(t), \quad t \in [-1,+1]. \tag{3.2.17}$$

In connection with (3.2.4) we find

$$P_n'(1) - P_{n-2}'(1) \;=\; 2n-1. \tag{3.2.18}$$

Therefore it follows that

$$P_n'(t) - P_{n-2}'(t) \;=\; (2n-1)P_{n-1}(t)\;, \quad t \in [-1,+1]. \tag{3.2.19}$$

Equivalently we have

$$(2n-1)\int_x^1 P_{n-1}(t)\,dt \;=\; P_{n-2}(x) - P_n(x), \quad n \geq 2, \tag{3.2.20}$$

for all $x \in [-1,+1]$. By similar arguments we are able to show that

$$P_{n+1}'(t) - tP_n'(t) \;=\; (n+1)P_n(t), \tag{3.2.21}$$

$$(t^2-1)P_n'(t) \;=\; ntP_n(t) - nP_{n-1}(t), \tag{3.2.22}$$

$$(n+1)P_{n+1}(t) + nP_{n-1}(t) - (2n+1)tP_n(t) \;=\; 0. \tag{3.2.23}$$

The formulae (3.2.21)–(3.2.23) are known as the *recurrence formulae for the Legendre polynomials.* Moreover we have

$$(t^2-1)P_n'(t) = \frac{n(n+1)}{2n+1}(P_{n+1}(t) - P_{n-1}(t)). \tag{3.2.24}$$

For later use we mention the following result.

Lemma 3.2.3. *The following identities are valid:*

(i) For $n = 0,1,\ldots$

$$P_{2n+1}'(t) = \sum_{k=0}^{n}(4k+1)P_{2k}(t)$$

(ii) For $n = 1,2,\ldots$

$$P_{2n}'(t) = \sum_{k=0}^{n-1}(4k+3)P_{2k+1}(t)$$

(iii) For $n = 1, 2, \dots$

$$(1+t)P_n'(t) = \sum_{k=0}^{n-1}(2k+1)P_k(t) + nP_n(t).$$

Proof. We prove statement (iii) only. It is clear that there exist coefficients $a_{n,0}, \dots, a_{n,n}$ such that

$$(1+t)P_n'(t) = \sum_{k=0}^{n} a_{n,k}P_k(t). \tag{3.2.25}$$

Now, by virtue of (3.2.1),

$$\int_{-1}^{+1} P_n'(t)(1+t)P_l(t)\,dt = a_{n,l}\frac{2}{2l+1} \tag{3.2.26}$$

for $l = 0, \dots, n$. Partial integration yields

$$\int_{-1}^{+1} P_n'(t)(1+t)P_l(t)\,dt = 2 - \delta_{n,l}\frac{2}{2l+1} - \int_{-1}^{+1} P_n(t)(1+t)P_l'(t)\,dt. \tag{3.2.27}$$

For $l = 0, \dots, n-1$ the last integral on the right-hand side vanishes, since $(1+t)P_l'(t)$ is of degree $\leq n-1$. Thus it follows that $a_{n,l} = 2l+1$ for $l = 0, \dots, n-1$. An easy calculation shows that

$$\begin{aligned}\int_{-1}^{+1} P_n(t)P_n'(t)(1+t)\,dt &= \frac{1}{2}P_n^2(t)(1+t)\Big|_{-1}^{+1} - \frac{1}{2}\int_{-1}^{+1} P_n^2(t)\,dt \\ &= 1 - \frac{1}{2n+1},\end{aligned} \tag{3.2.28}$$

hence, $a_{n,n} = n$. This gives the required result. $\square$

The power series

$$\phi(h) = \sum_{n=0}^{\infty} P_n(t)h^n, \quad t \in [-1,+1], \tag{3.2.29}$$

is absolutely and uniformly convergent for all h with $|h| \leq h_0, h_0 \in [0,1)$. By differentiation with respect to h and comparing coefficients according to (3.2.29) we find

$$(1+h^2-2ht)\phi'(h) = (t-h)\phi(h). \tag{3.2.30}$$

This differential equation is uniquely solvable under the initial condition $\phi(0) = 1$. Since it is not hard to show that $h \mapsto (1+h^2-2ht)^{1/2}$, $h \in (-1,+1)$, solves this initial value problem, we have the following *generating series expansion* of the Legendre polynomials.

Lemma 3.2.4. *For $t \in [-1,+1]$ and all $h \in (-1,+1)$*

$$\sum_{n=0}^{\infty} P_n(t)h^n = \frac{1}{\sqrt{1+h^2-2ht}}.$$

Remark. Among other areas of application, the subject of potential theory is concerned with forces of attraction due to the presence of a gravitational field. Central to the discussion of gravitational attraction is Newton's law of gravitation for the force field generated by a single particle: the gravitational force f at a free space (i.e. free of point masses) is related to the potential function F according to $f(x) = -\nabla_x F(x)$, $x \neq y$, when the potential between a mass point y and a point of free space x has the form $F(x) = k|x-y|^{-1}$, $x \neq y$ (k is the gravitational constant). Because of spherical symmetry of the gravitational field, the potential function of a single particle depends only upon the radial distance, i.e. the inner product of the direction vectors of x and y. In order to obtain this result, let us suppose for the sake of definition $x = |x|\xi$, $y = |y|\eta$, $\xi, \eta \in \Omega$, $|x| < |y|$. Then we find

$$\frac{1}{|x-y|} = \frac{1}{|y|}\left(1+\left(\frac{|x|}{|y|}\right)^2 - 2\frac{|x|}{|y|}\xi\cdot\eta\right)^{-1/2}. \tag{3.2.31}$$

Returning now to Lemma 3.2.4 with $t = \xi\cdot\eta$ and $h = |x|/|y|$ we find that the potential function has the series expansion

$$\frac{1}{|x-y|} = \frac{1}{|y|}\sum_{n=0}^{\infty}\left(\frac{|x|}{|y|}\right)^n P_n(\xi\cdot\eta). \tag{3.2.32}$$

Moreover, our considerations have shown that

$$\frac{1}{|x-y|} = \sum_{n=0}^{\infty}\frac{(-1)^n}{n!}|x|^n(\xi\cdot\nabla_y)^n\frac{1}{|y|}, \tag{3.2.33}$$

where

$$\frac{(-1)^n}{n!}(\xi\cdot\nabla_y)^n\frac{1}{|y|} = \frac{P_n(\xi\cdot\eta)}{|y|^{n+1}}, \quad n = 0,1,\ldots \tag{3.2.34}$$

Identity (3.2.34) is known as *Maxwell's representation formula.* It shows that the Legendre polynomials may be obtained by repeated differentiations of the '*fundamental solution*' $y \to |y|^{-1}, y \neq 0$, of the Laplace equation in the direction of the unit vector ξ. Thus the potential on the right-hand side of Maxwell's representation formula may be regarded as the potential of a pole of order n with the axis ξ at the origin.

The power series in Lemma 3.2.4 can be differentiated for all $h \in (-1,+1)$. Thus it follows that

$$-\frac{h-t}{(1+h^2-2ht)^{3/2}} = \sum_{n=1}^{\infty} nP_n(t)h^{n-1}. \tag{3.2.35}$$

Now it is easy to see that

$$\frac{1}{\sqrt{1+h^2-2ht}} - \frac{2h^2-2ht}{(1+h^2-2ht)^{3/2}} = \frac{1-h^2}{(1+h^2-2ht)^{3/2}}. \tag{3.2.36}$$

This gives us the following result.

Lemma 3.2.5. *For all $t \in [-1,+1]$ and $h \in (-1,1)$*

$$\frac{1-h^2}{(1+h^2-2ht)^{3/2}} = \sum_{n=0}^{\infty}(2n+1)h^n P_n(t).$$

Remark. Consider series representations of the form

$$\Phi_{\beta,k}(h,t) = \sum_{n=0}^{\infty}(n-\beta)^k P_n(t)h^n, \tag{3.2.37}$$

where $h \in (-1,+1)$, $t \in [-1,+1]$, $\beta \in \mathbb{R} \setminus \{0,1,\ldots\}$, $k = 0,1\ldots.$ Starting from $\Phi_{\beta,0}(h,t)$ elementary representations of a large class of sums $\Phi_{\beta,k}(h,t)$ can be ascertained by simple manipulations (see for example I. S. GRADSHTEYN, I. W. RYZHIK (1980), W. MAGNUS *et al.* (1966), A.P. PRUDNIKOV *et al.* (1980)).

Lemma 3.2.4 can be used to prove an integral representation for the Legendre polynomial. To this end we start from the well known elementary integral

$$\int_0^{\pi} \frac{d\varphi}{1+\gamma\cos\varphi} = \frac{\pi}{\sqrt{1-\gamma^2}} \quad (|\gamma|<1). \tag{3.2.38}$$

We put

$$\gamma = -\frac{h\sqrt{t^2-1}}{1-ht}. \tag{3.2.39}$$

On the one hand it follows that

$$\int_0^{\pi} \frac{1}{1+\gamma\cos\varphi}\,d\varphi = (1-ht)\int_0^{\pi}[1-h(t+\sqrt{t^2-1}\cos\varphi)]^{-1}\,d\varphi$$

$$= (1-ht)\int_0^{\pi}\sum_{n=0}^{\infty}(t+\sqrt{t^2-1}\cos\varphi)^n h^n \, d\varphi \quad (3.2.40)$$

On the other hand we obtain

$$\frac{\pi}{\sqrt{1-\gamma^2}} = \frac{\pi}{\sqrt{1-(h^2(t^2-1)/(1-ht))^2}} = \frac{\pi(1-ht)}{\sqrt{1+h^2-2ht}}. \quad (3.2.41)$$

In connection with Lemma 3.2.4 this yields

$$\frac{\pi}{\sqrt{1-\gamma^2}} = (1-ht)\pi\sum_{n=0}^{\infty}P_n(t)h^n. \quad (3.2.42)$$

By comparison we therefore obtain

$$\sum_{n=0}^{\infty}\int_0^{\pi}(t+\sqrt{t^2-1}\cos\varphi)^n \, d\varphi \; h^n = \pi\sum_{n=0}^{\infty}P_n(t)h^n. \quad (3.2.43)$$

This gives us the *Laplace representation of Legendre polynomials.*

Lemma 3.2.6. *For $t \in [-1,+1]$ and $n = 0, 1, \ldots$*

$$P_n(t) = \frac{1}{\pi}\int_0^{\pi}(t+\sqrt{t^2-1}\cos\varphi)^n \, d\varphi.$$

Next we list some inequalities for Legendre polynomials which will not be proved here, viz.

$$|P_n(t)| \leq \sqrt{\frac{2}{\pi(n+1/2)\sqrt{1-t^2}}}, \quad t \in (-1,+1), \quad (3.2.44)$$

and

$$|P_n'(t)| \leq \frac{49}{40}\sqrt{n+(1/2)}(1-t^2)^{-3/4}, \quad t \in (-1,+1). \quad (3.2.45)$$

The bound in (3.2.45) is already presented in a qualitative way in G. Szegö (1959), but the numerical constant is due to A. Martin (1991).

3.3 Fundamental Systems

When discussing spherical harmonics we need to understand more about the geometry of the sphere. It is well known that the spherical geometry is closely related to the *orthogonal group* $O(3)$ of real 3×3 matrices $\mathbf{t}$ such that $\mathbf{t}^T\mathbf{t} = \mathbf{i}$, where $\mathbf{i}$ denotes the 3×3 identity matrix, and $SO(3)$ the *special orthogonal group*, the subgroup of matrices $\mathbf{t} \in O(3)$ such that the determinant of $\mathbf{t}$ is 1. For any function $F \in \mathcal{L}^2(\Omega)$, the transformation $\xi \mapsto \mathbf{t}\xi$, $\xi \in \Omega$ produces a change in the functional values of F. The function $F(\mathbf{t}\cdot) : \xi \mapsto F(\mathbf{t}\xi), \xi \in \Omega$ is known as the $\mathbf{t}$-*transformed function.* A subspace $\mathcal{V} \subset \mathcal{L}^2(\Omega)$ is called an *invariant subspace* if, for all $F \in \mathcal{V}$, the $\mathbf{t}$-transformed function $F(\mathbf{t}\cdot)$ is of class $\mathcal{V}$ for all $\mathbf{t} \in SO(3)$. More precisely, we define the operator $R_{\mathbf{t}} : \mathcal{L}^2(\Omega) \to \mathcal{L}^2(\Omega)$ by letting $R_{\mathbf{t}}F(\xi) = F(\mathbf{t}\xi), \xi \in \Omega$. Let $S \subset SO(3)$ be a subgroup of $SO(3)$. A subspace $\mathcal{V} \subset \mathcal{L}^2(\Omega)$ is said to be invariant with respect to S or simply S-*invariant* if $F \in \mathcal{V}$ implies that $R_{\mathbf{t}}F \in \mathcal{V}$ for all $\mathbf{t} \in S$. Let $F, G \in \mathcal{L}^2(\Omega), \mathbf{t} \in SO(3)$. Observing the change of coordinates $\zeta = \mathbf{t}\xi$, we find

$$\int_\Omega F(\mathbf{t}\xi)G(\xi)\, d\omega(\xi) = (\det \mathbf{t})^2 \int_\Omega F(\zeta)G(\mathbf{t}^T\zeta)\, d\omega(\zeta) \tag{3.3.1}$$

(observe that $d\omega(\mathbf{t}\xi) = (\det \mathbf{t})\, d\omega(\xi)$). Thus it follows that

$$(R_{\mathbf{t}}F, G)_{\mathcal{L}^2(\Omega)} = (F, R_{\mathbf{t}^T}G)_{\mathcal{L}^2(\Omega)}. \tag{3.3.2}$$

An invariant subspace may itself contain one or more invariant subspaces. If this is the case $\mathcal{V}$ is said to be *reducible.* If there are no invariant subspaces of $\mathcal{V}$ (other than $\mathcal{V}$ itself), then $\mathcal{V}$ is said to be *irreducible.*

Let $\eta \in \Omega$ be fixed. We denote the subgroup of all rotations $\mathbf{t} \in SO(3)$ with $\mathbf{t}\eta = \eta$ by

$$SO_\eta(3) = \{\mathbf{t} \in SO(3) \mid \mathbf{t}\eta = \eta\}. \tag{3.3.3}$$

Our aim is to prove the irreducibility of the space $Harm_n$ of spherical harmonics of order n. As auxiliary point we use another important characterization of the Legendre polynomial: consider the function $L_n : \mathbb{R}^3 \to \mathbb{R}, n = 0, 1, \ldots$ defined by $(\mathrm{i} = \sqrt{-1})$

$$L_n(x) = \frac{1}{2\pi} \int_0^{2\pi} (x_3 + \mathrm{i}x_1 \cos\alpha + \mathrm{i}x_2 \sin\alpha)^n\, d\alpha, \quad x = (x_1, x_2, x_3)^T. \tag{3.3.4}$$

Clearly, L_n is a homogeneous polynomial of degree n which is harmonic and symmetric with respect to the x_3 axis. This function has the value 1 at ε^3. Moreover, after introducing the coordinates (1.2.24), we see that

$$\begin{aligned} L_n(x) &= \frac{r^n}{2\pi} \int_0^{2\pi} (t + \mathrm{i}\sqrt{1-t^2}\cos(\alpha - \varphi))^n\, d\alpha \qquad (3.3.5) \\ &= \frac{r^n}{2\pi} \int_0^{2\pi} (t + \mathrm{i}\sqrt{1-t^2}\cos\alpha)^n d\alpha \end{aligned}$$

$$= \quad r^n P_n(t)$$

(cf. Lemma 3.2.6). This implies that

$$L_n\left(\frac{x}{|x|}\right) = P_n(t), \quad x = |x|\left(t\varepsilon^3 + \sqrt{1-t^2}(\cos\varphi\varepsilon^1 + \sin\varphi\varepsilon^2)\right), \quad (3.3.6)$$

for all $x \in \mathbb{R}^3 \setminus \{0\}$. We are thus led to the following result.

Theorem 3.3.1. *Let H_n be a homogeneous, harmonic polynomial of degree n with the following properties:*

(i) $H_n(\mathbf{t}x) = H_n(x)$ for all orthogonal transformations $\mathbf{t} \in SO_{\varepsilon^3}(3)$

(ii) $H_n(\varepsilon^3) = 1$.

Then H_n is uniquely determined, and we have

$$H_n(x) = r^n P_n(t) = L_n(x), \quad x = r(t\varepsilon^3 + \sqrt{1-t^2}(\cos\varphi\varepsilon^1 + \sin\varphi\varepsilon^2)).$$

Proof. According to Theorem 2.2.1, H_n is uniquely determined by the homogeneous polynomials $A_n : (x_1, x_2) \mapsto A_n(x_1, x_2)$ and $A_{n-1} : (x_1, x_2) \mapsto A_{n-1}(x_1, x_2)$. The condition (i) implies that these polynomials depend only on $x_1^2 + x_2^2$. We thus get

$$A_n(x_1, x_2) = a(x_1^2 + x_2^2)^k, \qquad A_{n-1}(x_1, x_2) = 0 \quad \text{for } n = 2k \qquad (3.3.7)$$

and

$$A_{n-1}(x_1, x_2) = b(x_1^2 + x_2^2)^k, \qquad A_n(x_1, x_2) = 0 \quad \text{for } n = 2k+1. \qquad (3.3.8)$$

Apart from a multiplicative constant, A_n is therefore determined by condition (i). The value of the constant is then fixed by condition (ii). Using polar coordinates (1.2.4) we see that H_n depends on t only, as $x_1^2 + x_2^2 = r^2(1-t^2)$. Thus we have shown that there exists one and only one homogeneous harmonic polynomial H_n (dependent only on t) satisfying (i) and (ii). From the considerations given above, however, we know that L_n satisfies the required assumptions, hence, $L_n = H_n$. □

In other words, the only function $Y_n \in Harm_n$ satisfying (i) $Y_n(\mathbf{t}\xi) = Y_n(\xi), \xi \in \Omega$, for all orthogonal transformations $\mathbf{t} \in SO_\eta(3)$ (ii) $Y_n(\eta) = 1$, is given by $\xi \mapsto Y_n(\xi) = P_n(\xi \cdot \eta), \xi \in \Omega$.

After these preparations we now prove the following theorem.

Theorem 3.3.2. *The space $Harm_n$ of spherical harmonics of order n is irreducible.*

Proof. Assume that there exists an invariant subspace $\mathcal{Y}$ of dimension $d(\mathcal{Y}) < d(Harm_n) = 2n+1$. Then we would be able to show that the orthogonal complement $\mathcal{Y}^\perp$ of $\mathcal{Y}$ in $Harm_n$ (with respect to $(\cdot,\cdot)_{\mathcal{L}^2(\Omega)}$) is an invariant subspace, for

$$\begin{aligned}(F, F^\perp(\mathbf{t}\cdot))_{\mathcal{L}^2(\Omega)} &= \int_\Omega F(\xi)F^\perp(\mathbf{t}\xi)d\omega(\xi) \\ &= (\det \mathbf{t})^2 \int_\Omega F(\mathbf{t}^T\xi)F^\perp(\xi)\, d\omega(\xi) \\ &= 0\end{aligned}$$

holds for all $F \in \mathcal{Y}, F^\perp \in \mathcal{Y}^\perp$ (observe that $F(\mathbf{t}\cdot)$ is an element in $\mathcal{Y}$). But this means that $F^\perp(\mathbf{t}\cdot)$ is an element of $\mathcal{Y}^\perp$, i.e. $\mathcal{Y}^\perp$ is an invariant subspace. Now, because of the invariance of $\mathcal{Y}$ and $\mathcal{Y}^\perp$, G and $G^\perp$ being represented in terms of the $\mathcal{L}^2(\Omega)$-orthonormal system $\{Y_{n,j}\}$, respectively,

$$\begin{aligned}G &= \sum_{j=1}^{d(\mathcal{Y})} Y_{n,j}(\varepsilon^3)Y_{n,j} \in \mathcal{Y}, \\ G^\perp &= \sum_{j=d(\mathcal{Y})+1}^{d(Harm_n)} Y_{n,j}(\varepsilon^3)Y_{n,j} \in \mathcal{Y}^\perp\end{aligned}$$

satisfy $G(\mathbf{t}\cdot) = G, G^\perp(\mathbf{t}\cdot) = G^\perp$ for all $\mathbf{t} \in SO_{\varepsilon^3}(3)$. Moreover, $G, G^\perp$ do not vanish identically (note that there exist elements in $\mathcal{Y}, \mathcal{Y}^\perp$ different from zero at ε^3). Thus not all values of $Y_{n,j}(\varepsilon^3), j = 1, \ldots, d(\mathcal{Y})$ or $j = d(\mathcal{Y})+1, \ldots, d(Harm_n)$ are zero. From Theorem 3.3.1, therefore, there exists a constant $a \in \mathbb{R}\backslash\{0\}$ such that $G = aG^\perp$, in contradiction to our assumption. This proves Theorem 3.3.2. □

The irreducibility of $Harm_n$ leads us to simple representations of spherical harmonics (cf. T. Gervens (1989)).

Lemma 3.3.3. *Let Z_n be a member of $Harm_n$. Then there exist $2n+1$ orthogonal transformations $\mathbf{t}_1, \ldots, \mathbf{t}_{2n+1}$ such that any $Y_n \in Harm_n$ can be represented with real numbers $a_1, \ldots, a_{2n+1}$ in the form*

$$Y_n = \sum_{j=1}^{2n+1} a_j Z_n(\mathbf{t}_j\cdot).$$

Proof. Consider the set $\mathcal{V} = \{Z_n(\mathbf{t}\cdot) \mid \mathbf{t} \in O(3)\}$. Clearly, there exist $\mathbf{t}_1, \ldots, \mathbf{t}_{2n+1} \in O(3)$ such that

$$\mathcal{V} = \operatorname{span}\{Z_n(\mathbf{t}_1\cdot), \ldots, Z_n(\mathbf{t}_{2n+1}\cdot)\}. \tag{3.3.9}$$

Therefore, $Z_n \in \mathcal{V}$ implies $Z_n(\mathbf{t}\cdot) \in \mathcal{V}$. From the irreducibility of $Harm_n$ (cf. Theorem 3.3.2) it follows that $\mathcal{V}$ is an invariant non-void space of dimension $d(\mathcal{V}) = d(Harm_n)$, and the $2n+1$ linearly independent spherical harmonics $Z_n(\mathbf{t}_1\cdot), \ldots, Z_n(\mathbf{t}_{2n+1}\cdot)$ form a basis. □

Lemma 3.3.4. *There exist $2n+1$ points $\eta_1, \ldots, \eta_{2n+1}$ of the unit sphere Ω such that any $Y_n \in Harm_n$ can be respresented with real numbers $a_1, \ldots, a_{2n+1}$ in the form*

$$Y_n(\xi) = \sum_{j=1}^{2n+1} a_j P_n(\eta_j \cdot \xi), \quad \xi \in \Omega.$$

Proof. From Lemma 3.3.3 we know that there exist $2n+1$ orthogonal transformations $\mathbf{t}_1, \ldots, \mathbf{t}_{2n+1}$ such that any $Y_n \in Harm_n$ admits the form

$$Y_n(\xi) = \sum_{j=1}^{2n+1} a_j P_n(\varepsilon^3 \cdot \mathbf{t}_j \xi), \quad \xi \in \Omega. \tag{3.3.10}$$

But this means that Lemma 3.3.4 follows with $\eta_j = \mathbf{t}_j^T \varepsilon^3$, $j = 1, \ldots, 2n+1$. □

In connection with Maxwell's representation formula (3.2.34) it is possible to write the following.

Lemma 3.3.5. *For every $Y_n \in Harm_n$*

$$\frac{Y_n(\xi)}{|x|^{n+1}} = \frac{(-1)^n}{n!} \sum_{j=1}^{2n+1} a_j (\eta_j \cdot \nabla_x)^n \frac{1}{|x|}, \quad x = |x|\xi, \quad x \neq 0. \tag{3.3.11}$$

In other words, every *'outer harmonic'*

$$H_{-n-1} : x \mapsto H_{-n-1}(x) = |x|^{-(n+1)} Y_n(\xi), \quad x \in \mathbb{R}^3 \backslash \{0\}$$

may be regarded as the potential of a linear combination of multipoles with real axes. The system of points $\eta_1, \ldots, \eta_{2n+1}$ thus corresponds to a system of multipoles in Maxwell's interpretation of spherical harmonics.

Finally we come to the definition of *fundamental systems relative to* $Harm_n$.

Definition 3.3.6. A system X_{2n+1} of $2n+1$ points $\eta_1, \ldots, \eta_{2n+1}$ of the unit sphere Ω is called a fundamental system relative to $Harm_n$ if the matrix

$$\mathbf{matr}_{X_{2n+1}}(Y_{n,1}, \ldots, Y_{n,2n+1}) = \begin{pmatrix} Y_{n,1}(\eta_1) & \ldots & Y_{n,1}(\eta_{2n+1}) \\ \vdots & & \vdots \\ Y_{n,2n+1}(\eta_1) & \ldots & Y_{n,2n+1}(\eta_{2n+1}) \end{pmatrix} \tag{3.3.12}$$

is of maximal rank $2n+1$.

From Lemma 3.3.4 we obtain the following.

Lemma 3.3.7. *There exists a system* $X_{2n+1} = \{\eta_1, \ldots, \eta_{2n+1}\} \subset \Omega$ *satisfying*

$$\det(\mathbf{matr}_{X_{2n+1}}(Y_{n,1}, \ldots, Y_{n,2n+1})) \neq 0.$$

Proof. From Lemma 3.3.4 it follows that there exist $2n+1$ points $\eta_1, \ldots, \eta_{2n+1}$ such that the functions $P_n(\eta_1 \cdot), \ldots, P_n(\eta_{2n+1} \cdot)$ are linearly independent. Therefore the Gram matrix

$$\begin{aligned} &\left(\left(\frac{2n+1}{4\pi}\right)^2 \int_\Omega P_n(\eta_j \cdot \eta) P_n(\eta_k \cdot \eta)\, d\omega(\eta)\right)_{j,k=1,\ldots,2n+1} \\ &= \left(\frac{2n+1}{4\pi} P_n(\eta_j \cdot \eta_k)\right)_{j,k=1,\ldots,2n+1} \end{aligned} \tag{3.3.13}$$

is of maximal rank, and its determinant is positive. By virtue of the addition theorem we obtain

$$\begin{aligned} &(\mathbf{matr}_{X_{2n+1}}(Y_{n,1}, \ldots, Y_{n,2n+1}))^T \, \mathbf{matr}_{X_{2n+1}}(Y_{n,1}, \ldots, Y_{n,2n+1}) \\ &- (\frac{2n+1}{4\pi} P_n(\eta_j \cdot \eta_k))_{j,k=1,\ldots,2n+1} \\ &= \frac{2n+1}{4\pi} \mathbf{matr}_{X_{2n+1}}(P_n(\eta_1 \cdot), \ldots, P_n(\eta_{2n+1} \cdot)). \end{aligned} \tag{3.3.14}$$

Hence,

$$\begin{aligned} &\det \left(\frac{2n+1}{4\pi} P_n(\eta_j \cdot \eta_k)\right)_{j,k=1,\ldots,2n+1} \\ &= \det(\mathbf{matr}_{X_{2n+1}}(Y_{n,1}, \ldots, Y_{n,2n+1}))^2 > 0. \end{aligned} \tag{3.3.15}$$

This shows that $\det(\mathbf{matr}_{X_{2n+1}}(Y_{n,1}, \ldots, Y_{n,2n+1})) \neq 0$, as required. □

3.4 Closure and Completeness

Let, as usual, $\{Y_{n,j}\}, n = 0, 1, \ldots, j = 1, \ldots, 2n+1$, be an $\mathcal{L}^2(\Omega)$-orthonormal sequence of spherical harmonics. Let F be an arbitrary element of class $\mathcal{L}^2(\Omega)$ (or $\mathcal{C}(\Omega)$). The series

$$\sum_{n=0}^{\infty} \sum_{j=1}^{2n+1} (F, Y_{n,j})_{\mathcal{L}^2(\Omega)} Y_{n,j}$$

is the Fourier series (or orthogonal expansion) for F in terms of spherical harmonics. The constants $F^\wedge(n,j) = (F, Y_{n,j})_{\mathcal{L}^2(\Omega)}$ are known as the *Fourier coefficients of* F (or orthogonal coefficients of F with respect to $Y_{n,j}$). One frequently writes

$$F \sim \sum_{n=0}^{\infty} \sum_{j=1}^{2n+1} (F, Y_{n,j})_{\mathcal{L}^2(\Omega)} Y_{n,j}$$

to indicate that the right-hand sum is associated in a formal way with the left-hand side. In view of the fact that

$$\mathrm{Proj}_{Harm_n}(F) = \sum_{j=1}^{2n+1} (F, Y_{n,j})_{\mathcal{L}^2(\Omega)} Y_{n,j}$$

we may write

$$F \sim \sum_{n=0}^{\infty} \mathrm{Proj}_{Harm_n}(F),$$

and hence the *Fourier series* (orthogonal expansion in terms of spherical harmonics) of an element F is merely the sum of the projections of the element on the orthonormal system of spherical harmonics. The relation between an element and its Fourier series has been the object of many investigations. Of particular importance for practical purposes are results in the framework of the pre-Hilbert space $(\mathcal{C}(\Omega), (\cdot,\cdot)_{\mathcal{L}^2(\Omega)})$ or the Hilbert space $(\mathcal{L}^2(\Omega), (\cdot,\cdot)_{\mathcal{L}^2(\Omega)})$ which will be established below.

Starting point is Lemma 3.2.5 from which we obtain

$$\int_{-1}^{+1} \frac{1-h^2}{(1+h^2-2ht)^{3/2}}\, dt = \sum_{n=0}^{\infty} (2n+1) h^n \int_{-1}^{+1} P_n(t) P_0(t)\, dt \qquad (3.4.1)$$

for all $h \in (-1, 1)$. Hence, we have

$$\frac{1}{2} \int_{-1}^{+1} \frac{1-h^2}{(1+h^2-2ht)^{3/2}}\, dt = 1. \qquad (3.4.2)$$

Theorem 3.4.1 *(Poisson integral formula). If F is continuous on Ω, then*

$$\lim_{h\to 1, h<1} \sup_{\xi\in\Omega} \left|\frac{1}{4\pi}\int_\Omega \frac{(1-h^2)F(\eta)}{(1+h^2-2h(\xi\cdot\eta))^{3/2}}\,d\omega(\eta) - F(\xi)\right| = 0.$$

Proof. Observing the identity (3.4.2) we get

$$\begin{aligned}&\frac{1}{4\pi}\int_\Omega \frac{(1-h^2)F(\eta)}{(1+h^2-2h(\xi\cdot\eta))^{3/2}}\,d\omega(\eta) - F(\xi)\\ &= \frac{1}{4\pi}\int_\Omega \frac{(1-h^2)(F(\eta)-F(\xi))}{(1+h^2-2h(\xi\cdot\eta))^{3/2}}\,d\omega(\eta).\end{aligned} \tag{3.4.3}$$

For $h \in \left[\frac{1}{2}, 1\right)$ we split the integral into two parts:

$$\int_\Omega \cdots = \int\limits_{-1\le\xi\cdot\eta\le 1-\sqrt[3]{1-h}} \cdots + \int\limits_{1-\sqrt[3]{1-h}\le\xi\cdot\eta\le 1} \cdots . \tag{3.4.4}$$

On the one hand, we find

$$1+h^2-2ht = (1-h)^2 + 2h(1-t) \ge 2h\sqrt[3]{1-h} \tag{3.4.5}$$

and

$$\begin{aligned}\frac{1-h^2}{(1+h^2-2ht)^{3/2}} &\le \frac{1-h^2}{(2h\sqrt[3]{1-h})^{3/2}}\\ &= \frac{1+h}{(2h)^{3/2}}\frac{1-h}{\sqrt{1-h}}\\ &\le 2\sqrt{1-h}\end{aligned} \tag{3.4.6}$$

provided that $t \in \left[-1, 1-\sqrt[3]{1-h}\right]$. This leads us to

$$\left|\int\limits_{-1\le\xi\cdot\eta\le 1-\sqrt[3]{1-h}} \frac{(1-h^2)(F(\xi)-F(\eta))}{(1+h^2-2h(\xi\cdot\eta))^{3/2}}\,d\omega(\eta)\right| \tag{3.4.7}$$

$$\le 4\pi\|F\|_{C(\Omega)} \int_{-1}^{1-\sqrt[3]{1-h}} \frac{1-h^2}{(1+h^2-2ht)^{3/2}}\,dt \le 16\pi\|F\|_{C(\Omega)}\sqrt{1-h}.$$

On the other hand, F is uniformly continuous on Ω. Thus there exists a positive function $\mu : h \mapsto \mu(h)$ with

$$\lim_{\substack{h\to 1\\ h<1}} \mu(h) = 0 \tag{3.4.8}$$

such that

$$|F(\xi) - F(\eta)| \le \mu(h) \tag{3.4.9}$$

for all $\eta \in \Omega$ satisfying $1 - \sqrt[3]{1-h} \le \xi \cdot \eta \le 1$. Consequently, in connection with (3.4.2), we are able to deduce that

$$\left| \int\limits_{1-\sqrt[3]{1-h}\le\xi\cdot\eta\le 1} \frac{(1-h^2)(F(\xi)-F(\eta))}{(1+h^2-2h(\xi\cdot\eta))^{3/2}}\, d\omega(\eta) \right| \le \mu(h). \tag{3.4.10}$$

Letting h tend towards 1 we obtain the desired result. □

Combining Theorem 3.4.1 and Lemma 3.2.5 we find the following result.

Theorem 3.4.2. *Let F be of class $\mathcal{C}(\Omega)$. Then the series*

$$\sum_{n=0}^{\infty} h^n \sum_{j=1}^{2n+1} F^{\wedge}(n,j) Y_{n,j}(\xi)$$

with

$$F^{\wedge}(n,j) = (F, Y_{n,j})_{\mathcal{L}^2(\Omega)} = \int_{\Omega} F(\eta) Y_{n,j}(\eta)\, d\omega(\eta)$$

converges uniformly with respect to all $\xi \in \Omega$ for fixed $h \in (0, h_0), h_0 < 1$, and

$$\lim_{h\to 1, h<1} \sum_{n=0}^{\infty} h^n \sum_{j=1}^{2n+1} F^{\wedge}(n,j) Y_{n,j}(\xi) \;=\; F(\xi).$$

Theorem 3.4.2 enables us to prove the *closure of the system of spherical harmonics in the space $\mathcal{C}(\Omega)$* with respect to $\|\cdot\|_{\mathcal{C}(\Omega)}$.

Corollary 3.4.3. *The system $\{Y_{n,j}\}_{n=0,1,\ldots,\ j=1,\ldots,2n+1}$ is closed in $\mathcal{C}(\Omega)$, that is for any given $\varepsilon > 0$ and each $F \in \mathcal{C}(\Omega)$ there exists a linear combination*

$$\sum_{n=0}^{N} \sum_{j=1}^{2n+1} d_{n,j} Y_{n,j}$$

such that

$$\left\| F - \sum_{n=0}^{N} \sum_{j=1}^{2n+1} d_{n,j} Y_{n,j} \right\|_{\mathcal{C}(\Omega)} \leq \varepsilon.$$

Proof. Given $F \in \mathcal{C}(\Omega)$. Then, for any given $\varepsilon > 0$, there exists a real number $h = h(\varepsilon) < 1$ such that

$$\sup_{\xi \in \Omega} \left| F(\xi) - \sum_{n=0}^{\infty} h^n \sum_{j=1}^{2n+1} F^\wedge(n,j) Y_{n,j}(\xi) \right| \leq \frac{\varepsilon}{2}. \tag{3.4.11}$$

On the other hand, there exists an index $N = N(\varepsilon)$ such that

$$\sup_{\xi \in \Omega} \left| \sum_{n=0}^{\infty} h^n \sum_{j=1}^{2n+1} F^\wedge(n,j) Y_{n,j}(\xi) - \sum_{n=0}^{N} h^n \sum_{j=1}^{2n+1} F^\wedge(n,j) Y_{n,j}(\xi) \right| \leq \frac{\varepsilon}{2}. \tag{3.4.12}$$

But this means that

$$\sup_{\xi \in \Omega} \left| F(\xi) - \sum_{n=0}^{N} \sum_{j=1}^{2n+1} h^n F^\wedge(n,j) Y_{n,j}(\xi) \right| \leq \varepsilon, \tag{3.4.13}$$

which proves Corollary 3.4.3. □

Corollary 3.4.4. *The system $\{Y_{n,j}\}_{n=0,1,\ldots,\ j=1,\ldots,2n+1}$ is closed in the space $\mathcal{C}(\Omega)$ with respect to $\|\cdot\|_{\mathcal{L}^2(\Omega)}$, that is for any given $\varepsilon > 0$ and any given $F \in \mathcal{C}(\Omega)$, there exists a linear combination*

$$\sum_{n=0}^{N} \sum_{j=1}^{2n+1} b_{n,j} Y_{n,j}$$

such that

$$\left\| F - \sum_{n=0}^{N} \sum_{j=1}^{2n+1} b_{n,j} Y_{n,j} \right\|_{\mathcal{L}^2(\Omega)} \leq \varepsilon.$$

Proof. Corollary 3.4.4 follows immediately from Corollary 3.4.3 by using the norm estimate (1.2.10). □

Corollary 3.4.5. *The system $\{Y_{n,j}\}_{n=0,1,\ldots,\ j=1,\ldots,2n+1}$ is closed in the space $\mathcal{L}^2(\Omega)$ with respect to $\|\cdot\|_{\mathcal{L}^2(\Omega)}$.*

Proof. $\mathcal{C}(\Omega)$ is dense in $\mathcal{L}^2(\Omega)$, that is for every $F \in \mathcal{L}^2(\Omega)$ there exists a function $G \in \mathcal{C}(\Omega)$ with $\|F - G\|_{\mathcal{L}^2(\Omega)} \leq \varepsilon/2$. The function $G \in \mathcal{C}(\Omega)$ admits an arbitrarily close approximation by finite linear combinations of spherical harmonics. Therefore the proof of the closure is clear. □

Truncated spherical harmonic expansions admit the following *minimum property* which should be mentioned for the convenience of the reader.

Lemma 3.4.6. *Assume that* $F \in \mathcal{L}^2(\Omega)$. *Then*

$$\left\| F - \sum_{n=0}^{m} \sum_{j=1}^{2n+1} F^\wedge(n,j) Y_{n,j} \right\|_{\mathcal{L}^2(\Omega)} = \inf_{Y \in Harm_{0,\ldots,m}} \|F - Y\|_{\mathcal{L}^2(\Omega)} ,$$

i.e. the problem of finding the linear combination in $Harm_{0,\ldots,m}$ *which is minimal in the* $\mathcal{L}^2(\Omega)$*-norm is solved by the orthogonal projection* $Proj_{Harm_{0,\ldots,m}}(F)$ *of* F *onto* $Harm_{0,\ldots,m}$.

Proof. An easy calculation shows that

$$\begin{aligned} \left\| F - \sum_{n=0}^{m} \sum_{j=1}^{2n+1} L_{n,j} Y_{n,j} \right\|^2_{\mathcal{L}^2(\Omega)} &= (F,F)_{\mathcal{L}^2(\Omega)} \\ &\quad - \sum_{n=0}^{m} \sum_{j=1}^{2n+1} \mid F^\wedge(n,j) \mid^2 \\ &\quad + \sum_{n=0}^{m} \sum_{j=1}^{2n+1} \mid L_{n,j} - F^\wedge(n,j) \mid^2 \end{aligned} \tag{3.4.14}$$

for arbitrarily given coefficients $L_{n,j} \in \mathbb{R}$. Therefore, the minimum of the right-hand side of the equation (3.4.14) is achieved if and only if $L_{n,j} = F^\wedge(n,j)$ for $n = 0, \ldots, m,\ j = 1, \ldots, 2n+1$. Moreover,

$$\left\| F - \sum_{n=0}^{m} \sum_{j=1}^{2n+1} F^\wedge(n,j) Y_{n,j} \right\|^2_{\mathcal{L}^2(\Omega)} = (F,F)_{\mathcal{L}^2(\Omega)} - \sum_{n=0}^{m} \sum_{j=1}^{2n+1} |F^\wedge(n,j)|^2 .$$

□

We summarize our results in the *fundamental theorem of spherical harmonic expansions.*

Theorem 3.4.7. *The following seven statements are equivalent:*

(i) $\{Y_{n,j}\}$ *is closed in* $\mathcal{L}^2(\Omega)$ *(closure).*

(ii) *The orthogonal expansion of any element* $H \in \mathcal{L}^2(\Omega)$ *converges in norm to* H, *i.e.*

$$\lim_{m\to\infty} \left\| H - \sum_{n=0}^{m} \sum_{j=1}^{2n+1} (H, Y_{n,j})_{\mathcal{L}^2(\Omega)} Y_{n,j} \right\|_{\mathcal{L}^2(\Omega)} = 0.$$

(iii) *Parseval's identity holds. That is, for any* $H \in \mathcal{L}^2(\Omega)$,

$$\|H\|^2_{\mathcal{L}^2(\Omega)} = (H,H)_{\mathcal{L}^2(\Omega)} = \sum_{n=0}^{\infty} \sum_{j=1}^{2n+1} |\,(H, Y_{n,j})_{\mathcal{L}^2(\Omega)}\,|^2 .$$

(iv) *The extended Parseval identity holds. That is for any* $H, K \in \mathcal{L}^2(\Omega)$,

$$(H,K)_{\mathcal{L}^2(\Omega)} = \sum_{n=0}^{\infty} \sum_{j=1}^{2n+1} (H, Y_{n,j})_{\mathcal{L}^2(\Omega)} (K, Y_{n,j})_{\mathcal{L}^2(\Omega)}.$$

(v) *There is no strictly larger orthonormal system containing the orthonormal system* $\{Y_{n,j}\}_{n=0,1,\ldots,\ j=1,\ldots,2n+1}$.

(vi) *The system* $\{Y_{n,j}\}_{n=0,1,\ldots,\ j=1,\ldots,2n+1}$ *has the completeness property. That is,* $H \in \mathcal{L}^2(\Omega)$ *and* $(H, Y_{n,j})_{\mathcal{L}^2(\Omega)} = 0$ *for all* $n = 0,1,\ldots,$ $j = 1,\ldots,2n+1$, *implies* $H = 0$.

(vii) *An element* H *of* $\mathcal{L}^2(\Omega)$ *is determined uniquely by its orthogonal coefficients. That is, if* $(H, Y_{n,j})_{\mathcal{L}^2(\Omega)} = (K, Y_{n,j})_{\mathcal{L}^2(\Omega)}$, $n = 0,1,\ldots,$ $j = 1,\ldots,2n+1$, *then* $H = K$.

The proof of Theorem 3.4.7 is omitted (see for example P.J. Davis (1963)). The property (ii) is of great importance for practical purposes. In particular it tells us that any continuous function may be approximated (in the $\mathcal{L}^2(\Omega)$-sense) by finite truncations of its Fourier (orthogonal) expansion in terms of any $\mathcal{L}^2(\Omega)$-orthonormal system of spherical harmonics $\{Y_{n,j}\}$.

Definition 3.4.8. The *spherical Fourier transform* $FT : F \mapsto (FT)(F)$, $F \in \mathcal{L}^1(\Omega)$, is defined by

$$((FT)(F))(n,j) = F^\wedge(n,j) = \int_\Omega F(\eta) Y_{n,j}(\eta)\, d\omega(\eta).$$

As we have shown above, the restriction of the spherical Fourier transform to $\mathcal{L}^2(\Omega)$ forms a mapping from $\mathcal{L}^2(\Omega)$ into the space

$$\mathcal{L}^2_{FT}(\mathcal{J}) = \left\{ \{H(n,j)\} \left| \sum_{n=0}^{\infty} \sum_{j=1}^{2n+1} \mid H(n,j) \mid^2 < \infty \right. \right\}, \tag{3.4.15}$$

where we have used the abbreviation

$$\mathcal{J} = \{(n,j) \mid n = 0,1,\ldots, j = 1,\ldots, 2n+1\}. \tag{3.4.16}$$

From the considerations given above it is clear that any function $F \in \mathcal{L}^2(\Omega)$ is characterized by its sequence $\{F^\wedge(n,j)\} \in \mathcal{L}^2_{FT}(\mathcal{J})$.

Lemma 3.4.9 *(inverse transform). For $\{H(n,j)\} \in \mathcal{L}^2_{FT}(\mathcal{J})$ define the mapping*

$$(FT)^{-1} : \mathcal{L}^2_{FT}(\mathcal{J}) \to \mathcal{L}^2(\Omega)$$

by

$$(FT)^{-1}(\{H(n,j)\}) = \sum_{n=0}^{\infty} \sum_{j=1}^{2n+1} H(n,j) Y_{n,j}.$$

Then

$$\begin{aligned} (FT)^{-1}(FT) &= I_{\mathcal{L}^2(\Omega)}, \\ (FT)(FT)^{-1} &= I_{\mathcal{L}^2_{FT}(\mathcal{J})}. \end{aligned}$$

Moreover it should be noted that, for $F, G \in \mathcal{X}(\Omega)$, the relation

$$\lim_{N\to\infty} \left\| G - \sum_{n=0}^{N} \sum_{j=1}^{2n+1} F^\wedge(n,j) Y_{n,j} \right\|_{\mathcal{X}(\Omega)} = 0 \tag{3.4.17}$$

implies $F = G$ almost everywhere on Ω. But

$$\lim_{N\to\infty} \left\| F - \sum_{n=0}^{N} \sum_{j=1}^{2n+1} F^\wedge(n,j) Y_{n,j} \right\|_{\mathcal{X}(\Omega)} = 0 \tag{3.4.18}$$

does not hold for every $F \in \mathcal{X}(\Omega)$ provided that $\mathcal{X}(\Omega)$ is $\mathcal{C}(\Omega)$ or $\mathcal{L}^p(\Omega)$ with $p \in [1, \frac{4}{3}] \cup [4, \infty)$ (cf. R. Askey (1966), H. Pollard (1948)). In other words, finite truncations of Fourier expansions by means of $\mathcal{L}^2(\Omega)$-orthonormal spherical harmonics do not form an 'approximation process' when $\mathcal{X}(\Omega)$ is chosen as mentioned before. To the knowledge of the authors the convergence is unproved for $p \in [1, \frac{4}{3}] \cup [4, \infty)$. However, if F is assumed to be Lipschitz continuous, i.e. $F \in Lip(\Omega)$, T. GRONWALL (1914) has shown that F can be recovered by its Fourier expansion in uniform sense:

$$\lim_{N \to \infty} \left\| F - \sum_{n=0}^{N} \sum_{j=1}^{2n+1} F^{\wedge}(n,j) Y_{n,j} \right\|_{\mathcal{C}(\Omega)} = 0. \tag{3.4.19}$$

For the method of Cesàro summability to Fourier series the reader is referred to, for example, E. KOGBETLIANTZ (1924) or K.K. CHEN (1928).

3.5 Eigenfunctions of the Beltrami Operator

Next we discuss the role played by spherical harmonics as eigenfunctions of the Beltrami operator. From Lemma 3.1.2 we know that any spherical harmonic of order n is an infinitely often differentiable eigenfunction of the Beltrami operator corresponding to the eigenvalues $(\Delta^*)^{\wedge}(n) = -n(n+1)$, $n = 0, 1, \ldots$. On the other hand, the completeness of the system of spherical harmonics enables us to show that every eigenfunction $\xi \mapsto Y_n(\xi)$, $\xi \in \Omega$, of the Beltrami operator corresponding to the eigenvalue $-n(n+1)$ defines a homogeneous harmonic polynomial $x \mapsto H_n(x) = r^n Y_n(\xi)$, $x = r\xi$, $r = |x|$, $\xi \in \Omega$, of degree n.

We start our considerations with the following.

Lemma 3.5.1. *Assume that* $\lambda \neq n(n+1)$, $n = 0, 1, \ldots$. *Let* $K \in \mathcal{C}^{(\infty)}(\Omega)$ *satisfy the differential equation*

$$(\Delta^*_\xi + \lambda) K(\xi) = 0, \quad \xi \in \Omega.$$

Then $K = 0$.

Proof. Let $\{Y_{n,j}\}$ be an $\mathcal{L}^2(\Omega)$-orthonormal system of spherical harmonics. According to Green's second surface identity we get

$$\int_\Omega \Big(Y_{n,j}(\xi) \Delta^*_\xi K(\xi) - K(\xi) \Delta^*_\xi Y_{n,j}(\xi) \Big)\, d\omega(\xi) = 0. \tag{3.5.1}$$

Thus it follows that

$$\big(\lambda - n(n+1)\big) \int_\Omega K(\xi) Y_{n,j}(\xi)\, d\omega(\xi) = 0 \tag{3.5.2}$$

for $n = 0, 1, \ldots$, $j = 1, \ldots, 2n+1$. Since λ is assumed to be different from $n(n+1)$, $n = 0, 1, \ldots$, this implies

$$K^{\wedge}(n,j) = \int_\Omega K(\xi) Y_{n,j}(\xi)\, d\omega(\xi) = 0 \tag{3.5.3}$$

for $n = 0, 1, \ldots$, $j = 1, \ldots, 2n+1$. The completeness property of spherical harmonics, therefore, shows that $K = 0$, as required. □

Lemma 3.5.2. *Let $K \in \mathcal{C}^{(\infty)}(\Omega)$ satisfy the differential equation*

$$\left(\Delta_\xi^* + n(n+1)\right)K(\xi) = 0, \quad \xi \in \Omega.$$

Then $K \in Harm_n$.

Proof. By the same arguments as given above it follows that

$$(n(n+1) - k(k+1)) \int_\Omega K(\xi) Y_{k,j}(\xi)\, d\omega(\xi) = 0 \tag{3.5.4}$$

for $k = 0, 1, \ldots, j = 1, \ldots, 2k+1$. For all values k different from n we thus have $K^\wedge(k,j) = (K, Y_{k,j})_{\mathcal{L}^2(\Omega)} = 0$. But this means that K is a member of the span of $Y_{n,1}, \ldots, Y_{n,2n+1}$, as required. □

Summarizing our results we therefore obtain the following theorem.

Theorem 3.5.3. *The spherical harmonics $Y_n : \xi \mapsto Y_n(\xi)$ of order $n = 0, 1, \ldots$ are the everywhere on the unit sphere Ω infinitely often differentiable eigenfunctions of the Beltrami operator Δ^* corresponding to the eigenvalues $(\Delta^*)^\wedge(n) = -n(n+1)$, $n = 0, 1, \ldots$. The functions $x \mapsto H_n(x)$, $x \in \mathbb{R}^3$, with $H_n(x) = r^n Y_n(\xi)$, $x = r\xi, r = |x|$ are polynomials in cartesian coordinates which satisfy the Laplace equation and are homogeneous of degree n. Conversely, every homogeneous harmonic polynomial of degree n restricted to the unit sphere Ω is a spherical harmonic of order n.*

3.6 Funk–Hecke Formula

An outstanding result in the theory of spherical harmonics is the *Funk–Hecke formula* (cf. H. Funk (1916), E. Hecke (1918)).

Theorem 3.6.1. *Suppose that G is absolutely integrable on the interval $[-1,+1]$, i.e. $G \in \mathcal{L}^1[-1,+1]$. Then, for all $(\xi, \eta) \in \Omega^2$ and $n = 0, 1, \ldots,$*

$$\int_\Omega G(\xi \cdot \zeta) P_n(\eta \cdot \zeta)\, d\omega(\zeta) = G^\wedge(n) P_n(\xi \cdot \eta),$$

where

$$G^\wedge(n) = 2\pi \int_{-1}^{+1} G(t) P_n(t)\, dt = (G, P_n)_{\mathcal{L}^2[-1,+1]}.$$

Proof. For brevity we set

$$V_n(\xi,\eta) = \int_\Omega G(\xi\cdot\zeta)P_n(\eta\cdot\zeta)\,d\omega(\zeta). \tag{3.6.1}$$

Then, with any orthogonal matrix $\mathbf{t}$,

$$\begin{aligned} V_n(\mathbf{t}\xi,\mathbf{t}\eta) &= \int_\Omega G(\mathbf{t}\xi\cdot\zeta)P_n(\mathbf{t}\eta\cdot\zeta)\,d\omega(\zeta) \qquad (3.6.2)\\ &= (\det \mathbf{t})\int_\Omega G(\mathbf{t}\xi\cdot\mathbf{t}\zeta)P_n(\mathbf{t}\eta\cdot\mathbf{t}\zeta)\,d\omega(\mathbf{t}\zeta). \end{aligned}$$

But this shows us that

$$V_n(\mathbf{t}\xi,\mathbf{t}\eta) = (\det \mathbf{t})^2 V_n(\xi,\eta) = V_n(\xi,\eta). \tag{3.6.3}$$

On the other hand, according to (3.6.1) with ξ fixed, $V_n(\xi,\cdot)$ is a spherical harmonic of order n which, by virtue of (3.6.3), is invariant under orthogonal transformations. Therefore, by Theorem 3.3.1 there exists a constant $G^\wedge(n)$ such that $V_n(\xi,\eta) = G^\wedge(n)P_n(\xi\cdot\eta)$. In order to determine $G^\wedge(n)$ we set $\xi=\eta$ and find

$$\begin{aligned} G^\wedge(n) &= \int_\Omega G(\xi\cdot\zeta)P_n(\xi\cdot\zeta)\,d\omega(\zeta)\\ &= \int_\Omega G(\varepsilon^3\cdot\zeta)P_n(\varepsilon^3\cdot\zeta)\,d\omega(\zeta). \qquad (3.6.4) \end{aligned}$$

Using (1.2.24) it follows that $G^\wedge(n)$ admits the required representation. □

From Lemma 3.1.4 and Theorem 3.6.1 we get the following corollary.

Corollary 3.6.2. *Assume that $G \in \mathcal{L}^1[-1,1]$. Then, for every $Y_n \in Harm_n$,*

$$\int_\Omega G(\xi\cdot\eta)Y_n(\eta)d\omega(\eta) = G^\wedge(n)Y_n(\xi), \quad \xi\in\Omega,$$

where $G^\wedge(n)$ is given above.

The Funk–Hecke formula establishes the close connection between the orthogonal invariance of the sphere and the addition theorem. The principle of the Funk–Hecke formula is that of the integral operator mapping F to the '*convolution of G and F*' (cf. Definition 3.6.6). The kernel G depends only on the inner product $\xi\cdot\eta$, or equivalently on the distance between ξ and η. The spherical harmonics Y_n are the eigenfunctions of the integral operator corresponding to the eigenvalues $G^\wedge(n)$. Therefore, the Funk–Hecke formula greatly simplifies most manipulations with spherical harmonics.

Definition 3.6.3. The *Legendre transform* $G \mapsto (LT)(G), G \in \mathcal{L}^1[-1,+1]$, is defined by

$$((LT)(G))(n) = G^\wedge(n) = (G, P_n)_{\mathcal{L}^2[-1,+1]}.$$

The series

$$\sum_{n=0}^{\infty} \frac{2n+1}{4\pi} G^\wedge(n) P_n$$

is called *Legendre expansion* of G (with *Legendre coefficients* $G^\wedge(n), n = 0, 1, \ldots$).

The restriction of the Legendre transform to $\mathcal{L}^2[-1,+1]$ is a mapping from $\mathcal{L}^2[-1,+1]$ into the space $\mathcal{L}^2_{LT}(\mathbb{N}_0)$ of square-summable sequences $\{A_n\}$:

$$\mathcal{L}^2_{LT}(\mathbb{N}_0) = \left\{ \{A_n\} \left| \sum_{n=0}^{\infty} \frac{2n+1}{4\pi} |A_n|^2 < \infty \right. \right\}. \tag{3.6.5}$$

According to Parseval's identity we have

$$(G, H)_{\mathcal{L}^2[-1,+1]} = \sum_{n=0}^{\infty} \frac{2n+1}{4\pi} G^\wedge(n) H^\wedge(n) \tag{3.6.6}$$

for all $G, H \in \mathcal{L}^2[-1,+1]$. Moreover

$$\lim_{N \to \infty} \left\| G - \sum_{n=0}^{N} \frac{2n+1}{4\pi} G^\wedge(n) P_n \right\|_{\mathcal{L}^2[-1,+1]} = 0 \tag{3.6.7}$$

for all $G \in \mathcal{L}^2[-1,+1]$.

Lemma 3.6.4 *(inverse transform). For $\{A_n\} \in \mathcal{L}^2_{LT}(\mathbb{N}_0)$ define the mapping*

$$(LT)^{-1} : \mathcal{L}^2_{LT}(\mathbb{N}_0) \to \mathcal{L}^2[-1,+1]$$

by

$$(LT)^{-1}(\{A_n\}) = \sum_{n=0}^{\infty} \frac{2n+1}{4\pi} A_n P_n.$$

Then

$$\begin{aligned} (LT)^{-1}(LT) &= I_{\mathcal{L}^2[-1,+1]}, \\ (LT)(LT)^{-1} &= I_{\mathcal{L}^2_{LT}}(\mathbb{N}_0). \end{aligned}$$

Following H. POLLARD (1948) the convergence of the Legendre series of G is guaranteed for all $G \in \mathcal{L}^p[-1,+1]$ with $p \in (\frac{4}{3}, 4)$ (corresponding to $\|\cdot\|_{\mathcal{L}^p[-1,+1]}$). For $p \in [1,\infty] \setminus (\frac{4}{3}, 4)$ this assertion is generally false.

Observe that, for $G \in \mathcal{L}^2[-1,+1], G^\wedge(n) = O(n^{-2+\varepsilon}), \varepsilon > 0$, holds for $n \to \infty$. Furthermore it is not difficult to show that, for $G \in \mathcal{C}^{(2)}[-1,+1]$, $(L_t G)^\wedge(n) = -n(n+1)G^\wedge(n)$ so that $G^\wedge(n) = O(n^{-2(k+1)+\varepsilon}), \varepsilon > 0$, is valid for $G \in \mathcal{C}^{(2k)}[-1,+1], k \in \mathbb{N}_0$.

Lemma 3.6.5. *If $G \in \mathcal{C}^{(\infty)}[-1,+1]$, then*

$$\sum_{n=0}^{\infty} \frac{2n+1}{4\pi}(G^\wedge(n))^\alpha < \infty$$

holds for all $\alpha > 0$.

Convolutions on the sphere have been discussed by many authors (see, for example, S. BOCHNER (1954), A.P. CALDERON, A. ZYGMUND (1955)).

Definition 3.6.6. Assume that $G \in \mathcal{L}^1[-1,+1]$ and $F \in \mathcal{X}(\Omega)$. Then $G * F$ defined by

$$(G * F)(\xi) = \int_\Omega G(\xi \cdot \eta) F(\eta)\, d\omega(\eta)$$

is called the *convolution of G and F.*

Convolutions have the following properties:

(i) for $G \in \mathcal{L}^1[-1,+1], \quad Y_n \in Harm_n$

$$(G * Y_n)(\xi) = G^\wedge(n) Y_n(\xi), \quad \xi \in \Omega, \tag{3.6.8}$$

(ii) for $F \in \mathcal{L}^p(\Omega)$ and $G \in \mathcal{L}^q[-1,+1], 1 \le p, q < \infty$, $G * F$ exists almost everywhere on Ω, and we have *Young's inequality*

$$\|G * F\|_{\mathcal{L}^r(\Omega)} \le \|F\|_{\mathcal{L}^p(\Omega)} \|G\|_{\mathcal{L}^q[-1,+1]}, \quad \frac{1}{r} = \frac{1}{q} + \frac{1}{p} - 1 \ge 0. \tag{3.6.9}$$

In particular,

$$\begin{aligned} \|G * F\|_{\mathcal{L}^p(\Omega)} &\le \|F\|_{\mathcal{L}^p(\Omega)} \|G\|_{\mathcal{L}^1[-1,+1]}, \\ \|G * F\|_{\mathcal{L}^q(\Omega)} &\le \|F\|_{\mathcal{L}^1(\Omega)} \|G\|_{\mathcal{L}^q[-1,+1]}. \end{aligned}$$

(iii) for $F \in \mathcal{C}(\Omega)$ and $G \in \mathcal{L}^1[-1,+1]$, the convolution $G * F$ exists everywhere on Ω and is continuous, and we have

$$\|G * F\|_{\mathcal{C}(\Omega)} \le \|F\|_{\mathcal{C}(\Omega)} \|G\|_{\mathcal{L}^1[-1,+1]}.$$

(iv) for all $Y_n \in Harm_n$ (cf. Corollary 3.6.2)

$$\int_\Omega (G * F)(\eta) Y_n(\eta)\, d\omega(\eta) = G^\wedge(n) \int_\Omega F(\eta) Y_n(\eta)\, d\omega(\eta). \tag{3.6.10}$$

For later use we introduce the following definition.

Definition 3.6.7. Assume that $G \in \mathcal{L}^1[-1,+1]$ and $F \in \mathcal{L}^1(\Omega)$. For $(\xi, \zeta) \in \Omega^2$ we let

$$\begin{aligned} G^{(1)}(\xi \cdot \zeta) &= G(\xi \cdot \zeta) \\ G^{(k)}(\xi \cdot \zeta) &= \int_\Omega G^{(k-1)}(\xi \cdot \eta) G(\zeta \cdot \eta)\, d\omega(\eta), \quad k = 2, 3 \ldots . \end{aligned}$$

Then $G^{(k)} * F$ is called the kth *iterated convolution* of G and F.

Obviously it follows that

$$(G^{(k)})^\wedge(n) = (G^\wedge(n))^k, \quad n = 0, 1, \ldots, \quad k = 1, 2, \ldots . \tag{3.6.11}$$

Next we introduce the spherical counterpart T_h of the so-called *translation operator* (cf. RUDIN (1950))

$$T_h(F)(\xi) = \frac{1}{2\pi\sqrt{1-h^2}} \int_{\substack{\xi \cdot \eta = h \\ \eta \in \Omega}} F(\eta)\, d\sigma(\eta), \quad F \in \mathcal{X}(\Omega), \quad h \in (-1, +1), \tag{3.6.12}$$

where $d\sigma$ is the line element in $\mathbb{R}^3$. T_h is a bounded positive linear operator mapping $\mathcal{X}(\Omega)$ into $\mathcal{X}(\Omega)$ with the following properties:

$$||T_h(F)||_{\mathcal{X}(\Omega)} \le ||F||_{\mathcal{X}(\Omega)}, \tag{3.6.13}$$

$$\int_\Omega T_h(F)(\xi) Y_n(\xi)\, d\omega(\xi) = P_n(h) \int_\Omega F(\xi) Y_n(\xi)\, d\omega(\xi), \tag{3.6.14}$$

for $n = 0, 1, \ldots, h \in (-1, +1), F \in \mathcal{X}(\Omega)$. In particular,

$$T_h(Y_n)(\xi) = P_n(h) Y_n(\xi), \quad \xi \in \Omega, Y_n \in Harm_n. \tag{3.6.15}$$

Finally,

$$\lim_{\substack{h \to 1 \\ h < 1}} ||F - T_h(F)||_{\mathcal{X}(\Omega)} = 0. \tag{3.6.16}$$

In other words, $Y_n \in Harm_n$ implies $T_h(Y_n) \in Harm_n$, and $h \mapsto T_h(Y_n)$ is a polynomial of degree $\le n$. Furthermore it is not hard to see that for all $G \in \mathcal{L}^1[-1,+1]$ and all $F \in \mathcal{X}(\Omega)$

$$(G * F)(\xi) = 2\pi \int_{-1}^{+1} G(t) T_t(F)(\xi)\, dt \tag{3.6.17}$$

almost everywhere. Clearly, if $P : [-1, +1] \to \mathbb{R}$ is a polynomial of degree $\leq m$ and $F \in \mathcal{X}(\Omega)$, then $P * F$ is a member of the class of all spherical harmonics of order $\leq m$.

Bibliographical Notes

For more theoretical details on spherical harmonics the reader may want to consult some of the following references: R. COURANT, D. HILBERT (1924), W. FREEDEN (1979), E.W. HOBSON (1955), H. HOCHSTADT (1971), N.N. LEBEDEV (1973), J. LENSE (1953), C. MÜLLER (1952, 1966, 1969), R.T. SEELEY (1966), I.N. SNEDDON (1956), N.J. VILENKIN (1968), and many others.

4

GREEN'S FUNCTIONS WITH RESPECT TO ITERATED BELTRAMI OPERATORS

A fundamental question when approximating a function on the sphere by its truncated Fourier expansion (orthogonal expansion) in terms of spherical harmonics is the existence of a manageable error term. An adequate answer is the construction of integral formulae that will now be presented. It will be shown that the error term between a function and its truncated Fourier series expansion is explicitly available in integral form provided that sufficient smoothness is imposed on the function under consideration. An essential tool for the integral formulae is the theory of Green's function on the unit sphere Ω with respect to the operators $\partial_n = \Delta^* - (\Delta^*)^\wedge(n)$, $(\Delta^*)^\wedge(n) = -n(n+1)$, $n = 0, 1, 2, ...$ and their iterations (which also play a basic role in decompositions of spherical vector and tensor fields). In Chapter 5 integral formulae for more general operators will be considered in more detail.

4.1 Definition

We begin with the theory of Green's functions with respect to the operators $\partial_n, n = 0, 1, \ldots$.

Definition 4.1.1. $G(\partial_n;\cdot,\cdot) : (\xi, \eta) \mapsto G(\partial_n;\xi,\eta)$, $-1 \leq \xi \cdot \eta < 1$, is called Green's function with respect to the operator $\partial_n = \Delta^* - (\Delta^*)^\wedge(n)$ if it satisfies the following properties:

(i) *(differential equation)* for every point $\xi \in \Omega, \eta \mapsto G(\partial_n;\xi,\eta)$ is twice continuously differentiable on $\{\eta \in \Omega \mid -1 \leq \xi \cdot \eta < 1\}$, and we have

$$(\partial_n)_\eta G(\partial_n;\xi,\eta) = -\frac{2n+1}{4\pi} P_n(\xi \cdot \eta), \quad -1 \leq \xi \cdot \eta < 1,$$

where $(\partial_n)_\eta$ means that the operator ∂_n is applied to the variable η.

(ii) *(characteristic singularity)* for every $\xi \in \Omega$, the function

$$\eta \mapsto G(\partial_n;\xi,\eta) - \frac{1}{4\pi} \ln(1 - \xi \cdot \eta)$$

is continuously differentiable on Ω.

(iii) *(rotational symmetry)* for all orthogonal transformations $\mathbf{t}$,

$$G(\partial_n; \mathbf{t}\xi, \mathbf{t}\eta) = G(\partial_n; \xi, \eta).$$

(iv) *(normalization)* for every $\xi \in \Omega$,

$$\int_\Omega G(\partial_n; \xi, \eta) P_n(\xi \cdot \eta)\, d\omega(\eta) = 0.$$

We first prove the *uniqueness of Green's function* with respect to the operator ∂_n.

Lemma 4.1.2. *$G(\partial_n; \cdot, \cdot)$ is uniquely determined by its defining properties (i)–(iv).*

Proof. Denote by $D(\partial_n; \cdot, \cdot)$ the difference between two Green's functions satisfying (i)–(iv). Then we have the following properties:

(i) $D(\partial_n; \xi, \cdot)$ is twice continuously differentiable for all $\eta \in \Omega$ with $-1 \leq \xi \cdot \eta < 1$, and

$$(\partial_n)_\eta D(\partial_n; \xi, \eta) = 0, \tag{4.1.1}$$

(ii) $D(\partial_n; \xi, \cdot)$ is continuously differentiable for all $\eta \in \Omega$,

(iii) for all orthogonal transformations $\mathbf{t}$,

$$D(\partial_n; \mathbf{t}\xi, \mathbf{t}\eta) = D(\partial_n; \xi, \eta), \tag{4.1.2}$$

(iv)

$$\int_\Omega D(\partial_n; \xi, \eta) P_n(\xi \cdot \eta)\, d\omega(\eta) = 0 \tag{4.1.3}$$

for all $\xi \in \Omega$.

The properties (i)–(iii) show that $D(\partial_n; \xi, \cdot)$ is an everywhere on the unit sphere Ω infinitely often differentiable function satisfying the differential equation (i). Therefore $D(\partial_n; \xi, \cdot)$ must be a spherical harmonic of order n. $D(\partial_n; \xi, \eta)$ depends only on the scalar product of ξ and η, i.e.

$$D(\partial_n; \xi, \eta) = D(\partial_n; \xi \cdot \eta) = \alpha_n P_n(\xi \cdot \eta). \tag{4.1.4}$$

From (iv) we obtain

$$\int_\Omega D(\partial_n; \xi, \eta) P_n(\xi \cdot \eta)\, d\omega(\eta) = \alpha_n \int_\Omega P_n(\xi \cdot \eta) P_n(\xi \cdot \eta)\, d\omega(\eta) = 0. \tag{4.1.5}$$

Hence, $\alpha_n = 0$. But this means that Green's function $G(\partial_n; \cdot, \cdot)$ is uniquely determined by the defining properties (i)–(iv). □

Remark. Throughout this chapter we normally write $G(\partial_n;\xi,\eta)$ instead of $G(\partial_n;\xi\cdot\eta)$, $(\xi,\eta)\in\Omega^2$. But it should be kept in mind that $G(\partial_n;\xi,\eta)$ depends only on the scalar product of ξ and η, i.e. $G(\partial_n;\cdot)$ may be understood to be a function defined on the interval $[-1,+1]$.

An easy calculation shows that

$$-\frac{1}{4\pi}\int_0^1 \frac{1+r}{r}\left(\frac{1}{(1-2r(\xi\cdot\eta)+r^2)^{\frac{1}{2}}}-1\right)dr \tag{4.1.6}$$
$$=\frac{1}{4\pi}\ln(1-\xi\cdot\eta)+\frac{1}{4\pi}-\frac{1}{4\pi}\ln 2$$

satisfies all the defining properties (i)–(iv) of Green's function with respect to ∂_0. Hence, by virtue of the uniqueness, we have the explicit representation

$$G(\partial_0;\xi,\eta)=\frac{1}{4\pi}\ln(1-\xi\cdot\eta)+\frac{1}{4\pi}-\frac{1}{4\pi}\ln 2,\quad \xi,\eta\in\Omega,\ -1\le\xi\cdot\eta<1. \tag{4.1.7}$$

Observing the characteristic singularity of Green's function we see by applying Green's second surface identity that the spherical harmonics of order n, i.e. the eigenfunctions Y_n of the Beltrami operator Δ^* with respect to the eigenvalues $(\Delta^*)^\wedge(n)$, $n=0,1,\ldots$, are eigenfunctions of Green's (kernel) function $G(\partial_n;\cdot,\cdot)$ in the sense of the integral equation

$$(n(n+1)-k(k+1))\int_\Omega G(\partial_n;\xi,\eta)Y_k(\eta)\,d\omega(\eta)=(1-\delta_{n,k})Y_k(\xi). \tag{4.1.8}$$

Furthermore, if $\xi,\eta\in\Omega$ with $-1\le\xi\cdot\eta<1$, $G(\partial_n;\xi,\eta)$ allows the bilinear expansion

$$G(\partial_n;\xi,\eta)=\sum_{(\partial_n)^\wedge(k)\neq 0}\frac{1}{(\partial_n)^\wedge(k)}\sum_{j=1}^{2k+1}Y_{k,j}(\xi)Y_{k,j}(\eta), \tag{4.1.9}$$

where $(\partial_n)^\wedge(k)$ is given by

$$(\partial_n)^\wedge(k)=(\Delta^*)^\wedge(k)-(\Delta^*)^\wedge(n)=-k(k+1)+n(n+1). \tag{4.1.10}$$

The symbol $\sum_{(\partial_n)^\wedge(k)\neq 0}$ means that the sum is to be extended over all non-negative integers k for which the denominator $(\partial_n)^\wedge(k)$ is different from zero, i.e. $(\Delta^*)^\wedge(k)\neq(\Delta^*)^\wedge(n)$. Using the addition theorem we can rewrite the bilinear expansion of $G(\partial_n;\xi,\eta)$ in the form

$$G(\partial_n;\xi,\eta)=\sum_{(\partial_n)^\wedge(k)\neq 0}\frac{2k+1}{4\pi}\left((\partial_n)^\wedge(k)\right)^{-1}P_k(\xi\cdot\eta). \tag{4.1.11}$$

This gives rise to the following result.

Lemma 4.1.3. *For $n = 1, 2, \ldots$ and $\xi, \eta \in \Omega$ with $-1 \leq \xi \cdot \eta < 1$*

$$\begin{aligned} G(\partial_n; \xi, \eta) &= \frac{1}{4\pi} P_n(\xi \cdot \eta) \ln (1 - \xi \cdot \eta) \\ &\quad - \frac{1}{2\pi} \sum_{k=0}^{n-1} \frac{2k+1}{(\partial_n)^\wedge(k)} P_k(\xi \cdot \eta) \\ &\quad - \frac{1}{4\pi} \Big(\frac{2n+1}{2} \int_{-1}^{+1} P_n^2(t) \ln (1-t)\, dt \Big) P_n(\xi \cdot \eta). \end{aligned}$$

Proof. We have to show that $G(\partial_n; \cdot, \cdot)$ satisfies the defining properties (i)–(iv) of Definition 4.1.1. An easy calculation yields

$$\begin{aligned} &\left((1-t^2) \left(\frac{d}{dt} \right)^2 - 2t \frac{d}{dt} + n(n+1) \right) P_n(t) \ln (1-t) \\ &= -2(1+t) P_n'(t) - P_n(t). \end{aligned} \tag{4.1.12}$$

In connection with Lemma 3.2.3 we therefore obtain

$$\begin{aligned} &\left((1-t^2) \left(\frac{d}{dt} \right)^2 - 2t \frac{d}{dt} + n(n+1) \right) (-P_n(t) \ln(1-t)) \\ &+ \left((1-t^2) (\frac{d}{dt})^2 - 2t \frac{d}{dt} + n(n+1) \right) \left(-2 \sum_{k=0}^{n-1} \frac{2k+1}{n(n+1) - k(k+1)} P_k(t) \right) \\ &= 2(1+t) P_n'(t) + P_n(t) - 2 \sum_{k=0}^{n-1} (2k+1) P_k(t) \\ &= (2n+1) P_n(t). \end{aligned} \tag{4.1.13}$$

This shows that condition (i) of Definition 4.1.1 is valid. Because of $P_n(1) = 1, n = 1, 2, \ldots$, condition (ii) is certainly satisfied. Finally it is not difficult to see that

$$\int_\Omega G(\partial_n; \xi, \eta) P_n(\xi \cdot \eta)\, d\omega(\eta) = 0.$$

□

According to the classical Fredholm–Hilbert theory of linear integral equations (cf. D. HILBERT (1912), R. COURANT, D. HILBERT (1924)) we inductively define Green's functions with respect to iterated operators

$$\partial_{0,\ldots,m} = \partial_0 \cdot \ldots \cdot \partial_m, \tag{4.1.14}$$

where

$$\partial_n = \Delta^* - (\Delta^*)^\wedge(n), \quad n = 0, \ldots, m,$$

by the following convolutions

$$G(\partial_{0,\ldots,m}; \xi, \eta) = \int_\Omega G(\partial_{0,\ldots,m-1}; \xi, \zeta) G(\partial_m; \zeta, \eta)\, d\omega(\zeta), \; m = 1, 2\ldots \,. \tag{4.1.15}$$

$G(\partial_{0,\ldots,m}; \cdot, \cdot) : \Omega \times \Omega \to \mathbb{R}$ is called *Green's function with respect to the operator* $\partial_{0,\ldots,m}$. An illustration is given in Figure 4.1.1.

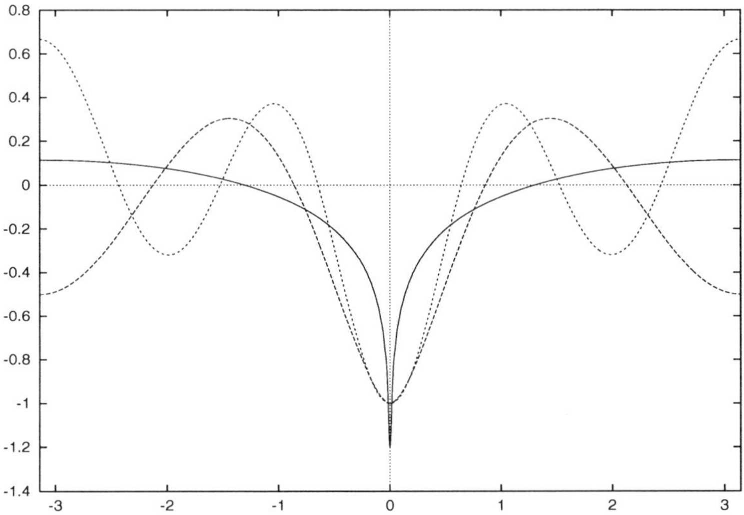

FIG. 4.1.1. Green's functions $G(\partial_{0,\ldots,m}; \cos\vartheta)$ for $m = 0,\ 1,\ 2$ (normalized)

For positive integers m, $G(\partial_{0,\ldots,m}; \xi, \eta)$ is continuous on the whole sphere Ω as a function of η with ζ fixed, or as a function of ζ with η fixed. On the other hand, the bilinear expansion of $G(\partial_{0,\ldots,m}; \xi, \eta), m \geq 1$,

$$\sum_{k=m+1}^{\infty} \sum_{l=1}^{2k+1} \frac{1}{(\partial_{0,\ldots,m})^\wedge(k)} Y_{k,l}(\xi) Y_{k,l}(\eta) = \frac{1}{4\pi} \sum_{k=m+1}^{\infty} \frac{2k+1}{(\partial_{0,\ldots,m})^\wedge(k)} P_k(\xi \cdot \eta)$$

with

$$(\partial_{0,\ldots,m})^\wedge(k) = (\partial_0)^\wedge(k) \cdot \ldots \cdot (\partial_m)^\wedge(k)$$

is absolutely and uniformly convergent both in ξ and η respectively and uniformly in ξ and η together. Hence, the representation theorem of the theory of orthogonal expansions yields

$$G(\partial_{0,\dots,m};\xi,\eta) = \frac{1}{4\pi}\sum_{k=m+1}^{\infty}\frac{2k+1}{(\partial_{0,\dots,m})^\wedge(k)}P_k(\xi\cdot\eta).$$

Let m be an integer with $m \geq 2$. Then the derivative

$$(\partial_2\cdot\ldots\cdot\partial_m)_\eta\ G(\partial_{0,\dots,m};\xi,\eta) = G(\partial_0\partial_1;\xi,\eta), \quad -1\leq\xi\cdot\eta\leq 1,$$

as a function of η for fixed ξ is a continuous function on Ω. For integers $m \geq 1$, the derivative

$$(\partial_1\cdot\ldots\cdot\partial_m)_\eta\ G(\partial_{0,\dots,m};\xi,\eta) = G(\partial_0;\xi,\eta), \quad -1\leq\xi\cdot\eta<1, \tag{4.1.16}$$

as a function of η possesses a logarithmic singularity in $\xi \in \Omega$ (cf. Definition 4.1.1). These properties are of basic interest in spherical spline settings corresponding to iterated Beltrami derivatives (cf. Chapter 6).

4.2 Integral Formulae

Let ξ be a fixed point of the unit sphere Ω. Assume that F is continuously differentiable on Ω. Then, for each sufficiently small ρ, Green's first surface identity gives

$$\int_{|\xi-\eta|\geq\rho,|\eta|=1}\left\{F(\eta)\Delta^*_\eta G(\Delta^*;\xi,\eta)+\nabla^*_\eta F(\eta)\cdot\nabla^*_\eta G(\Delta^*;\xi,\eta)\right\}d\omega(\eta)$$

$$= \int_{|\xi-\eta|=\rho,|\eta|=1} F(\eta)\frac{\partial}{\partial\nu_\eta}G(\Delta^*;\xi,\eta)\,d\sigma(\eta), \tag{4.2.1}$$

where ν is the unit normal to the circle consisting of all points $\eta\in\Omega$ with $|\ \xi-\eta\ |=\rho$, tangential to Ω, and directed exterior to the set of all points $\eta\in\Omega$ with $|\ \xi-\eta\ |\geq\rho$. Explicitly, for each point η, we have

$$\nu_\eta = -(1-(\xi\cdot\eta)^2)^{-\frac{1}{2}}\eta\wedge(\eta\wedge\xi). \tag{4.2.2}$$

In identity (4.2.1) we firstly observe the differential equation of Green's function

$$\int_{|\xi-\eta|\geq\rho,|\eta|=1} F(\eta)\Delta^*_\eta G(\Delta^*;\xi,\eta)\,d\omega(\eta) = -\frac{1}{4\pi}\int_{|\xi-\eta|\geq\rho,|\eta|=1} F(\eta)\,d\omega(\eta). \tag{4.2.3}$$

In coordinates $\eta = s\varepsilon^3+\sqrt{1-s^2}(\cos\lambda\varepsilon^1+\sin\lambda\varepsilon^2)$ the line element can be expressed as follows:

$$d\sigma^2(\eta) = (1-s^2)^{-1}\,ds^2 + (1-s^2)\,d\lambda^2. \tag{4.2.4}$$

Therefore, by virtue of the logarithmic singularity of Green's function $G(\Delta^*;\cdot,\cdot)$, we get in analogy to the well known results of potential theory (cf. O.D. Kellogg (1929)), on passing to the limit $\rho\to 0$, the following result.

Theorem 4.2.1 *(first fundamental theorem for ∇^*). Let ξ be a fixed point of the unit sphere Ω. Suppose that F is a continuously differentiable function on Ω. Then*

$$F(\xi) = \frac{1}{4\pi}\int_\Omega F(\eta)\,d\omega(\eta) - \int_\Omega (\nabla^*_\eta G(\Delta^*;\xi,\eta))\cdot(\nabla^*_\eta F(\eta))\,d\omega(\eta).$$

An immediate consequence is the following corollary.

Corollary 4.2.2 *(first fundamental theorem for L^*). Under the assumptions of Theorem 4.2.1*

$$F(\xi) = \frac{1}{4\pi}\int_\Omega F(\eta)\,d\omega(\eta) - \int_\Omega (L^*_\eta G(\Delta^*;\xi,\eta))\cdot(L^*_\eta F(\eta))\,d\omega(\eta).$$

From Green's second surface identity we get for each sufficiently small $\rho > 0$

$$\begin{aligned}&\int\limits_{|\xi-\eta|\geq\rho,|\eta|=1} \{G(\Delta^*;\xi,\eta)(\Delta^*)_\eta F(\eta) - F(\eta)(\Delta^*)_\eta G(\Delta^*;\xi,\eta)\}\,d\omega(\eta)\\ &= \int\limits_{|\xi-\eta|=\rho,|\eta|=1} \left\{G(\Delta^*;\xi,\eta)\frac{\partial}{\partial\nu_\eta}F(\eta) - F(\eta)\frac{\partial}{\partial\nu_\eta}G(\Delta^*;\xi,\eta)\right\}d\sigma(\eta),\end{aligned} \tag{4.2.5}$$

provided that F is twice continuously differentiable on Ω. By observing the characteristic properties of the Green function with respect to Δ^* we now obtain

Theorem 4.2.3 *(second fundamental theorem). Let ξ be a fixed point of the unit sphere Ω. Suppose that F is a twice continuously differentiable function on Ω. Then*

$$F(\xi) = \frac{1}{4\pi}\int_\Omega F(\eta)\,d\omega(\eta) + \int_\Omega G(\Delta^*;\xi,\eta)(\Delta^*_\eta F(\eta))\,d\omega(\eta).$$

The fundamental theorems as stated above compare the value of a function at a point $\xi \in \Omega$ with the integral mean of F relative to the unit sphere Ω under explicit representation of the error term in integral form. Observing the fact that the pth convolution of Green's function with respect to ∂_n coincides with the Green function with respect to ∂_n^p, i.e. $G^{(p)}(\partial_n;\xi,\eta) = G((\partial_n)^p;\xi,\eta)$, $p \geq 1, n \geq 0$, we obtain more generally

$$F(\xi) = \frac{2n+1}{4\pi}\int_\Omega F(\eta)P_n(\xi\cdot\eta)\,d\omega(\eta) + \int_\Omega G(\partial_n^p;\xi,\eta)((\partial_n^p)_\eta F(\eta))\,d\omega(\eta) \tag{4.2.6}$$

provided that F is of class $\mathcal{C}^{(2p)}(\Omega)$, where

$$G(\partial_n^p;\xi,y) = \frac{1}{4\pi} \sum_{(\partial_n)^\wedge(k)\neq 0} \frac{2k+1}{((\partial_n)^\wedge(k))^p} P_k(\xi\cdot\eta). \tag{4.2.7}$$

Hence we are now able to compare a function $F \in \mathcal{C}^{(2p)}(\Omega)$ with the nth order term of its orthogonal expansion in terms of spherical harmonics.

By use of Green's function $G(\partial_0\partial_1;\cdot,\cdot)$ with respect to the operator $\partial_0\partial_1$ we are able to generalize the second fundamental theorem. Observing the recursion property

$$(\partial_1)_\eta G(\partial_0\partial_1;\xi,\eta) = G(\partial_0;\xi,\eta) - \frac{3}{4\pi(\partial_0)^\wedge(1)} P_1(\xi\cdot\eta) \tag{4.2.8}$$

we obtain

$$\begin{aligned} &\int_\Omega G(\partial_0;\xi,\eta)\big((\partial_0)_\eta F(\eta)\big)\,d\omega(\eta) \\ &\quad = \int_\Omega (\partial_1)_\eta G(\partial_0\partial_1;\xi,\eta)\big((\partial_0)_\eta F(\eta)\big)\,d\omega(\eta) \\ &\qquad + \frac{3}{4\pi(\partial_0)^\wedge(1)} \int_\Omega P_1(\xi\cdot\eta)\big((\partial_0)_\eta F(\eta)\big)\,d\omega(\eta). \end{aligned} \tag{4.2.9}$$

Integration by parts, i.e. application of Green's second identity, yields for a function $F \in \mathcal{C}^{(4)}(\Omega)$

$$\begin{aligned} &\int_\Omega (\partial_1)_\eta G(\partial_0\partial_1;\xi,\eta)\big((\partial_0)_\eta F(\eta)\big)\,d\omega(\eta) \\ &\quad = \int_\Omega G(\partial_0\partial_1;\xi,\eta)\big((\partial_0\partial_1)_\eta F(\eta)\big)\,d\omega(\eta). \end{aligned} \tag{4.2.10}$$

In the same way we get

$$\frac{3}{4\pi(\partial_0)^\wedge(1)} \int_\Omega P_1(\xi\cdot\eta)\,((\partial_0)_\eta F(\eta))\,d\omega(\eta) = \frac{3}{4\pi}\int_\Omega P_1(\xi\cdot\eta)F(\eta)\,d\omega(\eta). \tag{4.2.11}$$

Therefore, we have

$$\begin{aligned} &\int_\Omega G(\partial_0;\xi,\eta)\big((\partial_0)_\eta F(\eta)\big)\,d\omega(\eta) \\ &\quad = \int_\Omega G(\partial_0\partial_1;\xi,\eta)\big((\partial_0\partial_1)_\eta F(\eta)\big)\,d\omega(\eta) + \frac{3}{4\pi}\int_\Omega F(\eta)P_1(\xi\cdot\eta)\,d\omega(\eta) \end{aligned} \tag{4.2.12}$$

provided that F is a four times continuously differentiable function on Ω. Thus, by combination of Theorem 4.2.3 and (4.2.12), we have, for $F \in \mathcal{C}^{(4)}(\Omega)$

$$F(\xi) = \sum_{n=0}^{1} \frac{2n+1}{4\pi} \int_\Omega F(\eta)P_n(\xi\cdot\eta)\,d\omega(\eta) \tag{4.2.13}$$

$$+\int_\Omega G(\partial_0\partial_1;\xi,\eta)\big((\partial_0\partial_1)_\eta F(\eta)\big)\,d\omega(\eta).$$

More generally, by successive integration by parts, we obtain in connection with the definition of $G(\partial_{0,\ldots,m};\cdot,\cdot)$ the following integral formulae.

Theorem 4.2.4. *Let m be a non-negative integer and ξ be a fixed point of the unit sphere Ω. Let F be a $(2m+2)$times continuously differentiable function on Ω. Then*

$$F(\xi)=\sum_{n=0}^{m}\frac{2n+1}{4\pi}\int_\Omega F(\eta)P_n(\xi\cdot\eta)\,d\omega(\eta)$$
$$+\int_\Omega G(\partial_{0,\ldots,m};\xi,\eta)\big((\partial_{0,\ldots,m})_\eta F(\eta)\big)\,d\omega(\eta). \qquad (4.2.14)$$

If F is $(2m+1)$times continuously differentiable on Ω, then

$$F(\xi)=\sum_{n=0}^{m}\frac{2n+1}{4\pi}\int_\Omega F(\eta)P_n(\xi\cdot\eta)\,d\omega(\eta) \qquad (4.2.15)$$
$$-\int_\Omega\big(\nabla^*_\eta G(\partial_{0,\ldots,m};\xi,\eta)\big)\cdot\big(\nabla^*_\eta(\partial_{1,\ldots,m})_\eta F(\eta)\big)\,d\omega(\eta),$$

where

$$\nabla^*_\eta G(\partial_{0,\ldots,m};\xi,\eta)=\Big(\frac{1}{4\pi}\sum_{k=m+1}^{\infty}\frac{2k+1}{(\partial_{0,\ldots,m})^\wedge(k)}P'_k(\xi\cdot\eta)\Big)(\xi-(\xi\cdot\eta)\eta).$$

Analogously, if F is $(2m+1)$times continuously differentiable on Ω, then

$$F(\xi)=\sum_{n=0}^{m}\frac{2n+1}{4\pi}\int_\Omega F(\eta)P_n(\xi\cdot\eta)\,d\omega(\eta)$$
$$-\int_\Omega L^*_\eta G(\partial_{0,\ldots,m};\xi,\eta)\cdot L^*_\eta(\partial_{1,\ldots,m})_\eta F(\eta)\,d\omega(\eta),$$

where

$$L^*_\eta G(\partial_{0,\ldots,m};\xi,\eta)=\left(\frac{1}{4\pi}\sum_{k=m+1}^{\infty}\frac{2k+1}{(\partial_{0,\ldots,m})^\wedge(k)}P'_k(\xi\cdot\eta)\right)(\xi\wedge\eta).$$

Inserting the addition theorem of spherical harmonics we find from (4.2.14) for functions $F\in C^{(2m+2)}(\Omega)$

$$F(\xi)=\sum_{n=0}^{m}\sum_{j=1}^{2n+1}Y_{n,j}(\xi)\int_\Omega F(\eta)Y_{n,j}(\eta)\,d\omega(\eta)$$

$$+ \int_{\Omega} G(\partial_{0,\ldots,m}; \xi, \eta) \big((\partial_{0,\ldots,m})_\eta F(\eta) \big) \, d\omega(\eta). \tag{4.2.16}$$

This formula gives a comparison between the mth partial sum of the Fourier expansion of F into spherical harmonics and the functional value of F with explicit knowledge of the remainder term.

More general, by iterated application of Green's second identity, we obtain

$$\begin{aligned} F(\xi) &= \sum_{n=0}^{m} \sum_{j=1}^{2n+1} Y_{n,j}(\xi) \int_{\Omega} F(\eta) Y_{n,j}(\eta) \, d\omega(\eta) \\ &\quad + \int_{\Omega} G(\partial_0^{p_0} \cdot \ldots \partial_m^{p_m}; \xi, \eta)((\partial_0^{p_0} \cdot \ldots \partial_m^{p_m}) F(\eta)) \, d\omega(\eta) \end{aligned}$$

provided that $p_l \geq 1$, $l = 0, \ldots, m$, and F is $2(p_0 + \ldots + p_m)$times continuously differentiable on Ω.

The identity (4.2.14) can be written as follows:

$$\begin{aligned} & F(\xi) - \sum_{n=0}^{m} \sum_{j=1}^{2n+1} Y_{n,j}(\xi) \int_{\Omega} F(\eta) Y_{n,j}(\eta) \, d\omega(\eta) \\ &= \int_{\Omega} (\partial_{0,\ldots,m})_\eta G(\partial^2_{0,\ldots,m}; \xi, \eta) \big((\partial_{0,\ldots,m})_\eta F(\eta) \big) \, d\omega(\eta), \end{aligned} \tag{4.2.17}$$

where $\partial^2_{0,\ldots,m} = \partial_{0,\ldots,m} \partial_{0,\ldots,m}$ and $G(\partial^2_{0,\ldots,m}; \cdot, \cdot)$ is given by the convolution integral (for a graphical impression, see Figure 4.2.1)

$$\begin{aligned} G(\partial^2_{0,\ldots,m}; \xi, \eta) &= G^{(2)}(\partial_{0,\ldots,m}; \xi, \eta) \\ &= \int_{\Omega} G(\partial_{0,\ldots,m}; \xi, \zeta) G(\partial_{0,\ldots,m}, \zeta, \eta) \, d\omega(\zeta). \end{aligned} \tag{4.2.18}$$

Clearly, $G(\partial^2_{0,\ldots,m}; \cdot, \cdot)$ allows the bilinear expansion

$$G(\partial^2_{0,\ldots,m}; \xi, \eta) = \frac{1}{4\pi} \sum_{k=m+1}^{\infty} \frac{2k+1}{(\partial^2_{0,\ldots,m})^\wedge(k)} P_k(\xi \cdot \eta), \quad \xi, \eta \in \Omega, \tag{4.2.19}$$

with

$$(\partial^2_{0,\ldots,m})^\wedge(k) = ((\partial_{0,\ldots,m})^\wedge(k))^2. \tag{4.2.20}$$

From (4.2.17) we obtain, for example,

$$\left\| F - \sum_{n=0}^{m} \sum_{j=1}^{2n+1} (F, Y_{n,j})_{\mathcal{L}^2(\Omega)} Y_{n,j} \right\|_{\mathcal{L}^2(\Omega)}$$

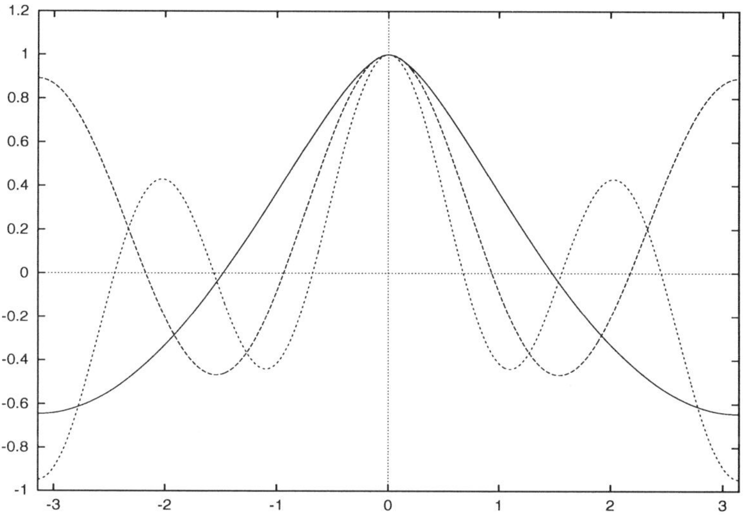

FIG. 4.2.1. Green's functions $G(\partial^2_{0,\ldots,m}; \cos\vartheta)$ for $m = 0,\ 1,\ 2$ (normalized)

$$\leq \left(\frac{1}{4\pi} \sum_{k=m+1}^{\infty} \frac{2k+1}{(\partial^2_{0,\ldots,m})^\wedge(k)} \right)^{\frac{1}{2}} \|(\partial_{0,\ldots,m})F\|_{\mathcal{L}^2(\Omega)}. \tag{4.2.21}$$

We omit $\mathcal{L}^p(\Omega)$-estimates for $p \neq 2$ and $\mathcal{C}(\Omega)$-estimates.

Remark. Let $\rho_0, \ldots, \rho_m$ be complex numbers. Suppose that $\Lambda_{0,\ldots,m}$ is the operator recursively given by

$$\Lambda_{0,\ldots,m} = (\Delta^* + \rho_m)\Lambda_{0,\ldots,m-1}, \quad m \geq 1, \tag{4.2.22}$$

$$\Lambda_0 = \Delta^* + \rho_0. \tag{4.2.23}$$

Then Green's function $G(\Lambda_{0,\ldots,m}; \cdot, \cdot)$ with respect to the operator $\Lambda_{0,\ldots,m}$ possesses the bilinear expansion

$$G(\Lambda_{0,\ldots,m}; \xi, \eta) = \frac{1}{4\pi} \sum_{(\Lambda_{0,\ldots,m})^\wedge(k) \neq 0} \frac{2k+1}{(\Lambda_{0,\ldots,m})^\wedge(k)} P_k(\xi \cdot \eta), \tag{4.2.24}$$

where $\sum_{(\Lambda_{0,\ldots,m})^\wedge(k)\neq 0}$ means that the summation is extended over all non-negative integers k for which

$$(\Lambda_{0,\ldots,m})^\wedge(k) = ((\Delta^*)^\wedge(k) + \rho_0) \cdot \ldots \cdot ((\Delta^*)^\wedge(k) + \rho_m) \neq 0. \tag{4.2.25}$$

The corresponding integral formula reads

$$\begin{aligned} F(\xi) &= \sum_{(\Lambda_{0,\ldots,m})^\wedge(k)=0} \frac{2k+1}{4\pi} \int_\Omega F(\eta) P_k(\xi\cdot\eta)\, d\omega(\eta) \\ &\quad + \int_\Omega G(\Lambda_{0,\ldots,m};\xi,\eta)((\Lambda_{0,\ldots,m})_\eta F(\eta))\, d\omega(\eta) \qquad (4.2.26) \end{aligned}$$

provided that F is of class $\mathcal{C}^{(2m+2)}(\Omega)$. As a consequence we obtain for an operator $\Lambda_{0,\ldots,m}$ satisfying $(\Lambda_{0,\ldots,m})^\wedge(k) \neq 0$ for all $k \geq 0$

$$F(\xi) = \int_\Omega G(\Lambda_{0,\ldots,m};\xi,\eta)((\Lambda_{0,\ldots,m})_\eta F(\eta))\, d\omega(\eta). \qquad (4.2.27)$$

For details the reader is referred to W. FREEDEN (1980B), W. FREEDEN, R. REUTER (1982). Later on we shall generalize the integral formula of the above type to pseudodifferential operators.

4.3 Beltrami Differential Equation

In what follows, let $Harm_{0,\ldots,m}, m \geq 0$, denote the space of all spherical harmonics of order $\leq m$ so that

$$Harm_{0,\ldots,m} = \oplus_{l=0}^{m} Harm_l. \qquad (4.3.1)$$

Obviously, $Harm_{0,\ldots,m}$ possesses the dimension

$$M = d(Harm_{0,\ldots,m}) = \sum_{j=0}^{m} d(Harm_j) = \sum_{j=0}^{m}(2j+1) = (m+1)^2. \qquad (4.3.2)$$

The space $Harm_{0,\ldots,m}$ equipped with the inner product $(\cdot,\cdot)_{\mathcal{L}^2(\Omega)}$ is an M-dimensional Hilbert space with the reproducing kernel $K_{Harm_{0,\ldots,m}}(\cdot,\cdot) : \Omega\times\Omega \to \mathbb{R}$ given by

$$K_{Harm_{0,\ldots,m}}(\xi,\eta) = \sum_{n=0}^{m}\sum_{j=1}^{2n+1} Y_{n,j}(\xi)Y_{n,j}(\eta) = \sum_{n=0}^{m} \frac{2n+1}{4\pi} P_n(\xi\cdot\eta). \qquad (4.3.3)$$

Note that the recursion relation

$$(n+1)(P_{n+1}(t) - P_n(t)) - n(P_n(t) - P_{n-1}(t)) = (2n+1)(t-1)P_n(t)$$

implies

$$(\xi\cdot\eta - 1)K_{Harm_{0,\ldots,m}}(\xi,\eta) = \frac{m+1}{4\pi}\left(P_{m+1}(\xi\cdot\eta) - P_m(\xi\cdot\eta)\right)$$

for all $(\xi,\eta) \in \Omega^2$.

Let Y be an element of class $Harm_{0,\ldots,m}$ of spherical harmonics of order $\leq m$:

$$Y(\eta) = \sum_{n=0}^{m} \sum_{j=1}^{2n+1} Y^{\wedge}(n,j) Y_{n,j}(\eta), \quad \eta \in \Omega. \tag{4.3.4}$$

Then, observing the differential equation

$$(\partial_n)_\eta Y_{n,j}(\eta) = 0, \quad \eta \in \Omega, \tag{4.3.5}$$

for $n = 0, \ldots, m,\ j = 1, \ldots, 2n+1$ and for all $\eta \in \Omega$ we have

$$\begin{aligned}(\partial_{0,\ldots,m})_\eta Y(\eta) &= (\partial_0 \cdot \ldots \cdot \partial_m)_\eta Y(\eta) \\ &= \sum_{n=0}^{m} \sum_{j=1}^{2n+1} Y^{\wedge}(n,j)(\partial_0 \cdot \ldots \cdot \partial_m)_\eta Y_{n,j}(\eta) \\ &= 0 \end{aligned} \tag{4.3.6}$$

for all $\eta \in \Omega$. On the other hand, we know from (4.2.16) that any solution $Y \in \mathcal{C}^{(2m+2)}(\Omega)$ of the homogeneous differential equation

$$(\partial_{0,\ldots,m})_\eta Y(\eta) = 0, \quad \eta \in \Omega, \tag{4.3.7}$$

is representable in the form

$$Y(\eta) = \sum_{n=0}^{m} \sum_{j=1}^{2n+1} Y^{\wedge}(n,j) Y_{n,j}(\eta), \quad \eta \in \Omega. \tag{4.3.8}$$

But this means that $Harm_{0,\ldots,m}$ is the null space of the operator $\partial_{0,\ldots,m}$.

For given $H \in \mathcal{C}(\Omega)$, the integral formula (Theorem 4.2.4) can be used to discuss the general differential equation

$$(\partial_{0,\ldots,m})_\eta F(\eta) = H(\eta), \quad \eta \in \Omega. \tag{4.3.9}$$

By virtue of Green's surface identity we first see that

$$\int_\Omega ((\partial_{0,..,m})_\eta F(\eta)) Y(\eta)\, d\omega(\eta) = \int_\Omega ((\partial_{0,..,m})_\eta Y(\eta)) F(\eta)\, d\omega(\eta) = 0 \tag{4.3.10}$$

for all elements $Y \in Harm_{0,\ldots,m}$. From the considerations given above it is clear that any function $Y \in Harm_{0,\ldots,m}$ can be added to F without changing the differential equation (4.3.9). However, if we require that F is orthogonal to the null space $Harm_{0,\ldots,m}$ of $\partial_{0,..,m}$, then the differential equation is uniquely solvable the following result.

This finally leads us to the following result.

Theorem 4.3.1. *Let H be a continuous function on Ω orthogonal to $Harm_{0,...,m}$, i.e.*

$$\int_{\Omega} H(\eta)Y_{n,j}(\eta)\, d\omega = 0 \tag{4.3.11}$$

for $n = 0, ..., m,\ j = 1, ..., 2n + 1$. Then the function F given by

$$F(\xi) = \int_{\Omega} G(\partial_{0,..,m}; \xi, \eta)H(\eta)\, d\omega(\eta),\ \ \xi \in \Omega, \tag{4.3.12}$$

represents the only $(2m + 2)$times continuously differentiable solution of the differential equation (4.3.9) on Ω which is orthogonal to $Harm_{0,...,m}$.

Bibliographical Notes

Green's functions with respect to iterated Beltrami operators have been investigated in detail by W. Freeden (1978, 1979, 1980b) for the three-dimensional case and R. Reuter (1982) in the higher-dimensional case. A more general approach has been presented by W. Freeden, R. Reuter (1982). The integral formulae are due to W. Freeden (1978, 1979, 1980b). Related formulae in $\mathbb{R}^n$ are multidimensional Euler summation and cubature formulae (cf. W. Freeden (1982a), W. Freeden, J. Fleck (1987) and the references therein). Later on Green's functions with respect to iterated Beltrami operators will be used in decomposition theorems for spherical vector and tensor fields.

5

SPHERICAL RADIAL BASIS FUNCTIONS

In this chapter we deal with spherical radial basis functions, i.e. kernel functions depending only on the (spherical) distance between unit vectors. The importance of our considerations lies in the development of general Sobolev spaces $\mathcal{H}$ and (invariant) pseudodifferential operators Λ in which it is possible that (i) the radial basis function is recognized as axisymmetric reproducing kernel function of $\mathcal{H}$ and (ii) the kernel admits a Legendre expansion determined by the symbol $\{\Lambda^\wedge(n)\}$ of a pseudodifferential operator Λ. Criteria are given for measuring the space-frequency localization of (radial basis) functions on the sphere. The trade off between space localization on the sphere and frequency localization in terms of spherical harmonics is described in form of an 'uncertainty principle'. We have to distinguish two classes of radial basis functions: the first class is characterized by the fact that all Legendre coefficients of the kernel are non-vanishing. Consequently the Sobolev space $\mathcal{H}$ contains *all* spaces $Harm_n, n = 0, 1, \ldots,$ as constituting members. This first class of radial basis functions includes Green's functions to certain (iterated) Beltrami derivatives, the Abel–Poisson kernel, the Gauß–Weierstraß kernel, etc. All these kernels have a support which covers the whole space, although some of them show good space localization in the sense of our criteria (for example, Abel–Poisson kernel, Gauß–Weierstraß kernel). The second class of radial basis functions consists of locally supported kernel functions (that means axisymmetric 'finite elements' generated by rotating onedimensional piecewise polynomials). It turns out that the zeros occurring in the Legendre coefficients for certain diameters of the support correspond to zeros of Gegenbauer polynomials. Correspondingly, our locally supported radial basis functions define reproducing kernels of Sobolev spaces which differ from the underlying Sobolev space of the first class in that *not all* spaces $Harm_n$ are contained as constituting members.

5.1 Sobolev Spaces and Pseudodifferential Operators

Starting point of our considerations (cf. W. Freeden (1984a,b)) is the linear space $\mathcal{A}$ consisting of all sequences $\{A_{n,j}\}$ of real numbers $A_{n,j}$, $(n,j) \in \mathcal{J}$:

$$\mathcal{A} = \{\{A_{n,j}\} \mid A_{n,j} \in \mathbb{R},\ n = 0, 1, ..., \ j = 1, ..., 2n+1\}. \qquad (5.1.1)$$

For simplicity, $\{A_n\} \in \mathcal{A}$ is understood to be equivalent to $\{A_{n,j}\} \in \mathcal{A}$ with $A_{n,j} = A_n$ for $j = 1, \ldots, 2n+1$.

Let $\{A_n\} \in \mathcal{A}$ be a sequence with $A_n \neq 0$ for all n. Consider the set $\mathcal{E} = \mathcal{E}(\{A_n\}; \Omega)$ of all infinitely differentiable functions F on Ω satisfying

$$\sum_{n=0}^{\infty} \sum_{j=1}^{2n+1} |A_n|^2 (F, Y_{n,j})^2_{\mathcal{L}^2(\Omega)} < \infty. \tag{5.1.2}$$

From the Cauchy–Schwarz inequality it follows that

$$\begin{aligned} &\left| \sum_{n=0}^{\infty} \sum_{j=1}^{2n+1} |A_n|^2 (F, Y_{n,j})_{\mathcal{L}^2(\Omega)} (G, Y_{n,j})_{\mathcal{L}^2(\Omega)} \right|^2 \\ &\leq \left(\sum_{n=0}^{\infty} \sum_{j=1}^{2n+1} |A_n|^2 (F, Y_{n,j})^2_{\mathcal{L}^2(\Omega)} \right) \left(\sum_{n=0}^{\infty} \sum_{j=1}^{2n+1} |A_n|^2 (G, Y_{n,j})^2_{\mathcal{L}^2(\Omega)} \right) \end{aligned} \tag{5.1.3}$$

for all $F, G \in \mathcal{E}$, hence, the left-hand side of (5.1.3) is finite whenever each member of the right-hand side is finite. We can therefore define an inner product $(\cdot,\cdot)_{\mathcal{H}(\{A_n\};\Omega)}$ on the space $\mathcal{E}$ by

$$(F, G)_{\mathcal{H}(\{A_n\};\Omega)} = \sum_{n=0}^{\infty} \sum_{j=1}^{2n+1} |A_n|^2 (F, Y_{n,j})_{\mathcal{L}^2(\Omega)} (G, Y_{n,j})_{\mathcal{L}^2(\Omega)}. \tag{5.1.4}$$

The associated norm is given by

$$\|F\|_{\mathcal{H}(\{A_n\};\Omega)} = \left((F, F)_{\mathcal{H}(\{A_n\};\Omega)} \right)^{1/2}. \tag{5.1.5}$$

Definition 5.1.1. The *Sobolev space* $\mathcal{H}(\{A_n\}; \Omega)$ is the completion of $\mathcal{E}(\{A_n\}; \Omega)$ under the norm (5.1.5):

$$\mathcal{H}(\{A_n\}; \Omega) = \overline{\mathcal{E}(\{A_n\}; \Omega)}^{\|\cdot\|_{\mathcal{H}(\{A_n\};\Omega)}}.$$

$\mathcal{H}(\{A_n\}; \Omega)$ equipped with the inner product corresponding to the norm (5.1.5) is a Hilbert space. It also follows from the Cauchy–Schwarz inequality (5.1.3) that $(F, G)_{\mathcal{H}(\{1\};\Omega)}$ exists if $F \in \mathcal{H}(\{A_n\}; \Omega)$ and $G \in \mathcal{H}(\{A_n^{-1}\}; \Omega)$. In addition,

$$|(F, G)_{\mathcal{H}(\{1\};\Omega)}| \leq \|F\|_{\mathcal{H}(\{A_n\};\Omega)} \|G\|_{\mathcal{H}(\{A_n^{-1}\};\Omega)}. \tag{5.1.6}$$

Hence, $(\cdot,\cdot)_{\mathcal{H}(\{1\};\Omega)}$ defines a duality of $\mathcal{H}(\{A_n\}; \Omega)$ and $\mathcal{H}(\{A_n^{-1}\}; \Omega)$.

Remark. $\mathcal{H}(\{A_n\}; \Omega)$ may be identified with a subspace of $\mathcal{A}$ by associating with each $F \in \mathcal{H}(\{A_n\}; \Omega)$ its sequence $\{F^\wedge(n, j)\}$, $F^\wedge(n, j) = (F, Y_{n,j})_{\mathcal{L}^2(\Omega)}$.

As shown before, $\Delta^* Y_{n,j} = -n(n+1)Y_{n,j}$ for all $n = 0, 1, ..., j = 1, ..., 2n+1$. Therefore we formally have

$$\left(\left(-\Delta^* + \frac{1}{4}\right)^{s/2} F\right)^{\wedge}(n,j) = \left(n + \frac{1}{2}\right)^{s} F^{\wedge}(n,j). \tag{5.1.7}$$

For brevity we let

$$\mathcal{H}_s(\Omega) = \mathcal{H}(\{ (n+1/2)^s \}; \Omega) \tag{5.1.8}$$

for each $s \in \mathbb{R}$. Hence, the norm in $\mathcal{H}_s(\Omega)$ reads as follows

$$\|F\|_{\mathcal{H}_s(\Omega)} = \|F\|_{\mathcal{H}(\{(n+\frac{1}{2})^s\};\Omega)} = \|(-\Delta^* + \frac{1}{4})^{s/2} F\|_{\mathcal{L}^2(\Omega)}. \tag{5.1.9}$$

In particular, $\mathcal{H}_0(\Omega) = \mathcal{L}^2(\Omega)$, and we define the norm $\|\cdot\|_{\mathcal{H}_0(\Omega)}$ to be simply the $\mathcal{L}^2$-norm. Moreover, according to our construction, $\mathcal{C}^{(\infty)}(\Omega)$ is a dense subspace of $\mathcal{H}_s(\Omega)$ for each s. If $t < s$, then $\|F\|_{\mathcal{H}_t(\Omega)} \leq \|F\|_{\mathcal{H}_s(\Omega)}$, and $\mathcal{H}_s(\Omega) \subset \mathcal{H}_t(\Omega)$. Thus the union of the $\mathcal{H}_s(\Omega)$-spaces over all integers s may be understood as a subspace of $\mathcal{A}$. If $F \in \mathcal{H}_s(\Omega)$ and $G \in \mathcal{H}_{-s}(\Omega)$, then

$$\begin{aligned}&(F,G)_{\mathcal{H}(\{1\};\Omega)} \\ &= \int_\Omega \left(\left(-\Delta^*_\xi + \frac{1}{4}\right)^{s/2} F(\xi)\right)\left(\left(-\Delta^*_\xi + \frac{1}{4}\right)^{-s/2} G(\xi)\right) d\omega(\xi).\end{aligned} \tag{5.1.10}$$

It should be noted that if F and G are both in $\mathcal{L}^2(\Omega)$, then

$$(F,G)_{\mathcal{H}(\{1\};\Omega)} = (F,G)_{\mathcal{L}^2(\Omega)} = \int_\Omega F(\xi) G(\xi)\, d\omega(\xi). \tag{5.1.11}$$

The operator $(-\Delta^* + \frac{1}{4})^{t/2}$ is a bounded operator from $\mathcal{H}_{s+t}(\Omega)$ to $\mathcal{H}_s(\Omega)$ for each s, indeed,

$$\|(-\Delta^* + \frac{1}{4})^{t/2} F\|_{\mathcal{H}_s(\Omega)} = \|F\|_{\mathcal{H}_{s+t}(\Omega)} \tag{5.1.12}$$

for all $F \in \mathcal{H}_{s+t}(\Omega)$.

Since pseudodifferential operators on the sphere do not belong to the traditional equipment, we make now a brief description of certain parts of them. For simplicity, we restrict ourselves to invariant pseudodifferential operators. A more detailed discussion (from the point of physical geodesy) can be found in S.L. SVENSSON (1983). The definitions below are due to J. CUI, W. FREEDEN (1997). General references to the mathematical theory of pseudodifferential operators are for example L. HÖRMANDER (1965) and G.I. ESKIN (1981).

Definition 5.1.2. Let $\{\Lambda^\wedge(n)\}_{n=0,1,\ldots}$ be a sequence of real numbers $\Lambda^\wedge(n)$ satisfying

$$\lim_{n\to\infty} \frac{|\Lambda^\wedge(n)|}{(n+\frac{1}{2})^t} = const. \neq 0$$

for some $t \in \mathbb{R}$. Then the operator $\Lambda : \mathcal{H}_s(\Omega) \to \mathcal{H}_{s-t}(\Omega)$ defined by

$$\Lambda(F) = \sum_{n=0}^{\infty} \sum_{j=1}^{2n+1} \Lambda^\wedge(n) F^\wedge(n,j) Y_{n,j}, \quad F \in \mathcal{H}_s(\Omega), \tag{5.1.13}$$

is called an (invariant) *pseudodifferential operator (PDO) of order* t. $\{\Lambda^\wedge(n)\}$ is called the *spherical symbol* of Λ. Moreover, if

$$\lim_{n\to\infty} \frac{|\Lambda^\wedge(n)|}{(n+\frac{1}{2})^t} = 0$$

for all $t \in \mathbb{R}$, then the operator $\Lambda : \mathcal{H}_s(\Omega) \to \mathcal{C}^{(\infty)}(\Omega)$ is called *pseudodifferential operator of order* $-\infty$ (note that equation (5.1.13) is understood in the $\mathcal{H}_{s-t}(\Omega)$-topology).

The spherical symbol has many appealing properties. It is easily seen that $(\Lambda'+\Lambda'')^\wedge(n) = (\Lambda')^\wedge(n)+(\Lambda'')^\wedge(n)$, $(\Lambda'\Lambda'')^\wedge(n) = (\Lambda')^\wedge(n)(\Lambda'')^\wedge(n)$ for all n. Obviously we have

$$\Lambda(Y_{n,j}) = \Lambda^\wedge(n) Y_{n,j}, \quad n = 0,1,\ldots, \quad j = 1,\ldots,2n+1, \tag{5.1.14}$$

(i.e. all pseudodifferential operators Λ considered here are invariant operators, so that the symbol is independent of j). Note that for all invertible Λ on $\mathcal{H}_s(\Omega)$ (i.e. $\Lambda^\wedge(n) \neq 0$ for all n)

$$\|U\|_{\mathcal{H}(\{\Lambda^\wedge(n)A_n\};\Omega)} = \|\Lambda(U)\|_{\mathcal{H}(\{A_n\};\Omega)}, \quad U \in \mathcal{H}(\{\Lambda^\wedge(n)A_n\};\Omega). \tag{5.1.15}$$

In this case, $\mathcal{H}(\{\Lambda^\wedge(n)A_n\};\Omega) = \Lambda^{-1}\mathcal{H}(\{A_n\};\Omega)$. Obviously, $\{r^n\}, 0 \leq r < 1$ (or $\{e^{-\frac{1}{2}n(n+1)h}\}, h > 0$) define pseudodifferential operators of order $-\infty$.

Some examples, which are of basic interest for us, are listed below:

(1) *Beltrami operator.* It is not invertible since $(\Delta^*)^\wedge(0) = 0$, but $-\Delta^* + \frac{1}{4}$ has the symbol $\{(-\Delta^* + \frac{1}{4})^\wedge(n)\}$ with $(-\Delta^* + \frac{1}{4})^\wedge(n) = (n+\frac{1}{2})^2$, $n = 0,1,\ldots$ and, hence, has an inverse $(-\Delta^* + \frac{1}{4})^{-1}$ which is a rational pseudodifferential operator of order -2, i.e. $n \mapsto ((-\Delta^* + \frac{1}{4})^{-1})^\wedge(n)$, $n \in \mathbb{N}_0$, is a rational function of order -2. More generally, $(-\Delta^* + \frac{1}{4})^s$ is a rational pseudodifferential operator of order $2s$ and has the spherical symbol $\{((-\Delta^* + \frac{1}{4})^s)^\wedge(n)\}$ with

$$\left(\left(-\Delta^* + \frac{1}{4}\right)^s\right)^\wedge(n) = \left(n + \frac{1}{2}\right)^{2s}, \quad n = 0,1,\ldots. \tag{5.1.16}$$

(2) *Integral operator* Λ *of the single layer potential on* Ω. The operator defined by

$$\Lambda(U)(\xi) = \frac{1}{2\pi} \int_\Omega U(\eta) \frac{1}{|\xi - \eta|}\, d\omega(\eta), \quad \xi \in \Omega, \tag{5.1.17}$$

is a rational pseudodifferential operator of order -1 possessing the symbol $\{(\Lambda)^\wedge(n)\}$ with $(\Lambda)^\wedge(n) = (n + \frac{1}{2})^{-1} = ((-\Delta^* + \frac{1}{4})^{-1/2})^\wedge(n)$. This can be easily seen by inserting the series expansion of $|\xi - \eta|^{-1}$ in terms of Legendre polynomials (cf. (3.2.32)). A consequence is that Λ is invertible.

(3) *Integral operator* Λ *of the double layer potential on* Ω. The operator defined by

$$\Lambda(U)(\xi) = \frac{1}{2\pi} \int_\Omega U(\eta) \frac{\partial}{\partial \nu_\eta} \frac{1}{|\xi - \eta|}\, d\omega(\eta), \quad \xi \in \Omega, \tag{5.1.18}$$

($\partial/\partial\nu$ is the derivative in the direction of the exterior normal) can be written in the form $\Lambda = (-\frac{1}{2})(-\Delta^* + \frac{1}{4})^{-1/2}$.

(4) *Stokes operator.* This operator is defined by

$$\Lambda(U)(\xi) = \frac{1}{4\pi} \int_\Omega St(\xi, \eta) U(\eta)\, d\omega(\eta), \quad \xi \in \Omega \tag{5.1.19}$$

where $St(\cdot, \cdot)$ is the Stokes kernel (cf. G.G. STOKES (1849)). It has the spherical symbol $\{(\Lambda)^\wedge(n)\}$:

$$(\Lambda)^\wedge(n) = \begin{cases} 0 & \text{for } \ n = 1 \\ 1/(n-1) & \text{for } \ n = 0, 2, 3, 4, \ldots . \end{cases} \tag{5.1.20}$$

The problem of determining the 'disturbing potential' U with $\Lambda(U) = F$ from prescribed 'gravity anomalies' F is the 'fundamental problem of classical physical geodesy' (see for example W.A. HEISKANEN, H. MORITZ (1967), P.A. MEISSL (1971), E. GROTEN (1979)).

(5) *Green's integral operator.* The operator (cf. Chapter 4) given by

$$\Lambda(U)(\xi) = \frac{1}{4\pi} \int_\Omega G(\partial_0; \xi, \eta) U(\eta)\, d\omega(\eta), \quad \xi \in \Omega \tag{5.1.21}$$

has the spherical symbol $\{(\Lambda)^\wedge(n)\}$, where (cf. (4.1.9))

$$(\Lambda)^\wedge(n) = \begin{cases} 0 & \text{for } \ n = 0 \\ -1/(n(n+1)) & \text{for } \ n = 1, 2, \ldots . \end{cases} \tag{5.1.22}$$

We see that all these listed operators form mappings from $\mathcal{H}_s(\Omega)$ into some $\mathcal{H}_{s-t}(\Omega)$, where t is the order of the pseudodifferential operator Λ under consideration. For example, the Beltrami operator is an operator from $\mathcal{H}_s(\Omega)$ into $\mathcal{H}_{s-2}(\Omega)$, etc. Sometimes an operator Λ satisfying $\Lambda^\wedge(n) \to 0$ is called a *smoothing operator*, because $\Lambda^\wedge(n) \to 0$ means that the higher order harmonics are subdued by the operator. The Beltrami operator has an opposite effect, because the higher order harmonics are amplified.

(6) *Poisson operator.* The operator given by

$$\Lambda(U)(\xi) = \frac{r^2-1}{4\pi} \int_\Omega \frac{U(\eta)}{|r\xi - \eta|^3}\, d\omega(\eta), \quad r \geq 1, \quad \xi \in \Omega \tag{5.1.23}$$

has the symbol $\{r^{-(n+1)}\}$.

(7) *Operator of first order radial derivative at the height r.* This operator given by

$$\Lambda(U)(\xi) = \int_\Omega D_r(\xi \cdot \eta) U(\eta)\, d\omega(\eta), \quad r > 1, \quad \xi \in \Omega \tag{5.1.24}$$

with

$$\begin{aligned} D_r(\xi\cdot\eta) & \\ &= \left(\frac{\partial}{\partial r}\right) \frac{1}{4\pi} \frac{r^2-1}{(r^2+1-2r(\xi\cdot\eta))^{3/2}} \\ &= \left(\frac{\partial}{\partial r}\right) \sum_{n=0}^{\infty} \frac{2n+1}{4\pi} \left(\frac{1}{r}\right)^{n+1} P_n(\xi\cdot\eta) \end{aligned} \tag{5.1.25}$$

has the symbol $\{-(n+1)r^{-(n+2)}\}$. It plays a particular role in satellite-to-satellite tracking problems (see for example R. RUMMEL (1979), F. SCHNEIDER (1996B, 1997), W. FREEDEN *et al.* (1997)).

(8) *Operator of second order radial derivative at the height r.* This operator given by

$$\Lambda(U)(\xi) = \int_\Omega \left((\frac{\partial}{\partial r}) D_r(\xi\cdot\eta) \right) U(\eta)\, d\omega(\eta), \quad r > 1, \quad \xi \in \Omega$$

has the symbol $\{(n+1)(n+2)r^{-(n+3)}\}$. It is of basic importance in satellite gradiometry (see for example R. RUMMEL (1986), R. RUMMEL, M. VAN GELDEREN (1992), M. SCHREINER (1994, 1995B, 1996A), M. BAYER (1996), F. SCHNEIDER (1996B, 1997), W. FREEDEN *et al.* (1997)).

Finally we discuss the transformation of a *classical inner boundary-value problem of potential theory* into a pseudodifferential equation. For given F determine U such that

$$\Delta U = 0 \quad \text{inside } \Omega, \tag{5.1.26}$$

$$\alpha \frac{\partial U}{\partial \nu} + \beta U = F \quad \text{on } \Omega, \quad |\alpha| + |\beta| > 0. \tag{5.1.27}$$

Suppose that U satisfies the assumption of the fundamental theorem

$$U(\xi) + \frac{1}{2\pi} \int_{\Omega} \left(\frac{1}{|\xi - \eta|} \frac{\partial U}{\partial \nu}(\eta) - U(\eta) \frac{\partial}{\partial \nu_\eta} \frac{1}{|\xi - \eta|} \right) d\omega(\eta) = 0. \tag{5.1.28}$$

Then, in connection with (5.1.17), (5.1.18), we find

$$U + \left(-\Delta^* + \frac{1}{4} \right)^{-1/2} \left(\frac{\partial U}{\partial \nu} \right) + \frac{1}{2} \left(-\Delta^* + \frac{1}{4} \right)^{-1/2} (U) = 0. \tag{5.1.29}$$

Hence it follows that

$$\frac{\partial U}{\partial \nu} = -\frac{1}{2} \left(I + 2 \left(-\Delta^* + \frac{1}{4} \right)^{1/2} U \right). \tag{5.1.30}$$

The boundary-value problem (5.1.26), (5.1.27) therefore, may be written as a pseudodifferential equation. Given $F \in \mathcal{H}_s(\Omega)$, find $U \in \mathcal{H}_{s+(1-\delta_{\alpha,0})}(\Omega)$ such that

$$\Lambda(U) = F, \; \Lambda = -\alpha \left(-\Delta^* + \frac{1}{4} \right)^{1/2} + \left(\beta - \frac{\alpha}{2} \right). \tag{5.1.31}$$

Obviously,

$$\Lambda^\wedge(n) = -\alpha \left(n + \frac{1}{2} \right) + \left(\beta - \frac{\alpha}{2} \right), \quad n = 0, 1, \ldots. \tag{5.1.32}$$

Examples are the *Dirichlet problem* ($\alpha = 0, \beta = 1$) and the *Neumann problem* ($\alpha = -1, \beta = 0$) corresponding to the boundary Ω. If Λ is invertible, i.e. $\Lambda^\wedge(n) \neq 0$ for all $n \in \mathbb{N}_0$, the solution of the inner boundary-value problem is given in the form

$$U(x) = \sum_{n=0}^{\infty} \sum_{j=1}^{2n+1} (\Lambda^\wedge(n))^{-1} F^\wedge(n,j) |x|^n Y_{n,j}(\xi), \quad x = |x|\xi.$$

5.2 Reproducing Kernel Sobolev Spaces

Next we consider elements of the Sobolev spaces $\mathcal{H}(\{A_n\};\Omega)$, $\{A_n\}$ given sequence in $\mathcal{A}$ with $A_n \neq 0$ for all n, as formal orthogonal expansions or 'generalized functions'. A fundamental lemma due to Sobolev says that under certain circumstances the formal orthogonal expansion actually converges uniformly to a function in ordinary sense. This lemma will be discussed now in more detail. We begin (cf. W. FREEDEN (1984B, 1987C)) with a definition.

Definition 5.2.1. A sequence $\{A_n\} \in \mathcal{A}$ is said to be $\{B_n\}$*-summable of order* m if the following conditions are satisfied:

(i) $A_n \neq 0$ for all $n \geq m+1$

(ii)

$$\Sigma_m(\{B_n\},\{A_n\}) = \sum_{n=m+1}^{\infty} \frac{2n+1}{4\pi}\frac{|B_n|^2}{|A_n|^2} < \infty, \quad \{B_n\} \in \mathcal{A}. \tag{5.2.1}$$

Furthermore, $\{A_n\} \in \mathcal{A}$ is said to be *summable of order* m if

(i) $A_n \neq 0$ for all $n \geq m+1$

(ii)

$$\Sigma_m(\{A_n\}) = \sum_{n=m+1}^{\infty} \frac{2n+1}{4\pi}\frac{1}{|A_n|^2} < \infty. \tag{5.2.2}$$

Remark. $\{B_n\}$-summability is formally understood as $\{B_n\}$-summability of order -1, and we write

$$\Sigma(\{B_n\},\{A_n\}) = \Sigma_{-1}(\{B_n\},\{A_n\}).$$

Associate with F the series $\sum_{n=0}^{\infty}\sum_{j=1}^{2n+1} F^\wedge(n,j)Y_{n,j}$. It is of fundamental importance to know when this series converges uniformly. This answer is supplied by the following *Sobolev lemma.*

Lemma 5.2.2. *Let the sequence* $\{A_n\}$ *be* $\{B_n\}$*-summable,* $B_n \neq 0$ *for all* n*. Then each* $F \in \mathcal{H}(\{B_n^{-1}A_n\};\Omega)$ *corresponds to a continuous function on* Ω*.*

Proof. For each positive integer N, we have

$$\left|\sum_{n=0}^{N}\sum_{j=1}^{2n+1} F^\wedge(n,j)Y_{n,j}(\xi)\right|^2 \tag{5.2.3}$$

$$
\begin{aligned}
&= \left| \sum_{n=0}^{N} \sum_{j=1}^{2n+1} B_n A_n^{-1} Y_{n,j}(\xi) A_n B_n^{-1} F^\wedge(n,j) \right|^2 \\
&\leq \left(\sum_{n=0}^{N} \sum_{j=1}^{2n+1} |B_n|^2 |A_n|^{-2} \left(Y_{n,j}(\xi)\right)^2 \right) \\
&\quad \times \left(\sum_{n=0}^{N} \sum_{j=1}^{2n+1} |B_n|^{-2} |A_n|^2 \left(F^\wedge(n,j)\right)^2 \right) \\
&\leq \Sigma\left(\{B_n\},\{A_n\}\right) \; \|F\|^2_{\mathcal{H}(\{B_n^{-1}A_n\};\Omega)}.
\end{aligned}
$$

□

In particular, we mention the following result.

Lemma 5.2.3. *If $F \in \mathcal{H}_s(\Omega)$, where $s > k+1$, then F corresponds to a function of class $\mathcal{C}^{(k)}(\Omega)$.*

It is well known that a continuously differentiable function is Lipschitz-continuous on Ω. The following lemma yields an estimate of the Lipschitz constant.

Lemma 5.2.4. *If $F \in \mathcal{H}_s(\Omega)$, $s > 2$, then F is Lipschitz-continuous. More explicitly,*

$$|F(\xi) - F(\eta)| \leq C_F(s)|\xi - \eta|,$$

where

$$C_F(s) = \left(\frac{1}{2} \sum_{n=1}^{\infty} \frac{2n+1}{4\pi} \frac{n(n+1)}{(n+\frac{1}{2})^{2s}} \right)^{1/2} \|F\|_{\mathcal{H}_s(\Omega)}.$$

Proof. The Cauchy–Schwarz inequality gives us in connection with the addition theorem of spherical harmonics and the estimate (3.2.6)

$$
\begin{aligned}
&|F(\xi) - F(\eta)|^2 \\
&\quad = \left| \sum_{n=0}^{\infty} \sum_{j=1}^{2n+1} F^\wedge(n,j)\left(Y_{n,j}(\xi) - Y_{n,j}(\eta)\right) \right|^2 \\
&\quad \leq 2 \left(\sum_{n=0}^{\infty} \sum_{j=1}^{2n+1} \left(n + \frac{1}{2}\right)^{2s} \left(F^\wedge(n,j)\right)^2 \right) \qquad (5.2.4) \\
&\quad\quad \times \left(\sum_{n=0}^{\infty} \frac{2n+1}{4\pi} \left(n+\frac{1}{2}\right)^{-2s} \left(1 - P_n(\xi\cdot\eta)\right) \right)
\end{aligned}
$$

$$\leq \quad \|F\|^2_{\mathcal{H}_s(\Omega)} \left(\sum_{n=0}^{\infty} \frac{2n+1}{4\pi} \frac{n(n+1)}{(n+\frac{1}{2})^{2s}} \right) |\xi - \eta|^2.$$

This proves Lemma 5.2.4. □

Let F be a member of the Sobolev space $\mathcal{H}_s(\Omega), s > 1$. Then it follows that

$$\left| F(\xi) - \sum_{n=0}^{N} \sum_{j=1}^{2n+1} F^\wedge(n,j) Y_{n,j}(\xi) \right|$$
$$\leq \left(\sum_{n=N+1}^{\infty} \sum_{j=1}^{2n+1} (F^\wedge(n,j))^2 (n+\frac{1}{2})^{2s} \right)^{1/2}$$
$$\left(\sum_{n=N+1}^{\infty} \sum_{j=1}^{2n+1} |Y_{n,j}(\xi)|^2 (n+\frac{1}{2})^{-2s} \right)^{1/2}$$

for each $\xi \in \Omega$. This enables us to derive an estimate between $F \in \mathcal{H}_s(\Omega)$ and its Nth truncated spherical harmonic expansion.

Lemma 5.2.5. *For $F \in \mathcal{H}_s(\Omega), s > 1$,*

$$\sup_{\xi \in \Omega} |F(\xi) - \sum_{n=0}^{N} \sum_{j=1}^{2n+1} F^\wedge(n,j) Y_{n,j}(\xi)| \leq \frac{C}{N^{s-1}} \|F\|_{\mathcal{H}_s(\Omega)},$$

where the constant C depends only on s.

Of great importance for our considerations are Sobolev spaces equipped with a reproducing kernel structure.

Theorem 5.2.6. *Let $\{A_n\}$ be summable. Then the space $\mathcal{H} = \mathcal{H}(\{A_n\};\Omega)$ (further on we write $\mathcal{H}$ instead of $\mathcal{H}(\{A_n\};\Omega)$ if no confusion is likely to arise), is a functional Hilbert subspace of $\mathcal{C}(\Omega)$. $\mathcal{H}$ has the reproducing kernel $K_{\mathcal{H}} : \Omega \times \Omega \to \mathbb{R}$ given by*

$$\begin{aligned} K_{\mathcal{H}}(\xi,\eta) &= \sum_{n=0}^{\infty} \sum_{j=1}^{2n+1} |A_n|^{-2} Y_{n,j}(\xi) Y_{n,j}(\eta) \\ &= \sum_{n=0}^{\infty} \frac{2n+1}{4\pi} |A_n|^{-2} P_n(\xi \cdot \eta), \quad \xi, \eta \in \Omega. \end{aligned}$$

The system $\{A_n^{-1} Y_{n,j}\}_{n=0,1,\ldots,j=1,\ldots,2n+1}$ represents a Hilbert basis in $\mathcal{H}$.

Proof. The Sobolev lemma implies that $\mathcal{H}$ is a subset of $\mathcal{C}(\Omega)$. A necessary and sufficient condition that $\mathcal{H}$ has a reproducing kernel function is that for each $\xi \in \Omega$, the linear functional $L_\xi : \mathcal{H} \to \mathbb{R}$ defined by $L_\xi F = F(\xi), \xi \in \Omega$, is bounded: $|L_\xi F| \leq C\|F\|_{\mathcal{H}}$ for some constant C and for all $F \in \mathcal{H}$. In fact, the Cauchy–Schwarz inequality shows that

$$|F(\xi)| \leq C\|F\|_{\mathcal{H}}, \tag{5.2.5}$$

where

$$C^2 = K_{\mathcal{H}}(\xi,\xi) \leq \Sigma(\{A_n\}) = \sum_{n=0}^{\infty} \frac{2n+1}{4\pi}|A_n|^{-2}. \tag{5.2.6}$$

This means C is uniformly bounded. Suppose, conversely, that $|L_\xi F| \leq C\|F\|_{\mathcal{H}}$ holds for each fixed $\xi \in \Omega$. Then one can find a function $G_\xi \in \mathcal{H}$ such that $F(\xi) = (F, G_\xi)_{\mathcal{H}}$. Now set $K_{\mathcal{H}}(\xi,\cdot) = G_\xi$, and it is clear that $K_{\mathcal{H}}(\cdot,\cdot)$ is the reproducing kernel (cf. N. ARONSZAJN (1950)). Clearly, the reproducing kernel is unique in $\mathcal{H}$. □

From our estimates for the Legendre polynomials (see Lemma 3.2.1) it is not difficult to verify the following result.

Lemma 5.2.7. *Let the sequence $\{A_n\} \in \mathcal{A}$ be $\{n^\sigma\}$-summable. Then*

$$|K_{\mathcal{H}}(\xi,\eta) - K_{\mathcal{H}}(\xi,\zeta)| \leq 2\sum_{n=0}^{\infty} \frac{2n+1}{4\pi}\frac{1}{|A_n|^2}\left(\frac{n(n+1)}{2}\right)^{\sigma} |\zeta-\eta|^{\sigma}$$

holds for all $(\xi,\eta,\zeta) \in \Omega^3$ provided that $\sigma \in [0,1]$.

Lemma 5.2.8. *Let the sequence $\{A_n\} \in \mathcal{A}$ be $\{n^{2\sigma+\frac{1}{2}}\}$-summable. Then*

$$\begin{aligned} &|K_{\mathcal{H}}(\xi,\xi) - 2K_{\mathcal{H}}(\xi,\eta) + K_{\mathcal{H}}(\eta,\eta)| \\ &\leq \ 4\sum_{n=0}^{\infty} \frac{(2n+1)^2}{4\pi}\frac{1}{|A_n|^2}\left(\frac{n(n+1)}{2}\right)^{2\sigma} |\xi-\eta|^{2\sigma} \end{aligned}$$

holds for all $(\xi,\eta) \in \Omega^2$ provided that $\sigma \in [0,1]$.

The space $\mathcal{H}_{0,\ldots,m} = Harm_{0,\ldots,m}, m \geq 0$, is a finite-dimensional Hilbert space ($d(\mathcal{H}_{0,\ldots,m}) = M = (m+1)^2$) with inner product

$$(F,G)_{\mathcal{H}_{0,\ldots,m}} = \sum_{n=0}^{m}\sum_{j=1}^{2n+1} |A_n|^2 (F, Y_{n,j})_{\mathcal{L}^2(\Omega)} (G, Y_{n,j})_{\mathcal{L}^2(\Omega)} \tag{5.2.7}$$

and reproducing kernel

$$K_{\mathcal{H}_{0,\ldots,m}}(\xi,\eta) = \sum_{n=0}^{m}\sum_{j=1}^{2n+1} |A_n|^{-2} Y_{n,j}(\xi) Y_{n,j}(\eta). \tag{5.2.8}$$

Let us denote by $\mathcal{H}_{0,\ldots,m}^{\perp} = (\mathcal{H}_{0,\ldots,m}^{\perp}(\{A_n\};\Omega))$ the orthogonal complement of $\mathcal{H}_{0,\ldots,m}$ in $\mathcal{H}(\{A_n\};\Omega)$. The linear space $\mathcal{H}_{0,\ldots,m}^{\perp}(\{A_n\};\Omega)$ is a Hilbert space with inner product $(\cdot,\cdot)_{\mathcal{H}_{0,\ldots,m}^{\perp}}$

$$(F,G)_{\mathcal{H}_{0,\ldots,m}^{\perp}} = \sum_{n=m+1}^{\infty}\sum_{j=1}^{2n+1} |A_n|^{2} (F,Y_{n,j})_{\mathcal{L}^2(\Omega)} (G,Y_{n,j})_{\mathcal{L}^2(\Omega)} \tag{5.2.9}$$

and reproducing kernel

$$K_{\mathcal{H}_{0,\ldots,m}^{\perp}}(\xi,\eta) = \sum_{n=m+1}^{\infty}\sum_{j=1}^{2n+1} |A_n|^{-2} Y_{n,j}(\xi) Y_{n,j}(\eta). \tag{5.2.10}$$

Hence, $\mathcal{H}$ is the orthogonal direct sum of $\mathcal{H}_{0,\ldots,m}$ and $\mathcal{H}_{0,\ldots,m}^{\perp}$ with the inner product

$$(\cdot,\cdot)_{\mathcal{H}} = (\cdot,\cdot)_{\mathcal{H}_{0,\ldots,m}} + (\cdot,\cdot)_{\mathcal{H}_{0,\ldots,m}^{\perp}} \tag{5.2.11}$$

and the reproducing kernel

$$K_{\mathcal{H}}(\cdot,\cdot) = K_{\mathcal{H}_{0,\ldots,m}}(\cdot,\cdot) + K_{\mathcal{H}_{0,\ldots,m}^{\perp}}(\cdot,\cdot). \tag{5.2.12}$$

The null space of the seminorm $\|\cdot\|_{\mathcal{H}_{0,\ldots,m}^{\perp}(\{A_n\};\Omega)}$ is the linear space $\mathcal{H}_{0,\ldots,m}$.

5.3 Integral Formulae

Consider now a pseudodifferential operator Λ with summable symbol $\{\Lambda^\wedge(n)\}$. Then it follows that

$$\begin{aligned} F(\xi) &= (F, K_{\mathcal{H}(\{\Lambda^\wedge(n)\};\Omega)}(\cdot,\xi))_{\mathcal{H}(\{\Lambda^\wedge(n)\};\Omega)} \\ &= (F, K_{\mathcal{H}_{0,\ldots,m}(\{\Lambda^\wedge(n)\};\Omega)}(\cdot,\xi))_{\mathcal{H}_{0,\ldots,m(\{\Lambda^\wedge(n)\};\Omega)}} \\ &\quad +(F, K_{\mathcal{H}_{0,\ldots,m}^{\perp}(\{\Lambda^\wedge(n)\};\Omega)}(\cdot,\xi))_{\mathcal{H}^{\perp}_{0,\ldots,m(\{\Lambda^\wedge(n)\};\Omega)}} \end{aligned}$$

holds for all $m \in \mathbb{N}_0$, $\xi \in \Omega$, and $F \in \mathcal{H}(\{\Lambda^\wedge(n)\};\Omega)$. By virtue of Parseval's identity it is not difficult to verify that

$$\begin{aligned} &(F, K_{\mathcal{H}_{0,\ldots,m(\{\Lambda^\wedge(n)\};\Omega)}}(\cdot,\xi))_{\mathcal{H}_{0,\ldots,m(\{\Lambda^\wedge(n)\};\Omega)}} \\ &\quad = \sum_{n=0}^{m}\sum_{j=1}^{2n+1} (F,Y_{n,j})_{\mathcal{L}^2(\Omega)} Y_{n,j}(\xi), \end{aligned}$$

whereas

$$(F, K_{\mathcal{H}_{0,\ldots,m}^{\perp}(\{\Lambda^\wedge(n)\};\Omega)}(\cdot,\xi))_{\mathcal{H}_{0,\ldots,m}^{\perp}(\{\Lambda^\wedge(n)\};\Omega)}$$

$$= \int_\Omega \Lambda_\eta K_{\mathcal{H}^\perp_{0,\ldots,m}(\{\Lambda^\wedge(n)\};\Omega)}(\xi,\eta)(\Lambda_\eta F(\eta))\, d\omega(\eta).$$

This leads us to the following *integral formula with respect to invariant pseudodifferential operators* with summable symbol.

Theorem 5.3.1. *Let Λ be a pseudodifferential operator with summable symbol $\{\Lambda^\wedge(n)\}$. Then*

$$F(\xi) = \sum_{n=0}^{m}\sum_{j=1}^{2n+1}(F,Y_{n,j})_{\mathcal{L}^2(\Omega)}Y_{n,j}(\xi) + \int_\Omega \Lambda_\eta K_{\mathcal{H}^\perp_{0,\ldots,m}(\{\Lambda^\wedge(n)\};\Omega)}(\xi,\eta)(\Lambda_\eta F(\eta))\, d\omega(\eta)$$

holds for every $\xi \in \Omega$ and every $F \in \mathcal{H}(\{\Lambda^\wedge(n)\};\Omega)$.

Consider the matrix

$$\mathbf{matr}_{X_N}(Y_{0,1},\ldots Y_{m,2m+1}) = \begin{pmatrix} Y_{0,1}(\eta_1) & \ldots & Y_{0,1}(\eta_N) \\ \vdots & & \vdots \\ Y_{m,2m+1}(\eta_1) & \ldots & Y_{m,2m+1}(\eta_N) \end{pmatrix}$$

corresponding to an $\mathcal{L}^2(\Omega)$-orthonormal system $\{Y_{n,j}\}$ of spherical harmonics. Unfortunately, according to Haar's theorem (see for example P.J. Davis (1963)), this matrix is not of maximal rank for all systems $X_N = \{\eta_1,\ldots,\eta_N\}$ of distinct points $\eta_1,\ldots,\eta_N \in \Omega$, $N \geq M, M = d(Harm_{0,\ldots,m})$. However, it is clear from Section 2.1 that there exists a large class of systems X_M having a non-degenerate matrix $\mathbf{matr}_{X_M}(Y_{0,1},\ldots,Y_{m,2m+1})$.

Definition 5.3.2. A system X_M of points $\eta_1,\ldots,\eta_M$ on Ω is called *fundamental system relative to $Harm_{0,\ldots,m}$* if, for (an $\mathcal{L}^2(\Omega)$-orthonormal basis) $\{Y_{0,1},\ldots,Y_{m,1},\ldots,Y_{m,2m+1}\}$ of $Harm_{0,\ldots,m}$, $\mathbf{matr}_{X_M}(Y_{0,1},\ldots,Y_{m,2m+1})$ is regular. A system $X_N = \{\eta_1,\ldots,\eta_N\}$ of $N \geq M$ points $\eta_1,\ldots,\eta_N$ on Ω is called *admissible system relative to $Harm_{0,\ldots,m}$* if it contains a fundamental subset relative to $Harm_{0,\ldots,m}$.

Remark. For convenience we always assume that $X_M \subset X_N$ is a fundamental system relative to $Harm_{0,\ldots,m}$. This is always achievable by reordering an admissible system.

It should be noted that via the addition theorem,

$$(\mathbf{matr}_{X_M}(Y_{0,1},\ldots,Y_{m,2m+1}))^T\mathbf{matr}_{X_M}(Y_{0,1},\ldots,Y_{m,2m+1}) \quad (5.3.1)$$

$$= \mathbf{matr}_{X_M}\left(\sum_{n=0}^{m}\frac{2n+1}{4\pi}P_n(\eta_1\cdot),\ldots,\sum_{n=0}^{m}\frac{2n+1}{4\pi}P_n(\eta_M\cdot)\right).$$

Consequently, the property of X_M being a fundamental system relative to the set $Harm_{0,\ldots,m}$ is independent of the choice of the orthonormal basis $\{Y_{0,1},\ldots,Y_{m,2m+1}\}$.

Let $X_N = \{\eta_1,...,\eta_N\}$ be an admissible system relative to $Harm_{0,\ldots,m}$. Assume that $F:\Omega\to\mathbb{R}$ is a function of class $\mathcal{H}(\{\Lambda^\wedge(n)\};\Omega)$. Let Y be an element of $Harm_{0,\ldots,m}$ of the decomposition

$$Y = \sum_{n=0}^{m}\sum_{j=1}^{2n+1} c_{n,j}Y_{n,j},\ c_{n,j} = Y^\wedge(n,j). \tag{5.3.2}$$

Then, for all solutions $a\in\mathbb{R}^N, a=(a_1,...,a_N)^T$, of the linear system

$$\mathbf{matr}_{X_N}(Y_{0,1,\ldots,}Y_{m,2m+1})a = c, \tag{5.3.3}$$

that means for all $a\in\mathbb{R}^N$ satisfying

$$\sum_{k=1}^{N} a_kY_{n,j}(\eta_k) = c_{n,j};\quad n=0,...,m,\quad j=1,...,2n+1, \tag{5.3.4}$$

we have

$$Y = \sum_{k=1}^{N} a_k \sum_{n=0}^{m}\sum_{j=1}^{2n+1} Y_{n,j}(\eta_k)Y_{n,j}. \tag{5.3.5}$$

Observing this fact we obtain the following result.

Theorem 5.3.3. *Let Λ be a pseudodifferential operator with summable symbol $\{\Lambda^\wedge(n)\}$. Assume that $X_N = \{\eta_1,\ldots,\eta_N\}$ is an admissible system relative to $Harm_{0,\ldots,m}$ and that $Y\in Harm_{0,\ldots,m}$. Then*

$$\int_\Omega Y(\eta)F(\eta)d\omega(\eta) = \sum_{k=1}^{N} a_kF(\eta_k)$$
$$-\sum_{k=1}^{N} a_k\int_\Omega\left(\Lambda_\eta(K_{\mathcal{H}^\perp_{0,\ldots,m}(\{\Lambda^\wedge(n)\};\Omega)}(\eta_k,\eta))\right)(\Lambda_\eta F(\eta))\,d\omega(\eta)$$

holds for all $F\in\mathcal{H}(\{\Lambda^\wedge(n)\};\Omega)$ and for all solutions $a\in\mathbb{R}^N$, $a = (a_1,\ldots,a_N)^T$ satisfying (5.3.4) with $c=(c_{0,1},\ldots,c_{m,1},\ldots,c_{m,2m+1})^T$, $c_{n,j} = Y^\wedge(n,j) = (Y,Y_{n,j})_{\mathcal{L}^2(\Omega)}$.

As an immediate consequence of Theorem 5.3.3 we mention the following.

Corollary 5.3.4. *If $F \in Harm_{0,\ldots,m}$ and $Y_{n,j} \in Harm_{0,\ldots,m}$, then*

$$F^{\wedge}(n,j) = \int_{\Omega} Y_{n,j}(\eta)F(\eta)d\omega(\eta) = \sum_{k=1}^{N} a_k^{n,j} F(\eta_k) \tag{5.3.6}$$

holds for all $a_1^{n,j}, \ldots, a_N^{n,j} \in \mathbb{R}$ with

$$\sum_{k=1}^{N} a_k^{n,j} Y_{p,q}(\eta_k) = \delta_{p,n}\delta_{q,j}, \quad p = 0, \ldots, m, \quad q = 1, \ldots, 2p+1. \tag{5.3.7}$$

Thus we are led to a spherical counterpart to *Shannon's sampling rule.*

Corollary 5.3.5 *(spherical Shannon sampling theorem in $Harm_{0,\ldots,m}$). Let F be in $Harm_{0,\ldots,m}$ and $X_N = \{\eta_1, \ldots, \eta_N\}$ be an admissible system relative to $Harm_{0,\ldots,m}$. Then F can be reconstructed from its samples at points out of X_N by the following interpolation formula:*

$$F(\xi) = \sum_{k=1}^{N} F(\eta_k)L_k(\xi), \quad \xi \in \Omega,$$

where $L_k \in Harm_{0,\ldots,m}$ is given by

$$L_k(\xi) = \sum_{n=0}^{m}\sum_{j=1}^{2n+1} a_k^{n,j} Y_{n,j}(\xi)$$

and the coefficients $a_k^{n,j}$ are determined by (5.3.7).

Proof. From Corollary 5.3.4 we are able to deduce that

$$\begin{aligned}\sum_{k=1}^{N} F(\eta_k)L_k(\xi) &= \sum_{n=0}^{m}\sum_{j=1}^{2n+1}\left(\sum_{k=1}^{N} a_k^{n,j}F(\eta_k)\right)Y_{n,j}(\xi)\\ &= \sum_{n=0}^{m}\sum_{j=1}^{2n+1} F^{\wedge}(n,j)Y_{n,j}(\xi)\\ &= F(\xi).\end{aligned}$$

This proves Corollary 5.3.5. □

Next we are interested in further consequences of Theorem 5.3.3.

Corollary 5.3.6. *The summation formula*

$$\sum_{k=1}^{N} a_k \int_{\Omega} \Lambda_\eta K_{\mathcal{H}^{\perp}_{0,\ldots,m}(\{\Lambda^\wedge(n)\};\Omega)}(\eta_k,\eta)(\Lambda_\eta F(\eta))\, d\omega(\eta) = \sum_{k=1}^{N} a_k F(\eta_k) \tag{5.3.8}$$

holds for every selection of vectors $a \in \mathbb{R}^N$ satisfying

$$\mathbf{matr}_{X_N}(Y_{0,1,\ldots,}Y_{m,2m+1})a = 0 \tag{5.3.9}$$

and every function $F \in \mathcal{H}(\{\Lambda^\wedge(n)\};\Omega)$.

Corollary 5.3.7. *The formula*

$$\int_{\Omega} (\Lambda_\eta S(\eta))(\Lambda_\eta F(\eta))\, d\omega(\eta) = \sum_{k=1}^{N} a_k F(\eta_k) \tag{5.3.10}$$

holds for all $S : \Omega \to \mathbb{R}$ of the form

$$S(\eta) \quad = \quad Y(\eta) + \sum_{k=1}^{N} a_k K_{\mathcal{H}^{\perp}_{0,\ldots,m}(\{\Lambda^\wedge(n)\};\Omega)}(\eta,\eta_k), \tag{5.3.11}$$

$\eta \in \Omega$, $Y \in Harm_{0,\ldots,m}$, satisfying the linear system

$$\sum_{k=1}^{N} a_k Y_{n,j}(\eta_k) = 0, \quad n = 0,\ldots,m, \quad j = 1,\ldots,2n+1. \tag{5.3.12}$$

In particular, for $N = M$, $a_1 = \ldots = a_m = 0$ so that

$$S(\eta) = Y(\eta), \quad \eta \in \Omega, \quad Y \in Harm_{0,\ldots,m}. \tag{5.3.13}$$

The function S occurring in Corollary 5.3.7 will be recognized in Chapter 6 as spherical spline function of order m relative to X_N, and the identity (5.3.10) is the so-called spline integration formula which enables us to justify minimum norm properties in spherical spline interpolation (cf. W. FREEDEN (1981A)).

Theorem 5.3.1 allows the following reformulation.

Theorem 5.3.8. *For a given function $W \in \mathcal{L}^2(\Omega)$ orthogonal to $Harm_{0,\ldots,m}$ denote by V the function of class $\mathcal{H}(\{\Lambda^\wedge(n)\};\Omega)$ satisfying $\Lambda(V) = W$ and V orthogonal to $Harm_{0,\ldots,m}$. For a given admissible system $X_N =$*

$\{\eta_1, \ldots, \eta_N\}$ *and weights* $a_1, \ldots, a_N$ *with* $\mathbf{matr}_{X_N}(Y_{0,1,\ldots,}Y_{m,2m+1})a = c$, *let* $G^W_{Harm_{0,\ldots,m}} : \Omega \to \mathbb{R}$ *designate the function*

$$G^W_{Harm_{0,\ldots,m}}(\eta) = -\sum_{k=1}^{N} a_k \Lambda_\eta K_{\mathcal{H}^{\perp}_{0,\ldots,m}}(\eta_k, \eta) - V(\eta), \quad \eta \in \Omega. \tag{5.3.14}$$

Then the identity

$$\int_\Omega (Y(\eta) + W(\eta))F(\eta)\, d\omega(\eta) = \sum_{k=1}^{N} a_k F(\eta_k) + R^W_{Harm_{0,\ldots,m}}(F) \tag{5.3.15}$$

holds for $Y \in Harm_{0,\ldots,m}$ *with* $(Y, Y_{n,j})_{\mathcal{L}^2(\Omega)} = c_{n,j}$ *and every* $F \in \mathcal{H}(\{\Lambda^\wedge(n)\}; \Omega)$, *where the remainder term is given by*

$$R^W_{Harm_{0,\ldots,m}}(F) = \int_\Omega G^W_{Harm_{0,\ldots,m}}(\eta)(\Lambda_\eta F(\eta))\, d\omega(\eta). \tag{5.3.16}$$

The Cauchy–Schwarz inequality yields the error estimate

$$\left|R^W_{Harm_{0,\ldots,m}}(F)\right| \leq \left(\int_\Omega |G^W_{Harm_{0,\ldots,m}}(\eta)|^2\, d\omega(\eta)\right)^{1/2} \times \left(\sum_{n=m+1}^{\infty} \sum_{j=1}^{2n+1} (\Lambda^\wedge(n))^2 (F^\wedge(n,j))^2\right)^{1/2}.$$

The last estimate will be exploited later on to establish best integration formulae in approximate integration.

5.4 Peano's Theorem

Again suppose that Λ is a pseudodifferential operator with summable symbol $\{\Lambda^\wedge(n)\}$. For brevity we write $\mathcal{H} = \mathcal{H}(\{\Lambda^\wedge(n)\}; \Omega)$ throughout this section. The integral formulae developed in Section 5.3 enable us to formulate a spherical analogue (cf. W. FREEDEN (1981A)) of the well known Peano theorem (G. PEANO (1913)).

Theorem 5.4.1 *(Peano's theorem). Let* $\mathcal{I} : \mathcal{H} \to \mathbb{R}$ *be a bounded linear functional on* $\mathcal{H}$ *such that* $\mathcal{I}Y = 0$ *whenever* $Y \in Harm_{0,\ldots,m}$. *Then, for* $F \in \mathcal{H}$,

$$\mathcal{I}F = \int_\Omega (\Lambda_\eta K(\eta))(\Lambda_\eta F(\eta))\, d\omega(\eta), \tag{5.4.1}$$

where

$$K(\eta) = \mathcal{I}_\xi K_{\mathcal{H}^{\perp}_{0,\ldots,m}}(\xi, \eta), \quad \eta \in \Omega. \tag{5.4.2}$$

The function $K : \Omega \to \mathbb{R}$ *is called the Peano kernel for the linear functional* $\mathcal{I}$. *(*$\mathcal{I}_\xi$ *means that the linear functional is applied to the* ξ*-variable.)*

Proof. We apply $\mathcal{I}$ to both sides of the integral formula (Theorem 5.3.1) and use the fact that $\mathcal{I}$ vanishes on elements of class $Harm_{0,\ldots,m}$:

$$\mathcal{I}F = \mathcal{I}_\xi \int_\Omega (\Lambda_\eta K_{\mathcal{H}^\perp_{0,\ldots,m}}(\xi,\eta))(\Lambda_\eta F(\eta))\, d\omega(\eta). \tag{5.4.3}$$

Now, the linear functional allows the interchange of $\mathcal{I}$ and the integral. This yields the desired result. □

Consider an approximation of a linear functional $\mathcal{I}$ by a linear functional $\mathcal{L}$ (cf. A. SARD (1949)) of the form

$$\mathcal{L}F = \sum_{k=1}^{N} a_k F(\eta_k), \tag{5.4.4}$$

where $a_1, \ldots, a_N$ are real numbers, and $X_N = \{\eta_1, \ldots, \eta_N\}$ is an admissible system relative to $Harm_{0,\ldots,m}$. $\mathcal{L}$ is called *exact for the degree* m, if $\mathcal{I}F = \mathcal{L}F$ whenever $F \in Harm_{0,\ldots,m}$. The remainder, when $\mathcal{L}$ is used to approximate $\mathcal{I}$, is a linear functional $\mathcal{R}$ defined by $\mathcal{R} = \mathcal{I} - \mathcal{L}$. If the approximation of $\mathcal{I}$ by $\mathcal{L}$ is exact for the degree m and F is a function of class $\mathcal{H}$, then $\mathcal{R}$ satisfies the assumptions of Peano's theorem. Hence it follows that

$$\mathcal{R}F = \int_\Omega (\Lambda_\eta K(\eta))(\Lambda_\eta F(\eta))\, d\omega(\eta), \tag{5.4.5}$$

where

$$K(\eta) = \mathcal{R}_\xi K_{\mathcal{H}^\perp_{0,\ldots,m}}(\xi,\eta), \quad \eta \in \Omega. \tag{5.4.6}$$

Applying the Cauchy–Schwarz inequality to (5.4.5) we obtain

$$|\mathcal{R}F| \leq \left(\int_\Omega |\Lambda_\eta K(\eta)|^2\, d\omega(\eta)\right)^{1/2} \left(\sum_{n=m+1}^{\infty} \sum_{j=1}^{2n+1} (\Lambda^\wedge(n))^2 (F^\wedge(n,j))^2\right)^{1/2}, \tag{5.4.7}$$

where equality in formula (5.4.7) holds only if K and the series expansion $\sum_{n=m+1}^{\infty}\sum_{j=1}^{2n+1} \Lambda^\wedge(n) F^\wedge(n,j) Y_{n,j}$ are linearly dependent. We note that the quantity J given by

$$J = \int_\Omega |\Lambda_\eta K(\eta)|^2\, d\omega(\eta) \tag{5.4.8}$$

depends on the knots $\eta_1, \ldots, \eta_N$ and the operator $\mathcal{I}$, but not on the function $F \in \mathcal{H}$. Inserting (5.4.6) into (5.4.8) we have

$$J = \mathcal{R}_\xi \mathcal{R}_\zeta \int_\Omega \Lambda_\eta K_{\mathcal{H}^\perp_{0,\ldots,m}}(\xi,\eta) \Lambda_\eta K_{\mathcal{H}^\perp_{0,\ldots,m}}(\zeta,\eta)\, d\omega(\eta), \tag{5.4.9}$$

i.e.

$$J = \mathcal{R}_\xi \mathcal{R}_\zeta\ K_{\mathcal{H}^\perp_{0,\ldots,m}}(\xi,\zeta). \tag{5.4.10}$$

Collecting our results we therefore obtain the following *a priori estimate.*

Theorem 5.4.2. *Let $\mathcal{I}$ be a bounded linear functional on $\mathcal{H}$, and let $\mathcal{L}$ be an approximation to $\mathcal{I}$ of the form (5.4.4), exact for the degree m. Then, for each $F \in \mathcal{H}$,*

$$|\mathcal{R}F| \\ \leq \left((\mathcal{I}_\xi \mathcal{I}_\zeta - 2\mathcal{I}_\xi \mathcal{L}_\zeta + \mathcal{L}_\xi \mathcal{L}_\zeta) K_{\mathcal{H}^\perp_{0,\ldots,m}}(\xi,\zeta) \right)^{1/2} \left(\int_\Omega |\Lambda_\eta F(\eta)|^2\, d\omega(\eta) \right)^{1/2}.$$

5.5 Localization

Of particular interest for our purposes are kernel functions localizing sufficiently well in space domain as well as in frequency domain. The chief measure for an $\mathcal{L}^2$-function defined on Euclidean space is the variance. In what follows we adopt this notion for functions defined on the sphere Ω.

Let $F : \Omega \to \mathbb{R}$ be in class $\mathcal{L}^2(\Omega)$ satisfying $\|F\|_{\mathcal{L}^2(\Omega)} = 1$. We associate to F the normal vector field $o^{(1)}F : \eta \mapsto o^{(1)}_\eta F(\eta) = \eta F(\eta), \eta \in \Omega$ (cf. Section 12.1). The 'centre of gravity of the spherical window' in space domain is defined by the expectation value of the operator $o^{(1)}$

$$g_F^{o^{(1)}} = \int_\Omega (o^{(1)}_\eta F(\eta)) F(\eta)\, d\omega(\eta) = \int_\Omega \eta |F(\eta)|^2\, d\omega(\eta) \in \mathbb{R}^3$$

thereby interpreting $|F(\eta)|^2 d\omega(\eta)$ as surface mass distribution over the sphere Ω embedded in Euclidean space $\mathbb{R}^3$. It is clear that $g_F^{o^{(1)}}$ lies in $\overline{\Omega_{\text{int}}}$, i.e. $|g_F^{o^{(1)}}| \leq 1$. The *variance in space domain* is understood in canonical sense as the variance of the operator $o^{(1)}$

$$\begin{aligned} \sigma_F^{o^{(1)}} &= \int_\Omega |(o^{(1)} - g_F^{o^{(1)}})\ F(\eta)|^2\, d\omega(\eta) \\ &= \int_\Omega |\eta - g_F^{o^{(1)}}|^2\ |F(\eta)|^2\, d\omega(\eta) \in \mathbb{R}. \end{aligned}$$

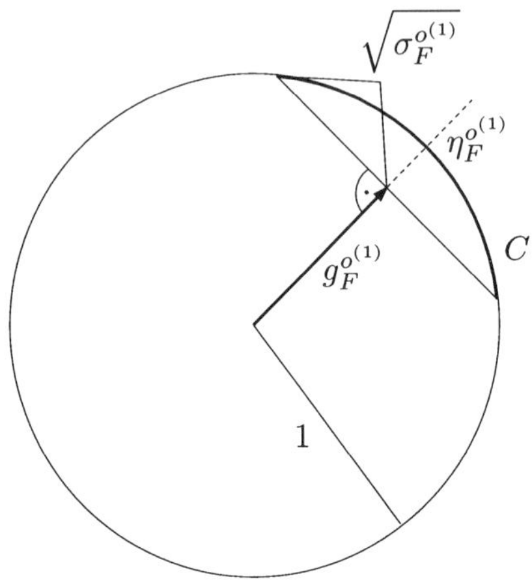

FIG. 5.5.1. Localization in a spherical cap

From the identity $|\eta - g_F^{o^{(1)}}|^2 = 1 + |g_F^{o^{(1)}}|^2 - 2\eta \cdot g_F^{o^{(1)}}, \eta \in \Omega$, it follows immediately that

$$\begin{aligned}&\int_\Omega |\eta - g_F^{o^{(1)}}|^2 \, |F(\eta)|^2 \, d\omega(\eta) \\ &= \int_\Omega (1 + (g_F^{o^{(1)}})^2 - 2\eta \cdot g_F^{o^{(1)}})|F(\eta)|^2 \, d\omega(\eta),\end{aligned}$$

hence, $\sigma_F^{o^{(1)}} = 1- \mid g_F^{o^{(1)}} \mid^2$. Obviously, $0 \leq \sigma_F^{o^{(1)}} \leq 1$. Figure 5.5.1 illustrates a geometric interpretation of $g_F^{o^{(1)}}$ and $\sigma_F^{o^{(1)}}$. We associate to $g_F^{o^{(1)}}, g_F^{o^{(1)}} \neq 0$, and its projection $\eta_F^{o^{(1)}} = g_F^{o^{(1)}} / \mid g_F^{o^{(1)}} \mid$ onto the sphere Ω the spherical cap

$$C = C(\eta_F^{o^{(1)}}; \mid g_F^{o^{(1)}} \mid) = \{\eta \in \Omega \mid 1 - \eta \cdot \eta_F^{o^{(1)}} \leq 1 - \mid g_F^{o^{(1)}} \mid\}.$$

Then the boundary ∂C is a circle with radius $(\sigma_F^{o^{(1)}})^{1/2}$. According to our construction the mass may be regarded to be essentially localized in the cap C. As one thinks of a radial basis function F to be a window function on Ω, the window is determined by $C(\eta_F^{o^{(1)}}; \mid g_F^{o^{(1)}} \mid)$, and its width is given by $(\sigma_F^{o^{(1)}})^{1/2}$.

Since we are particularly interested in axisymmetric functions (i.e. 'radial basis functions' on the sphere), some simplifications can be done. Let

G be in $\mathcal{L}^2[-1,+1]$ and $\|G\|_{\mathcal{L}^2[-1,+1]} = 1$. Then the corresponding 'centre of gravity' can be computed readily by

$$g^{o^{(1)}}_{G(\cdot\varepsilon^3)} = \int_\Omega \eta |G(\eta\cdot\varepsilon^3)|^2 \, d\omega(\eta) = \left(2\pi \int_{-1}^{+1} t|G(t)|^2 \, dt\right) \varepsilon^3.$$

Letting

$$t^{o^{(1)}}_G = |g^{o^{(1)}}_{G(\cdot\varepsilon^3)}| = 2\pi \int_{-1}^{+1} t|G(t)|^2 \, dt$$

we find for the corresponding variance

$$\sigma^{o^{(1)}}_G = 2\pi \int_{-1}^{+1} (1 + (t^{o^{(1)}}_G)^2 - 2t^{o^{(1)}}_G t)|G(t)|^2 \, dt = 1 - (t^{o^{(1)}}_G)^2.$$

Next, our purpose is to calculate expectation value and variance of the surface curl gradient $o^{(3)} = L^*$. Assume that F is of class $\mathcal{H}(\{(-\Delta^*)^\wedge(n)\};\Omega)$ such that $\|F\|_{\mathcal{L}^2(\Omega)} = 1$. Elementary calculations show that

$$g^{o^{(3)}}_F = \int_\Omega (o^{(3)}_\eta F(\eta))F(\eta) \, d\omega(\eta) = 0$$

for all $F \in \mathcal{H}(\{(-\Delta^*)^\wedge(n)\};\Omega)$. Moreover, from the surface integral theorem of Gauß we are able to deduce that

$$\begin{aligned} \sigma^{o^{(3)}}_F &= \int_\Omega |(o^{(3)}_\eta - g^{o^{(3)}}_F)F(\eta)|^2 \, d\omega(\eta) \\ &= \int_\Omega o^{(3)}_\eta F(\eta) o^{(3)}_\eta F(\eta) \, d\omega(\eta) \\ &= \int_\Omega (-\Delta^*_\eta F(\eta))F(\eta) \, d\omega(\eta) = g^{-\Delta^*}_F. \end{aligned}$$

This expression, which gives $\sigma^{o^{(3)}}_F$ in terms of the expectation value of the operator $-\Delta^*$, helps us to find a measure of the 'angular frequency' in F. Indeed, it becomes clear from the spherical harmonic expansion

$$g^{-\Delta^*}_F = \sum_{n=0}^{\infty} \sum_{j=1}^{2n+1} (n(n+1))(F^\wedge(n,j))^2$$

that a large value of $g^{-\Delta^*}_F$ is caused by a slow decay in the series expansion, but this means that the higher order spherical harmonics are contributing significantly. But the presence of high order harmonics indicates high frequency behaviour in F. Thus, although the expectation value of the operator vanishes for the whole class of functions $F \in \mathcal{H}(\{(-\Delta^*)^\wedge(n)\};\Omega)$ with

$\|F\|_{\mathcal{L}^2(\Omega)} = 1$, the variance $\sigma_F^{o^{(3)}}$ may be interpreted to be a measure in frequency domain. Again we achieve a simplification by confining ourselves to axisymmetric functions G. Then we have

$$\begin{aligned} \sigma_G^{o^{(3)}} &= \sum_{n=0}^{\infty} \frac{2n+1}{4\pi} n(n+1)|G^\wedge(n)|^2 \\ &= -\int_\Omega G(\eta\cdot\varepsilon^3)\Delta_\eta^* G(\eta\cdot\varepsilon^3)\, d\omega(\eta) \\ &= -2\pi \int_{-1}^{+1} G(t) L_t G(t)\, dt, \end{aligned}$$

where L_t is the Legendre operator (cf. Lemma 3.2.2).

All functions including windows obey an uncertainty relation which states that sharp localization in space and in frequency are mutually exclusive. In other words, an uncertainty relation measures the trade-off between space localization on Ω and frequency localization.

Theorem 5.5.1. *Let $F \in \mathcal{H}(\{(-\Delta^*)^\wedge(n)\};\Omega)$ satisfy $\|F\|_{\mathcal{L}^2(\Omega)} = 1$. Then*

$$\sigma_F^{o^{(1)}} \sigma_F^{o^{(3)}} \geq | g_F^{o^{(1)}} |^2 . \tag{5.5.1}$$

If $g_F^{o^{(1)}}$ is non-vanishing, then

$$\Delta_F^{o^{(1)}} \Delta_F^{o^{(3)}} \geq 1, \tag{5.5.2}$$

where we have introduced the abbreviations

$$\Delta_F^{o^{(1)}} = \frac{(\sigma_F^{o^{(1)}})^{1/2}}{| g_F^{o^{(1)}} |} \quad \textit{and} \quad \Delta_F^{o^{(3)}} = (\sigma_F^{o^{(3)}})^{1/2}.$$

Proof. We essentially follow F.J. NARCOWICH, J.D. WARD (1996). Applying the Cauchy–Schwarz inequality we obtain

$$(\sigma_F^{o^{(1)}})^{1/2} (\sigma_F^{o^{(3)}})^{1/2} \geq g_F,$$

where

$$g_F = \left| \int_\Omega |\eta - g_F^{o^{(1)}}|\, |F(\eta)|\, |(\eta \wedge \nabla_\eta^*) F(\eta)|\, d\omega(\eta) \right| .$$

Observing the definition of the vector product we find

$$g_F \geq \left| \int_\Omega F(\eta)(\eta - g_F^{o^{(1)}}) \wedge (\eta \wedge \nabla_\eta^*) F(\eta)\, d\omega(\eta) \right| .$$

We use the fact that

$$\int_\Omega F(\eta) \left((\eta - g_F^{o^{(1)}}) \wedge (\eta \wedge \nabla_\eta^*) F(\eta) \right) d\omega(\eta)$$

$$= \int_{\Omega} F(\eta)\left((\eta \wedge \nabla^*_\eta) \wedge (\eta - g_F^{o^{(1)}})F(\eta)\right)\, d\omega(\eta)$$

and the identity

$$(\eta - g_F^{o^{(1)}}) \wedge (\eta \wedge \nabla^*_\eta) + (\eta \wedge \nabla^*_\eta) \wedge (\eta - g_F^{o^{(1)}}) = -2\eta.$$

This yields

$$g_F \geq \left| \int_{\Omega} F(\eta)(-\eta)F(\eta)\, d\omega(\eta)\right| = \mid g_F^{o^{(1)}} \mid .$$

But this is the desired result. □

For simplicity, we write $t_G^{o^{(1)}}$ instead of $t_{G(\varepsilon^3\cdot)}^{o^{(1)}}$ throughout the remainder of this book.

Corollary 5.5.2. *Let $G(\varepsilon^3\cdot) \in \mathcal{H}(\{(-\Delta^*)^\wedge(n)\};\Omega)$ satisfy the condition $\|G\|_{\mathcal{L}^2[-1,+1]} = 1$. If $t_G^{o^{(1)}}$ is non-vanishing, then*

$$\Delta_G^{o^{(1)}} \Delta_G^{o^{(3)}} \geq 1,$$

where

$$\Delta_G^{o^{(1)}} = \frac{(\sigma_G^{o^{(1)}})^{1/2}}{t_G^{o^{(1)}}} \quad \text{and} \quad \Delta_G^{o^{(3)}} = (\sigma_G^{o^{(3)}})^{1/2}.$$

Next we are interested in the localization of the so-called *Gaussian kernel*

$$G(t) = \mathrm{e}^{-\frac{\lambda}{2}(1-t)}, \quad \lambda > 0, \quad t \in [-1,+1].$$

G takes its name from the bell-shaped (Gaussian) probability density function (note that $\mathrm{e}^{-\frac{\lambda}{2}(1-\cos\,\psi)} \simeq \mathrm{e}^{-\frac{\lambda}{2}(\frac{\psi^2}{2})}$ for small $\psi \geq 0$). It is easy to see that

$$\tilde{G}(t) = \gamma(\lambda)\mathrm{e}^{-\frac{\lambda}{2}(1-t)}, \quad \gamma(\lambda) = \frac{1}{\sqrt{4\pi}}\left(\frac{1}{2\lambda}(1-\mathrm{e}^{-2\lambda})\right)^{-1/2},$$

satisfies $\|\tilde{G}\|_{\mathcal{L}^2[-1,+1]} = 1$. Furthermore we get

$$\begin{aligned} t_{\tilde{G}}^{o^{(1)}} &= \coth\lambda - \frac{1}{\lambda}, \\ \sigma_{\tilde{G}}^{o^{(1)}} &= 1 - (\coth\lambda - \frac{1}{\lambda})^2, \end{aligned}$$

and

$$\sigma_{\tilde{G}}^{o^{(3)}} = \frac{\lambda}{2}\coth\lambda - \frac{1}{2}.$$

It is not difficult to deduce that $t_{\tilde{G}}^{o^{(1)}} \to 1$ and $\sigma_{\tilde{G}}^{o^{(1)}}\sigma_{\tilde{G}}^{o^{(3)}} \to 1$ as $\lambda \to \infty$. This shows us that the best value of the uncertainty principle is 1.

It should be mentioned that the interpretation of $\sigma_F^{o^{(3)}}$ as variance in 'total angular frequency' also shows some drawbacks: firstly the expectation value of the surface curl gradient is of vectorial nature which seems to be inadequate in frequency localization by means of scalar spherical harmonics, secondly the value of $g_F^{o^{(3)}}$ vanishes for all candidates F. In other words, the 'centre of gravity of the spherical window' in frequency domain is independent of the choice of the function F under consideration. This is the reason why we are finally interested in the variance of the operator $-\Delta^*$

$$\sigma_F^{-\Delta^*} = \int_\Omega |((-\Delta_\eta^*) - g_F^{-\Delta^*})F(\eta)|^2 d\omega(\eta)$$

as being a measure for the spread in frequency (note that the corresponding expectation value $g_F^{-\Delta^*}$ is now scalar-valued). An easy calculation yields

$$\sigma_F^{-\Delta^*} = g_F^{(-\Delta^*)^2} - (g_F^{-\Delta^*})^2. \tag{5.5.3}$$

In connection with Theorem 5.5.1 and (5.5.3) this leads us to the following result (cf. W. FREEDEN, U. WINDHEUSER (1997))

Corollary 5.5.3. *Let $F \in \mathcal{H}(\{((-\Delta^*)^\wedge(n))^2\};\Omega)$ satisfy $\|F\|_{\mathcal{L}^2(\Omega)} = 1$. Then*

$$\sigma_F^{o^{(1)}} \sigma_F^{-\Delta^*} \geq |g_F^{o^{(1)}}|^2 \, \frac{g_F^{(-\Delta^*)^2} - (g_F^{-\Delta^*})^2}{g_F^{-\Delta^*}} \tag{5.5.4}$$

provided that $g_F^{-\Delta^*} \neq 0$, i.e. $F^\wedge(n,j) \neq 0$ for some $(n,j), n > 0$.

If the right-hand side of (5.5.4) is non-vanishing, then

$$\Delta_F^{o^{(1)}} \Delta_F^{-\Delta^*} \geq 1, \tag{5.5.5}$$

where

$$\Delta_F^{-\Delta^*} = \left(\frac{\sigma_F^{-\Delta^*}}{(g_F^{(-\Delta^*)^2} - (g_F^{-\Delta^*})^2)/g_F^{-\Delta^*}} \right)^{1/2} = \left(g_F^{-\Delta^*}\right)^{1/2}.$$

The best value of the uncertainty principle (5.5.5) is 1.

Finally, we are looking at the localization of spherical harmonics. Consider the members $Y_{n,j}$ of an $\mathcal{L}^2(\Omega)$-orthonormal system of spherical harmonics. Then, by definition, $\|Y_{n,j}\|_{\mathcal{L}^2(\Omega)} = 1$ for all n, j. Furthermore,

$$g_{Y_{n,j}}^{o^{(1)}} = 0, g_{Y_{nj}}^{-\Delta^*} = n(n+1), g_{Y_{n,j}}^{(-\Delta^*)^2} = (n(n+1))^2.$$

As expected we have

$$\sigma_{Y_{n,j}}^{o^{(1)}} = 1, \ \sigma_{Y_{n,j}}^{-\Delta^*} = 0.$$

Remark. The diploma thesis A. CLEMENS (1997) gives a graphical impression of the localization properties for a large class of radial basis functions on the sphere.

5.6 Examples of Radial Basis Functions

Up to this point, our theoretical conclusions have been given under predefined Hilbert space structure. However, the choice of the norm is an important problem in constructive approximation. In what follows we give a first list of possible choices of the norm. The list is divided into four parts: (1) Green's functions with respect to certain (iterated) Beltrami differential operators (2) kernels related to (invertible) pseudodifferential operators (3) Abel–Poisson kernel and related radial basis functions (4) Gauß–Weierstraß kernel and related radial basis functions. All radial basis functions discussed here share the property of having a global support, i.e. their support covers the whole sphere Ω.

5.6.1 *Green's Kernels Corresponding to Iterated Beltrami Operators*

Let us construct the Hilbert space $\mathcal{H} = \mathcal{H}(\{A_n\};\Omega)$ by choosing the inner product $(\cdot,\cdot)_{\mathcal{H}(\{A_n\};\Omega)}$ corresponding to the norm $\|\cdot\|_{\mathcal{H}(\{A_n\};\Omega)}$, given by

$$\begin{aligned}\|F\|^2_{\mathcal{H}(\{A_n\};\Omega)} &= \left|\int_\Omega F(\xi)Y_{0,1}(\xi)\,d\omega(\xi)\right|^2 \\ &\quad + \sum_{n=1}^{\infty}\sum_{j=1}^{2n+1}((\Delta^*)^\wedge(n))^2\left|\int_\Omega F(\xi)Y_{n,j}(\xi)\,d\omega(\xi)\right|^2.\end{aligned}$$

That means the sequence $\{A_n\}$ is given by $A_0 = 1$, $A_n = (\Delta^*)^\wedge(n)$, $n = 1, 2, \ldots$, where the sequence $\{(\Delta^*)^\wedge(n)\}$ is the spherical symbol of the Beltrami operator $\partial_0 = \Delta^*$. Obviously, $\{A_n\}$ is $\{(n+\frac{1}{2})^\tau\}$-summable for all $\tau < 1$.

Observing the definition of the spherical harmonics we get via Green's surface identity

$$\begin{aligned}&(\Delta^*)^\wedge(n)\int_\Omega F(\xi)Y_{n,j}(\xi)\,d\omega(\xi)\\ &= \int_\Omega F(\xi)(\Delta^*_\xi Y_{n,j}(\xi))\,d\omega(\xi) = \int_\Omega(\Delta^*_\xi F(\xi))Y_{n,j}(\xi)\,d\omega(\xi).\end{aligned}$$

Consequently, according to Parseval's identity, we obtain

$$\|F\|^2_{\mathcal{H}(\{A_n\};\Omega)} = \frac{1}{4\pi}\int_\Omega |F(\xi)|^2\, d\omega(\xi) + \int_\Omega |\Delta^*_\xi F(\xi)|^2\, d\omega(\xi). \tag{5.6.1}$$

Remark. The integral taken over the square of the (distributional) Beltrami derivative on Ω

$$\int_\Omega |\Delta^*_\xi F(\xi)|^2\, d\omega(\xi) \tag{5.6.2}$$

can be seen in parallel to the one-dimensional integral over an interval $[a,b]$

$$\int_a^b |F''(t)|^2\, dt \tag{5.6.3}$$

that may be physically interpreted (at least in the linearized sense) as the bending energy of a statically deflected thin beam which indeed is proportional to the integral taken over the square of the (linearized) curvature of the elastica of the beam. The physical interpretation of (5.6.2) will turn out to be of basic importance in variational characterization of spherical splines.

For the reproducing kernel in $\mathcal{H} = \mathcal{H}(\{A_n\};\Omega)$ we find (cf. (4.2.7))

$$K_{\mathcal{H}}(\xi,\eta) = \frac{1}{4\pi} + \frac{1}{4\pi}\sum_{n=1}^{\infty} \frac{2n+1}{(n(n+1))^2} P_n(\xi\cdot\eta). \tag{5.6.4}$$

Apart from an additive constant the series on the right-hand side coincides with Green's function on the unit sphere corresponding to the differential operator ∂_0^2 (cf. W. FREEDEN (1978, 1979, 1980B)).

Consider $\{A_n\}$ with $A_n = 1$ for $n = 0,\ldots,m$, $A_n = (\partial_{0,\ldots,m})^\wedge(n)$ for $n = m+1, m+2,\ldots$, where the sequence $\{(\partial_{0,\ldots,m})^\wedge(n)\}$ is the spherical symbol of the operator $\partial_{0,\ldots,m} = \partial_0 \cdot \ldots \cdot \partial_m$ with $\partial_n = \Delta^* - (\Delta^*)^\wedge(n)$. The kernel of the operator $\partial_{0,\ldots,m}$ is the space $Harm_{0,\ldots,m}$. The norm is given by

$$\|F\|^2_{\mathcal{H}(\{A_n\};\Omega)} = \sum_{n=0}^{m}\sum_{j=1}^{2n+1} \left|\int_\Omega F(\xi) Y_{n,j}(\xi)\, d\omega(\xi)\right|^2 + \int_\Omega |(\partial_{0,\ldots,m})_\xi F(\xi)|^2\, d\omega(\xi), \tag{5.6.5}$$

whereas the reproducing kernel in $\mathcal{H}(\{A_n\};\Omega)$ reads

$$K_{\mathcal{H}}(\xi,\eta) = \sum_{n=0}^{m} \frac{2n+1}{4\pi} P_n(\xi\cdot\eta) + \frac{1}{4\pi}\sum_{k=m+1}^{\infty} \frac{2k+1}{((\partial_{0,\ldots,k})^\wedge(n))^2} P_k(\xi\cdot\eta). \tag{5.6.6}$$

The infinite series on the right-hand side is the bilinear expansion of Green's function of the operator $\partial^2_{0,\ldots,m}$ (cf. W. FREEDEN (1981A)). The sequence

$\{A_n\}$ is $\{(n+\frac{1}{2})^\tau\}$-summable for all $\tau < 2m+1$. An illustration of these kernels is given in Figure 5.6.1.

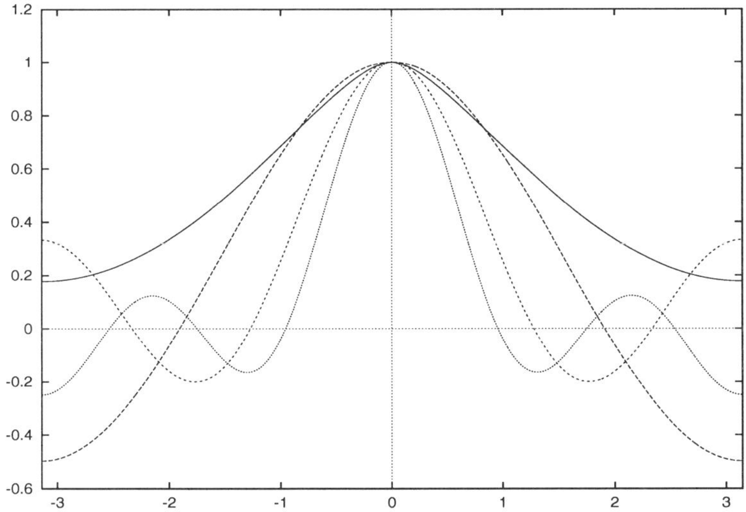

FIG. 5.6.1. The kernel $K_{\mathcal{H}}(\cos\vartheta)$ of (5.6.6) for $m = 0, \ldots, 3$

5.6.2 *Kernels Corresponding to Invertible Pseudodifferential Operators*

Let $\{A_n\} = \{A^\wedge(n)\}$, where A is a pseudodifferential operator on Ω with spherical symbol $\{A^\wedge(n)\}$ such that (i) $n \mapsto |A^\wedge(n)|^2$, $n \in \mathbb{N}_0$, is a real rational function (ii) there exist two positive constants C, C' with

$$C\left(n+\frac{1}{2}\right)^{2+\varepsilon} \le (A^\wedge(n))^2 \le C'\left(n+\frac{1}{2}\right)^{\sigma}, \tag{5.6.7}$$

for some $\varepsilon > 0, \sigma \ge 2+\varepsilon$. Then the norm reads

$$\|F\|^2_{\mathcal{H}(\{A_n\};\Omega)} = \int_\Omega |A_\eta F(\eta)|^2 \, d\omega(\eta) = \sum_{n=0}^{\infty}\sum_{j=1}^{2n+1} (A^\wedge(n))^2 (F^\wedge(n,j))^2. \tag{5.6.8}$$

For the reproducing kernel in $\mathcal{H} = \mathcal{H}(\{A^\wedge(n)\};\Omega)$ we find

$$K_{\mathcal{H}}(\xi,\eta) = \sum_{n=0}^{\infty} \frac{2n+1}{4\pi} \frac{1}{(A^\wedge(n))^2} P_n(\xi\cdot\eta). \tag{5.6.9}$$

The sequence $\{A_n\}$ is $\{(n+\frac{1}{2})^\tau\}$-summable for all $\tau < \varepsilon/2$. It should be mentioned that examples can be found for which the kernel representations are expressible as elementary functions. An example is $A^\wedge(0) = 1, A^\wedge(n) = ((2n+1)\ n(n+1))^{1/2}$, $n = 1,2,\ldots$. Note that the Legendre transform of the function $t \mapsto 1 - 2\ln\left(1+\sqrt{(1-t)/2}\right)$, $t \in [-1,+1]$ is given by $4\pi((2n+1)n(n+1))^{-1}, n = 1,2,\ldots$ (cf. W. FREEDEN (1990B)).

An alternative to come to a good candidate for numerical purposes is to replace $\{A^\wedge(n)\}$ by the sequence $\{h^{-n/2}A^\wedge(n)\}, 0 < h < 1$, for which the correspondig kernel is available as elementary function and to use this kernel for parameters h close to the value 1.

5.6.3 *Abel–Poisson Kernel and Related Functions*

Let $\{A_n\} = \{h^{-n/2}\Lambda^\wedge(n)\}$, $0 < h < 1$, where $n \mapsto |\Lambda^\wedge(n)|^2$, $n \in \mathbb{N}_0$, is a real rational function with

$$0 < |\Lambda^\wedge(n)|^2 \leq C' \left(n+\frac{1}{2}\right)^\tau \tag{5.6.10}$$

for some τ. Of course, the parameter h gives additional flexibility in choosing the norm, and also results in more general sequences $\{A_n\}$ being possible since the factor h^n appears in the infinite sum for the reproducing kernel. Now, the norm reads

$$\|F\|^2_{\mathcal{H}(\{A_n\};\Omega)} = \sum_{n=0}^{\infty}\sum_{j=1}^{2n+1} \frac{1}{h^n}\left|\int_\Omega (\Lambda_\xi F(\xi))Y_{n,j}(\xi)\,d\omega(\xi)\right|^2, \tag{5.6.11}$$

while for the reproducing kernel in $\mathcal{H}(\{A_n\};\Omega)$ we find

$$K_{\mathcal{H}(\{A_n\};\Omega)}(\xi,\eta) = \sum_{n=0}^{\infty} \frac{2n+1}{4\pi}(\Lambda^\wedge(n))^{-2}h^n P_n(\xi\cdot\eta). \tag{5.6.12}$$

The sequence $\{A_n\}$ is $\{(n+\frac{1}{2})^\tau\}$-summable for all τ. Moreover, different types of kernel representations are available in terms of elementary functions. We only mention here (cf. W. FREEDEN (1984B)):

(i) *Abel–Poisson kernel:* $\Lambda^\wedge(n) = 1$, $n = 0,1,\ldots$

$$K_{\mathcal{H}(\{A_n\};\Omega)}(\xi,\eta) = Q_h(\xi\cdot\eta) = \frac{1}{4\pi}\frac{1-h^2}{(L_h(\xi\cdot\eta))^{3/2}} \tag{5.6.13}$$

(ii) *'Singularity kernel':* $\Lambda^\wedge(n) = (n+\frac{1}{2})^{1/2}, n = 0,1,\ldots$

$$K_{\mathcal{H}(\{A_n\};\Omega)}(\xi,\eta) = \frac{1}{2\pi}\frac{1}{(L_h(\xi\cdot\eta))^{1/2}} \tag{5.6.14}$$

(iii) *'Logarithmic kernel'*: $\Lambda^\wedge(n) = ((2n+1)(n+1))$, $n = 0, 1, \ldots$

$$K_{\mathcal{H}(\{A_n\};\Omega)}(\xi,\eta) = \frac{1}{2\pi h}\ln\Big(1 + \frac{2h}{(L_h(\xi\cdot\eta))^{1/2}+1-h}\Big), \tag{5.6.15}$$

where we have used the abbreviation

$$L_h(\xi\cdot\eta) = 1 + h^2 - 2h(\xi\cdot\eta). \tag{5.6.16}$$

Graphical illustrations are given in the Figures 5.6.2–5.6.4.

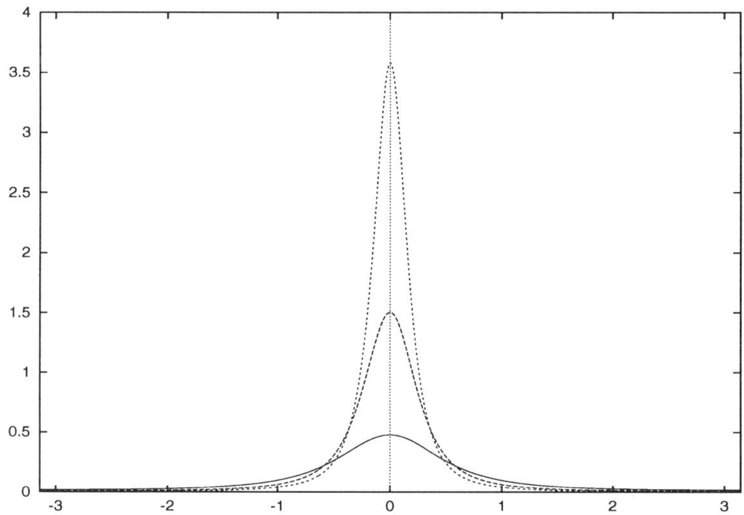

FIG. 5.6.2. Abel–Poisson kernel $Q_h(\cos\vartheta)$ for $h = 0.5,\ 0.7,\ 0.8$

Of particular significance for our further investigations is the *Abel–Poisson kernel* $Q_h : [-1,+1] \to \mathbb{R}$, $h \in (0,1)$, which should be discussed in more detail (cf. W. FREEDEN, M. SCHREINER (1995A)):

$$Q_h(t) = \frac{1}{4\pi}\frac{1-h^2}{(1+h^2-2ht)^{3/2}} = \frac{1}{4\pi}\sum_{n=0}^{\infty}(2n+1)h^n P_n(t). \tag{5.6.17}$$

From $1 + h^2 - 2ht = (1-h)^2 + 2h(1-t)$ it follows that $Q_h(t) > 0$ for all $t \in [-1,+1]$. By virtue of the Funk–Hecke formula in connection with (5.6.17) we get

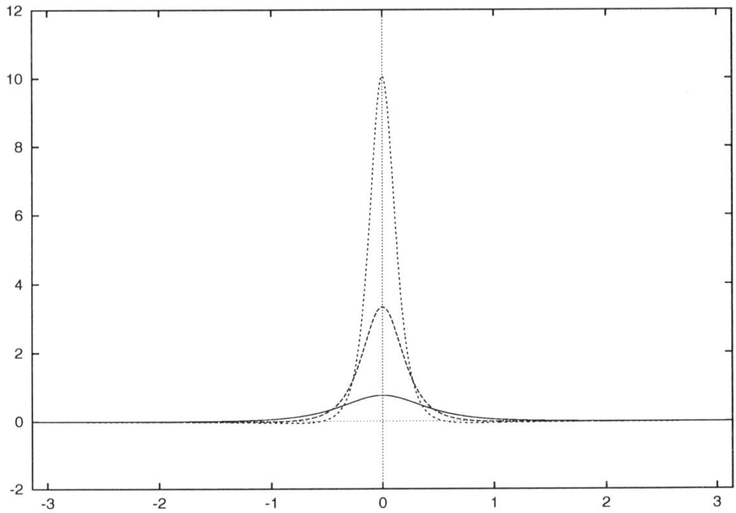

FIG. 5.6.3. Singularity kernel $K_{\mathcal{H}(\{A_n\};\Omega)}(\cos\vartheta)$ for $h = 0.5,\ 0.7,\ 0.8$

$$\int_{\Omega} Q_h(\xi \cdot \eta)\, d\omega(\eta) = 1 \tag{5.6.18}$$

for all $\xi \in \Omega$ and $h < 1$. For $t \in [-1, +1]$, we have

$$|Q_h(t)| \leq Q_h(1) = \frac{1}{4\pi} \frac{1+h}{(1-h)^2}. \tag{5.6.19}$$

Hence,

$$|Q_h(t)| \leq \frac{1}{2\pi} \frac{1}{(1-h)^2}, \quad t \in [-1, 1]. \tag{5.6.20}$$

Observing the first derivative

$$\frac{d}{dt} Q_h(t) = \frac{3}{4\pi} \frac{h(1-h^2)}{(1+h^2-2ht)^{5/2}}, \tag{5.6.21}$$

we obtain from the mean value theorem that

$$|Q_h(t) - Q_h(t')| \leq \frac{3}{2\pi} \frac{1}{(1-h)^4} |t - t'| \tag{5.6.22}$$

for all $t, t' \in [-1, +1]$. It is not hard to see that

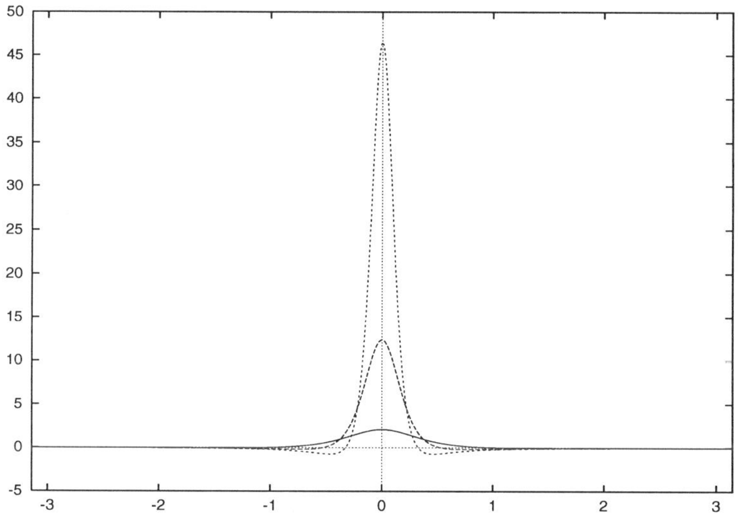

FIG. 5.6.4. Logarithmic kernel $K_{\mathcal{H}(\{A_n\};\Omega)}(\cos\vartheta)$ for $h = 0.5,\ 0.7,\ 0.8$

$$\int_\Omega Q_h(\xi\cdot\zeta)Q_{h'}(\eta\cdot\zeta)\,d\omega(\zeta) = Q_{hh'}(\xi\cdot\eta) \tag{5.6.23}$$

for all $\xi,\eta\in\Omega$ and $h,h'<1$. Therefore, in view of (5.6.20) and (5.6.22), the iterated Abel–Poisson kernel fulfils

$$\left|Q_h^{(2)}(t)\right| \leq \frac{1}{2\pi}\frac{1}{(1-h^2)^2}, \quad t\in[-1,+1], \tag{5.6.24}$$

and

$$\left|Q_h^{(2)}(t) - Q_h^{(2)}(t')\right| \leq \frac{3}{2\pi}\frac{1}{(1-h^2)^4}|t-t'| \tag{5.6.25}$$

for all $t,t'\in[-1,+1]$.

In order to investigate the *localization properties* of the Abel–Poisson kernel, consider the function

$$\tilde{Q}_h(t) = \|Q_h\|^{-1}_{\mathcal{L}^2[-1,+1]}Q_h(t), \quad t\in[-1,+1].$$

An easy calculation (cf. W. FREEDEN, U. WINDHEUSER (1997)) gives us

$$\|Q_h\|_{\mathcal{L}^2[-1,+1]} = (Q_{h^2}(1))^{1/2} = \left(\frac{1+h^2}{4\pi}\right)^{1/2}\frac{1}{1-h^2},$$

$$t_{\tilde{Q}_h}^{o^{(1)}} = \frac{2h}{1+h^2}, \quad \sigma_{\tilde{Q}_h}^{o^{(3)}} = \frac{6h^2}{(1-h^2)^2}.$$

This shows us that

$$\Delta_{\tilde{Q}_h}^{o^{(1)}} = \frac{1-h^2}{2h}, \quad \Delta_{\tilde{Q}_h}^{o^{(3)}} = \sqrt{6}\frac{h}{1-h^2}.$$

Thus we finally obtain

$$\Delta_{\tilde{Q}_h}^{o^{(1)}} \Delta_{\tilde{Q}_h}^{o^{(3)}} = \frac{\sqrt{6}}{2} = \sqrt{\frac{3}{2}} > 1.$$

Note that the value of $\Delta_{\tilde{Q}_h}^{o^{(1)}} \Delta_{\tilde{Q}_h}^{o^{(3)}}$ is independent of h.

5.6.4 *Gauß–Weierstraß Kernel and Related Functions*

Set $\{A_n\} = \{\mathrm{e}^{-n(n+1)\rho}\Lambda^\wedge(n)\}$, $\rho > 0$, where Λ is a pseudodifferential operator satisfying the property that $n \mapsto (\Lambda^\wedge(n))^2, n = 0, 1, \ldots$ is a real rational function with $0 < |\Lambda^\wedge(n)|^2 \leq C'(n+\frac{1}{2})^\tau$ for some τ. The simplest example is the *Gauß–Weierstraß kernel* ($\Lambda^\wedge(n) = 1, n = 0, 1, \ldots$)

$$K_{\mathcal{H}(\{A_n\};\Omega)}(\xi,\eta) = W_\rho(\xi\cdot\eta) = \sum_{n=0}^{\infty} \mathrm{e}^{-n(n+1)\rho}\frac{2n+1}{4\pi}P_n(\xi\cdot\eta), \tag{5.6.26}$$

which will be discussed below in more detail.

It is known from results due to S. Bochner (1950, 1954) that $W_\rho(t) \geq 0$ for all $t \in [-1,+1]$. Clearly, $(W_\rho)^\wedge(n) = \mathrm{e}^{-n(n+1)\rho}, n = 0, 1, \ldots$. Furthermore,

$$\int_\Omega W_\rho(\xi\cdot\eta)\,d\omega(\eta) = 1 \tag{5.6.27}$$

for all $\xi \in \Omega$. Since $|P_n(t)| \leq P_n(1) = 1$ for $t \in [-1,+1]$, we have

$$|W_\rho(t)| \leq \frac{1}{4\pi}\sum_{n=0}^{\infty} \mathrm{e}^{-n(n+1)\rho}(2n+1), \quad t \in [-1,1]. \tag{5.6.28}$$

To estimate the infinite sum on the right-hand side, we mention (cf. W. Freeden, M. Schreiner (1995A)) first that the function $A : [0,\infty) \to \mathbb{R}$, defined by

$$A(x) = \mathrm{e}^{-x(x+1)\rho}(2x+1) \tag{5.6.29}$$

is positive, monotone increasing for $x \in [0, x^*)$, and monotone decreasing for $x \in (x^*, \infty)$, where

$$x^* = -\frac{1}{2} + \frac{1}{\sqrt{2}}\frac{1}{\sqrt{\rho}}. \tag{5.6.30}$$

Therefore we can estimate

$$\sum_{n=0}^{\infty} \mathrm{e}^{-n(n+1)\rho}(2n+1) \quad \leq \quad 2A(x^*) + \int_0^\infty A(x)\,dx$$

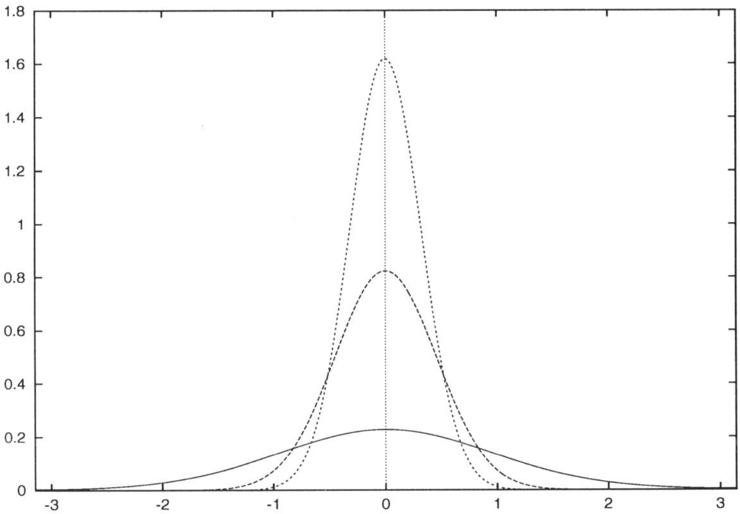

FIG. 5.6.5. Gauß–Weierstraß kernel $W_\rho(\cos\vartheta)$ for $\rho = 0.4,\ 0.1,\ 0.05$

$$
\begin{aligned}
&= 2\sqrt{\frac{2}{\rho}}\mathrm{e}^{(\rho-2)/4} + \int_0^\infty \mathrm{e}^{-y\rho}\,dy \\
&\leq 2\sqrt{\frac{2}{\rho}} + \frac{1}{\rho} \\
&\leq (2\sqrt{2}+1)\frac{1}{\rho}. \qquad (5.6.31)
\end{aligned}
$$

Hence, we obtain

$$|W_\rho(t)| \leq \frac{1}{\pi}\frac{1}{\rho}, \quad t \in [-1,+1]. \qquad (5.6.32)$$

Furthermore, in connection with the estimate (3.2.5), we find for $t, t' \in [-1,+1]$

$$|W_\rho(t) - W_\rho(t')| \leq |t-t'| \sum_{n=0}^{\infty} \mathrm{e}^{-n(n+1)\rho}\frac{2n+1}{4\pi}\frac{n(n+1)}{2}. \qquad (5.6.33)$$

Using that the function $B : [0,\infty) \to \mathbb{R}$, defined by

$$B(x) = \mathrm{e}^{-x(x+1)\rho}(2x+1)\frac{x(x+1)}{2} \qquad (5.6.34)$$

is positive, monotone increasing for $x \in [0, x^*)$, and monotone decreasing for $x \in (x^*, \infty)$, where

$$x^* = -\frac{1}{2} + \frac{1}{4}\sqrt{\frac{2}{\rho}}\sqrt{\rho + 6 + \sqrt{\rho^2 + 4\rho + 36}}, \tag{5.6.35}$$

we get by similar considerations as before

$$\begin{aligned}\sum_{n=0}^{\infty} e^{-n(n+1)\rho}(2n+1)\frac{n(n+1)}{2} &\leq 2B(x^*) + \int_0^{\infty} B(x)\,dx \\ &\leq 6\frac{1}{\rho^{3/2}} + \frac{1}{2\rho^2}\Gamma(2) \\ &\leq \frac{13}{2}\frac{1}{\rho^2}. \end{aligned} \tag{5.6.36}$$

Hence, we end up with

$$|W_\rho(t) - W_\rho(t')| \leq \frac{13}{8\pi}\frac{1}{\rho^2}|t - t'|, \quad t, t' \in [-1, +1]. \tag{5.6.37}$$

Obviously, we have

$$\int_\Omega W_\rho(\xi \cdot \eta) W_{\rho'}(\eta \cdot \zeta)\, d\omega(\eta) = W_{\rho+\rho'}(\xi \cdot \zeta), \tag{5.6.38}$$

for all $\xi, \zeta \in \Omega$ and $\rho, \rho' > 0$. For the iterated kernel $W_\rho^{(2)}$, we find by virtue of (5.6.32) and (5.6.37) that

$$\left|W_\rho^{(2)}(t)\right| \leq \frac{1}{2\pi}\frac{1}{1-h}, \quad t \in [-1, +1], \tag{5.6.39}$$

and

$$\left|W_\rho^{(2)}(t) - W_\rho^{(2)}(t')\right| \leq \frac{13}{32\pi}\frac{1}{(1-h)^2}|t - t'|, \quad t, t' \in [-1, +1]. \tag{5.6.40}$$

In order to investigate the *localization* of the Gauß–Weierstraß kernel, consider

$$\tilde{W}_\rho(t) = \|W_\rho\|^{-1}_{\mathcal{L}^2[-1,+1]} W_\rho(t), \quad t \in [-1, +1],$$

where

$$\|W_\rho\|_{\mathcal{L}^2[-1,+1]} = \left(\sum_{n=0}^{\infty} \frac{2n+1}{4\pi} e^{-n(n+1)2\rho}\right)^{1/2}.$$

The values $t_{\tilde{W}_\rho}^{o^{(1)}}$ and $\sigma_{\tilde{W}_\rho}^{o^{(3)}}$ are calculated numerically (more explicitly, $t_{\tilde{W}_\rho}^{o^{(1)}}$ via approximate integration and $\sigma_{\tilde{W}_\rho}^{o^{(3)}}$ by series truncation). Table 5.6.1 shows results for different values of $\rho \in [0.05, 0.8]$. It seems that $\Delta_{\tilde{W}\rho}^{o^{(1)}} \Delta_{\tilde{W}\rho}^{o^{(3)}}$ tends to the value 1 as $\rho \to 0$.

Table 5.6.1

ρ	$\Delta^{o^{(3)}}_{\tilde{W}_\rho}$	$\Delta^{o^{(1)}}_{\tilde{W}_\rho}$	$\Delta^{o^{(1)}}_{\tilde{W}_\rho}\Delta^{o^{(3)}}_{\tilde{W}_\rho}$
0.05	3.108	0.325	1.012
0.1	2.159	0.475	1.025
0.2	1.468	0.717	1.053
0.4	0.942	1.177	1.110
0.8	0.468	2.545	1.193

5.7 Positive Definiteness

The sequence $\{A_n\}_{n=0,1,\ldots}$ defining the corresponding reproducing kernel Hilbert spaces $\mathcal{H}(\{A_n\};\Omega)$ has been considered under the assumption that $A_n \neq 0$ for all n. In practice, however, one often has to deal with the more general case, where some A_n are allowed to be zero. We now study this situation. We start with the following definition.

Definition 5.7.1. Let $\{A_n\}_{n=0,1,\ldots} \in \mathcal{A}$ be a sequence. We split $\mathbb{N}_0$ into two parts such that $\mathbb{N}_0 = \mathcal{N} \cup \mathcal{N}_0$, $\mathcal{N} \cap \mathcal{N}_0 = \emptyset$, where

$$\begin{aligned}\mathcal{N} &= \{n \in \mathbb{N}_0 |\ A_n \neq 0\},\\ \mathcal{N}_0 &= \{n \in \mathbb{N}_0 |\ A_n = 0\}.\end{aligned}$$

The sequence $\{A_n\}$ is called $\{B_n\}$*-summable with respect to* $\mathcal{N}$ if

$$\Sigma^{\mathcal{N}}(\{B_n\},\{A_n\}) = \sum_{n\in\mathcal{N}} \frac{2n+1}{4\pi}\frac{|B_n|^2}{|A_n|^2} < \infty.$$

As above, a $\{1\}$-summable sequence with respect to $\mathcal{N}$ is simply called *summable with respect to* $\mathcal{N}$:

$$\Sigma^{\mathcal{N}}(\{A_n\}) = \sum_{n\in\mathcal{N}} \frac{2n+1}{4\pi}\frac{1}{|A_n|^2} < \infty.$$

For a given sequence $\{A_n\}$, consider the linear space $\mathcal{E}^{\mathcal{N}}(\{A_n\};\Omega)$ of all functions $F \in \mathcal{C}^{(\infty)}(\Omega)$ satisfying $(F, Y_{n,j})_{\mathcal{L}^2(\Omega)} = 0$ for all $n \in \mathcal{N}_0$ and $\sum_{n\in\mathcal{N}}\sum_{j=1}^{2n+1}|A_n|^2(F,Y_{n,j})^2_{\mathcal{L}^2(\Omega)} < \infty$. On $\mathcal{E}^{\mathcal{N}}(\{A_n\};\Omega)$ we introduce an inner product by

$$(F,G)_{\mathcal{H}^{\mathcal{N}}(\{A_n\};\Omega)} = \sum_{n\in\mathcal{N}} |A_n|^2 \sum_{j=1}^{2n+1} (F,Y_{n,j})_{\mathcal{L}^2(\Omega)}(G,Y_{n,j})_{\mathcal{L}^2(\Omega)}$$

and define the space $\mathcal{H}^{\mathcal{N}}(\{A_n\};\Omega)$ to be the completion of $\mathcal{E}^{\mathcal{N}}(\{A_n\};\Omega)$ with respect to the topology induced by $(\cdot,\cdot)_{\mathcal{H}^{\mathcal{N}}(\{A_n\};\Omega)}$. Then we end up with a Hilbert space. If there is no confusion likely to arise we simply write in the following $\mathcal{H}$ instead of $\mathcal{H}^{\mathcal{N}}(\{A_n\};\Omega)$. Note that in this situation not all $Harm_n$ are contained in $\mathcal{H}$. Indeed we have $Harm_n \subset \mathcal{H}$ if and only if $n \in \mathcal{N}$.

The Hilbert space $\mathcal{H}$ corresponding to a summable sequence $\{A_n\}$ with respect to $\mathcal{N}$ possesses a reproducing kernel $K_{\mathcal{H}}(\cdot,\cdot)$, since the evaluation functional $F \mapsto F(\xi), F \in \mathcal{H}$, is continuous for every $\xi \in \Omega$ (cf. N. Aronszajn (1950)). It can be even seen, that $K_{\mathcal{H}}(\cdot,\cdot)$ admits a uniformly convergent series representation of the form

$$K_{\mathcal{H}}(\xi,\eta) = \sum_{n\in\mathcal{N}} \frac{2n+1}{4\pi}\frac{1}{|A_n|^2}P_n(\xi\cdot\eta), \quad (\xi,\eta)\in\Omega^2.$$

Since the function $K_{\mathcal{H}}(\cdot,\eta)$ is an η–zonal function we are allowed to write $K_{\mathcal{H}}(\xi,\eta) = K_{\mathcal{H}}(\xi\cdot\eta)$ for all $(\xi,\eta)\in\Omega^2$. Note that the Legendre coefficients are given by

$$K_{\mathcal{H}}^{\wedge}(n) = \begin{cases} |A_n|^{-2} & \text{for} \quad n \in \mathcal{N} \\ 0 & \text{for} \quad n \in \mathcal{N}_0. \end{cases}$$

All $K_{\mathcal{H}}^{\wedge}(n)$ are non-negative. Thus it turns out that each kernel $K_{\mathcal{H}}(\cdot,\cdot)$ is positive definite. The details will be investigated below.

Assume now that the summable sequence $\{A_n\}$ with respect to $\mathcal{N}$ satisfies, in addition, $A_n \neq 0$ for $n = 0,\ldots,m$, i.e. $\{0,\ldots,m\}\subset\mathcal{N}$. Then the space $\mathcal{H} = \mathcal{H}(\{A_n\};\Omega)$ can be decomposed into the orthogonal direct sum

$$\mathcal{H} = \mathcal{H}_{0,\ldots,m} \oplus \mathcal{H}_{0,\ldots,m}^{\perp},$$

where $\mathcal{H}_{0,\ldots,m} = Harm_{0,\ldots,m}$ (note that we exclude the case $\mathcal{N} = \emptyset$). $\mathcal{H}_{0,\ldots,m}$ and $\mathcal{H}_{0,\ldots,m}^{\perp}$ defined in canonical way are Hilbert spaces possessing the reproducing kernels

$$K_{\mathcal{H}_{0,\ldots,m}}(\xi,\eta) = \sum_{n=0}^{m} \frac{2n+1}{4\pi}\frac{1}{|A_n|^2}P_n(\xi\cdot\eta), \quad (\xi,\eta)\in\Omega^2, \tag{5.7.1}$$

$$K_{\mathcal{H}_{0,\ldots,m}^{\perp}}(\xi,\eta) = \sum_{\substack{n\in\mathcal{N}\\ n>m}} \frac{2n+1}{4\pi}\frac{1}{|A_n|^2}P_n(\xi\cdot\eta), \quad (\xi,\eta)\in\Omega^2, \tag{5.7.2}$$

respectively. In the same way as described above we have

$$K_{\mathcal{H}}(\cdot,\cdot) = K_{\mathcal{H}_{0,\ldots,m}}(\cdot,\cdot) + K_{\mathcal{H}_{0,\ldots,m}^{\perp}}(\cdot,\cdot)$$

and

$$(\cdot,\cdot)_{\mathcal{H}} = (\cdot,\cdot)_{\mathcal{H}_{0,\ldots,m}} + (\cdot,\cdot)_{\mathcal{H}_{0,\ldots,m}^{\perp}}.$$

The norm $\|\cdot\|_{\mathcal{H}_{0,\ldots,m}^{\perp}}$ of $\mathcal{H}_{0,\ldots,m}^{\perp}$ can be understood to be a seminorm in $\mathcal{H}$ with kernel $Harm_{0,\ldots,m}$. This fact will be of importance for the forthcoming definition of spherical splines (cf. Chapter 6). The reproducing kernels (5.7.1) and (5.7.2) can be seen to satisfy the condition of positive definiteness.

Definition 5.7.2. Let $K : [-1,+1] \to \mathbb{R}$ be continuous. K is said to be *conditionally positive definite of order* m, if for any (admissible) system $X_N = \{\eta_1, \ldots, \eta_N\}$ on Ω and all real scalars $a_1, \ldots, a_N$ satisfying

$$\mathbf{matr}_{X_N}(Y_{0,1}, \ldots, Y_{m,2m+1})a = 0$$

the quadratic form

$$\sum_{k=1}^{N}\sum_{l=1}^{N} a_k a_l K(\eta_k \cdot \eta_l)$$

is non-negative.

K is said to be *conditionally strictly positive definite of order* m on Ω if the quadratic form is positive for all $a \neq 0$. *(Strict) positive definiteness* is formally understood as conditionally (strict) positive definiteness of order -1.

It is a remarkable fact that positive definite functions can be characterized by their Legendre expansion. First important results go back to I.J. SCHOENBERG (1942). In our approach we first mention the following result.

Lemma 5.7.3. *Each Legendre polynomial P_n is positive definite.*

Proof. It follows from the addition theorem that for all $\{\eta_1, \ldots, \eta_N\} \subset \Omega$ and $(a_1, \ldots, a_N)^T \in \mathbb{R}^N \backslash \{0\}$

$$\sum_{k=1}^{N}\sum_{l=1}^{N} a_k a_l P_n(\eta_k \cdot \eta_l) = \frac{4\pi}{2n+1} \sum_{j=1}^{2n+1} \left| \sum_{k=1}^{N} a_k Y_{n,j}(\eta_k) \right|^2 \geq 0$$

This proves the assertion of Lemma 5.7.3. □

Now we come to a more detailed characterization of positive definiteness (cf. W. FREEDEN *et al.* (1994), M. SCHREINER (1997A,B)).

Theorem 5.7.4. *Assume that the Legendre coefficients of $K : [-1,+1] \to \mathbb{R}$ satisfy*

$$\sum_{n=0}^{\infty} \frac{2n+1}{4\pi} K^\wedge(n) < \infty,$$

Then the following statements are equivalent:

(i) for all $\varphi \in \mathcal{C}^{(\infty)}(\Omega)$

$$\int_\Omega \int_\Omega K(\xi \cdot \eta)\varphi(\xi)\varphi(\eta)\, d\omega(\xi)\, d\omega(\eta) \geq 0$$

(ii) for all $Y_{n,j}$ *with* $n = 0, 1, \ldots, j = 1, \ldots, 2n+1$

$$\int_\Omega \int_\Omega K(\xi \cdot \eta) Y_{n,j}(\xi) Y_{n,j}(\eta)\, d\omega(\xi)\, d\omega(\eta) \geq 0.$$

(iii) $K^\wedge(n) \geq 0$ *for all* $n \geq 0$.

(iv) K *is positive definite.*

Proof. That (i) implies (ii) is clear. We therefore turn to (ii) $\Rightarrow$ (iii). (iii) follows immediately from (ii) since for all $n = 0, 1, \ldots, j = 1, \ldots, 2n+1$

$$K^\wedge(n) = \sum_{j=1}^{2n+1} \int_\Omega \int_\Omega K(\xi \cdot \eta) Y_{n,j}(\xi) Y_{n,j}(\eta)\, d\omega(\xi) d\omega(\eta).$$

In order to develop (iv) from (iii) suppose that $\eta_1, \ldots, \eta_N \in \Omega$ are pairwise distinct and assume that $(a_1, \ldots, a_n)^T$ is different from $(0, \ldots, 0)^T$. Then

$$\begin{aligned} 0 &\leq \sum_{n=0}^{\infty} h^{2n} \sum_{j=1}^{2n+1} \left(\sum_{k=1}^{N} a_k Y_{n,j}(\eta_k) \right)^2 K^\wedge(n) \\ &= \int_\Omega \int_\Omega K(\xi \cdot \eta) \Phi_h(\xi) \Phi_h(\eta)\, d\omega(\xi)\, d\omega(\eta), \end{aligned}$$

where

$$\Phi_h(\xi) = \sum_{k=1}^{N} a_k Q_h(\eta_k \cdot \xi), \quad \xi \in \Omega,$$

and Q_h is the Abel–Poisson kernel (cf. (5.6.13)). Letting $h \to 1$ we see that

$$\sum_{k=1}^{N} \sum_{l=1}^{N} a_k a_l K(\eta_k \cdot \eta_l) \geq 0,$$

as required. Finally, assume that (iv) is true and suppose that $\varphi \in \mathcal{C}^{(\infty)}(\Omega)$. Since K is continuous and therefore Riemann-integrable there exists a sequence of numbers r_k^N, $N = 1, 2, \ldots, k = 1, \ldots, N$, and a sequence of point sets $X_N = \{\eta_1^N, \ldots, \eta_N^N\} \subset \Omega$, $N \in \mathbb{N}$, with pairwise distinct points $\eta_1^N, \ldots, \eta_N^N$ such that

$$\int_\Omega \int_\Omega K(\xi \cdot \eta)\varphi(\xi)\varphi(\eta)\, d\omega(\xi)\, d\omega(\eta)$$
$$= \lim_{N\to\infty} \sum_{k=1}^{N}\sum_{l=1}^{N} r_k^N r_l^N K(\eta_k \cdot \eta_l)\varphi(\eta_k)\varphi(\eta_l).$$

By assumption,

$$\sum_{k=1}^{N}\sum_{l=1}^{N} K(\eta_k \cdot \eta_l)(r_k^N \varphi(\eta_k))(r_l^N \varphi(\eta_l))$$

is non-negative. Thus, assertion (i) is verified. □

Corollary 5.7.5. *Assume that $\{A_n\}$ is summable with respect to $\mathcal{N}$. Then $K_{\mathcal{H}^{\mathcal{N}}(\{A_n\};\Omega)}$ satisfies (i)–(iv) of Theorem 5.7.4.*

For the purpose of interpolation the property of strict conditionally positive definiteness is of importance. It is not hard to derive the following result from Definition 5.7.1.

Lemma 5.7.6. *If $K \in \mathcal{C}[-1,+1]$ is conditionally strictly positive definite of order m, then $K^{\perp}_{0,\ldots,m}$ defined by*

$$K^{\perp}_{0,\ldots,m}(t) = K(t) - \sum_{n=0}^{m} \frac{2n+1}{4\pi} K^{\wedge}(n) P_n(t), \quad t \in [-1,+1]$$

is conditionally strictly positive definite of order m.

Remark. In the case that $m = -1$, the sum $\Sigma_{n=0}^{m}$ is assumed to be zero.

Unfortunately, we do not know equivalent conditions comparable to those of Theorem 5.7.4 for the case of strict positive definiteness. That is the reason why we are interested in sufficient conditions for a kernel to be strictly positive definite. A first result is due to Y. Xu and E.W. Cheney (1992), who proved that if all the Legendre coefficients of a kernel are positive, then the kernel is strictly positive definite. By using locally supported kernels we are able to improve this result (cf. Theorem 5.9.3).

An equivalent condition for strict positive definiteness is the following.

Lemma 5.7.7. *Let $K : [-1, +1] \to \mathbb{R}$ satisfy*

$$\sum_{n=0}^{\infty} \frac{2n+1}{4\pi} K^{\wedge}(n) < \infty.$$

Then the kernel K is strictly positive definite if and only if the functions $K(\eta_1 \cdot), \ldots, K(\eta_N \cdot)$ are linearly independent for any choice of pairwise distinct points $\eta_1, \ldots, \eta_N \in \Omega$.

Proof. According to our assumptions K is the reproducing kernel of a Hilbert space $\mathcal{H}$. Therefore, $K(\eta_k \cdot \eta_l) = (K(\eta_k \cdot), K(\eta_l \cdot))_{\mathcal{H}}$. Hence,

$$\begin{pmatrix} K(\eta_1 \cdot \eta_1) & \ldots & K(\eta_1 \cdot \eta_N) \\ \vdots & & \vdots \\ K(\eta_N \cdot \eta_1) & \ldots & K(\eta_N \cdot \eta_N) \end{pmatrix}$$

is as Gram matrix positive definite if and only if the functions $K(\eta_1 \cdot), \ldots,$ $K(\eta_N \cdot)$ are linearly independent (see for example P.J. Davis (1963)). □

As an immediate consequence we find the following.

Corollary 5.7.8. *Let $\{A_n\}$ be summable. Moreover, assume that the points $\eta_1, \ldots, \eta_N \in \Omega$ are pairwise distinct. Then the functions $K_{\mathcal{H}(\{A_n\};\Omega)}(\eta_1, \cdot), \ldots, K_{\mathcal{H}(\{A_n\};\Omega)}(\eta_N, \cdot)$ are linearly independent.*

5.8 Further Examples of Radial Basis Functions

In this section we complete our list of radial basis functions. Thereby we drop the condition that all the Legendre coefficients of a kernel have to be positive. We firstly come to kernels related to radial basis functions in Euclidean space $\mathbb{R}^3$ and concentrate afterwards on locally supported kernels (i.e. axisymmetric 'finite elements') on the sphere.

5.8.1 *Kernels Related to Radial Basis Functions of Euclidean Spaces*

C.A. Micchelli (1986) has made a description of existence and uniqueness of the solution to the interpolation problem for radial (basis) functions in n-dimensional Euclidean space $\mathbb{R}^n$. This analysis relates the radial functions to the classical notion of positive definite functions and extends a study due to I.J. Schoenberg (1938). Refinements of these results were developed by W.R. Madych, S.A. Nelson (1988) based on Fourier transform techniques in $\mathbb{R}^n$. In what follows we are interested in the most popular radial basis functions in $\mathbb{R}^3$ and their reformulations in spherical language.

The sequence $\{A_n\}$, $K^{\wedge}(n) = A_n^{-2}$, generating the *Gaussian kernel* (cf. Section 5.5)

$$K(t) = 2\lambda(1 - e^{-2\lambda})e^{-\lambda(1-t)}, \ \lambda > 0, t \in [-1, +1] \tag{5.8.1}$$

satisfies the following recursion formula

$$A_0^{-1} = 1, \quad A_1^{-1} = \frac{1 + \lambda^{-2\lambda}}{1 - \lambda^{-2\lambda}} - \frac{1}{\lambda}, \tag{5.8.2}$$

$$A_{n+1}^{-1} = -\frac{2n+1}{\lambda} A_n^{-1} + A_{n-1}^{-1}, \quad n \geq 1. \tag{5.8.3}$$

The values A_n^{-1} are all positive. To prove this it is enough to show that

$$J_n = \int_{-1}^{+1} e^{\lambda t} P_n(t)\, dt > 0, \quad n > 0, \quad \lambda > 0. \tag{5.8.4}$$

Substituting the uniformly convergent series of the exponential function into the integral above yields

$$J_n = \sum_{k=n}^{\infty} \frac{1}{k!} \lambda^k \left((-1)^{k+1} + 1\right) \int_0^1 t^k P_n(t)\, dt, \tag{5.8.5}$$

where we note that (cf. W. MAGNUS *et al.* (1966)) the integrals occurring in the series (5.8.5) are always positive for $k \geq n$.

A function $G \in \mathcal{C}(0, \infty)$ is termed conditionally strictly positive definite of order m in $\mathbb{R}^3$, if for any system $X_N = \{x_1, \ldots, x_N\} \subset \mathbb{R}^3$ of pairwisely distinct points and scalars $a_1, \ldots, a_N$ satisfying $\sum_{i=1}^N a_i P(x_i) = 0$ for all polynomials P of degree $\leq m$ in $\mathbb{R}^3$ the quadratic form

$$\sum_{k=1}^{N} \sum_{i=1}^{N} a_k a_i G(|x_k - x_i|) \tag{5.8.6}$$

is positive. Comparing this definition (in Euclidean space $\mathbb{R}^3$) with Definition 5.7.2 we obtain the following statement: if $G \in \mathcal{C}[0, \infty)$ is conditionally strictly positive definite of order m in $\mathbb{R}^3$, then $t \mapsto K(t) = G(\sqrt{2 - 2t}), t \in [-1, +1]$ is continuous on $[-1, +1]$ and conditionally strictly positive definite of order m on Ω (in the sense of Definition 5.7.2).

A result due to I.J. SCHOENBERG (1938) shows us that the kernel $t \mapsto K(t) = G(\sqrt{2 - 2t})$, $t \in [-1, +1]$, is strictly positive definite whenever $G \in \mathcal{C}^{(\infty)}[0, \infty)$ satisfies $(-1)^j (d^j/d\tau^j) G(\sqrt{\tau}) > 0$, $\tau > 0, j = 0, 1, \ldots$. An extension of this result is proved by C.A. MICCHELLI (1986): let $G \in \mathcal{C}^{(\infty)}(0, \infty) \cap \mathcal{C}^{(0)}[0, \infty)$ satisfy $(-1)^j (d^j/d\tau^j) G(\sqrt{\tau}) \geq 0, \tau > 0, j > m$ and $(d^{m+1}/d\tau^{m+1}) G(\sqrt{\tau}) \neq const.$ Then $t \mapsto K(t) = G(\sqrt{2 - 2t}), t \in [-1, +1]$

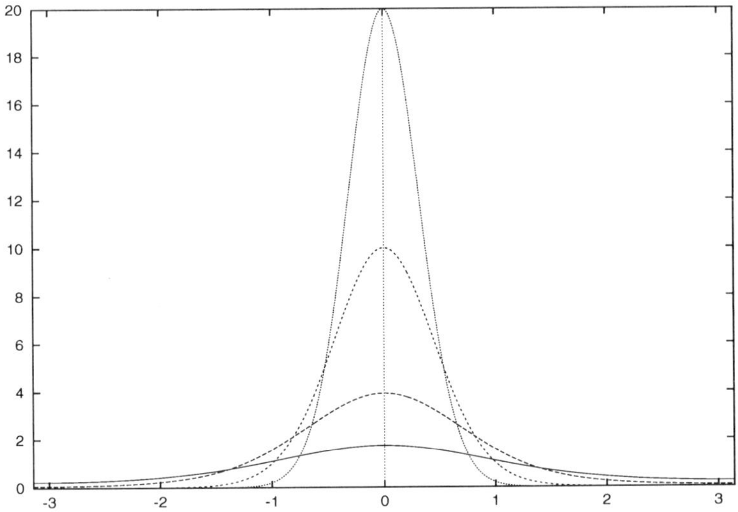

FIG. 5.8.1. Gaussian kernel $K(\cos\vartheta)$ for $\lambda = 1,\ 2,\ 5,\ 10$

is conditionally strictly positive definite of order m. A list of functions satisfying one of these conditions includes (cf. N. DYN (1989)):

$$
\begin{array}{rl}
\text{(i)} & G(\tau) = (c^2+\tau^2)^{-1/2}, \ c>0, \\
\text{(ii)} & G(\tau) = (-1)^{m+1}\tau^{\beta}, \quad 2m<\beta<2m+2, \quad m\geq 0, \\
\text{(iii)} & G(\tau) = (-1)^{m+1}\tau^{2m}\ln\tau, \quad m\geq 1, \\
\text{(iv)} & G(\tau) = (-1)^{m+1}(c^2+\tau^2)^{\beta/2}, \quad 2m<\beta<2m+2, \quad c>0, \quad m\geq 0, \\
\text{(v)} & G(\tau) = (-1)^{m+1}(c^2+\tau^2)^{m}\ln(c^2+\tau^2)^{1/2}, \quad c>0, \quad m\geq 0.
\end{array}
$$

The *inverse multiquadric* (i) satisfies Schoenberg's condition (see also Figure 5.8.2), while (ii)–(v) satisfy Micchelli's condition.

Another criterion that guarantees the regularity of the matrix

$$
\begin{pmatrix}
G(\sqrt{2-2\eta_1\cdot\eta_1}) & \dots & G(\sqrt{2-2\eta_N\cdot\eta_1}) \\
\vdots & & \vdots \\
G(\sqrt{2-2\eta_1\cdot\eta_N}) & \dots & G(\sqrt{2-2\eta_N\cdot\eta_N})
\end{pmatrix}
\tag{5.8.7}
$$

was conjectured in R. FRANKE (1983) and proved in C.A. MICCHELLI (1986): let G be in $\mathcal{C}^{(\infty)}(0,\infty)\cap\mathcal{C}^{(0)}[0,\infty)$ satisfying the conditions $G(\tau)>0$ and $(-1)^{j+1}(d^j/d\tau^j)G(\sqrt{\tau})\geq 0, \tau>0,\ j\geq 1$, then

$$
(-1)^{N-1}\det\left((G(\sqrt{2-2\eta_i\cdot\eta_j}))_{i,j=1}^{N}\right) > 0. \tag{5.8.8}
$$

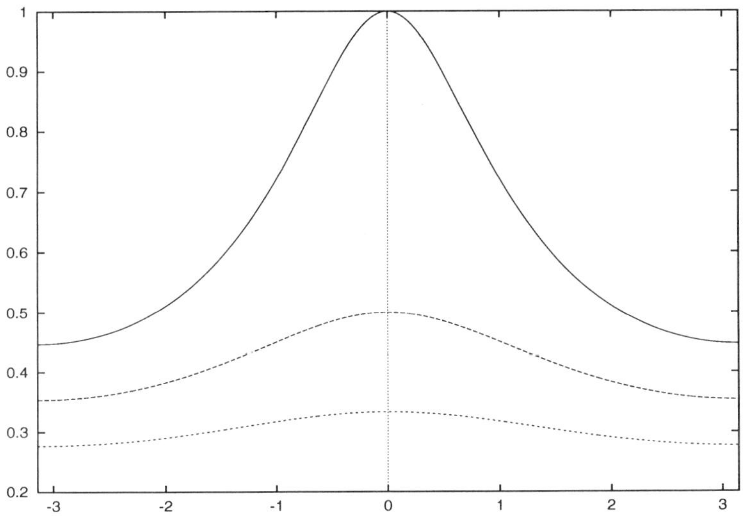

FIG. 5.8.2. Inverse multiquadric $G((2-2\cos\vartheta)^{1/2})$ for $c = 1,\ 2,\ 3$

A function satisfying (5.8.8) is the well known multiquadric (cf. Figure 5.8.3)

$$K(t) = G(\sqrt{2-2t}) = (c^2 + 2 - 2t)^{\nu}, \quad 0 < \nu < 1, \quad c > 0.$$

By calculating a simple example it can be shown that the multiquadric without approximation order (i.e. $m = -1$) is not conditionally strictly positive definite (in contrast to the situation in Euclidean space $\mathbb{R}^2$). But if we take approximation order 0 (cf. example (v)), the corresponding kernel defines a reproducing kernel in the space $\mathcal{H}_0^{\perp}$.

5.8.2 *Locally Supported Kernels*

Consider the piecewise polynomial function $B_h^{(k)} : [-1,1] \to \mathbb{R}$, $k = 0, 1, \ldots$, $h \in (0,1)$, given by

$$B_h^{(k)}(t) = \begin{cases} 0 & \text{for} \quad -1 \le t \le h \\ (t-h)^k/(1-h)^k & \text{for} \quad h < t \le 1. \end{cases} \tag{5.8.9}$$

Let $\eta \in \Omega$ be fixed. Then the η-zonal function $B_h^{(k)}(\eta\cdot\ ) : \Omega \to \mathbb{R}$ has a local support. More explicitly,

$$\operatorname{supp} B_h^{(k)}(\eta\cdot\ ) = \{\xi \in \Omega \mid\ h \le \xi \cdot \eta \le 1\}.$$

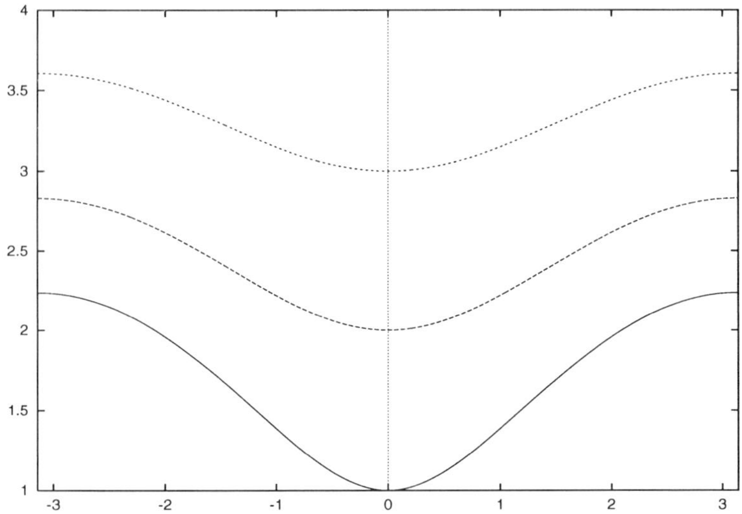

FIG. 5.8.3. Multiquadric $G((2-2\cos\vartheta)^{1/2})$ for $\nu = 1/2$ and $c = 1,\ 2,\ 3$

An easy calculation shows that the iterated η-zonal function $(B_h^{(k)})^{(2)}(\eta\cdot\ )$ has the support

$$\text{supp }(B_h^{(k)})^{(2)}(\eta\cdot\ ) = \{\xi \in \Omega \mid 2h^2 - 1 \leq \xi\cdot\eta \leq 1\}.$$

Graphical impressions of the locally supported kernels are given in Figure 5.8.4 and Figure 5.8.5, respectively.

Remark. For non-positive values of h, supp $(B_h^{(k)})^{(2)}(\eta\cdot\ ) = \Omega$. That is the reason why we restrict ourselves to parameter values $h \in (0,1)$.

It follows from the Funk–Hecke formula that

$$(B_h^{(k)})^\wedge(n) = 2\pi \int_h^1 \frac{(t-h)^k}{(1-h)^k} P_n(t)\, dt.$$

A recursion formula can be derived by considering the integrals (cf. P.A. MEISSL (1971), L. SJÖBERG (1980), W. FREEDEN, J.C. MASON (1990))

$$I_{n,k} = \int_h^1 (t-h)^k P_n(t)\, dt. \tag{5.8.10}$$

Straightforward integration yields

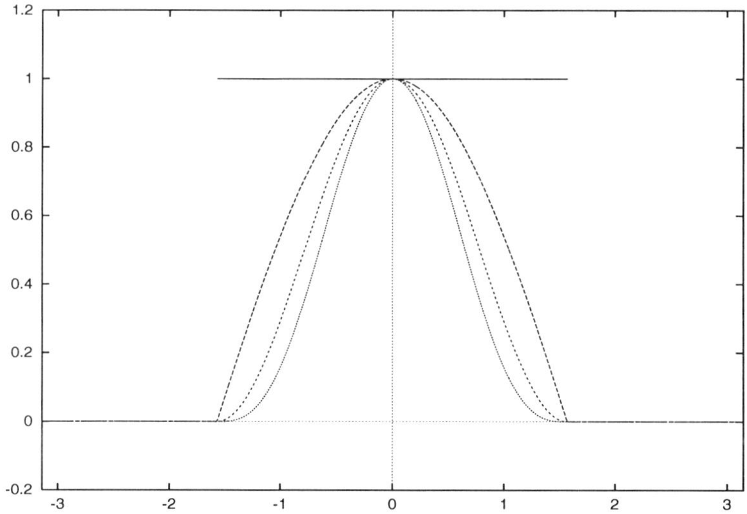

FIG. 5.8.4. Locally supported kernel $B_h^{(k)}(\cos\vartheta)$ for $h = 0$ and $k = 0, \ldots, 3$

$$I_{0,k} = \frac{(1-h)^{k+1}}{k+1}, \qquad I_{1,k} = I_{0,k}\frac{k+1+h}{k+2}.$$

From the recurrence formula (3.2.23) we obtain ($n \geq 1$)

$$(n+1)I_{n+1,k} + nI_{n-1,k} - (2n+1)I_{n,k+1} - (2n+1)hI_{n,k} = 0.$$

Furthermore, we find by partial integration and (3.2.19) that

$$(2n+1)I_{n,k+1} = -(k+1)(I_{n+1,k} - I_{n-1,k}). \tag{5.8.11}$$

Combining these results we arrive at the following recursion formula:

$$(n+k+2)I_{n+1,k} = (2n+1)hI_{n,k} + (k+1-n)I_{n-1,k}.$$

This gives us the following result.

Lemma 5.8.1. *For $k \geq 0$, $n \geq 1$ we have*

$$\begin{aligned}
(B_h^{(k)})^\wedge(0) &= 2\pi\frac{1-h}{k+1} \\
(k+2)(B_h^{(k)})^\wedge(1) &= (k+1+h)(B_h^{(k)})^\wedge(0) \\
(n+k+2)(B_h^{(k)})^\wedge(n+1) &= (2n+1)h(B_h^{(k)})^\wedge(n) \\
&\quad +(k+1-n)(B_h^{(k)})^\wedge(n-1).
\end{aligned}$$

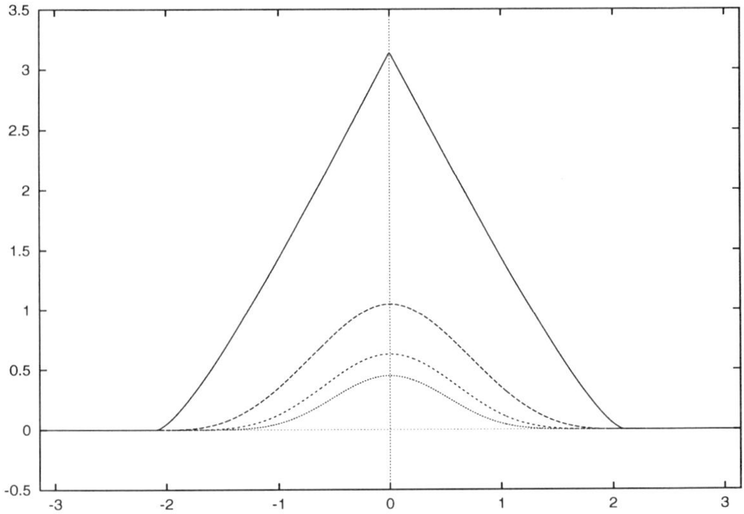

FIG. 5.8.5. Iterated locally supported kernel $(B_h^{(k)})^{(2)}(\cos\vartheta)$ for $h = 0.5$ and $k = 0, \ldots, 3$

Next we consider the asymptotic behaviour of $(B_h^{(k)})^\wedge(n)$ for fixed $h \in (0,1)$ as $n \to \infty$. For $k = 0$ and $n \geq 1$ we get from (3.2.19)

$$I_{n,0} = \int_h^1 P_n(t)dt = \frac{1}{2n+1}(P_{n+1}(h) - P_{n-1}(h)).$$

Using the estimate (3.2.44) we see that $|(B_h^{(0)})^\wedge(n)| = O(n^{-3/2})$ as n tends to infinity. From (5.8.11) we conclude recursively that $|(B_h^{(k)})^\wedge(n)| = O(n^{-(3/2)-k})$ as $n \to \infty$. Hence, the sequence $\{A_n\}$ defined by

$$A_n = \begin{cases} ((B_h^{(k)})^\wedge(n))^{-1} & \text{if} \qquad (B_h^{(k)})^\wedge(n) \neq 0 \\ 0 & \text{otherwise} \end{cases}$$

is summable with respect to $\mathcal{N}$. Therefore, the kernel $(B_h^{(k)})^{(2)}$ can be regarded as reproducing kernel of the space $\mathcal{H} = \mathcal{H}^{\mathcal{N}}(\{A_n\};\Omega)$. For a detailed characterization of $\mathcal{H}$ it is important to know which numbers $(B_h^{(k)})^\wedge(n)$ are zero. This investigation will be performed below. Before we come to this problem we note that the Lipschitz constant of $B_h^{(k)}$ can be estimated as follows: for $k = 1, 2, \ldots$ and $t, t' \in [1-h, 1]$

$$B_h^{(k)}(t) - B_h^{(k)}(t') = (1-h)^{-k}(t-t')\sum_{l=0}^{k-1}(t-h)^l(t'-h)^{k-1-l}$$

so that for all $t, t' \in [-1,1]$ and $k = 1, 2, \ldots$

$$\left|B_h^{(k)}(t) - B_h^{(k)}(t')\right| \le \frac{2^{k-1}}{(1-h)^k}|t-t'|. \tag{5.8.12}$$

Remark. Locally supported kernels of the above type ('*finite elements on the sphere*') have been discussed in more detail by S.L. Svensson (1984), H.J. Schaffeld (1988), W. Freeden, J. Mason (1990), J. Cui *et al.* (1992), M. Schreiner (1997a). Instead of the trial functions of type (5.8.9) we may consider many other trial functions, for example, certain B-spline expressions (cf. W. Freeden, J. Mason (1990)).

In order to characterize the underlying Hilbert space corresponding to a kernel $B_h^{(k)}$, $h \in (0,1)$, it is important to know which Legendre coefficients $(B_h^{(k)})^\wedge(n)$ are different from zero. Lemma 5.8.1 shows that each $(B_h^{(k)})^\wedge(n)$ is a polynomial of degree $n+1$ in the variable h which is zero for $h = 1$. Thus, the linear factor $(1-h)$ may be divided, and we define

$$b_n^{(k)}(h) = \frac{1}{2\pi(1-h)}(B_h^{(k)})^\wedge(n), \quad n = 0, 1, \ldots, \; h \in (0,1).$$

$b_n^{(k)}(h), h \in [-1,+1]$, is a polynomial of degree n in the variable h. From the recurrence relation for the $(B_h^{(k)})^\wedge(n)$ (cf. Lemma 5.8.1) we immediately get the following recursion formula for $b_n^{(k)}(h)$ $(n \ge 1)$:

$$\begin{aligned} b_0^{(k)}(h) &= \frac{1}{k+1}, \\ (k+2)b_1^{(k)}(h) &= (k+1+h)b_0^{(k)}(h), \\ (n+k+2)b_{n+1}^{(k)}(h) &= (2n+1)hb_n^{(k)}(h) + (k+1-n)b_{n-1}^{(k)}(h). \end{aligned} \tag{5.8.13}$$

The following is an immediate first result.

Lemma 5.8.2. *For all $k \ge 1$ we have*

$$\begin{aligned} b_0^{(k)}(-1) &= \frac{1}{k+1}, \\ b_n^{(k)}(-1) &= \frac{(k+1-n)\cdot\ldots\cdot(k-1)k}{(k+1)\cdot\ldots\cdot(k+1+n)}, \quad n \ge 1. \end{aligned}$$

Proof. The proof follows directly by induction by means of (5.8.13). □

Lemma 5.8.3. *For the lth derivatives of $b_n^{(k)}$ at the point $h = -1$ we have:*

(i) *if $l > n$:* $\quad (b_n^{(k)})^{(l)}(-1) = 0,$

(ii) *if $l = n$:* $\quad (b_n^{(k)})^{(l)}(-1) = \dfrac{(2l)!}{2^l} \dfrac{1}{(k+1)\cdot \ldots \cdot (k+1+l)},$

(iii) *if $l < n$:*

$$(b_n^{(k)})^{(l)}(-1) = \frac{(n-l+1)\cdot \ldots \cdot (n+l)}{2^l} \frac{(k+1-n)\cdot \ldots \cdot (k-l)}{(k+1)\cdot \ldots \cdot (k+1+n)}.$$

Proof. Since $b_n^{(k)}$ is a polynomial of degree n, part (i) is obvious. We prove part (ii) and (iii) by induction. For $l = 0$ the assertion is just Lemma 5.8.2. An easy calculation shows also the validity for $l = 1$. Assume now, (ii) and (iii) are true for all $l' \leq l$. Then we obtain from the recursion formula and the Leibniz rule

$$\begin{aligned}(b_{l+1}^{(k)})^{(l+1)}(h) &= \frac{2l+1}{l+k+2} h (b_l^{(k)})^{(l+1)}(h) + (l+1)\frac{2l+1}{l+k+2}(b_l^{(k)})^{(l)}(h) \\ &\quad + \frac{k+1-l}{l+k+2}(b_{l-1}^{(k)})^{(l)}(h) \\ &= (l+1)\frac{2l+1}{l+k+2}(b_l^{(k)})^{(l)}(h).\end{aligned}$$

Part (ii) follows then immediately by setting $h = -1$.

In order to prove the result for $(b_n^{(k)})^{(l+1)}(-1)$ if $n \geq l+1$ we apply an induction with respect to the variable n. For $n = l+1$ the result has just been proved. A similar calculation yields the case $n = l+2$. Under the assumption that (iii) is true for all n' with $l+1 \leq n' \leq n$ we have

$$\begin{aligned}(b_{n+1}^{(k)})^{(l+1)}(h) &= \frac{2n+1}{n+k+2} h (b_n^{(k)})^{(l+1)}(h) + (l+1)\frac{2n+1}{n+k+2}(b_n^{(k)})^{(l)}(h) \\ &\quad + \frac{k+1-n}{n+k+2}(b_{n-1}^{(k)})^{(l+1)}(h).\end{aligned}$$

Setting $h = -1$ and inserting the known formulae for $(b_n^{(k)})^{(l+1)}$, $(b_n^{(k)})^{(l)}$ and $(b_{n-1}^{(k)})^{(l+1)}$ the result follows after some lengthy but straightforward calculations. □

Lemma 5.8.3 shows in particular that $(b_{k+1}^{(k)})^{(l)}(-1) = 0$ for $l = 0, \ldots, k$. Since $b_{k+1}^{(k)}$ is a polynomial of degree $k+1$ it is clear that $b_{k+1}^{(k)}$ is of the form

$$b_{k+1}^{(k)} = \alpha_k (h+1)^{k+1} \tag{5.8.14}$$

with a constant α_k. This constant can be determined by taking the $(k+1)$ th derivative in (5.8.14) and substituting $h = -1$. With the results of Lemma 5.8.3 we arrive at

$$\frac{(2k+2)!}{2^{k+1}} \frac{1}{(k+1) \cdot \ldots \cdot (2k+2)} = \alpha_k (k+1)!.$$

Thus, $\alpha_k = 1/(2^{k+1}(k+1))$.

Using the recursion formula (5.8.13) we see that $b_{k+2}^{(k)}(h) = h b_{k+1}^{(k)}(h)$. As a consequence, all $b_n^{(k)}(h)$, $n \geq k+1$ have a zero of order $k+1$ at $h = -1$. Hence the definition

$$c_n^{(k)}(h) = \frac{1}{\alpha_k (h+1)^{k+1}} b_{n+k+1}^{(k)}(h), \quad n \geq 0, \tag{5.8.15}$$

yields a polynomial in h of degree n. Note that for $n \geq k+1$, $(B_h^{(k)})^\wedge(n) = 0$ if and only if $c_{n-k-1}^{(k)}(h) = 0$.

Next, we show that $c_n^{(k)}$ coincides up to a multiplicative constant with the Gegenbauer polynomial C_n^λ, $\lambda = k + 3/2$ (for the definition of the Gegenbauer polynomials see, e.g., G. SZEGÖ (1959)).

Lemma 5.8.4. *For $k \geq 0$, $n \in \mathbb{N}_0$, and $\lambda = k + 3/2$*

$$\begin{aligned} C_0^\lambda(h) &= c_0^{(k)}(h), \\ C_n^\lambda(h) &= \frac{1}{n!} \prod_{j=0}^{n-1} (2k+3+j)\, c_n^{(k)}(h), \quad n \geq 1. \end{aligned}$$

Proof. Set

$$\begin{aligned} \tilde{C}_0^\lambda(h) &= c_0^{(k)}(h), \\ \tilde{C}_n^\lambda(h) &= \frac{1}{n!} \prod_{j=0}^{n-1} (2k+3+j)\, c_n^{(k)}(h), \quad n \geq 1. \end{aligned}$$

The proof follows if we show that the polynomials $\tilde{C}_n^\lambda$ satisfy the recurrence relation of the Gegenbauer polynomials

$$\begin{gathered} C_0^\lambda(h) = 1, \quad C_1^\lambda(h) = 2\lambda h, \\ (n+1) C_{n+1}^\lambda(h) - 2(\lambda+n) h C_n^\lambda(h) + (2\lambda+n-1) C_{n-1}^\lambda(h) = 0, \qquad n > 1, \end{gathered} \tag{5.8.16}$$

for $\lambda = k + 3/2$ (see for example L.C. ANDREWS (1985) or G. SZEGÖ (1959)).

It is obvious that the initial values satisfy

$$\tilde{C}_0^\lambda(h) = c_0^{(k)}(h) = \frac{1}{\alpha_k} b_{k+1}^{(k)}(h) = 1$$

and

$$\begin{aligned}\tilde{C}_1^\lambda(h) &= (2k+3)c_1^{(k)}(h) = (2k+3)\frac{1}{\alpha_k(h+1)^{k+1}} b_{k+2}^{(k)}(h) \\ &= (2k+3)h\frac{1}{\alpha_k(h+1)^{k+1}} b_{k+1}^{(k)}(h) = (2k+3)h = 2\lambda h.\end{aligned}$$

Using (5.8.13) and (5.8.15) we obtain from

$$(n+2k+3)b_{n+k+2}^{(k)}(h) - (2n+2k+3)hb_{n+k+1}^{(k)}(h) + nb_{n+k}^{(k)}(h) = 0$$

that

$$\begin{aligned}&(n+2k+3)\tilde{C}_{n+1}^\lambda(h) - \frac{(2k+3+n)(2n+2k+3)}{n+1} h\tilde{C}_n^\lambda(h) \\ &+ \frac{(2k+3+n)(2k+2+n)n}{n(n+1)} \tilde{C}_{n-1}^\lambda(h) \\ &= 0.\end{aligned}$$

Thus,

$$(n+1)\tilde{C}_{n+1}^\lambda(h) - 2(\lambda+n)\tilde{C}_n^\lambda(h) + (2\lambda+n-1)\tilde{C}_{n-1}^\lambda(h) = 0,$$

which is just (5.8.16), as required. □

Theorem 5.8.5. *For $h \in (0,1)$ and $k = 0, 1, \ldots,$ the following statements are valid:*

(i) $(B_h^{(k)})^\wedge(n) \neq 0$ for $n = 0, 1, \ldots, k+2$.

(ii) For $n > k+2$, $(B_h^{(k)})^\wedge(n) = 0$ if and only if $C_{n-k-1}^{k+3/2}(h) = 0$, where $C_{n-k-1}^{k+3/2}$ denotes the Gegenbauer polynomial of order $n-k-1$.

Proof. We know from Lemma 5.8.1 that both $(B_h^{(k)})^\wedge(0)$ and $(B_h^{(k)})^\wedge(1)$ are positive. Furthermore it follows from the recurrence formula that $(B_h^{(k)})^\wedge(n+1) > 0$ as long as $n \leq k+1$. This proves (i). The second statement is a consequence of Lemma 5.8.4. □

Corollary 5.8.6. *We assume again that $k \geq 0$. The following statements are valid:*

(i) There exist parameters $h \in (0,1)$ such that $(B_h^{(k)})^\wedge(n) \neq 0$ for all $n \in \mathbb{N}_0$,

(ii) Let $m \in \mathbb{N}_0$ be given. Then there exists a number $h_0 \in (0,1)$ such that for all $h \in (h_0, 1)$ and all $n \leq m$, $(B_h^{(k)})^\wedge(n) \neq 0$.

Proof. Part (i) follows immediately from Theorem 5.8.5, since there exist only countably infinite points being a zero of a Gegenbauer polynomial.

It can be deduced from the theory of orthogonal polynomials (see for example F.G. TRICOMI (1970), L.C. ANDREWS (1985)) that the largest number $h \in (0,1)$ for which $C_{n-k-1}^{k+3/2}(h) = 0$ for an integer n with $k+2 \leq n \leq m$ is the largest zero of $C_{m-k-1}^{k+3/2}$. So just take this value as h_0. This proves part (ii). □

Remark. The first statement of Corollary 5.8.6 is of more theoretical concern than practical use, since it is also known that the set of all zeros of the Gegenbauer polynomials $C_n^{k+3/2}$, $n = 0, \ldots$, is dense in $[-1,1]$. But the second statement is very useful, since it shows us that we can ensure by a suitable choice of the parameter $h \in (0,1)$ that all spherical harmonics up to a certain order are contained in the Hilbert space $\mathcal{H}$, possessing $(B_h^{(k)})^{(2)}(\cdot,\cdot)$ as reproducing kernel.

Summarizing our results we therefore obtain the following.

Theorem 5.8.7. *Let $k \geq 0$ and $h \in (0,1)$. Suppose that the sets $\mathcal{N} = \{n \in \mathbb{N}_0 |\ (B_h^{(k)})^\wedge(n) \neq 0\}$ and $\mathcal{N}_0 = \{n \in \mathbb{N}_0 |\ (B_h^{(k)})^\wedge(n) = 0\}$ are determined by Theorem 5.8.5. Then the sequence*

$$A_n = \begin{cases} 1/(B_h^{(k)})^\wedge(n) & \text{for} \quad n \in \mathcal{N} \\ 0 & \text{for} \quad n \in \mathcal{N}_0 \end{cases} \tag{5.8.17}$$

is summable and defines a reproducing kernel Hilbert space $\mathcal{H} = \mathcal{H}^{\mathcal{N}}(\{A_n\}; \Omega)$. The reproducing kernel of $\mathcal{H}$ is given by

$$K_{\mathcal{H}}(\xi, \eta) = (B_h^{(k)})^{(2)}(\xi \cdot \eta), \quad \xi, \eta \in \Omega,$$

where

$$\text{supp } K_{\mathcal{H}}(\cdot, \eta) = \{\xi \in \Omega \mid\ 2h^2 - 1 \leq \xi \cdot \eta \leq 1\}.$$

This theorem shows how reproducing kernel Hilbert spaces with locally supported reproducing kernels can be constructed. However, taking $m \geq 0$, the modified kernels $K_{\mathcal{H}^{\perp}_{0,\ldots,m}}(\cdot,\cdot)$, are now globally supported. Thus, we are finally interested in developing a method to construct also kernels $K_{\mathcal{H}^{\perp}_{0,\ldots,m}}(\cdot,\cdot)$ that still remain locally supported.

The idea is simple: let $m \in \mathbb{N}_0$ be fixed. Choose $m+2$ values $h_1, \ldots, h_{m+2}$ out of the interval $(0,1)$ such that $h_1 < \ldots < h_{m+2}$. Then determine real numbers $j_1, \ldots, j_{m+2}$ in such a way that

$$J(t) = \sum_{i=1}^{m+2} j_i B_{h_i}^{(k)}(t), \quad t \in [-1,1] \tag{5.8.18}$$

satisfies

$$J^{\wedge}(n) = 0, \quad n = 0, \ldots, m. \tag{5.8.19}$$

If we define

$$K_{\mathcal{H}}(\xi,\eta) = \sum_{n=0}^{m} \frac{2n+1}{4\pi} P_n(\xi\cdot\eta) + J^{(2)}(\xi\cdot\eta), \quad \xi,\eta \in \Omega, \tag{5.8.20}$$

we end up with the reproducing kernel of a space $\mathcal{H} = \mathcal{H}^{\mathcal{N}}(\{A_n\};\Omega)$, where the sets $\mathcal{N}, \mathcal{N}_0$ and $\{A_n\}$, respectively, are given by

$$\begin{aligned} \mathcal{N} &= \left\{ n \in \mathbb{N}_0 \,\middle|\, \sum_{i=1}^{m+2} j_i (B_{h_i}^{(k)})^{\wedge}(n) \neq 0 \right\}, \\ \mathcal{N}_0 &= \left\{ n \geq m+1 \,\middle|\, \sum_{i=1}^{m+2} j_i (B_{h_i}^{(k)})^{\wedge}(n) = 0 \right\}, \end{aligned}$$

and

$$A_n = \begin{cases} 1/(j_1(B_{h_1}^{(k)})^{\wedge}(n) + \ldots + j_{m+2}(B_{h_{m+2}}^{(k)})^{\wedge}(n)) & \text{for } n \in \mathcal{N}, \\ & \phantom{\text{for }} n \geq m+1 \\ 1 & \text{for } n = 0,\ldots,m \\ 0 & \text{for } n \in \mathcal{N}_0. \end{cases}$$

Furthermore, it follows that $K_{\mathcal{H}^{\perp}_{0,\ldots,m}}(\cdot,\cdot) = J^{(2)}(\cdot,\cdot)$, and hence

$$\operatorname{supp} K_{\mathcal{H}^{\perp}_{0,\ldots,m}}(\eta\cdot\,) = \{\xi \in \Omega \mid 2h_1^2 - 1 \leq \xi\cdot\eta \leq 1\} \tag{5.8.21}$$

for all $\eta \in \Omega$. Note that we have used in definition (5.8.20) the iterated kernel $J^{(2)}$. This ensures that all the Legendre coefficients of $K_{\mathcal{H}}$ are non-negative.

So it remains to find $j_1, \ldots, j_{m+2} \in \mathbb{R}$ such that (5.8.19) is fulfiled (cf. M. Schreiner (1997a)).

Theorem 5.8.8. *Suppose that $m \in \mathbb{N}_0$, $0 < h_1 < \ldots < h_{m+2} < 1$. Then there exist values $j_1, \ldots, j_{m+2} \in \mathbb{R}$ such that (5.8.18) satisfies (5.8.19). If we require, in addition, that*

$$j_1 + \ldots + j_{m+2} = 1, \tag{5.8.22}$$

then the numbers j_i are uniquely determined. The support of $J^{(2)}(\eta\cdot\ )$ is, for fixed $\eta \in \Omega$, given by (cf. (5.8.21))

$$\operatorname{supp} J^{(2)}(\eta\cdot\ ) = \{\xi \in \Omega \mid\ 2h_1^2 - 1 \le \xi \cdot \eta \le 1\}. \tag{5.8.23}$$

Proof. We only have to show that the system of linear equations defined by (5.8.19) and (5.8.22) is uniquely solvable. It can be written down as follows:

$$\begin{pmatrix} 1 & \cdots & 1 \\ (B_{h_1}^{(k)})^\wedge(0) & \cdots & (B_{h_{m+2}}^{(k)})^\wedge(0) \\ \vdots & \ddots & \vdots \\ (B_{h_1}^{(k)})^\wedge(m) & \cdots & (B_{h_{m+2}}^{(k)})^\wedge(m) \end{pmatrix} \begin{pmatrix} j_1 \\ j_2 \\ \vdots \\ j_{m+2} \end{pmatrix} = \begin{pmatrix} 1 \\ 0 \\ \vdots \\ 0 \end{pmatrix}. \tag{5.8.24}$$

But we have seen in Lemma 5.8.1 that $\{1, (B_h^{(k)})^\wedge(0), \ldots, (B_h^{(k)})^\wedge(m)\}$ is a system of polynomials in h of degree $0, \ldots, m+1$, respectively. Thus, they form a unisolvent system for interpolation on the interval $[-1, 1]$ (cf. P.J. Davis (1963)). Therefore, the system (5.8.24) is uniquely solvable, as required. The property (5.8.23) is clear from our previous discussion. □

Figure 5.8.6 shows an example for the kernel J and its iteration $J^{(2)}$.

Finally, we shall prove that the locally supported kernels introduced above are strictly positive definite.

Theorem 5.8.9. *Let $k \ge 0$. The following statements hold true.*

(i) If $h \in (0, 1)$ then $(B_h^{(k)})^{(2)}$ is strictly positive definite.

(ii) Let $m \in \mathbb{N}_0$ and $0 < h_1 < \ldots < h_{m+2} < 1$, and assume that $J : [-1, 1] \to \mathbb{R}$ is defined as in Theorem 5.8.8. Then $J^{(2)}$ is strictly positive definite.

Proof. Assertion (i) is equivalent to the positive definiteness of the symmetric matrix

$$\begin{pmatrix} (B_h^{(k)})^{(2)}(\eta_1 \cdot \eta_1) & \cdots & (B_h^{(k)})^{(2)}(\eta_1 \cdot \eta_N) \\ \vdots & \ddots & \vdots \\ (B_h^{(k)})^{(2)}(\eta_N \cdot \eta_1) & \cdots & (B_h^{(k)})^{(2)}(\eta_N \cdot \eta_N) \end{pmatrix} \tag{5.8.25}$$

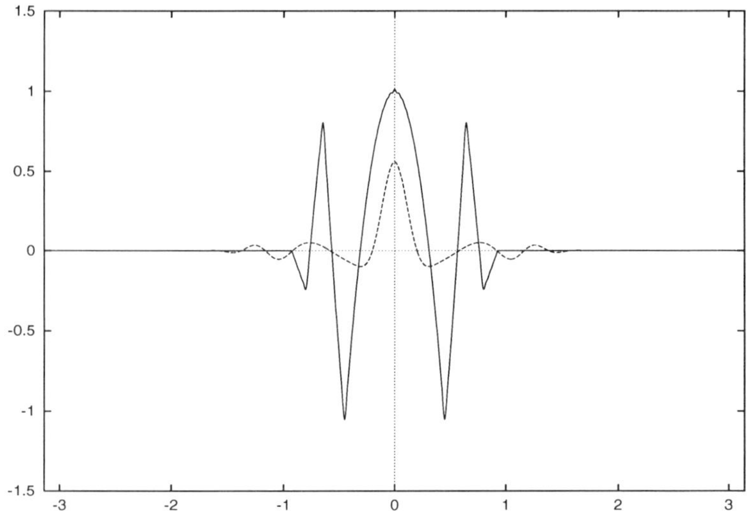

FIG. 5.8.6. The kernels $J(\cos\vartheta)$ and $J^{(2)}(\cos\vartheta)$ of Theorem 5.8.8 with $m=2$, $(h_1,\ldots,h_4)=(0.6,0.7,0.8,0.9)$, $(j_1,\ldots,j_4)=(-1,4,-6,4)$

for any pointset $X_N=\{\eta_1,\ldots,\eta_N\}$ of distinct points. However, since

$$(B_h^{(k)})^{(2)}(\eta_i\cdot\eta_k) \quad = \quad \int_\Omega B_h^{(k)}(\eta_i\cdot\zeta)B_h^{(k)}(\zeta\cdot\eta_k)\,d\omega(\zeta),$$

the matrix (5.8.25) can be recognized as a Gram matrix, and is therefore positive definite if and only if the functions $B_h^{(k)}(\eta_i\cdot\ )$, $i=1,\ldots,N$, are linearly independent. But this can be concluded from the discontinuity of the kth derivative of $B_h^{(k)}(\cdot,\cdot)$ in the following way: assume that $F=\sum_{i=1}^N a_iB_h^{(k)}(\eta_i\cdot\ )=0$. Thus, F is trivially of class $\mathcal{C}^{(\infty)}$. On the other hand, the directional derivative of order k of the function $\xi\mapsto B_h^{(k)}(\eta_i\cdot\xi)$ in direction $\eta_i-(\eta_i\cdot\xi)\xi$ does not exist for all $\xi\in M(\eta_i;h)$. Suppose now, that an a_i (without loss of generality, say a_1) is different from zero. Since there exists a point $\xi_1\in M(\eta_1;h)\setminus\cup_{j=2}^N M(\eta_j;h)$, it follows that the described kth derivative of F at the point ξ does not exist, a contradiction. Thus, $a_1=\ldots=a_N=0$ is proved, and therefore part (i) is shown. The second statement follows by similar arguments. $\square$

5.9 Strict Positive Definiteness

It is known from Y. Xu and E.W. Cheney (1992) that a sufficient condition for a positive definite kernel K to be strictly positive definite is $K^\wedge(n) > 0$ for all $n = 0, 1, \ldots$. In this section we prove a more general result by use of locally supported kernels and approximation properties of the Abel–Poisson kernel.

Let Q_h be the Abel–Poisson kernel (cf. (5.6.17)). We consider for $m \in \mathbb{N}_0$

$$(Q_h)^{\perp}_{0,\ldots,m}(t) = Q_h(t) - \sum_{n=0}^{m} \frac{2n+1}{4\pi} h^n P_n(t), \; t \in [-1,1].$$

A slight generalization of Theorem 3.4.1 is the following.

Theorem 5.9.1. *Let $F \in \mathcal{C}(\Omega)$ satisfy $(F, Y_{n,j})_{\mathcal{L}^2(\Omega)} = 0$ for all $n = 0, \ldots, m$, $j = 1, \ldots, 2n+1$. Then*

$$\lim_{\substack{h \to 1 \\ h<1}} \sup_{\xi \in \Omega} \left| F(\xi) - \int_\Omega (Q_h)^{\perp}_{0,\ldots,m}(\xi \cdot \eta) F(\eta)\, d\omega(\eta) \right| = 0.$$

This gives rise to the following formulation.

Theorem 5.9.2. *Suppose that $h \in (0,1)$, $m \in \mathbb{N}_0$. Then the function $(Q_h)^{\perp}_{0,\ldots,m}$ is strictly positive definite.*

Proof. Let $\eta_1, \ldots, \eta_N \in \Omega$ be pairwise distinct. According to Lemma 5.7.7 we have to show that the functions $(Q_h)^{\perp}_{0,\ldots,m}(\eta_1 \cdot\ ), \ldots, (Q_h)^{\perp}_{0,\ldots,m}(\eta_N \cdot\ )$ are linearly independent. For that purpose we assume that $a_1, \ldots, a_N \in \mathbb{R}$ satisfy

$$\sum_{i=1}^{N} a_i (Q_h)^{\perp}_{0,\ldots,m}(\eta_i \cdot\ ) = 0.$$

Since the series of $(Q_h)^{\perp}_{0,\ldots,m}$ in terms of the Legendre polynomials is uniformly convergent, it follows for all $\xi \in \Omega$ that

$$\begin{aligned} 0 &= \Big(\sum_{i=1}^{N} a_i (Q_h)^{\perp}_{0,\ldots,m}(\eta_i \cdot\ ), \frac{2n+1}{4\pi} P_n(\ \cdot \xi)\Big)_{\mathcal{L}^2(\Omega)} \\ &= \sum_{i=1}^{N} a_i h^n P_n(\eta_i \cdot \xi) \\ &= h^n \sum_{i=1}^{N} a_i P_n(\eta_i \cdot \xi). \end{aligned}$$

Thus, $\sum_{i=1}^{N} a_i P_n(\eta_i \cdot\ ) = 0$. The same calculation performed in the backward direction implies then

$$\sum_{i=1}^{N} a_i (Q_h)^{\perp}_{0,\dots,m}(\eta_i \cdot\ ) = 0 \tag{5.9.1}$$

for all $h \in (0,1)$.

Now, let $i^* \in \{1,\dots,N\}$ be fixed, and choose $\varepsilon < \Theta_N$, where Θ_N is given by $\Theta_N = \max_{j=1,\dots,N} \sup_{j \neq i} |\eta_i - \eta_j|$. We conclude from Theorem 5.8.8 that there exists a continuous function $H \in \mathcal{C}(\Omega)$ with the properties:

(i) $H(\eta_{i*}) = 1$,

(ii) $\operatorname{supp} H \subset \{\xi \in \Omega |\ |\xi - \eta| \leq \varepsilon\}$,

(iii) $(H, Y_n)_{\mathcal{L}^2(\Omega)} = 0$ for all $Y_n \in Harm_n$, $n = 0,\dots,m$.

Thus, we are able to deduce from Theorem 5.9.1 and the fact that $H(\eta_i) = 0$ if $i \neq i^*$,

$$a_{i^*} = \sum_{i=1}^{N} a_i H(\eta_i) = \lim_{\substack{h \to 1 \\ h < 1}} (\sum_{i=1}^{N} a_i (Q_h)^{\perp}_{0,\dots,m}(\eta_i \cdot\ ), H)_{\mathcal{L}^2(\Omega)}.$$

On the other hand, it follows from (5.9.1) for all $h \in (0,1)$ that

$$(\sum_{i=1}^{N} a_i (Q_h)^{\perp}_{0,\dots,m}(\eta_i \cdot\ ), H) = 0.$$

Thus, $a_{i^*} = 0$ and hence the functions $(Q_h)^{\perp}_{0,\dots,m}(\eta_1 \cdot\ ), \dots, (Q_h)^{\perp}_{0,\dots,m}(\eta_N \cdot\ )$ are linearly independent, as required. □

This result finally allows us to establish the following.

Theorem 5.9.3. *Let $K : [-1,1] \to \mathbb{R}$ be a positive definite kernel with uniformly convergent series expansion*

$$K(t) = \sum_{n=0}^{\infty} \frac{2n+1}{4\pi} K^\wedge(n) P_n(t), \quad t \in [-1,+1].$$

Assume that only finitely many $K^\wedge(n)$ are zero. Then K is strictly positive definite.

Proof. Let m denote the index of the largest integer n for that $K^\wedge(n) = 0$. Then $K^\wedge(n) > 0$ for all $n > m$. Assume that $\eta_1, \dots, \eta_N \in \Omega$ are pairwisely distinct. We show that $K(\eta_1 \cdot\ ), \dots, K(\eta_N \cdot\ )$ are linearly independent.

If $\sum_{i=1}^{N} a_i K(\eta_i \cdot\ ) = 0$, it follows as in the proof of Theorem 5.9.2 that

$$\sum_{i=1}^{N} a_i P_n(\eta_i \cdot\ ) = 0 \tag{5.9.2}$$

for all n for which $K^\wedge(n) > 0$. In particular, (5.9.2) is true for all $n > m$. But then we obtain that for every $h \in (0,1)$

$$\sum_{i=1}^{N} a_i (Q_h)^{\perp}_{0,\ldots,m}(\eta_i \cdot\ ) = 0.$$

Since we know from Theorem 5.9.2 that $(Q_h)^{\perp}_{0,\ldots,m}$ is strictly positive definite, $a_1 = \ldots = a_N = 0$. Hence, K is strictly positive definite. This is the desired result. □

Bibliographical Notes

The theory of spherical radial basis functions goes back to the celebrated paper of I.J. SCHOENBERG (1942), where positive definite functions on the sphere have been investigated. The starting point of their use for approximation methods on the sphere is W. FREEDEN (1978, 1979). From the literature in physical geodesy there is to mention the paper by S.L. SVENSSON (1984), where locally supported kernels were used for finite element methods on the sphere. For the theory of radial basis functions in Euclidean spaces, we mention the overview article by M.J.D. POWELL (1992) and the references therein. A list of radial basis (kernel) functions on the sphere is given in W. FREEDEN, J.C. MASON (1990), W. FREEDEN *et al.* (1994, 1996c). A sufficient condition for a spherical function to be strictly positive definite is due to Y. XU, E.W. CHENEY (1992). The described improvement is based on W. FREEDEN *et al.* (1994), M. SCHREINER (1997A,B).

6

SPHERICAL SPLINES

In this chapter we are concerned with spherical splines. Based on the results of the last chapter we are able to introduce spherical splines in close orientation to the one-dimensional polynomial spline approach. In Section 6.1 we start with a variational characterization of spline interpolation. Section 6.2 shows error estimates, while in Section 6.3 some convergence properties are considered. In Section 6.4 we investigate complete spline bases, while Section 6.5 deals with smoothing by spherical splines.

6.1 Variational Characterization of Spline Interpolation

Given arbitrary points $\eta_1, \ldots, \eta_N$, the problem that we address now is the construction of a 'smooth' function defined over Ω which interpolates the given data points $(\eta_i, y_i) \in \Omega \times \mathbb{R}$, $i = 1, \ldots, N$. Applications include modelling closed surfaces in computer aided geometric design and representing functions which estimate physical quantities such as temperature, pressure, gravitational or magnetic potentials at all points on the surface of the spherical earth based on a discrete sample of values taken at arbitrary locations.

Only a few solutions to this scattered data interpolation problem exist in the literature. Two methods seem to be useful: spherical splines and multiquadric surfaces. The purpose of this chapter is to set these methods in a unified context so that we can draw upon ideas from the theory of positive definite functions (see Chapter 5). It turns out that our approach can be based on any strictly positive definite kernel (of order m), so that the list given in the last chapter provides the user with a variety of possible interpolants. We distinguish two cases: (i) $m = -1$, i.e. no polynomial precision (ii) $m \geq 0$, i.e. polynomial precision of order m. *Polynomial precision* means that if the data come from a spherical harmonic of order $\leq m$ this polynomial is reproduced exactly. Clearly, as we shall see below, polynomial precision corresponds intimately to the norm to be minimized in the variational characterization of spline interpolation.

6.1.1 *Spherical Spline Interpolation, $m = -1$*

Let $\{A_n\} = \{A^\wedge(n)\}$ be the symbol of a pseudodifferential operator A being summable with respect to $\mathcal{N}$. From the summability we know that

$\mathcal{H} = \mathcal{H}^{\mathcal{N}}(\{A_n\};\Omega)$ admits the reproducing kernel $K_{\mathcal{H}}(\cdot,\cdot) : \Omega \times \Omega \to \mathbb{R}$, given by

$$K_{\mathcal{H}}(\xi,\eta) = \sum_{n\in\mathcal{N}} \frac{2n+1}{4\pi} \frac{1}{|A^\wedge(n)|^2} P_n(\xi\cdot\eta), \quad \xi,\eta\in\Omega. \tag{6.1.1}$$

We assume that $K_{\mathcal{H}}$ is strictly positive definite. According to the results of the last chapter we know that a sufficient condition for $K_{\mathcal{H}}$ to be strictly positive definite is if $\mathcal{N}_0$ consists of only finitely many elements.

Let there be given N data points $(\eta_1, y_1), \ldots, (\eta_N, y_N) \in \Omega \times \mathbb{R}$ corresponding to a system $X_N = \{\eta_1, \cdots, \eta_N\}$, $\eta_i \neq \eta_k$ for $i \neq k$. We firstly consider the interpolation problem of finding the *$\mathcal{H}$-smallest norm interpolant* to the prescribed data:

$$\inf_{F\in\mathcal{I}_N(y)} \|F\|_{\mathcal{H}} = \inf_{F\in\mathcal{I}_N(y)} \|A(F)\|_{\mathcal{L}^2(\Omega)}, \tag{6.1.2}$$

where

$$\mathcal{I}_N(y) = \{F\in\mathcal{H} \mid F(\eta_i) = y_i,\ i = 1,\ldots,N\}. \tag{6.1.3}$$

Though the smallest norm problem is standard in interpolation theory, the solution process should be characterized briefly. We start with the following definition.

Definition 6.1.1. Any function $S \in \mathcal{H}$ of the form

$$S(\xi) = \sum_{i=1}^{N} a_i K_{\mathcal{H}}(\eta_i,\xi), \quad a_i\in\mathbb{R}, \quad \xi\in\Omega$$

is called *spherical spline* in $\mathcal{H}$ relative to the system $X_N = \{\eta_1, ..., \eta_N\}$ in Ω. The class of all spherical splines is denoted by $Spline(\{A_n\}; X_N)$.

First we formulate the following lemmata.

Lemma 6.1.2. *If $F \in \mathcal{H}$ and $S \in Spline(\{A_n\}; X_N)$, then*

$$(S,F)_{\mathcal{H}} = \sum_{i=1}^{N} a_i F(\eta_i). \tag{6.1.4}$$

Lemma 6.1.3 *(uniqueness of interpolation). For given $y \in \mathbb{R}^N$, $y = (y_1, \ldots, y_N)^T$, there exists a unique $S \in Spline(\{A_n\}; X_N) \cap \mathcal{I}_N(y)$, denoted briefly by S_N.*

As spline, $S \in Spline(\{A_n\}; X_N)$ contains a total of N coefficients

$$S(\xi) = \sum_{i=1}^{N} a_i K_{\mathcal{H}}(\xi, \eta_i), \quad \xi \in \Omega. \tag{6.1.5}$$

The constraints $S(\eta_k) = y_k$ give N linear equations in these coefficients

$$\sum_{i=1}^{N} a_i K_{\mathcal{H}}(\eta_k, \eta_i) = y_k, \quad k = 1, \ldots, N. \tag{6.1.6}$$

From the strict positive definiteness of $K_{\mathcal{H}}$ it follows that the coefficient matrix is positive definite. This implies Lemma 6.1.3. □

The coefficient matrix of (6.1.6) is positive definite and symmetric, hence, powerful routines are available for numerical implementation (see for example J.J. DONGARRA *et al.* (1979), etc).

Next we prove the following.

Lemma 6.1.4 *(first minimum property). If $F \in \mathcal{I}_N(y)$, then*

$$\|F\|^2_{\mathcal{H}} = \|S_N\|^2_{\mathcal{H}} + \|S_N - F\|^2_{\mathcal{H}}.$$

Proof. Obviously, we have $(S_N - F, S_N)_{\mathcal{H}} = 0$. Consequently,

$$\begin{aligned}(F, F)_{\mathcal{H}} &= \big(S_N - (S_N - F), S_N - (S_N - F)\big)_{\mathcal{H}} \\ &= (S_N, S_N)_{\mathcal{H}} + (S_N - F, S_N - F)_{\mathcal{H}}.\end{aligned}$$

This proves Lemma 6.1.4. □

Lemma 6.1.5 *(second minimum property). If $S \in Spline(\{A_n\}; X_N)$ and $F \in \mathcal{I}_N(y)$, then*

$$\|S - F\|^2_{\mathcal{H}} = \|S_N - F\|^2_{\mathcal{H}} + \|S - S_N\|^2_{\mathcal{H}}.$$

Summarizing our results we therefore obtain the following result.

Theorem 6.1.6.. *The interpolation problem*

$$\|S_N\|_{\mathcal{H}} = \inf_{F \in \mathcal{I}_N(y)} \|F\|_{\mathcal{H}}$$

is well-posed in the sense that its solution exists, is unique, and depends continuously on the data $y_1, \ldots, y_N$. The uniquely determined solution is given in the explicit form

$$S_N(\xi) - \sum_{i=1}^{N} a_i K_{\mathcal{H}}(\eta_i, \xi), \quad \xi \in \Omega,$$

where the coefficients $a_i \in \mathbb{R}$ satisfy the linear equations (6.1.6).

Finally we mention the following result.

Corollary 6.1.7 *(Shannon sampling theorem in Spline $(\{A_n\}; X_N)$). Let F be in Spline$(\{A_n\}; X_N)$. Then F can be reconstructed from its samples at points of X_N by the following interpolation formula*

$$F(\xi) = \sum_{k=1}^{N} F(\eta_k) L_k(\xi), \quad \xi \in \Omega,$$

where $L_k \in$ Spline$(\{A_n\}; X_N)$ is given by

$$L_k(\xi) = \sum_{l=1}^{N} a_l^k K_{\mathcal{H}}(\xi, \eta_l)$$

and the coefficients a_j^k satisfy the linear equations

$$\sum_{l=1}^{N} a_l^k K_{\mathcal{H}}(\eta_i, \eta_l) = \delta_{ik}.$$

Proof. As member of *Spline*$(\{A_n\}; X_N)$, the function F is uniquely determined by

$$F(\xi) = \sum_{k=1}^{N} a_k K_{\mathcal{H}}(\xi, \eta_k), \quad \xi \in \Omega,$$

where

$$\sum_{k=1}^{N} a_k K_{\mathcal{H}}(\eta_i, \eta_k) = F(\eta_i), \quad i = 1, \ldots, N.$$

Now we have

$$\sum_{k=1}^{N} F(\eta_k) L_k(\xi) = \sum_{l=1}^{N} a_l K_{\mathcal{H}}(\xi, \eta_l),$$

where

$$a_l = \sum_{k=1}^{N} a_l^k F(\eta_k).$$

This gives us

$$\sum_{l=1}^{N} a_l K_{\mathcal{H}}(\eta_i, \eta_l) = \sum_{k=1}^{N} F(\eta_k) \sum_{l=1}^{N} a_l^k K_{\mathcal{H}}(\eta_i, \eta_l) = F(\eta_i),$$

$i = 1, \ldots, N.$ □

6.1.2 *Spherical Spline Interpolation, $m \geq 0$*

Let $m \geq 0$. We assume that the sequence $\{A_n\}$ is summable with respect to $\mathcal{N}$ and satisfies, in addition, $\{0, \ldots, m\} \subset \mathcal{N}$. This leads to an orthogonal decomposition of $\mathcal{H} = \mathcal{H}^{\mathcal{N}}(\{A_n\}; \Omega)$ by $\mathcal{H} = \mathcal{H}_{0,\ldots,m} \oplus \mathcal{H}^{\perp}_{0,\ldots,m}$ (as described in Section 5.7). The corresponding reproducing kernels are $K_{\mathcal{H}_{0,\ldots,m}}(\cdot,\cdot)$ and $K^{\perp}_{\mathcal{H}_{0,\ldots,m}}(\cdot,\cdot)$, and $\|\cdot\|_{H^{\perp}_{0,\ldots,m}}$ is a semi-norm in $\mathcal{H}$ with kernel $\mathcal{H}_{0,\ldots,m}$. We assume that $K_{\mathcal{H}}(\cdot,\cdot)$ is conditionally strictly positive definite of order m.

Let there be given N data points $(\eta_1, y_1), \ldots, (\eta_N, y_N)$ corresponding to an admissible system $X_N = \{\eta_1, \cdots, \eta_N\}$ relative to $Harm_{0,\ldots,m}$. We consider the interpolation problem of finding the $\mathcal{H}^{\perp}_{0,\ldots,m}$ *-smallest norm interpolant* to the prescribed data:

$$\inf_{F \in \mathcal{I}_N(y)} \|F\|_{\mathcal{H}^{\perp}_{0,\ldots,m}} = \inf_{F \in \mathcal{I}_N(y)} \left(\sum_{\substack{n \geq m+1 \\ n \in \mathcal{N}}} \sum_{j=1}^{2n+1} |A_n|^2 (F, Y_{n,j})^2_{\mathcal{L}^2(\Omega)} \right)^{1/2}. \tag{6.1.7}$$

Definition 6.1.8. Let $X_N = \{\eta_1, \ldots, \eta_N\}$ be an admissible system relative to the space $Harm_{0,\ldots,m}$. Then any function $S \in \mathcal{H}(\{A_n\}; \Omega)$ of the form

$$S(\xi) = P(\xi) + \sum_{i=1}^{N} a_i K_{\mathcal{H}^{\perp}_{0,\ldots,m}}(\eta_i, \xi), \quad \xi \in \Omega, \quad P \in Harm_{0,\ldots,m}, \tag{6.1.8}$$

is called *spherical spline in $\mathcal{H}$ of order m* relative to the system $X_N = \{\eta_1, \ldots, \eta_N\} \subset \Omega$, if $a \in \mathbb{R}^N$, $a = (a_1, \ldots, a_N)^T$, satisfies the linear system

$$\mathbf{matr}_{X_N}(Y_{0,1}, \ldots, Y_{m,2m+1})a = 0. \tag{6.1.9}$$

Correspondingly, the class of all spherical splines of order m in $\mathcal{H}$ is denoted by $Spline_{0,\ldots,m}(\{A_n\}; X_N)$.

Evidently, $Spline_{0,\ldots,m}(\{A_n\}; X_N)$ contains the class $Harm_{0,\ldots,m}$ as subset. Again we list four lemmata.

Lemma 6.1.9. *If $F \in \mathcal{H}$ and $S \in Spline_{0,\ldots,m}(\{A_n\}; X_N)$ is of the form (6.1.8), then*

$$(S, F)_{\mathcal{H}^{\perp}_{0,\ldots m}} = \sum_{i=1}^{N} a_i F(\eta_i).$$

Proof. First it follows that

$$\sum_{i=1}^{N} a_i F(\eta_i)$$

$$= \sum_{i=1}^{N} a_i \big(F, K_{\mathcal{H}_{0,\ldots,m}}(\cdot,\eta_i)\big)_{\mathcal{H}} + \sum_{i=1}^{N} a_i \big(F, K_{\mathcal{H}^{\perp}_{0,\ldots,m}}(\cdot,\eta_i)\big)_{\mathcal{H}}.$$

According to our assumption $a \in \mathbb{R}^n$, $a = (a_1, \ldots, a_N)^T$ satisfies the linear system (6.1.9). Therefore, we have

$$\sum_{i=1}^{N} a_i F(\eta_i) = (F, \sum_{i=1}^{N} a_i K_{\mathcal{H}^{\perp}_{0,\ldots,m}}(\cdot,\eta_i))_{\mathcal{H}^{\perp}_{0,\ldots,m}}.$$

This proves Lemma 6.1.9. □

Lemma 6.1.10 *(uniqueness of interpolation). For given $y \in \mathbb{R}^N$, $y = (y_1, \ldots, y_N)^T$, there exists a unique $S \in Spline_{0,\ldots,m}(\{A_n\}; X_N) \cap \mathcal{I}_N(y)$, denoted briefly by S_N.*

Proof. Any $S \in Spline_{0,\ldots,m}(\{A_n\}; X_N)$ contains a total of $(M + N)$ coefficients

$$S(\xi) = \sum_{n=0}^{m} \sum_{j=1}^{2n+1} c_{n,j} Y_{n,j}(\xi) + \sum_{i=1}^{N} a_i K_{\mathcal{H}^{\perp}_{0,\ldots,m}}(\xi,\eta_i), \quad \xi \in \Omega. \tag{6.1.10}$$

The constraints $S(\eta_k) = y_k$ give N linear equations in these coefficients:

$$\sum_{n=0}^{m} \sum_{j=1}^{2n+1} c_{n,j} Y_{n,j}(\eta_k) + \sum_{i=1}^{N} a_i K_{\mathcal{H}^{\perp}_{0,\ldots,m}}(\eta_k,\eta_i) = y_k, \quad k = 1, \ldots, N. \tag{6.1.11}$$

The linear system (6.1.9) provides M further equations:

$$\sum_{i=1}^{N} a_i Y_{n,j}(\eta_i) = 0, \quad n = 0, \ldots, m, \quad j = 1, \ldots, 2n+1. \tag{6.1.12}$$

Our assertion will be proved if it can be shown that the overall linear system is non-singular. This will be established if it can be shown that the corresponding homogeneous system has only the trivial solution in which all $N + M$ coefficients vanish. In other words, we only have to show that the only $\tilde{S} \in Spline_{0,\ldots,m}(\{A_n\}; X_N) \cap \mathcal{I}_N(0)$ is the trivial one. To prove this we obtain from Lemma 6.1.9 $\|\tilde{S}\|_{\mathcal{H}^{\perp}_{0,\ldots,m}} = 0$, i.e. $\tilde{S} \in Harm_{0,\ldots,m} \cap \mathcal{I}_N(0)$. Consequently, because of the admissibility assumption imposed on X_N, $\tilde{S} = 0$, as required. □

Lemma 6.1.11 *(first minimum property). If $F \in \mathcal{I}_N(y)$, then*

$$\|F\|^2_{\mathcal{H}^{\perp}_{0,\ldots,m}} = \|S_N\|^2_{\mathcal{H}^{\perp}_{0,\ldots,m}} + \|S_N - F\|^2_{\mathcal{H}^{\perp}_{0,\ldots,m}}. \tag{6.1.13}$$

Proof. Lemma 6.1.11 follows easily from Lemma 6.1.9. □

Lemma 6.1.12 *(second minimum property). If $S \in Spline_{0,\ldots,m}(\{A_n\}; X_N)$ and $F \in \mathcal{I}_N(y)$, then*

$$\|S - F\|^2_{\mathcal{H}^{\perp}_{0,\ldots,m}} = \|S_N - F\|^2_{\mathcal{H}^{\perp}_{0,\ldots,m}} + \|S - S_N\|^2_{\mathcal{H}^{\perp}_{0,\ldots,m}}.$$

Summarizing the results given above we therefore obtain the following result.

Theorem 6.1.13. *The spline interpolation problem*

$$\|S_N\|_{\mathcal{H}^{\perp}_{0,\ldots,m}} = \inf_{F \in \mathcal{I}_N(y)} \|F\|_{\mathcal{H}^{\perp}_{0,\ldots,m}} \tag{6.1.14}$$

is well-posed in the sense that its solution exists, is unique, and depends continuously on the data $y_1, \ldots, y_N$. The uniquely determined solution S_N is given in the explicit form

$$S_N(\xi) = \sum_{n=0}^{m} \sum_{j=1}^{2n+1} c_{n,j} Y_{n,j}(\xi) + \sum_{i=1}^{N} a_i K_{\mathcal{H}^{\perp}_{0,\ldots,m}}(\eta_i, \xi), \quad \xi \in \Omega,$$

where the coefficients $c_{n,j}, a_i$ satisfy the linear equations

$$\sum_{i=1}^{N} a_i K_{\mathcal{H}^{\perp}_{0,\ldots,m}}(\eta_k, \eta_i) + \sum_{n=0}^{m} \sum_{j=1}^{2n+1} c_{n,j} Y_{n,j}(\eta_k) = y_k, \quad k = 1, \ldots, N$$

and

$$\sum_{i=1}^{N} a_i Y_{n,j}(\eta_i) = 0, \quad n = 0, \ldots, m, \quad j = 1, \ldots, 2n+1.$$

Next we mention the following corollary.

Corollary 6.1.14 *(Shannon sampling theorem in $Spline_{0,\ldots,m}(\{A_n\}; X_N)$). Let F be an element of $Spline_{0,\ldots,m}(\{A_n\}; X_N)$. Then F can be reconstructed from its samples at points of X_N by the following interpolation formula (cf.* W. FREEDEN (1981A)*)*

$$F(\xi) = \sum_{k=1}^{N} F(\eta_k) L_k(\xi), \quad \xi \in \Omega,$$

where $L_k \in Spline_{0,\ldots,m}(\{A_n\}; X_N)$ is given by

$$L_k(\xi) = \sum_{n=0}^{m} \sum_{j=1}^{2n+1} c_{n,j}^k Y_{n,j}(\xi) + \sum_{l=1}^{N} a_l^k K_{\mathcal{H}_{0,\ldots,m}^{\perp}}(\xi, \eta_l)$$

and the coefficients $c_{n,j}^k, a_l^k$ satisfy the linear equations

$$\begin{aligned} \sum_{l=1}^{N} a_l^k K_{\mathcal{H}_{0,\ldots,m}^{\perp}}(\eta_i, \eta_l) + \sum_{n=0}^{m} \sum_{j=1}^{2n+1} c_{n,j}^k Y_{n,j}(\eta_i) &= \delta_{ik}, \\ \sum_{l=1}^{N} a_l^k Y_{p,q}(\eta_l) &= 0, \end{aligned}$$

$i, k = 1, \ldots, N$, $p = 0, 1, \ldots, m$, $\quad q = 1, \ldots, 2p+1$.

Proof. As an element class of $Spline_{0,\ldots,m}(\{A_n\}; X_N)$, the function F is uniquely determined by

$$F(\xi) = \sum_{n=0}^{m} \sum_{j=1}^{2n+1} c_{n,j} Y_{n,j}(\xi) + \sum_{l=1}^{N} a_l K_{\mathcal{H}_{0,\ldots,m}^{\perp}}(\xi, \eta_l), \quad \xi \in \Omega,$$

where

$$\begin{aligned} \sum_{l=1}^{N} a_l K_{\mathcal{H}_{0,\ldots,m}^{\perp}}(\eta_i, \eta_l) + \sum_{n=0}^{m} \sum_{j=1}^{2n+1} c_{n,j} Y_{n,,j}(\eta_i) &= F(\eta_i), \\ \sum_{l=1}^{N} a_l Y_{p,q}(\eta_l) &= 0, \end{aligned}$$

$i = 1, \ldots, N$, $p = 0, \ldots, m$, $q = 1, \ldots, 2p+1$. Now it follows that

$$\sum_{k=1}^{N} F(\eta_k) L_k(\xi) = \sum_{n=0}^{m} \sum_{j=1}^{2n+1} c_{n,j} Y_{n,j}(\xi) + \sum_{l=1}^{N} a_l K_{\mathcal{H}_{0,\ldots,m}^{\perp}}(\xi, \eta_l),$$

where

$$c_{n,,j} = \sum_{k=1}^{N} c_{n,j}^k F(\eta_k)$$

and

$$a_l = \sum_{k=1}^{N} a_l^k F(\eta_k).$$

This gives us

$$\begin{aligned}
&\sum_{n=0}^{m}\sum_{j=1}^{2n+1} c_{n,j} Y_{n,j}(\eta_i) + \sum_{l=1}^{N} a_l K(\eta_i,\eta_l) \\
&\quad = \sum_{k=1}^{N} F(\eta_k)\left(\sum_{n=0}^{m}\sum_{j=1}^{2n+1} c_{n,j}^k Y_{n,j}(\eta_i) + \sum_{l=1}^{N} a_l^k K_{\mathcal{H}_{0,\ldots,m}^{\perp}}(\eta_i,\eta_l)\right) \\
&\quad = F(\eta_i)
\end{aligned}$$

and

$$\sum_{l=1}^{N} a_l Y_{p,q}(\eta_l) = \sum_{k=1}^{N} F(\eta_k) \sum_{l=1}^{N} a_l^k Y_{p,q}(\eta_l) = 0.$$

This proves Corollary 6.1.14. □

Numerical experiences with the linear systems of equations stated in Theorem 6.1.13 have shown that the systems tend to be ill-conditioned unless m and N are not too large. However, it turns out that the condition is strongly dependent on the reproducing kernel function used for the interpolation problem. W. FREEDEN (1982C, 1984A,B, 1990B) proposes a combined Lagrange and Newton technique which also opens the advantage of a permanence property in spline interpolation. A general procedure for solving systems of equations arising from reproducing kernel functions has been given by C. GU (1989) (see also C. GU, G. WAHBA (1991)).

For approximation order $m \geq 0$ there is also the possibility of establishing a Lagrange expansion of the interpolant, in combination with a Newton part of the approximation when additional points are being added over time (see W. FREEDEN (1982C, 1984A,B)). To this end we remember that there exists in $Harm_{0,\ldots,m}$ a unique Lagrange basis $\{L_1,\ldots,L_M\}$, $M=(m+1)^2$, of the form

$$L_r(\xi) = \sum_{k=1}^{M} b_k^r \sum_{n=0}^{m}\sum_{j=1}^{2n+1} Y_{n,j}(\eta_k) Y_{n,j}(\xi), \quad \xi \in \Omega, \tag{6.1.15}$$

satisfying $L_r(\eta_k) = \delta_{r,k}$, $r,k=1,\ldots,M$, provided that $X_M = \{\eta_1,\ldots,\eta_M\}$ is a fundamental system on Ω relative to $Harm_{0,\ldots,m}$. Therefore, for every

$F \in \mathcal{H}$, the unique $Harm_{0,\ldots,m}$-interpolant pF of F for the $Harm_{0,\ldots,m}$-fundamental system X_M is given by the 'Lagrange form'

$$pF = \sum_{i=1}^{M} F(\eta_i) L_i. \tag{6.1.16}$$

The mapping $p : \mathcal{H} \to \mathcal{H}$ is a continuous linear projector of $\mathcal{H} \subset \mathcal{C}(\Omega)$ onto $Harm_{0,\ldots,m}$. Hence, p determines the following direct sum decomposition of $\mathcal{H}$:

$$\mathcal{H} = Harm_{0,\ldots,m} \oplus \overset{\circ}{\mathcal{H}}{}^{\perp}_{0,\ldots,m}, \tag{6.1.17}$$

where

$$\overset{\circ}{\mathcal{H}}{}^{\perp}_{0,\ldots,m} = \overset{\circ}{\mathcal{H}}{}^{\perp}_{0,\ldots,m}(\{A_n\};\Omega) = \{F \in \mathcal{H} | F(\eta_j) = 0,\ j = 1, \ldots, M\}. \tag{6.1.18}$$

Consequently, any $F \in \mathcal{H}$ can be written uniquely as

$$F = pF + \overset{\circ}{F}, \qquad \overset{\circ}{F} \in \overset{\circ}{\mathcal{H}}{}^{\perp}_{0,\ldots,m}. \tag{6.1.19}$$

$\overset{\circ}{\mathcal{H}}{}^{\perp}_{0,\ldots,m}$, as defined by (6.1.18) equipped with $(\cdot,\cdot)_{\mathcal{H}^{\perp}_{0,\ldots,m}}$ is a Hilbert subspace of $\mathcal{C}(\Omega)$. Furthermore, the reproducing structure in $\mathcal{H}$ enables us to show (cf. W. Freeden (1984a,b)) that

$$(\overset{\circ}{F}, K_{\overset{\circ}{\mathcal{H}}{}^{\perp}_{0,\ldots,m}}(\cdot,\eta))_{\mathcal{H}^{\perp}_{0,\ldots,m}} = \overset{\circ}{F}(\eta) \tag{6.1.20}$$

for all $\eta \in \Omega$ and $\overset{\circ}{F} \in \overset{\circ}{\mathcal{H}}{}^{\perp}_{0,\ldots,m}$, where the reproducing kernel $K_{\overset{\circ}{\mathcal{H}}{}^{\perp}_{0,\ldots,m}} : \Omega \times \Omega \to \mathbb{R}$ is given by

$$K_{\overset{\circ}{\mathcal{H}}{}^{\perp}_{0,\ldots,m}}(\xi,\eta) = K_{\mathcal{H}^{\perp}_{0,\ldots,m}}(\xi,\eta) - \sum_{i=1}^{M} K_{\mathcal{H}^{\perp}_{0,\ldots,m}}(\xi,\eta_i) L_i(\eta) \tag{6.1.21}$$
$$- \sum_{i=1}^{M} L_i(\xi) K_{\mathcal{H}^{\perp}_{0,\ldots,m}}(\eta,\eta_i) + \sum_{i=1}^{M}\sum_{j=1}^{M} L_i(\xi) K_{\mathcal{H}^{\perp}_{0,\ldots,m}}(\eta_i,\eta_j) L_j(\eta).$$

Of course,

$$\|\sum_{k=1}^{M} y_k L_k\|_{\mathcal{H}^{\perp}_{0,\ldots,m}} = 0. \tag{6.1.22}$$

But this means that the spline interpolation problem (6.1.7) reduces to the problem

$$\|\overset{\circ}{S}_{M+1,N}\|_{\mathcal{H}^{\perp}_{0,\ldots,m}} = \inf_{\overset{\circ}{F} \in \mathcal{I}_{M+1,N}(y)} \|\overset{\circ}{F}\|_{\mathcal{H}^{\perp}_{0,\ldots,m}}, \tag{6.1.23}$$

where

$$\mathcal{I}_{M+1,N}(y) = \{F \in \mathring{\mathcal{H}}^{\perp}_{0,\ldots,m} \mid F(\eta_k) = y_k,\ k = M+1,\ldots,N\}. \quad (6.1.24)$$

From the arguments leading to Theorem 6.1.6, however, we know that

$$\mathring{S}_{M+1,N}(\xi) = \sum_{k=M+1}^{N} d_k K_{\mathring{\mathcal{H}}^{\perp}_{0,\ldots,m}}(\xi,\eta_k), \quad \xi \in \Omega, \quad (6.1.25)$$

where the coefficients $d_{M+1},\ldots,d_N$ satisfy the linear equations

$$\sum_{k=M+1}^{N} d_k K_{\mathring{\mathcal{H}}^{\perp}_{0,\ldots,m}}(\eta_i,\eta_k) = y_i - \sum_{k=1}^{M} y_k L_k(\eta_i), \quad i = M+1,\ldots,N. \quad (6.1.26)$$

Corollary 6.1.15. *The uniquely determined solution S_N of the spline interpolation problem*

$$\|S_N\|_{\mathcal{H}^{\perp}_{0,\ldots,m}} = \inf_{F\in\mathcal{I}_N(y)} \|F\|_{\mathcal{H}^{\perp}_{0,\ldots,m}} \quad (6.1.27)$$

is given in the explicit form

$$S_N(\xi) = \sum_{k=1}^{M} y_k L_k(\xi) + \sum_{k=M+1}^{N} d_k K_{\mathring{\mathcal{H}}^{\perp}_{0,\ldots,m}}(\xi,\eta_k), \quad \xi \in \Omega, \quad (6.1.28)$$

where the coefficients $d_{M+1},\ldots,d_N$ satisfy the linear equations (6.1.26).

The coefficient matrix of the linear system (6.1.26) of $N-M$ equations in the unknowns d_k is, in fact, a Gram matrix consisting of $N-M$ linearly independent elements out of $\mathring{\mathcal{H}}^{\perp}_{0,\ldots,m}$. Thus it is symmetric and positive definite. This implies the unique solvability of the linear system (6.1.26).

6.2 Error Estimates

Suppose that $X_N = \{\eta_1,\ldots,\eta_N\} \subset \Omega$ is a system of N points ($\eta_i \neq \eta_j$ for $i \neq j$). We define the *X_N-width* Θ_{X_N} by setting

$$\Theta_{X_N} = \max_{\xi\in\Omega}\left(\min_{\eta\in X_N} |\xi-\eta|\right). \quad (6.2.1)$$

We first consider estimates for spherical spline interpolation with $m=-1$.

Theorem 6.2.1. *Let F be of class $\mathcal{H} = \mathcal{H}^{\mathcal{N}}(\{A_n\};\Omega)$. Suppose that $K_{\mathcal{H}}(\cdot,\cdot)$ satisfies the estimate*

$$|K_{\mathcal{H}}(\xi,\zeta) - K_{\mathcal{H}}(\xi,\eta)| \leq E^2_{\{A_n\}}|\zeta - \eta|^{\sigma} \tag{6.2.2}$$

for all $(\xi,\eta,\zeta) \in \Omega^3$, some $\sigma \geq 0$ and some positive constant $E_{\{A_n\}}$. Denote by $S_N \in \mathcal{H}$ the uniquely determined solution of the interpolation problem

$$\|S_N\|_{\mathcal{H}} = \inf_{G \in \mathcal{I}_N(y)} \|G\|_{\mathcal{H}}, \tag{6.2.3}$$

where

$$\mathcal{I}_N(y) = \{G \in \mathcal{H} | G(\eta_i) = F(\eta_i) = y_i, i = 1,\ldots,N\}.$$

Then

$$\sup_{\xi\in\Omega} |F(\xi) - S_N(\xi)| \leq 2\sqrt{2}E_{\{A_n\}}\Theta^{\sigma/2}_{X_N}\|F\|_{\mathcal{H}}. \tag{6.2.4}$$

Proof. For any $\xi \in \Omega$, there exists a point $\eta_i \in X_N$ with $|\xi - \eta_i| \leq \Theta_{X_N}$. Observing the fact that $S_N(\eta_i) = F(\eta_i)$ we obtain

$$S_N(\xi) - F(\xi) = S_N(\xi) - S_N(\eta_i) + F(\eta_i) - F(\xi). \tag{6.2.5}$$

From Theorem 5.2.6 we see that

$$\begin{aligned} S_N(\xi) - S_N(\eta_i) &= (K_{\mathcal{H}}(\xi,\cdot) - K_{\mathcal{H}}(\eta_i,\cdot), S_N)_{\mathcal{H}} \\ F(\xi) - F(\eta_i) &= (K_{\mathcal{H}}(\xi,\cdot) - K_{\mathcal{H}}(\eta_i,\cdot), F)_{\mathcal{H}}. \end{aligned} \tag{6.2.6}$$

The Cauchy–Schwarz inequality gives

$$\begin{aligned} |S_N(\xi) - S_N(\eta_i)| &\leq (\kappa_{\mathcal{H}}(\xi,\eta_i))^{1/2}\|S_N\|_{\mathcal{H}}, \\ |F(\xi) - F(\eta_i)| &\leq (\kappa_{\mathcal{H}}(\xi,\eta_i))^{1/2}\|F\|_{\mathcal{H}}, \end{aligned} \tag{6.2.7}$$

where

$$\kappa_{\mathcal{H}}(\xi,\eta_i) = (K_{\mathcal{H}}(\xi,\cdot) - K_{\mathcal{H}}(\eta_i,\cdot), K_{\mathcal{H}}(\xi,\cdot) - K_{\mathcal{H}}(\eta_i,\cdot))_{\mathcal{H}}. \tag{6.2.8}$$

As interpolating spline, S_N is the smoothest interpolant. Therefore we have

$$\|S_N\|_{\mathcal{H}} \leq \|F\|_{\mathcal{H}}. \tag{6.2.9}$$

But this shows us that

$$|S_N(\xi) - F(\xi)| \leq 2\left(\kappa_{\mathcal{H}}(\xi,\eta_i)\right)^{1/2}\|F\|_{\mathcal{H}}. \tag{6.2.10}$$

By virtue of our assumption we find

$$|\kappa_{\mathcal{H}}(\xi,\eta_i)| = |2(K_{\mathcal{H}}(\xi,\xi) - K_{\mathcal{H}}(\xi,\eta_i))| \leq 2E^2_{\{A_n\}}|\xi - \eta_i|^{\sigma}. \tag{6.2.11}$$

Summarizing our results we get

$$|S_N(\xi) - F(\xi)| \leq 2\sqrt{2}E_{\{A_n\}}|\xi - \eta_i|^{\sigma/2}\|F\|_{\mathcal{H}}. \tag{6.2.12}$$

This yields the desired result. □

Combining Theorem 6.2.1 and Lemma 5.2.7 we find the following.

Corollary 6.2.2. *Let F be of class $\mathcal{H}^{\mathcal{N}}(\{A_n\};\Omega)$. Suppose that $\{A_n\}$ is $\{n^\sigma\}$-summable, $\sigma \in [0,1]$. Then*

$$\sup_{\xi\in\Omega} |F(\xi) - S_N(\xi)| \leq 4D_{\sigma;\{A_n\}}\Theta_{X_N}^{\sigma/2}\|F\|_{\mathcal{H}},$$

where

$$D_{\sigma;\{A_n\}} = \left(\sum_{n\in\mathcal{N}} \frac{2n+1}{4\pi}|A_n|^{-2}(\frac{n(n+1)}{2})^\sigma\right). \tag{6.2.13}$$

Combining Theorem 6.2.1 and Lemma 5.2.8 we obtain the following.

Corollary 6.2.3. *Let F be of class $\mathcal{H}^{\mathcal{N}}(\{A_n\};\Omega)$. Suppose that $\{A_n\}$ is $\{n^{(\sigma+\frac{1}{2})}\}$-summable with respect to $\mathcal{N}$, $\sigma \in [0,1]$. Then*

$$\sup_{\xi\in\Omega} |F(\xi) - S_N(\xi)| \leq 4\sqrt{2}E_{\sigma;\{A_n\}}\Theta_{X_N}^{\sigma}\|F\|_{\mathcal{H}},$$

where

$$E_{\sigma;\{A_n\}} = \left(\sum_{n\in\mathcal{N}} \frac{(2n+1)^2}{4\pi}|A_n|^{-2}\left(\frac{n(n+1)}{2}\right)^{2\sigma}\right)^{1/2}. \tag{6.2.14}$$

Next we prove an error estimate for spline interpolation with polynomial precision $m \geq 0$ (cf. W. FREEDEN (1984B)).

Theorem 6.2.4. *Let F be of class $\mathcal{H}^{\mathcal{N}}(\{A_n\};\Omega)$. Suppose that $X_N = \{\eta_1,\ldots,\eta_N\}$ is an admissible system relative to $Harm_{0,\ldots,m}$. Furthermore, suppose that $\{A_n\}$ is $\{n^{\tau/2}\}$-summable with respect to $\mathcal{N}$, $\tau \geq 2\sigma+1$, $\sigma \in [0,1]$. Denote by $S_N \in \mathcal{H}$ the uniquely determined solution of the interpolation problem*

$$\|S_N\|_{\mathcal{H}^\perp_{0,\ldots,m}} = \inf_{G\in\mathcal{I}_N(y)} \|G\|_{\mathcal{H}^\perp_{0,\ldots,m}},$$

where

$$\mathcal{I}_N(y) = \{G \in \mathcal{H} \mid G(\eta_i) = F(\eta_i) = y_i, i = 1,\ldots,N\}.$$

Then

$$\sup_{\xi\in\Omega} |F(\xi) - S_N(\xi)| \leq 4\sqrt{2}E_{m;\sigma;\{A_n\}}\,\Theta_{X_N}^{\sigma}\,\|F\|_{\mathcal{H}^\perp_{0,\ldots,m}},$$

where

$$E_{m;\sigma;\{A_n\}} = \left(\sum_{\substack{n\in\mathcal{N}\\ n\geq m+1}} \frac{2n+1}{4\pi} |A_n|^{-2}((2n+1)c_n^{2\sigma} + 2c_n^{\sigma} C_{M,\sigma} + (C_{M,\sigma})^2) \right)^{1/2} \tag{6.2.15}$$

and

$$C_{M,\sigma} = \left(\sum_{k=1}^{M}\sum_{l=1}^{M} |b_l^k| \right) \left(\sum_{n=0}^{m} \frac{2n+1}{4\pi} c_n^{\sigma} \right). \tag{6.2.16}$$

Herein, c_n is equal to $n(n+1)/2$, and b_l^k are the constituting coefficients of L_k (cf. (6.1.15)).

Proof. For any given point $\xi \in \Omega$, there exists a member $\eta_i \in X_N$ satisfying $|\xi - \eta_i| \leq \Theta_N$. Observing the fact that $pS_N = pF$ and $S_N(\eta_i) = F(\eta_i)$ we obtain

$$S_N(\xi) - F(\xi) = \mathring{S}_N(\xi) - \mathring{S}_N(\eta_i) + \mathring{F}(\eta_i) - \mathring{F}(\xi). \tag{6.2.17}$$

Thus, by the triangle inequality, we have

$$| S_N(\xi) - F(\xi) | \leq | \mathring{S}_N(\xi) - \mathring{S}_N(\eta_i) | + | \mathring{F}(\xi) - \mathring{F}(\eta_i) | . \tag{6.2.18}$$

From Theorem 5.2.6 it can be deduced that

$$\mathring{S}_N(\xi) - \mathring{S}_N(\eta_i) = \left((K_{\mathring{\mathcal{H}}^{\perp}_{0,\ldots,m}}(\xi,\cdot) - K_{\mathring{\mathcal{H}}^{\perp}_{0,\ldots,m}}(\eta_i,\cdot)), \mathring{S}_N \right)_{\mathcal{H}^{\perp}_{0,\ldots,m}} \tag{6.2.19}$$

and

$$\mathring{F}(\xi) - \mathring{F}(\eta_i) = \left((K_{\mathring{\mathcal{H}}^{\perp}_{0,\ldots,m}}(\xi,\cdot) - K_{\mathring{\mathcal{H}}^{\perp}_{0,\ldots,m}}(\eta_i,\cdot)), \mathring{F} \right)_{\mathcal{H}^{\perp}_{0,\ldots,m}} .$$

Therefore the Cauchy–Schwarz inequality yields

$$| \mathring{S}_N(\xi) - \mathring{S}_N(\eta_i) | \leq \left(\kappa_{\mathring{\mathcal{H}}^{\perp}_{0,\ldots,m}}(\xi,\eta_i) \right)^{1/2} \|S_N\|_{\mathcal{H}^{\perp}_{0,\ldots,m}} \tag{6.2.20}$$

and

$$| \mathring{F}(\xi) - \mathring{F}(\eta_i) | \leq \left(\kappa_{\mathring{\mathcal{H}}^{\perp}_{0,\ldots,m}}(\xi,\eta_i) \right)^{1/2} \|F\|_{\mathcal{H}^{\perp}_{0,\ldots,m}},$$

where we have used the abbreviation

$$\kappa_{\mathring{\mathcal{H}}^{\perp}_{0,\ldots,m}}(\xi,\eta_i) = \|K_{\mathring{\mathcal{H}}^{\perp}_{0,\ldots,m}}(\xi,\cdot) - K_{\mathring{\mathcal{H}}^{\perp}_{0,\ldots,m}}(\eta_i,\cdot)\|^2_{\mathcal{H}^{\perp}_{0,\ldots,m}} . \tag{6.2.21}$$

As interpolating spline, S_N is the smoothest interpolant. Therefore we have

$$\|S_N\|_{\mathcal{H}^{\perp}_{0,\ldots,m}} \leq \|F\|_{\mathcal{H}^{\perp}_{0,\ldots,m}}. \tag{6.2.22}$$

Consequently we find

$$| S_N(\xi) - F(\xi) | \leq 2\Big(\kappa_{\overset{\circ}{\mathcal{H}}^{\perp}_{0,\ldots,m}}(\xi, \eta_i)\Big)^{\frac{1}{2}} \|F\|_{\mathcal{H}^{\perp}_{0,\ldots,m}}. \tag{6.2.23}$$

Elementary calculations yield

$$\begin{aligned}
&\kappa_{\overset{\circ}{\mathcal{H}}^{\perp}_{0,\ldots,m}}(\xi, \eta_i) \qquad\qquad (6.2.24)\\
&= K_{\mathcal{H}^{\perp}_{0,\ldots,m}}(\xi, \xi) - 2K_{\mathcal{H}^{\perp}_{0,\ldots,m}}(\xi, \eta_i) + K_{\mathcal{H}^{\perp}_{0,\ldots,m}}(\eta_i, \eta_i)\\
&\quad -2\sum_{k=1}^{M}(K_{\mathcal{H}^{\perp}_{0,\ldots,m}}(\xi, \eta_k) - K_{\mathcal{H}^{\perp}_{0,\ldots,m}}(\eta_i, \eta_k))(L_k(\xi) - L_k(\eta_i))\\
&\quad +\sum_{k=1}^{M}\sum_{j=1}^{M}(L_k(\xi) - L_k(\eta_i))K_{\mathcal{H}^{\perp}_{0,\ldots,m}}(\eta_i, \eta_k)(L_j(\xi) - L_j(\eta_i)).
\end{aligned}$$

Observing our estimates for Legendre polynomials we get

$$| \kappa_{\overset{\circ}{\mathcal{H}}^{\perp}_{0,\ldots,m}}(\xi, \eta_i) | \leq 4(E_{m;\sigma;\{A_n\}})^2 \mid \xi - \eta_i \mid^{2\sigma} \tag{6.2.25}$$

provided that $\sigma \in [0, 1]$. This yields Theorem 6.2.4. □

6.3 Convergence

As an immediate consequence of Theorem 6.2.4 we find the following.

Corollary 6.3.1. *Let F be of class $\mathcal{H}^{\mathcal{N}}(\{A_n\}; \Omega)$. Suppose that $\{X_N\}$ is a sequence of admissible systems $X_N = \{\eta_1^{(N)}, \ldots, \eta_N^{(N)}\}$ relative to $Harm_{0,\ldots,m}$ such that $\Theta_{X_N} \to 0$ as $N \to \infty$. Furthermore, suppose that $\{A_n\}$ is summable with respect to $\mathcal{N}$. Then the sequence $\{S_N\}$ of the uniquely determined solutions S_N corresponding to the data points $(\eta_i^{(N)}, y_i^{(N)}) \in \Omega \times \mathbb{R}, i = 1, \ldots, N$, satisfies the limit relation*

$$\lim_{N\to\infty} \sup_{\xi\in\Omega} |F(\xi) - S_N(\xi)| = 0.$$

Proof. Because of our assumptions and the continuity of the members of $\mathcal{H}^{\mathcal{N}}(\{A_n\};\Omega)$ there exists to every $\varepsilon > 0$ and to every $\xi \in \Omega$ a point η out of the union of all X_N such that both

$$|K_{\mathcal{H}^{\perp}_{0,\ldots,m}}(\xi,\xi) - K_{\mathcal{H}^{\perp}_{0,\ldots,m}}(\xi,\eta)| \leq \varepsilon \tag{6.3.1}$$

and

$$|L_k(\xi) - L_k(\eta)| \leq \varepsilon, \quad k = 1,\ldots,M, \tag{6.3.2}$$

are valid. Hence, Corollary 6.3.1 can be guaranteed by obvious conclusions. □

Next we mention the following result.

Theorem 6.3.2. *Let $\{X_N\}$ be a sequence of admissible systems $X_N = \{\eta_1^N,\ldots,\eta_N^N\}$ relative to $Harm_{0,\ldots,m}$ such that $\Theta_{X_N} \to 0$ as $N \to \infty$. Then, to any $F \in \mathcal{H}$ and to every $\varepsilon > 0$, there exist an integer $N = N(\varepsilon)$ and a spherical spline S_N of the form*

$$S_N = \sum_{n=1}^{M} F(\eta_n^{(N)})L_n + \sum_{n=M+1}^{N} d_n^F K^{\circ}_{\mathcal{H}^{\perp}_{0,\ldots,m}}(\cdot,\eta_n^{(N)})$$

uniquely determined by

$$\begin{aligned}\sum_{n=M+1}^{N} & d_n^F K^{\circ}_{\mathcal{H}^{\perp}_{0,\ldots,m}}(\eta_i^{(N)},\eta_n^{(N)}) \\ &= F(\eta_i^{(N)}) - \sum_{n=1}^{M} F(\eta_n^{(N)})L_n(\eta_i^{(N)}), \quad i = M+1,\ldots,N,\end{aligned}$$

such that

$$S_N(\eta_i^{(N)}) = F(\eta_i^{(N)}), \quad i = 1,\ldots,N,$$

and

$$\sup_{\xi\in\Omega} |S_N(\xi) - F(\xi)| \leq \varepsilon.$$

That means we are able to approximate any function $F \in \mathcal{H}$ uniformly on Ω in a constructive way using spline interpolation provided that the X_N-widths Θ_{X_N} tend to zero as $N \to \infty$.

For the next theorem we require, in addition, that all *constituting coefficients* A_n of $\mathcal{H} = \mathcal{H}(\{A_n\};\Omega)$ are non-zero, i.e. all spherical harmonics are elements of the space $\mathcal{H}$. Since the set of all finite linear combinations of spherical harmonics is uniformly dense in the space $\mathcal{C}(\Omega)$,

$\mathcal{H}$ (with $A_n \neq 0$ for all $n \in \mathbb{N}_0$) is a dense subset of $\mathcal{C}(\Omega)$, too. Furthermore, an extended version of Helly's theorem due to H. YAMABE (1950) shows us that to any $G \in \mathcal{C}(\Omega)$ and any (admissible) system $X_T = \{\eta_1, \ldots, \eta_T\}$, there exists an element $F \in \mathcal{H}$ in an ε-neighbourhood of G with $F(\eta_i) = G(\eta_i), \quad i = 1, \ldots, T$. Thus we finally obtain the following theorem, in connection with Theorem 6.2.4.

Theorem 6.3.3. *Assume that $\{A_n\}$ is a summable sequence, i.e. in particular $A_n \neq 0$ for all $n \in \mathbb{N}_0$. Let $X_T = \{\eta_1, \ldots, \eta_T\}$ be an admissible system on Ω relative to $Harm_{0,\ldots,m}$. Suppose that $\{X_N\}$, $N \to \infty$, is a sequence of admissible systems $X_N = \{\eta_1^{(N)}, \ldots, \eta_N^{(N)}\}$ relative to $Harm_{0,\ldots,m}$ such that $X_T \subset X_N$ for all N and $\Theta_{X_N} \to 0$ as $N \to \infty$. Then, to any $G \in \mathcal{C}(\Omega)$ and to every $\varepsilon > 0$, there exist an integer $N = N(\varepsilon)$ and a spherical spline $S_N \in Spline_{0,\ldots,m}(\{A_n\}; X_N)$ such that*

$$S_N(\eta_i) = G(\eta_i), \quad i = 1, \ldots, T$$

and

$$\sup_{\xi \in \Omega} |S_N(\xi) - G(\xi)| \leq \varepsilon.$$

Although Theorem 6.3.3 is a pure existence theorem, it is of importance both from theoretical and practical point of view. It justifies the use of spherical spline functions as basis system in problems of uniform approximation on the sphere.

6.4 Complete Spline Bases

Let $\{A_n\} \in \mathcal{A}$ be a summable sequence (in the sense of Definition 5.2.1). Let the reproducing kernel $K_{\mathcal{H}}(\cdot, \cdot)$ in the Sobolev space $\mathcal{H} = \mathcal{H}(\{A_n\}; \Omega)$ be given by (6.1.1). In $\mathcal{H}$ closed and complete systems can be generated conveniently from countable dense subsets $X \subset \Omega$ as follows (cf. W. FREEDEN (1984B, 1990B)).

Theorem 6.4.1. *Assume that*

$$X = \bigcup_{N=1}^{\infty} X_N,$$

with $X_N = \{\eta_1, \ldots, \eta_N\}$, $\eta_i \neq \eta_j$ for $i \neq j$, is a countable, dense set of points on Ω. Then

$$\overline{\operatorname{span}_{\eta \in X} K_{\mathcal{H}}(\eta, \cdot)} = \mathcal{H},$$

where completion is understood in the sense of $\| \cdot \|_{\mathcal{H}}$.

Proof. Our purpose is to show that $F \in \mathcal{H}$ and $(F, K_{\mathcal{H}}(\eta,\cdot))_{\mathcal{H}} = 0$ for all $\eta \in X$ implies that $F = 0$. Clearly, $(F, K_{\mathcal{H}}(\eta,\cdot))_{\mathcal{H}} = 0$ is equivalent to $F(\eta) = 0$ for all $\eta \in X$. However, according to our construction, F is continuous on Ω. Hence, if $F(\xi) \neq 0$ for some $\xi \in \Omega$, then F is different from zero for a whole neighbourhood of ξ on Ω. But this is impossible because of the density of X. □

According to the Gram–Schmidt orthonormalization procedure there corresponds to every system $X_N = \{\eta_1, \ldots, \eta_N\}$ an $\mathcal{H}$-orthonormal system in $\mathrm{span}_{\eta\in X_N} K_{\mathcal{H}}(\eta,\cdot)$. Because of the reproducing structure in $\mathcal{H}$ orthonormalization has the great advantage that any kind of approximate integration can be avoided. As a matter of fact, the determination of the orthogonal projection of a function $F \in \mathcal{H}$ onto $\mathrm{span}_{\eta\in X_N} K_{\mathcal{H}}(\eta,\cdot)$ (by solving the normal equations) is equal to $Spline(\{A_n\}; X_N)$-interpolation (i.e. $m = -1$) corresponding to the data points $(\eta, F(\eta)) \in \Omega \times \mathbb{R}$, $\eta \in X_N$.

As it is well known, the set of all finite linear combinations of spherical harmonics is uniformly dense in $\mathcal{C}(\Omega)$. Hence, $\mathcal{H}$ is a dense subset of $\mathcal{C}(\Omega)$, too. This leads us to the following.

Corollary 6.4.2. *Under the assumptions of Theorem 6.4.1*

$$\overline{\mathrm{span}_{\eta\in X} K_{\mathcal{H}}(\eta,\cdot)}^{\|\cdot\|_{\mathcal{C}(\Omega)}} = \mathcal{C}(\Omega).$$

Observing the norm estimate (1.2.10) and the density of $\mathcal{C}(\Omega)$ in $\mathcal{L}^2(\Omega)$ (with respect to $(\cdot,\cdot)_{\mathcal{L}^2(\Omega)}$) we get the following.

Corollary 6.4.3. *Under the assumptions of Theorem 6.4.1*

$$\overline{\mathrm{span}_{\eta\in X} K_{\mathcal{H}}(\eta,\cdot)}^{\|\cdot\|_{\mathcal{L}^2(\Omega)}} = \mathcal{L}^2(\Omega).$$

In a similar way we obtain the following.

Corollary 6.4.4. *Under the assumptions of Theorem 6.4.1*

$$\begin{aligned}
\bigoplus_{n=0}^{m} Harm_n \oplus \overline{\mathrm{span}_{\eta\in X} K_{\mathcal{H}^{\perp}_{0,\ldots,m}}(\eta,\cdot)}^{\|\cdot\|_{\mathcal{H}}} &= \mathcal{H},\\
\bigoplus_{n=0}^{m} Harm_n \oplus \overline{\mathrm{span}_{\eta\in X} K_{\mathcal{H}^{\perp}_{0,\ldots,m}}(\eta,\cdot)}^{\|\cdot\|_{\mathcal{C}(\Omega)}} &= \mathcal{C}(\Omega),\\
\bigoplus_{n=0}^{m} Harm_n \oplus \overline{\mathrm{span}_{\eta\in X} K_{\mathcal{H}^{\perp}_{0,\ldots,m}}(\eta,\cdot)}^{\|\cdot\|_{\mathcal{L}^2(\Omega)}} &= \mathcal{L}^2(\Omega).
\end{aligned}$$

Remark. If we drop the assumption that $A_n \neq 0$ for all $n \in \mathbb{N}_0$, the situation is slightly different: let $\{A_n\}$ be summable with respect to $\mathcal{N}$ and assume that $\{0, \ldots, m\} \in \mathcal{N}$. Then it is true that

$$\bigoplus_{n=0}^{m} Harm_n \oplus \operatorname*{span}_{\eta \in X} K_{\mathcal{H}^{\perp}_{0,\ldots,m}}(\eta, \cdot)$$

is dense in the spaces $\mathcal{H}$, $\{F \in \mathcal{C}(\Omega) \mid F^{\wedge}(n,j) = 0,\ n \in \mathcal{N}_0\}$, $\{F \in \mathcal{L}^2(\Omega) | F^{\wedge}(n,j) = 0,\ n \in \mathcal{N}_0\}$ with respect to the topologies $\|\cdot\|_{\mathcal{H}}$, $\|\cdot\|_{\mathcal{C}(\Omega)}$, $\|\cdot\|_{\mathcal{L}^2(\Omega)}$, respectively.

6.5 Combined Interpolation and Smoothing

It is often the case that the data are of heterogeneous character and (partially) affected with errors or irregularities. In this case interpolation is inappropriate and a *combined interpolating–smoothing method* should be used. As usual we introduce a reproducing Hilbert subspace $\mathcal{H} = \mathcal{H}^{\mathcal{N}}(\{A_n\}; \Omega)$ of $\mathcal{C}(\Omega)$, corresponding to a sequence $\{A_n\}$, summable with respect to $\mathcal{N}$. Again, we let $m \geq 0$ and assume that $\{0, \ldots, m\} \subset \mathcal{N}$. An observed quantity may be interpreted as the value of a linear functional applied on an element of the Hilbert space $\mathcal{H}$. For the smoothing procedure we use the well known functional which includes a trade-off between fidelity to the data and smoothness of the approximating function. The expression to be minimized is along the same lines as suggested by E.T. WHITTAKER (1923) and I.J. SCHOENBERG (1964A,B). More explicitly, let us suppose that as a result of observation or experience we have obtained the set of real numbers $A_i(F) = \alpha_i, i = 1, \ldots, p; B_j(F) = \beta_j, j = 1, \ldots, q$, corresponding to a system of $(p+q)$-functionals $A_1, \ldots, A_p; B_1, \ldots, B_q$ of the dual space $\mathcal{H}^*$ (i.e. the totality of bounded linear functionals defined on $\mathcal{H}$). A method (cf. W. FREEDEN, B. WITTE (1982)) is presented of giving as approximation to F a function $S \in \mathcal{H}$ 'smooth' with respect to the predefined seminorm $\|\cdot\|_{\mathcal{H}^{\perp}_{0,\ldots,m}}$ in $\mathcal{H}$ and satisfying the following properties:

$$\begin{aligned} A_i(S) \text{ is 'close' to } A_i(F) &= \alpha_i, \quad i = 1, \ldots, p, \\ B_i(S) \text{ is equal to } B_i(F) &= \beta_i, \quad i = 1, \ldots, q. \end{aligned}$$

Consequently this approximation procedure may be regarded, in fact, as a spline method combining interpolation and smoothing.

Theorem 6.5.1. *Suppose that δ and $\gamma_1^2, \ldots, \gamma_p^2$ are prescribed positive weights. Let $A_1, \ldots, A_p$ and $B_1, \ldots, B_q$ be systems of bounded linear functionals on $\mathcal{H}$ such that the matrix*

$$\begin{pmatrix} \mathbf{a} & \mathbf{b} & \mathbf{y} \\ \mathbf{b}^T & \mathbf{c} & \mathbf{z} \\ \mathbf{y}^T & \mathbf{z}^T & \mathbf{0} \end{pmatrix} \tag{6.5.1}$$

is non-singular, where the sub-matrices $\mathbf{a}, \mathbf{b}, \mathbf{c}, \mathbf{y}, \mathbf{z}$, are given as follows:

$$\mathbf{a} = \left(A_i A_j K_{\mathcal{H}^\perp_{0,\ldots,m}}(\cdot,\cdot) + \delta\gamma_i^2 \delta_{ij}\right)_{\substack{i=1,\ldots,p\\ j=1,\ldots,p}} \tag{6.5.2}$$

$$\mathbf{b} = \left(A_i B_j K_{\mathcal{H}^\perp_{0,\ldots,m}}(\cdot,\cdot)\right)_{\substack{i=1,\ldots,p\\ j=1,\ldots,q}} \tag{6.5.3}$$

$$\mathbf{c} = \left(B_i B_j K_{\mathcal{H}^\perp_{0,\ldots,m}}(\cdot,\cdot)\right)_{\substack{i=1,\ldots,q\\ j=1,\ldots,q}} \tag{6.5.4}$$

$$\mathbf{y} = \left(A_i Y_{n,j}\right)_{\substack{i=1,\ldots,p\\ n=0,\ldots,m,\, j=1,\ldots,2n+1}} \tag{6.5.5}$$

$$\mathbf{z} = \left(B_i Y_{n,j}\right)_{\substack{i=1,\ldots,q\\ n=0,\ldots,m,\, j=1,\ldots,2n+1}} . \tag{6.5.6}$$

Suppose that $\alpha \in \mathbb{R}^p, \alpha^T = (\alpha_1, \ldots, \alpha_p)$ and $\beta \in \mathbb{R}^q, \beta^T = (\beta_1, \ldots, \beta_q)$ are given vectors, then the function $S \in \mathcal{H}$ of the form

$$\begin{aligned} S(\xi) &= \sum_{i=1}^p a_i A_i K_{\mathcal{H}^\perp_{0,\ldots,m}}(\cdot,\xi) + \sum_{j=1}^q b_j B_j K_{\mathcal{H}^\perp_{0,\ldots,m}}(\cdot,\xi) \\ &\quad + \sum_{n=0}^m \sum_{l=1}^{2n+1} c_{n,l} Y_{n,l}(\xi),\ \xi \in \Omega, \end{aligned} \tag{6.5.7}$$

with coefficients

$$\begin{aligned} a \in \mathbb{R}^p, a^T &= (a_1, \ldots, a_p), \\ b \in \mathbb{R}^q, b^T &= (b_1, \ldots, b_q), \\ c \in \mathbb{R}^M, c^T &= (c_{0,1}, \ldots, c_{m,1}, \ldots, c_{m,2m+1}), \end{aligned}$$

uniquely determined by the linear equations

$$A_i S + \delta\gamma_i^2 a_i = \alpha_i\ , \quad i = 1, \ldots, p \tag{6.5.8}$$

$$B_j S = \beta_j\ , \quad j = 1, \ldots, q \tag{6.5.9}$$

$$\sum_{i=1}^p a_i A_i Y_{n,l} + \sum_{j=1}^q b_j B_j Y_{n,l} = 0, \quad n = 0, \ldots, m, \quad l = 1, \ldots, 2n+1, \tag{6.5.10}$$

represents the only element of $\mathcal{H}$ satisfying

$$\sum_{i=1}^p \left(\frac{A_i(S) - \alpha_i}{\gamma_i}\right)^2 + \delta(S,S)_{\mathcal{H}^\perp_{0,\ldots,m}}$$

$$\leq \sum_{i=1}^{p} \left(\frac{A_i(F) - \alpha_i}{\gamma_i} \right)^2 + \delta(F,F)_{\mathcal{H}^{\perp}_{0,\ldots,m}} \tag{6.5.11}$$

for all $F \in \mathcal{H}$ *with* $B_j(F) = \beta_j, j = 1, \ldots, q$.

Proof. Inserting the respresentation (6.5.7) into the equations (6.5.8), (6.5.9), (6.5.10) results in a linear system in the unknowns a_i, b_j, $c_{n,l}$. The matrix of the linear system is (6.5.1). Since the matrix (6.5.1) is assumed to be non-singular, the coefficients are uniquely determined.

Our purpose is to show that the function S determined by the conditions (6.5.8), (6.5.9), (6.5.10) satisfies the property (6.5.11).

For any function $F \in \mathcal{H}$ satisfying $B_j(F) = \beta_j, j = 1, \ldots, q$, it is easy to see that

$$(S,F)_{\mathcal{H}^{\perp}_{0,\ldots,m}} = \sum_{i=1}^{p} a_i A_i(F) + \sum_{j=1}^{q} b_j \beta_j. \tag{6.5.12}$$

Solving (6.5.8) for a_i, multiplying by $A_i(F)$, and summing over $i = 1, \ldots, p$ gives

$$\sum_{i=1}^{p} a_i A_i(F) = \frac{1}{\delta} \sum_{i=1}^{p} A_i(F) \left(\frac{\alpha_i - A_i(S)}{\gamma_i^2} \right). \tag{6.5.13}$$

Combining with (6.5.12) yields

$$\delta(S,F)_{\mathcal{H}^{\perp}_{0,\ldots,m}} = \sum_{i=1}^{p} A_i(F) \left(\frac{\alpha_i - A_i(S)}{\gamma_i^2} \right) + \delta \sum_{j=1}^{q} b_j \beta_j. \tag{6.5.14}$$

Now an elementary calculation shows us that

$$\begin{aligned} &\sum_{i=1}^{p} \left(\frac{A_i(F) - \alpha_i}{\gamma_i} \right)^2 + \delta(F,F)_{\mathcal{H}^{\perp}_{0,\ldots,m}} \qquad (6.5.15)\\ &= \sum_{i-1}^{p} \left(\frac{A_i(S) - \alpha_i}{\gamma_i} \right)^2 + \delta(S,S)_{\mathcal{H}^{\perp}_{0,\ldots,m}} \\ &\quad + \sum_{i=1}^{p} \left(\frac{A_i(F) - A_i(S)}{\gamma_i} \right)^2 + 2 \sum_{i=1}^{p} \frac{(A_i(F) - A_i(S))(A_i(S) - \alpha_i)}{\gamma_i^2} \\ &\quad + \delta(F-S, F-S)_{\mathcal{H}^{\perp}_{0,\ldots,m}} + 2\delta(S, F-S)_{\mathcal{H}^{\perp}_{0,\ldots,m}} \end{aligned}$$

for every $F \in \mathcal{H}$. According to (6.5.14) if follows that

$$2 \sum_{i=1}^{p} \frac{(A_i(F) - A_i(S))(A_i(S) - \alpha_i)}{\gamma_i^2} + 2\delta(S, F-S)_{\mathcal{H}^{\perp}_{0,\ldots,m}} = 0 \tag{6.5.16}$$

provided that $B_j(F) = \beta_j$. Therefore we have

$$\sum_{i=1}^{p} \left(\frac{A_i(F) - \alpha_i}{\gamma_i} \right)^2 + \delta(F,F)_{\mathcal{H}^{\perp}_{0,\ldots,m}} \tag{6.5.17}$$

$$= \sum_{i=1}^{p} \left(\frac{A_i(S) - \alpha_i}{\gamma_i} \right)^2 + \delta(S, S)_{\mathcal{H}^{\perp}_{0,\ldots,m}}$$
$$+ \sum_{i=1}^{p} \left(\frac{A_i(F) - A_i(S)}{\gamma_i} \right)^2 + \delta(F - S, F - S)_{\mathcal{H}^{\perp}_{0,\ldots,m}}$$

for all $F \in \mathcal{H}$ with $B_j(F) = \beta_j, j = 1, \ldots, q$. But this implies that

$$\sum_{i=1}^{p} \left(\frac{A_i(S) - \alpha_i}{\gamma_i} \right)^2 + \delta(S, S)_{\mathcal{H}^{\perp}_{0,\ldots,m}}$$
$$\leq \sum_{i=1}^{p} \left(\frac{A_i(F) - \alpha_i}{\gamma_i} \right)^2 + \delta(F, F)_{\mathcal{H}^{\perp}_{0,\ldots,m}} \tag{6.5.18}$$

for all $F \in \mathcal{H}$ satisfying the interpolation constraints $B_j(F) = \beta_j, j = 1, \ldots, q$. Equality in (6.5.18) holds if and only if

$$\sum_{i=1}^{p} \left(\frac{A_i(F) - A_i(S)}{\gamma_i} \right)^2 + \delta(F - S, F - S)_{\mathcal{H}^{\perp}_{0,\ldots,m}} = 0,$$

i.e. $F - S \in Harm_{0,\ldots,m}$ with $A_i(F - S) = 0, i = 1, \ldots, p$, and $B_j(F - S) = 0, j = 1, \ldots, q$. As a function of the space $Harm_{0,\ldots,m}$, $F - S$ admits the representation

$$F - S = \sum_{n=0}^{m} \sum_{l=1}^{2n+1} d_{n,l} Y_{n,l}, \tag{6.5.19}$$

where the coefficients $d \in \mathbb{R}^M, d^T = (d_{0,1}, \ldots, d_{m,1}, \ldots, d_{m,2m+1})$ satisfy the linear system

$$\begin{pmatrix} \mathbf{a} & \mathbf{b} & \mathbf{y} \\ \mathbf{b}^T & \mathbf{c} & \mathbf{z} \\ \mathbf{y}^T & \mathbf{z}^T & \mathbf{0} \end{pmatrix} \begin{pmatrix} 0 \\ 0 \\ d \end{pmatrix} = \begin{pmatrix} 0 \\ 0 \\ 0 \end{pmatrix}. \tag{6.5.20}$$

Since the matrix (6.5.1) is invertible it is obvious that $d = 0$. Therefore we have $F = S$. The element $S \in \mathcal{H}$ is the only function in $\mathcal{H}$ having the minimal property (6.5.18) and satisfying interpolating constraints. □

The values $\alpha_1, \ldots, \alpha_p$ and $\beta_1, \ldots, \beta_q$ are the observed quantities. The 'distance' of the values $A_i(S)$ to $\alpha_i, i = 1, \ldots, q$, can be controlled by choosing the constant δ in a suitable way. A small value of δ emphasizes fidelity to the observed data at the expense of smoothness, while a large value does the opposite. Taking $\delta = 0$ yields $A_i(S) = \alpha_i, i = 1, \ldots, p$, i.e.

the combined smoothing and interpolating procedure leads back to strict interpolation. It is of advantage to adapt the quantities $\gamma_1^2, \ldots, \gamma_p^2$ to the standard deviations of the measured values.

Remark. The smoothing parameter δ controls the trade off between 'approximation to the data' as measured by the first term and 'smoothness' of the solution as measured by the second term. C. Reinsch (1975) proposed a method to estimate a good value for the smoothing parameter δ. This simple method leads G. Wahba (1975), P. Craven, G. Wahba (1979), G. Wahba, J. Wendelberger (1980) to propose the method of generalized cross-validation (GCV). The method of generalized cross-validation was successfully improved by F. Utreras (1981), L. Elden (1984), M.F. Hutchinson, F.R. de Hoog (1987), and more recently by F. Utreras (1990) and others. Some theoretical aspects of GCV were analyzed by P. Craven, G. Wahba (1979), F. Utreras (1990), and D. Ragozin (1983).

Bibliographical Notes

Spline approximation on the sphere is becoming more and more important in recent time. The obvious applications involve approximation of data from environmental sources, such as meteorology, oceanography, and pollution. Applications include representing functions which model temperature, pressure, ozone, gravitational and magnetic forces, elastic deformation etc. at all points on the spherical surface of the earth based on a discrete sample of values taken at arbitrary points. But there are also applications in biomedical engineering, for example the electroencephalographic description of scalp potentials and scalp current densities from discretely placed electrodes (cf. F. Perrin *et al.* (1989)).

A major field of interest is gravity field and geoid determination in physical geodesy (see for example W. Freeden *et al.* (1996b)) which enjoys a renewed popularity in many positioning, mapping, and exploration applications with the advent of satellite based techniques (such as the global positioning system (GPS) and satellite altimetry). In former geophysical prospecting, which was dominated by seismic reflection surveying, gravity methods were mostly used as complements when difficulties with seismic methods arose. Nowadays, however, satellite gravity methods have brought a new concept into prospecting. High precision and resolution of the gravity field, obtained with space-borne satellite techniques, will change the ordinary routine in future prospecting. From being a secondary prospecting tool, the gravity field or the geoid, computed from (scattered) terrestrial and satellite data, will be used to locate prospective regions as well as individual prospects.

If the data are localized approximation problems on the sphere can be solved through application of methods designed for two-dimensional Euclidean space (see for example T.A. FOLEY, H. HAGEN (1994) and the references therein). However, problems like gravity determination involve essentially the entire surface of the sphere, or a sufficiently large part that modelling the data as arising in two-space is no longer appropriate. Since there is no differential mapping of the entire sphere to a bounded planar region, there is a need to develop approximation methods over the sphere itself.

Global approximation on the sphere has been studied by a number of authors, among them R. BARNHILL *et al.* (1987), T. FOLEY (1990), R.L. HARDY (1971), R. FRANKE (1983, 1985, 1987), J. HOSCHEK, D. LASSER (1992), J. HOSCHEK, G. SEEMANN (1992), C. LAWSON (1984), G. NIELSON, R. RAMERAJ (1987), H. POTTMANN, M. ECK (1990) and R. RENKA (1984) mostly seen from the viewpoint of computer aided geometric design. Spherical spline approximation, however, has its roots in physically motivated problems. It was independently developed by W. FREEDEN (1981A) and G. WAHBA (1981). A best approximate integration technique based on spherical spline interpolation (of order 0) has been presented already in W. FREEDEN (1978, 1979). Generalizations to multidimensional hyperspheres are due to W. FREEDEN, P. HERMANN (1986), H.J. TAIJERON *et al.* (1994), while generalizations to vectorial and tensorial problems of geosciences and satellite-based disciplines have been discussed by, e.g., R. FRANKE *et al.* (1988), W. FREEDEN, T. GERVENS (1988, 1991), W. FREEDEN *et al.* (1990, 1994, 1996B), G. WAHBA (1982, 1984, 1990). The list of applications in geophysics and physical geodesy is long (see for example J. CUI (1995), H. EULER *et al.* (1985, 1986), W. FREEDEN (1979, 1981A,B, 1982B,C, 1987A, 1990B), W. FREEDEN, M. SCHREINER (1994) H. v. GYSEN, L. MERRY (1990), T. KLING *et al.* (1987), R. REUTER (1982), F. SCHNEIDER (1994, 1996A,B, 1997), M. SCHREINER (1994, 1995) and many others).

The importance of our approach lies in the development of a general setting in which it is possible that the approximation can be chosen to embody desirable characteristics according to the source of the data. This is achieved through the idea of approximation by spline (cf. W. FREEDEN (1984A,B, 1990)) that minimize a weighted Sobolev norm of the approximant, with a large class of weights being at the disposal of the user. A survey paper on spherical spline theory is W. FREEDEN *et al.* (1996C).

The idea of spherical spline interpolation can be generalized to nonspherical surfaces (such as ellipsoid, (regular) earth's surface etc.). Recently, N. DYN *et al.* (1996) have studied variational principles and Sobolev-type estimates for interpolation on a $\mathcal{C}^{(\infty)}$-Riemannian manifold. Moreover, boundary-value problems corresponding to elliptic partial differential equations become successfully solvable by the generalized concept of splines even if the boundary is non-smooth. For the Laplace equation of potential theory we mention W. FREEDEN (1981B, 1985C, 1987A,B), W. FREEDEN *et al.* (1996B), R. REUTER (1982), L. SHURE *et al.* (1982), F. SCHNEIDER (1996B, 1997), for the Helmholtz equation we refer to W. FREEDEN (1987C), for the Cauchy–Navier equation of linear elasticity we list J.M. BRUDOWSKY (1992), J.M. BRUDOWSKY *et al.* (1992), W. FREEDEN (1990A), W. FREEDEN *et al.* (1990), W. FREEDEN *et al.* (1996A), M. TÜCKS (1996).

7

APPROXIMATE INTEGRATION

The method of expanding a function into a Fourier series in terms of trial functions (for example, spherical harmonics, certain types of radial basis functions) amounts to the problem of evaluating the expansion coefficients by integration. In practical applications (e.g. in many problems of geodesy or geophysics), however, the integrand is often not known continuously over all of Ω. Instead we are confronted with the situation that experimental data or measured values yielding the function only at prescribed points must be integrated. This gives rise to approximate integration. The aim of this chapter is to present in more detail some procedures of numerical integration on the sphere. The first one is a low discrepancy method (Section 7.1). It is based on the idea of equidistribution. The conceptual background is a spherical version of the Koksma–Hlawka formula corresponding to invariant pseudodifferential operators. Because of their simple structure and flexibility low discrepancy methods can be recommended for problems involving a large number of densely spaced data points, and under this assumption the method is satisfactory. The implementation on a computer is very easy. The second method (Section 7.2) deals with the standard procedure of integrating spherical interpolating polynomials. The third method combines the advantage of an optimal approximation formula with the requirement of an explicit (a priori) error estimation. It is a best approximation technique, since it provides a way of minimizing the error. This method essentially uses the properties of the reproducing kernels associated to summable pseudodifferential operators (as we shall see in Section 7.3). It can be applied to problems with regular as well as irregular spacing of the data. However, it requires the solution of a (time consuming) linear system of equations. Thus it can be recommended for practical use only in the case where other methods fail, particularly, for scattered data distribution. The fourth method (Section 7.4) deals with integration exact for spherical splines of a given index. It is shown that the integration of an interpolating spline is equivalent to the preceding best approximation technique.

7.1 Low Discrepancy Method

The idea of the low discrepancy method (cf., e.g. P.J. DAVIS, P.M. RABINOWITZ (1975)) is simple: approximate the integral of a function $F : \Omega \to \mathbb{R}$ by a mean of functional values at prescribed points. This is a reasonable approach if the data set is somehow 'equidistributed' over the sphere and if the function shows 'good' behaviour. These statements can be made rigorous by the concept of equidistribution and the notion of discrepancy. The starting point of the theory of equidistributed pointsets is the work of H. WEYL (1916). Recent results for Euclidean spaces $\mathbb{R}^n$ can be found in L. KUIPERS, H. NIEDERREITER (1974) and H. NIEDERREITER (1992). An extension of the Koksma–Hlawka formula to the spherical context has been proved by W. FREEDEN (1979). More detailed work about equidistribution on the sphere is due to E. HLAWKA (1981) and W. FREEDEN, M. SCHREINER (1995A), J. CUI, W. FREEDEN (1997).

For our purposes, the reason of investigating systems (ensembles) of points on the unit sphere is twofold. On the one hand, we use them for numerical evaluation of integrals, for example the inner product of two functions. To this end, 'irregularly equidistributed' systems seem to be the better advice (cf. Examples 7.1.12 and 7.1.13). On the other hand, we need systems of points as 'axes of the axisymmetric kernel functions' in Gabor, Toeplitz, and wavelet transforms. Here, hierarchical sequences of systems (possibly with 'regular structure') are needed (cf. Example 7.1.11).

Assume that a system $X_N = \{\eta_1^N, \ldots, \eta_N^N\}$ of points $\eta_k^N \in \Omega, k = 1, \ldots, N$, is associated with a partition $\Omega_{X_N} = \{\Omega_{\eta_1^N}, \ldots, \Omega_{\eta_N^N}\}$ of the unit sphere Ω into N mutually distinct parts $\Omega_{\eta_k^N}, \eta_k^N \in \Omega_{\eta_k^N}$, satisfying

$$0 < \|\Omega_{\eta_k^N}\| = \int_{\Omega_{\eta_k^N}} d\omega = 4\pi a_k^N, \quad k = 1, \ldots, N, \tag{7.1.1}$$

and

$$\sum_{k=1}^{N} a_k^N = 1. \tag{7.1.2}$$

Hence it makes sense to consider the number

$$\max_{k=1,\ldots,N} \sup_{\xi \in \Omega_{\eta_k^N}} |\xi - \eta_k^N|. \tag{7.1.3}$$

Consequently we are able to define the *partition size* of the system X_N by setting

$$\sigma(X_N) = \inf \max_{k=1,\ldots,N} \sup_{\xi \in \Omega_{\eta_k^N}} |\xi - \eta_k^N|, \tag{7.1.4}$$

where the infimum is taken over all partitions Ω_{X_N} of Ω satisfying (7.1.1) and (7.1.2). It follows that $a_k^N \to 0$ for $N \to \infty$ (k fixed) if $\sigma(X_N) \to 0$ so that (a_k^N) forms a *Toeplitz-triangular matrix*, i.e.

$$\begin{pmatrix} a_1^1 & & & & & \\ a_1^2 & a_2^2 & & & & \\ a_1^3 & a_2^3 & a_3^3 & & & \\ \vdots & \vdots & \vdots & & & \\ a_1^N & a_2^N & a_3^N & \ldots & a_N^N & \\ \vdots & \vdots & \vdots & & \vdots & \ddots \end{pmatrix}$$

is a triangular matrix satisfying $a_k^N \to 0$ and $N \to \infty$, k fixed, and $a_k^N > 0$ for all $N, 1 \leq k \leq N$,

$$\sum_{k=1}^{N} a_k^N = 1.$$

Let $X_N = \{\eta_1^N, \ldots, \eta_N^N\}$ be a system on Ω. Then it follows that for every partition $\Omega_{X_N} = \{\Omega_{\eta_1^N}, \ldots, \Omega_{\eta_N^N}\}$ of the above type,

$$\begin{aligned} \frac{1}{4\pi}\int_\Omega F(\eta)d\omega(\eta) &= \frac{1}{4\pi}\sum_{k=1}^{N}\int_{\Omega_{\eta_k^N}} F(\eta)d\omega(\eta) \qquad (7.1.5)\\ &= \sum_{k=1}^{N} a_k^N F(\eta_k^N) + \frac{1}{4\pi}\sum_{k=1}^{N}\int_{\Omega_{\eta_k^N}} (F(\eta) - F(\eta_k^N))d\omega(\eta). \end{aligned}$$

Thus we get the following result.

Lemma 7.1.1. *Let $X_N = \{\eta_1^N, \ldots, \eta_N^N\}$ be a system of points on Ω and let F be Lipschitz continuous on Ω with Lipschitz constant C_F. Then*

$$\left|\frac{1}{4\pi}\int_\Omega F(\eta)d\omega(\eta) - \sum_{k=1}^{N} a_k^N F(\eta_k^N)\right| \leq C_F\sigma(X_N).$$

In other words, the '*integral mean*'

$$\mu(F) = \frac{1}{4\pi}\int_\Omega F(\eta)d\omega(\eta) \qquad (7.1.6)$$

may be replaced by the '*approximate sum*' (Toeplitz mean)

$$\mu_{X_N}(F) = \sum_{k=1}^{N} a_k^N F(\eta_k^N) \qquad (7.1.7)$$

provided that the partition size of X_N is sufficiently small. In particular, if $a_k^N = 1/N$ for $k = 1, \ldots, N$, $\mu_{X_N}(F)$ is the *arithmetical mean of the function values of F at the points $\eta_k^N, k = 1, \ldots, N$.*

An immediate consequence of Theorem 5.3.3 is the following integral formula.

Theorem 7.1.2 *(Koksma–Hlawka formula for pseudodifferential operators). Let Λ be a pseudodifferential operator with summable symbol $\{\Lambda^\wedge(n)\}$. Assume $X_N = \{\eta_1^N, \ldots, \eta_N^N\}$ is a system of points $\eta_1^N, \ldots, \eta_N^N$ on Ω. Then*

$$\begin{aligned} \frac{1}{4\pi}\int_\Omega F(\eta)d\omega(y) &= \sum_{k=1}^N a_k^N F(\eta_k^N) \\ &- \sum_{k=1}^N a_k^N \int_\Omega \Lambda_\eta K_{\mathcal{H}_0^\perp(\{\Lambda^\wedge(n)\};\Omega)}(\eta,\eta_k^N)(\Lambda_\eta F(\eta))d\omega(\eta) \end{aligned}$$

holds for all $F \in \mathcal{H}(\{\Lambda^\wedge(n)\};\Omega)$ and all $a_1^N, \ldots, a_N^N$ with $\sum_{k=1}^N a_k^N = 1$.

For the convenience of the reader we mention a specialization of Theorem 7.1.2 to equidistributed pointsets (i.e. $a_k^N = 1/N, k = 1, \ldots, N$) and a pseudodifferential operator such that the corresponding reproducing kernel is available in elementary form (cf. J. CUI, W. FREEDEN (1997)).

Corollary 7.1.3. *Let Λ be the pseudodifferential operator with spherical symbol $\{\Lambda^\wedge(n)\}$ given by*

$$\Lambda^\wedge(n) = \begin{cases} ((2n+1)n(n+1))^{1/2} & , \quad n = 1, 2, \ldots \\ 1 & , \quad n = 0 \ . \end{cases}$$

Then

$$\begin{aligned} &\left| \frac{1}{4\pi}\int_\Omega F(\xi)d\omega(\xi) - \frac{1}{N}\sum_{k=1}^N F(\eta_k^N) \right| \qquad (7.1.8) \\ &\leq \frac{1}{N}\left(\frac{1}{4\pi}\sum_{r=1}^N\sum_{k=1}^N (1 - 2\ln(1 + \sqrt{\frac{1-\eta_r^N\cdot\eta_k^N}{2}})) \right)^{1/2} \|\Lambda(F)\|_{\mathcal{L}^2(\Omega)}. \end{aligned}$$

Proof. The estimate easily follows from the fact that the Legendre transform of $t \mapsto G(t) = 1 - 2\ln(1+\sqrt{(1-t)/2}), t \in [-1,+1]$, is given by (see for example W. MAGNUS *et al.* (1966)): $G^\wedge(n) = (4\pi(2n+1)n(n+1))^{-1}$, $n = 1, 2, \ldots$. □

Using the abbreviation

$$\varepsilon_{X_N}(F) = \mu(F) - \mu_{X_N}(F)$$

we get by virtue of Theorem 7.1.2 that

$$\varepsilon_{X_N}(F) = \int_{\Omega} \varepsilon_{X_N}(\Lambda_\eta K_{\mathcal{H}_0^{\perp}(\{\Lambda^\wedge(n)\};\Omega)}(\cdot,\eta))(\Lambda_\eta F(\eta))\, d\omega(\eta) \tag{7.1.9}$$

provided that F is a member of $\mathcal{H}(\{\Lambda^\wedge(n)\};\Omega)$. Using (7.1.9) we obtain from the Cauchy–Schwarz inequality

$$|\varepsilon_{X_N}(F)| \leq \left(\sum_{k=1}^{N}\sum_{l=1}^{N} a_k^N a_l^N K_{\mathcal{H}_0^{\perp}(\{\Lambda^\wedge(n)\};\Omega)}(\eta_k^N,\eta_l^N)\right)^{1/2} \left(\int_{\Omega} |\Lambda_\eta F(\eta)|^2\, d\omega(\eta)\right)^{1/2}.$$

The expression

$$D(X_N) = \left(\sum_{k=1}^{N}\sum_{l=1}^{N} a_k^N a_l^N K_{\mathcal{H}_0^{\perp}(\{\Lambda^\wedge(n)\};\Omega)}(\eta_k^N,\eta_l^N)\right)^{1/2}$$

is called *discrepancy of* X_N (more accurately, $\mathcal{L}^2(\Omega)$*-discrepancy of* X_N).

Lemma 7.1.4. *Assume that the function F is of class $\mathcal{H}(\{\Lambda^\wedge(n)\};\Omega)$. Then the following estimate is valid:*

$$|\varepsilon_{X_N}(F)| \leq D(X_N)\; V(F),$$

where the '$\mathcal{L}^2(\Omega)$-variance' $V(F)$ is given by

$$V(F) = \left(\int_{\Omega} |\Lambda_\eta F(\eta)|^2\, d\omega(\eta)\right)^{1/2}.$$

Next we consider sequences $\{X_N\}$ of systems $X_N = \{\eta_1^N,\ldots,\eta_N^N\}$. We call $\{X_N\}$ a *hierarchical sequence*, whenever $X_N \subset X_{N+1}$ for all N. One can always assume that $\eta_k^N = \eta_k^{N+1}$ holds for all $k = 1,\ldots,N$ and all positive integers N.

A sequence (hierarchical sequence) $\{X_N\}$ is called (a_k^N)*-distribution* (*hierarchical* (a_k^N)*-distribution*), if

$$\frac{1}{4\pi}\int_{\Omega} F(\xi)\, d\omega(\xi) = \lim_{N\to\infty} \sum_{k=1}^{N} a_k^N F(\eta_k^N) \tag{7.1.10}$$

holds for all $F \in \mathcal{H}_s(\Omega), s > 2$. When constructing hierarchical (a_k^N)-distributions it is clear that the convergence of (7.1.10) cannot be as fast as in the non-hierarchical case. This is the prize to be paid for the additional structure. In particular, an a_k^N-distribution with $a_k^N = 1/N, k = 1, \ldots, N$, is simply called *equidistribution*.

An easy consequence is the following.

Lemma 7.1.5. *If $\sigma(X_N) \to 0$ as $N \to \infty$, then $\{X_N\}$ is an (a_k^N)-distribution.*

Moreover, we also have the following lemma.

Lemma 7.1.6. *Suppose that $\{X_N\}$ is an (a_k^N)-distribution. Then (7.1.10) holds for all $F \in \mathcal{C}(\Omega)$.*

Proof. We know already that (7.1.10) is true for all continuously differentiable functions on Ω. Take now a continuous function F on Ω. For $\varepsilon > 0$ arbitrary, the closure of the system of spherical harmonics tells us that there exists a function $P \in \mathcal{H}_s(\Omega), s > 2$, such that

$$\sup_{\xi \in \Omega} |F(\xi) - P(\xi)| \leq \varepsilon.$$

Therefore it follows via the triangle inequality that

$$\begin{aligned}
& \left| \frac{1}{4\pi} \int_\Omega F(\eta)\, d\omega(\eta) - \sum_{k=1}^N a_k^N F(\eta_k^N) \right| \qquad (7.1.11)\\
\leq\ & \frac{1}{4\pi} \int_\Omega |F(\eta) - P(\eta)|\, d\omega(\eta) \\
& + \left| \frac{1}{4\pi} \int_\Omega P(\eta)\, d\omega(\eta) - \sum_{k=1}^N a_k^N P(\eta_k^N) \right| \\
& + \sum_{k=1}^N |a_k^N|\ |P(\eta_k^N) - F(\eta_k^N)| \\
\leq\ & 2\varepsilon + \left| \frac{1}{4\pi} \int_\Omega P(\eta)\, d\omega(\eta) - \sum_{k=1}^N a_k^N P(\eta_k^N) \right|.
\end{aligned}$$

For N sufficiently large we thus obtain $|\varepsilon_{X_N}(F)| \leq 3\varepsilon$, as required. □

It is not very complicated to extend Lemma 7.1.6 to all Riemann-integrable functions. We omit these considerations.

Next we want to adopt a famous criterion due to H. Weyl (1916). We consider the system $\{\mu_{X_N}(Y_{n,j})\}$ of '*Weyl-sums*' $\mu_{X_N}(Y_{n,j}), n = 0, 1, \ldots$, $j = 1, \ldots, 2n + 1$.

Theorem 7.1.7. *Let $\{X_N\}$ be a sequence of systems $X_N = \{\eta_1^N, \ldots, \eta_N^N\}$. Then the following statements are equivalent:*

(i) $\lim_{N\to\infty} \sum_{k=1}^{N} a_k^N Y_{n,j}(\eta_k^N) = 0$ *for* $n = 1, 2, \ldots, j = 1, \ldots, 2n+1$,

(ii) $\lim_{N\to\infty} \sum_{k=1}^{N} a_k^N F(\eta_k^N) = \frac{1}{4\pi} \int_\Omega F(\eta)\, d\omega(\eta)$ *for all* $F \in \mathcal{C}(\Omega)$.

Proof. Assume that (i) is true. Then we know from Theorem 7.1.2 that for all $F \in \mathcal{H}(\{\Lambda^\wedge(n)\}; \Omega)$

$$\lim_{N\to\infty} \int_\Omega \sum_{k=1}^{N} a_k^N K_{\mathcal{H}_0^\perp(\{\Lambda^\wedge(n)\};\Omega)}(\eta, \eta_k^N) Y_{n,j}(\eta)\, d\omega(\eta) = 0$$

for all $n = 1, 2, \ldots, j = 1, \ldots, 2n+1$. Hence,

$$\lim_{N\to\infty} \sum_{k=1}^{N} a_k^N K_{\mathcal{H}_0^\perp(\{\Lambda^\wedge(n)\};\Omega)}(\cdot, \eta_k^N) = 0.$$

Then, by virtue of Lemma 7.1.4, condition (ii) is true for all $F \in \mathcal{C}(\Omega)$. That condition (ii) implies (i) is clear from the orthogonality of spherical harmonics. □

As an immediate consequence we obtain the following corollary.

Corollary 7.1.8. $\{X_N\}$ *is an* (a_k^N)*-distribution if and only if*

$$\lim_{N\to\infty} D(X_N) = 0.$$

Finally we list some examples of equidistributed sequences ($a_k^N = 1/N$ for $k = 1, \ldots, N$) which are of practical importance. Thereby, any point of the below described ensembles is given in terms of the standard representation (cf. (1.2.4)):

$$(\sin(\vartheta)\cos(\varphi), \sin(\vartheta)\sin(\varphi), \cos(\vartheta))^T.$$

Example 7.1.9. *An ensemble $X_{N(\gamma)}$ on the unit sphere dependent on the choice of a control parameter $\gamma \in \mathbb{N}$ is given by the points $(\varphi_{ij}, \vartheta_i)$ as follows:*

(i) $\vartheta_0 = 0$, $\varphi_{01} = 0$ *(North Pole)*

(ii) $\Delta\vartheta = \pi/\gamma$

(iii) $\vartheta_i = i\Delta\vartheta, \quad 1 \le i \le \gamma - 1$

(iv) $\gamma_i = \left[2\pi/\arccos\left((cos\Delta\vartheta - \cos^2\vartheta_i)/\sin^2\vartheta_i\right)\right]$

(v) $\varphi_{ij} = (j - \frac{1}{2})(2\pi/\gamma_i),\ 1 \le j \le \gamma_i$

(vi) $\vartheta_\gamma = \pi,\ \varphi_{\gamma 1} = 0 \quad$ *(South Pole).*

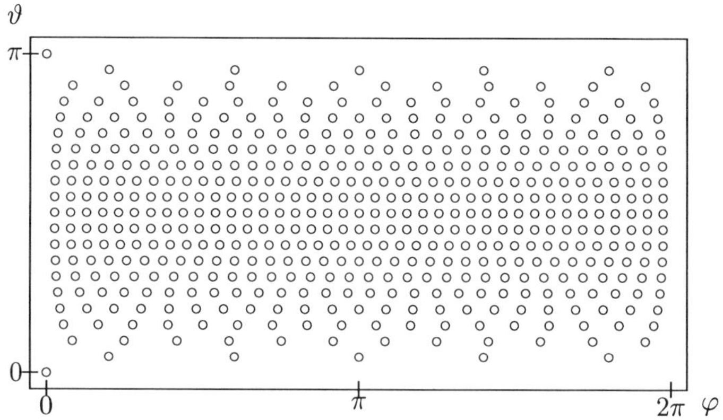

FIG. 7.1.1. Ensemble of Example 7.1.9. $\gamma = 20$, $N = 501$

Lemma 7.1.10. *The number of points $N(\gamma)$ for a given $\gamma \in \mathbb{N}$ can be estimated by*

$$N(\gamma) \le 2 + \frac{4}{\pi}\gamma^2.$$

Proof. For $1 \le i \le \gamma - 1$ we have the estimate

$$\frac{\gamma_i}{\gamma} = \frac{1}{\gamma}\left[2\pi/\arccos\left(\frac{\cos\pi/\gamma - \cos^2 i\pi/\gamma}{\sin^2 i\pi/\gamma}\right)\right] \le 2\sin\frac{i\pi}{\gamma}.$$

Hence, remembering the trapezoidal rule of numerical integration and observing that $\sin\pi t$ is concave for $t \in (-1, 1)$ we get

$$N(\gamma) \quad = \quad 2 + \sum_{i=1}^{\gamma-1}\gamma_i = 2 + \gamma^2\sum_{i=1}^{\gamma-1}\frac{1}{\gamma}\frac{\gamma_i}{\gamma} \tag{7.1.12}$$

$$\leq \quad 2 + 2\gamma^2 \sum_{i=1}^{\gamma-1} \frac{1}{\gamma} \sin \frac{i\pi}{\gamma}$$

$$\leq \quad 2 + 2\gamma^2 \int_0^1 \sin \pi t \, dt = 2 + \frac{4}{\pi}\gamma^2,$$

as required. □

Notice that we get an equidistribution but not a hierarchical equidistribution by increasing the parameter γ.

Example 7.1.11. *Again dependent on a parameter $\gamma \in \mathbb{N}$, we define the ensemble $X_{N(\gamma)}$ by relations of type*

(i) $\vartheta_0 = 0, \varphi_{01} = 0$ *(North Pole)*

(ii) $\Delta\vartheta = \pi/\gamma$

(iii) $\vartheta_i = i\Delta\vartheta, \quad 1 \leq i \leq \gamma - 1$

(iv) $i \leq \gamma/2 : \gamma_i = 4i$
$i > \gamma/2 : \gamma_i = 4(\gamma - i)$

(v) $\varphi_{ij} = (j - \frac{1}{2})(2\pi/\gamma_i), \ 1 \leq j \leq \gamma_i$

(vi) $\vartheta_\gamma = \pi, \ \varphi_{\gamma_1} = 0$ *(South Pole).*

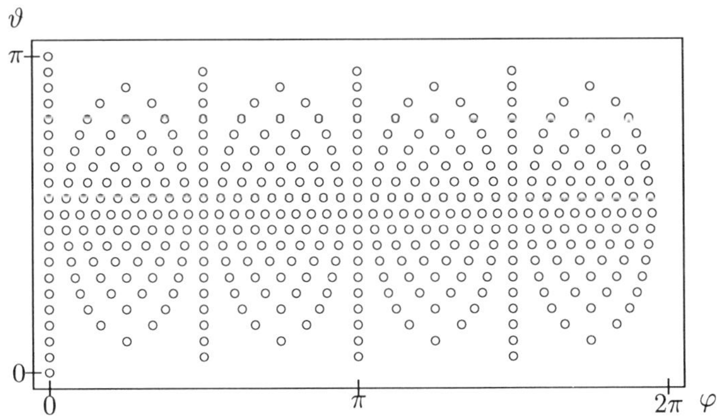

FIG. 7.1.2. Ensemble of Example 7.1.11. $\gamma = 20$, $N = 402$

The total number $N(\gamma)$ of points is determined by the choice of the value γ. We have

$$N(\gamma) = 2 + 4\left[\frac{\gamma+1}{2}\right]\left[\frac{\gamma}{2}\right].$$

By doubling the parameter γ one gets a hierarchical equidistribution.

The generation of the following two systems is based on the theory of uniform distributions in Euclidean spaces $\mathbb{R}^n$ (cf. L. KUIPERS, H. NIEDERREITER (1974)). More explicitly, the idea is to use a sequence in a rectangle which is uniformly distributed and then to transform it via spherical coordinates to the unit sphere Ω. This approach yields more 'irregular' point distributions for numerical integration.

Based on the idea of J.G. VAN DER CORPUT (1935) we consider for $p \in \{2,3,\ldots\}$ the unique expansion of $n-1$ $(n \in \mathbb{N})$ of the form

$$n - 1 = \sum_{j=0}^{s} a_j p^j, \quad a_j \in \{0,\ldots,p-1\}$$

and define

$$\Phi_p(n) = \sum_{j=0}^{s} a_j p^{-j-1}.$$

Clearly, $\Phi_p(n) \in [0,1)$ for all $n \in \mathbb{N}$. The mapping Φ_p is essentially involved in the next two examples.

Example 7.1.12. *Given $N \in \mathbb{N}$, we define the system $X_N = (\eta_1^N,\ldots,\eta_N^N)$ using standard coordinates (φ_n, ϑ_n) by*

$$(\varphi_n, \vartheta_n) = \left(2\pi\frac{n-1}{N}, \arccos(2\Phi_2(n)-1)\right), \quad n = 1,\ldots,N. \tag{7.1.13}$$

It is known for this so-called *Hammersley sequence* that for all $F \in \mathcal{C}(\Omega)$

$$\lim_{N\to\infty} \frac{1}{N}\sum_{n=1}^{N} F^*(\varphi_n, t_n) = \frac{1}{4\pi}\int_0^{2\pi}\int_{-1}^{1} F^*(\varphi,t)\,dt\,d\varphi, \tag{7.1.14}$$

where

$$t_n = \cos\vartheta_n$$

and

$$F^*(\varphi,t) = F(t\varepsilon^3 + \sqrt{1-t^2}(\cos\varphi\varepsilon^1 + \sin\varphi\varepsilon^2))$$

(cf. H. NIEDERREITER (1992)). Hence, according to the proof of Theorem 7.1.7, we have the limit relations

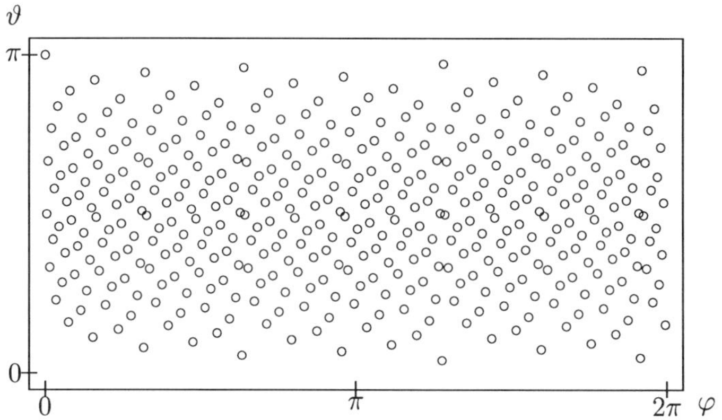

FIG. 7.1.3. Ensemble of Example 7.1.12. $N = 400$

$$\lim_{N\to\infty} D(X_N) = 0. \tag{7.1.15}$$

It is easy to see that we do not get a hierarchical sequence of systems by increasing N. In order to get a hierarchical sequence of ensembles we change the last definition slightly.

Example 7.1.13. *Let for* $n = 1, \ldots, N$,

$$(\varphi_n, \vartheta_n) = (2\pi\Phi_2(n), \arccos(2\Phi_3(n) - 1)). \tag{7.1.16}$$

This Corput–Halton sequence also satisfies (7.1.14), (7.1.15). It is obvious that a hierarchical sequence of systems is generated for increasing values of N.

Example 7.1.14. *The construction principle of the equidistribution*

$$\{(1/\sqrt{n})(x_1, x_2, x_3)^T | x_1^2 + x_2^2 + x_3^2 = n, n \in \mathbb{N}, x_i \in \mathbb{Z}, \}$$

is well known in analytic number theory (cf. A.V. MALYSEV (1957)*). The number of grid points* $x = (x_1, x_2, x_3)^T, x_i \in \mathbb{Z}$, *satisfying* $1 \le n \le m, m \in \mathbb{N}$, *can be deduced from arguments due to* R.F. ARENSTORF, D. JOHNSON (1979)*. It is clear that we get a hierarchical equidistribution for increasing* m.

Besides the equidistributions mentioned above there are many other examples. For more details the interested reader is referred to the literature.

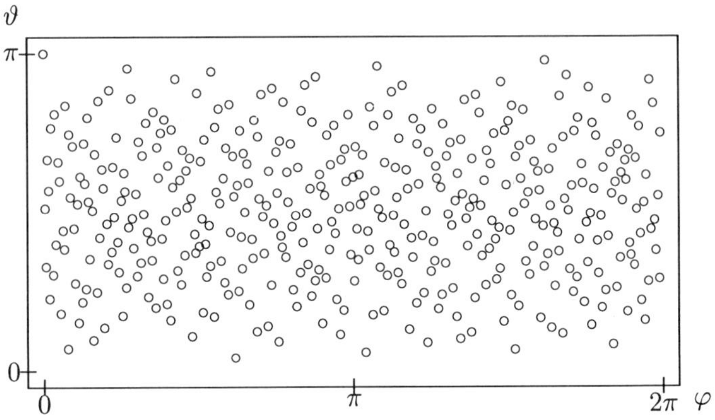

FIG. 7.1.4. Ensemble of Example 7.1.13. $N = 400$

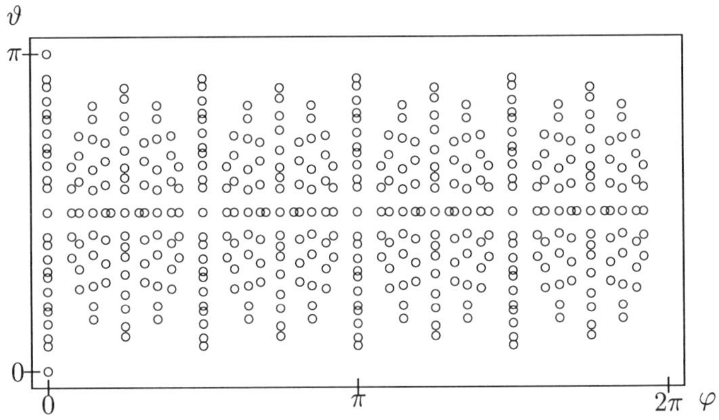

FIG. 7.1.5. Ensemble of Example 7.1.14. $N = 410$

Remark. The subdivision effected by grid lines of standard coordinates

(i) $\varphi_n = n(2\pi/\gamma), \quad 0 \leq n \leq 2\gamma - 1$

(ii) $\vartheta_i = i(\pi/\gamma), \quad 0 \le i \le \gamma$

forms rectangular blocks (with $2\gamma(\gamma+1)$ nodes). These blocks *do not generate* an equidistribution on the sphere Ω. Nevertheless, the 'trapezoidal rule'

$$\sum_{i=0}^{\gamma} d_i \sum_{n=0}^{2\gamma-1} F(\cos\vartheta_i\, \varepsilon^3 + \sin\vartheta_i(\cos\varphi_n\varepsilon^1 + \sin\varphi_n\varepsilon^2))$$

plays a particular role in numerical integration. It enables us to compute 'discrete convolutions' on the sphere for example, discretized wavelet transform (cf. Chapter 10) by use of fast Fourier transform techniques (see for example M. Bäcker (1995), U. Windheuser (1995, 1996)).

7.2 Interpolatory Polynomial Integration

Assume that $X_M = \{\eta_1, ..., \eta_M\}$, $M = (m+1)^2$, is a fundamental system relative to $Harm_{0,...,m}$. Then the matrix $\mathbf{matr}_{X_M}(Y_{0,1}, \ldots, Y_{m,2m+1})$ is regular. The addition theorem of spherical harmonics yields

$$\mathbf{h}^T\mathbf{h} = \mathbf{k}, \tag{7.2.1}$$

where

$$\mathbf{h} = \mathbf{matr}_{X_M}(Y_{0,1}, \ldots, Y_{m,2m+1}), \tag{7.2.2}$$

$$\mathbf{k} = \mathbf{matr}_{X_M}\left(K_{Harm_{0,\ldots,m}}(\eta_1, \cdot), \ldots, K_{Harm_{0,\ldots,m}}(\eta_M, \cdot)\right) \tag{7.2.3}$$

and

$$K_{Harm_{0,\ldots,m}}(\xi, \eta) = \sum_{n=0}^{m}\sum_{j=1}^{2n+1} Y_{n,j}(\xi)Y_{n,j}(\eta), \quad \xi, \eta \in \Omega. \tag{7.2.4}$$

Thus $\mathbf{k}$ is a positive definite, symmetric matrix. Hence there exists an orthogonal matrix $\mathbf{q}$ such that

$$\mathbf{qkq}^T = \mathbf{d}, \tag{7.2.5}$$

where

$$\mathbf{d} = \mathrm{diag}\,(\delta_1, \ldots, \delta_M) = \begin{pmatrix} \delta_1 & & 0 \\ & \ddots & \\ 0 & & \delta_M \end{pmatrix}, \tag{7.2.6}$$

and

$$\delta_{\min} = \delta_1 \le \delta_2 \le \ldots \le \delta_M = \delta_{\max}.$$

For a fundamental system X_M relative to $Harm_{0,...,m}$, the *Lagrangians* $L_1, ..., L_m \in Harm_{0,...,m}$ defined by

$$L_1(\xi) = \frac{\begin{vmatrix} Y_{0,1}(\xi) & \dots & Y_{m,1}(\xi) & \dots & Y_{m,2m+1}(\xi) \\ Y_{0,1}(\eta_2) & \dots & Y_{m,1}(\eta_2) & \dots & Y_{m,2m+1}(\eta_2) \\ \vdots & & \vdots & & \vdots \\ Y_{0,1}(\eta_M) & \dots & Y_{m,1}(\eta_M) & \dots & Y_{m,2m+1}(\eta_M) \end{vmatrix}}{\begin{vmatrix} Y_{0,1}(\eta_1) & \dots & Y_{m,1}(\eta_1) & \dots & Y_{m,2m+1}(\eta_1) \\ Y_{0,1}(\eta_2) & \dots & Y_{m,1}(\eta_2) & \dots & Y_{m,2m+1}(\eta_2) \\ \vdots & & \vdots & & \vdots \\ Y_{0,1}(\eta_M) & \dots & Y_{m,1}(\eta_M) & \dots & Y_{m,2m+1}(\eta_M) \end{vmatrix}},$$

$$L_2(\xi) = \frac{\begin{vmatrix} Y_{0,1}(\eta_1) & \dots & Y_{m,1}(\eta_1) & \dots & Y_{m,2m+1}(\eta_1) \\ Y_{0,1}(\xi) & \dots & Y_{m,1}(\xi) & \dots & Y_{m,2m+1}(\xi) \\ \vdots & & \vdots & & \vdots \\ Y_{0,1}(\eta_M) & \dots & Y_{m,1}(\eta_M) & \dots & Y_{m,2m+1}(\eta_M) \end{vmatrix}}{\begin{vmatrix} Y_{0,1}(\eta_1) & \dots & Y_{m,1}(\eta_1) & \dots & Y_{m,2m+1}(\eta_1) \\ Y_{0,1}(\eta_2) & \dots & Y_{m,1}(\eta_2) & \dots & Y_{m,2m+1}(\eta_2) \\ \vdots & & \vdots & & \vdots \\ Y_{0,1}(\eta_M) & \dots & Y_{m,1}(\eta_M) & \dots & Y_{m,2m+1}(\eta_M) \end{vmatrix}},$$

etc. satisfy $L_j(\eta_k) = \delta_{jk}$, $j, k = 1, ..., M$. Moreover, we have

$$|L_1(\xi)|^2 = \frac{\begin{vmatrix} K_{Harm_{0,\dots,m}}(\xi,\xi) & K_{Harm_{0,\dots,m}}(\xi,\eta_2) & \dots & K_{Harm_{0,\dots,m}}(\xi,\eta_M) \\ \vdots & \vdots & & \vdots \\ K_{Harm_{0,\dots,m}}(\eta_M,\xi) & K_{Harm_{0,\dots,m}}(\eta_M,\eta_2) & \dots & K_{Harm_{0,\dots,m}}(\eta_M,\eta_M) \end{vmatrix}}{\begin{vmatrix} K_{Harm_{0,\dots,m}}(\eta_1,\eta_1) & K_{Harm_{0,\dots,m}}(\eta_1,\eta_2) & \dots & K_{Harm_{0,\dots,m}}(\eta_1,\eta_M) \\ \vdots & \vdots & & \vdots \\ K_{Harm_{0,\dots,m}}(\eta_M,\eta_1) & K_{Harm_{0,\dots,m}}(\eta_M,\eta_2) & \dots & K_{Harm_{0,\dots,m}}(\eta_M,\eta_M) \end{vmatrix}}$$

$$|L_2(\xi)|^2 = \frac{\begin{vmatrix} K_{Harm_{0,\dots,m}}(\eta_1,\eta_1) & K_{Harm_{0,\dots,m}}(\eta_1,\xi) & \dots & K_{Harm_{0,\dots,m}}(\eta_1,\eta_M) \\ \vdots & \vdots & & \vdots \\ K_{Harm_{0,\dots,m}}(\eta_M,\eta_1) & K_{Harm_{0,\dots,m}}(\eta_M,\xi) & \dots & K_{Harm_{0,\dots,m}}(\eta_M,\eta_M) \end{vmatrix}}{\begin{vmatrix} K_{Harm_{0,\dots,m}}(\eta_1,\eta_1) & K_{Harm_{0,\dots,m}}(\eta_1,\eta_2) & \dots & K_{Harm_{0,\dots,m}}(\eta_1,\eta_M) \\ \vdots & \vdots & & \vdots \\ K_{Harm_{0,\dots,m}}(\eta_M,\eta_1) & K_{Harm_{0,\dots,m}}(\eta_M,\eta_2) & \dots & K_{Harm_{0,\dots,m}}(\eta_M,\eta_M) \end{vmatrix}}$$

etc. We are interested in the interpolation operator

$$T_{Harm_{0,\dots,m}} : \mathcal{C}(\Omega) \to Harm_{0,\dots,m} \tag{7.2.7}$$

defined by

$$T_{Harm_{0,\dots,m}}(F) = \sum_{j=1}^{M} F(\eta_j) L_j \tag{7.2.8}$$

for $F \in \mathcal{C}(\Omega)$. $Harm_{0,\dots,m}$ is finite-dimensional, so there is some member $Y \in Harm_{0,\dots,m}$ with

$$\|F - Y\|_{\mathcal{C}(\Omega)} = \inf_{P \in Harm_{0,\dots,m}} \|F - P\|_{\mathcal{C}(\Omega)} \tag{7.2.9}$$

(see for example G.G. LORENTZ (1966)). Furthermore, it is not difficult to see that $T_{Harm_{0,\dots,m}} Y = Y$. Now assume that $\|F\|_{\mathcal{C}(\Omega)} \leq 1$. Then we obtain

$$|T_{Harm_{0,\dots,m}}(F)(\xi)| \leq L(\xi), \quad \xi \in \Omega, \tag{7.2.10}$$

where the *Lebesgue function* $L \in \mathcal{C}(\Omega)$ is given by

$$L(\xi) = \sum_{j=1}^{M} |L_j(\xi)|, \quad \xi \in \Omega. \tag{7.2.11}$$

L attains its maximum value, i.e. there exists a point $\xi_0 \in \Omega$ with

$$| T_{Harm_{0,\dots,m}}(F)(\xi_0) | \leq L(\xi_0) = \max_{\xi \in \Omega} | L(\xi) | . \tag{7.2.12}$$

We let

$$\varepsilon_j = \operatorname{sign} L_j(\xi_0) \in \{-1, 0, +1\} \tag{7.2.13}$$

for $j = 1, \dots, M$ and choose $d \in \mathbb{R}$ such that

$$0 < d < \frac{1}{2} \min \left(|\eta_j - \eta_k|, j, k \in \{1, \dots, M\}, j \neq k\right). \tag{7.2.14}$$

Then there exists a function $H \in \mathcal{C}(\Omega)$ satisfying

$$H(\xi) = \begin{cases} \varepsilon_j(1 - (1/d)|\xi - \eta_j|) & \text{for } |\xi - \eta_j| \leq d, \quad j = 1, \dots, M, \\ 0 & \text{otherwise.} \end{cases} \tag{7.2.15}$$

Clearly, $\|H\|_{\mathcal{C}(\Omega)} \leq 1$. Thus it follows that

$$\|T_{Harm_{0,\dots,m}}\|_{\mathcal{C}(\Omega)} \geq \|T_{Harm_{0,\dots,m}}(H)\|_{\mathcal{C}(\Omega)} \tag{7.2.16}$$

$$
\begin{aligned}
&\geq T_{Harm_{0,\ldots,m}}(H)(\xi_0) && (7.2.17)\\
&= \sum_{j=1}^{M} \varepsilon_j L_j(\xi_0) && (7.2.18)\\
&= L(\xi_0). && (7.2.19)
\end{aligned}
$$

This shows that

$$
\|T_{Harm_{0,\ldots,m}}\|_{\mathcal{C}(\Omega)} = \max_{\xi\in\Omega} \sum_{j=1}^{M} |L_j(\xi)|. \tag{7.2.20}
$$

$\|T_{Harm_{0,\ldots,m}}\|_{\mathcal{C}(\Omega)}$ is called the *Lebesgue constant.* It follows that

$$
\begin{aligned}
\|F - T_{Harm_{0,\ldots,m}}(F)\|_{\mathcal{C}(\Omega)} &= \|(F-Y) + T_{Harm_{0,\ldots,m}}(Y-F)\|_{\mathcal{C}(\Omega)}\\
&\leq \left(1 + \|T_{Harm_{0,\ldots,m}}\|_{\mathcal{C}(\Omega)}\right) \|F - Y\|_{\mathcal{C}(\Omega)}.
\end{aligned}
$$

The determination of the *Lebesgue constant* is an important problem in interpolation (cf. D.L. RAGOZIN (1972)). In general, the explicit calculation of $\|T_{Harm_{0,\ldots,m}}\|_{\mathcal{C}(\Omega)}$ becomes a severe problem. Therefore it is of high value to know upper bounds. A first step is an easy, but reasonable estimate obtained by the Cauchy–Schwarz inequality, namely

$$
\|T_{Harm_{0,\ldots,m}}\|^2_{\mathcal{C}(\Omega)} \leq M \max_{\xi\in\Omega} \sum_{j=1}^{M} |L_j(\xi)|^2. \tag{7.2.21}
$$

In other words we are led to the question of finding an estimate for the Lagrangian square sum.

Lemma 7.2.1. *Let X_M be a fundamental system relative to $Harm_{0,\ldots,m}$. Then*

$$
\frac{M}{4\pi} \cdot \delta_{\max}^{-1} \leq \sum_{j=1}^{M} \mid L_j(\xi) \mid^2 \leq \frac{M}{4\pi} \cdot \delta_{\min}^{-1} \tag{7.2.22}
$$

holds for all $\xi \in \Omega$.

Proof. Let $l(\xi) = (L_1(\xi), .., L_M(\xi))^T, \xi \in \Omega$. Furthermore, define $u(\xi) = (U_1(\xi), ..., U_M(\xi))^T$ by

$$
u(\xi) = \mathbf{q}\ l(\xi), \quad \xi \in \Omega. \tag{7.2.23}
$$

Then, for $j = 1, ..., M$, U_j is an element of $Harm_{0,\ldots,m}$, and it follows that

$$
\sum_{j=1}^{M} |U_j(\xi)|^2 = \sum_{j=1}^{M} |L_j(\xi)|^2, \quad \xi \in \Omega. \tag{7.2.24}
$$

Moreover, for any pair $(\xi, \eta) \in \Omega^2$, we have

$$K_{Harm_{0,\ldots,m}}(\xi,\eta) = \sum_{j=1}^{M}\sum_{k=1}^{M} L_j(\xi) K_{Harm_{0,\ldots,m}}(\eta_j,\eta_k) L_k(\eta) \tag{7.2.25}$$

so that

$$K_{Harm_{0,\ldots,m}}(\xi,\eta) = l^T(\xi)\ \mathbf{k}\ l(\eta) = u^T(\xi)\ \mathbf{d}\ u(\eta). \tag{7.2.26}$$

But this means that

$$K_{Harm_{0,\ldots,m}}(\xi,\eta) = \sum_{j=1}^{M} \delta_j U_j(\xi) U_j(\eta), \quad \xi,\eta \in \Omega \tag{7.2.27}$$

and

$$K_{Harm_{0,\ldots,m}}(\xi,\xi) = \sum_{j=1}^{M} \delta_j |U_j(\xi)|^2 = \frac{M}{4\pi}, \quad \xi \in \Omega. \tag{7.2.28}$$

Consequently, in connection with (7.2.6), we obtain

$$\delta_1 \sum_{j=1}^{M} |L_j(\xi)|^2 \le \frac{M}{4\pi} \le \delta_M \sum_{j=1}^{M} |L_j(\xi)|^2. \tag{7.2.29}$$

This is the required result. □

From estimate (7.2.21) we get as immediate consequences the following lemmata.

Lemma 7.2.2. *Suppose that X_M is a fundamental system relative to $Harm_{0,\ldots,m}$. Then*

$$\|T_{Harm_{0,\ldots,m}}\|_{C(\Omega)} \le \frac{M^2}{4\pi}\delta_{\min}^{-1}. \tag{7.2.30}$$

Lemma 7.2.3. *Suppose that X_M is a fundamental system relative to $Harm_{0,\ldots,m}$. Then $\delta_{\min} = \delta_{\max}$ (i.e. $\delta_1 = \ldots = \delta_M$) if and only if*

$$\max_{\eta\in\Omega} \sum_{j=1}^{M} |L_j(\eta)|^2 = \sum_{j=1}^{M} |L_j(\xi)|^2 = 1, \quad \xi \in \Omega.$$

An easy consequence is

Lemma 7.2.4. *Let $\delta_{\min} = \delta_{\max}$ (i.e. $\delta_1 = \ldots = \delta_M$). Then*

$$\|T_{Harm_{0,\ldots,m}}\|_{\mathcal{C}(\Omega)} \le \sqrt{M}.$$

Next we give a charaterization of $\mathbf{k}^{-1}$ and its associated greatest eigenvalue $\delta_{\min}^{-1}$ in the context of the Lagrangians $L_1, ..., L_m$. We start with the fact that

$$\begin{aligned} \delta_{jk} &= L_k(\eta_j) = \int_\Omega K_{Harm_{0,\ldots,m}}(\eta_j,\xi) L_k(\xi)\, d\omega(\xi) \qquad (7.2.31) \\ &= \int_\Omega \sum_{p=1}^M K_{Harm_{0,\ldots,m}}(\eta_j,\eta_p) L_p(\xi) L_k(\xi)\, d\omega(\xi), \end{aligned}$$

i.e.

$$\sum_{p=1}^M K_{Harm_{0,\ldots,m}}(\eta_j,\eta_p)(L_p, L_k)_{\mathcal{L}^2(\Omega)} = \delta_{jk}. \qquad (7.2.32)$$

Hence we have verified that

$$\mathbf{k}^{-1} = ((L_j, L_k)_{\mathcal{L}^2(\Omega)})_{j,k=1,\ldots,M}. \qquad (7.2.33)$$

Especially, if $\delta_{\max} = \delta_{\min}$, we obtain

$$(L_j, L_j)_{\mathcal{L}^2(\Omega)} = \frac{4\pi}{M} \qquad (7.2.34)$$

for $j = 1, ..., M$. Moreover, from (7.2.33), we have

$$\delta_{\min}^{-1} = \max_{F \in Harm_{0,\ldots,m}, \sum_{j=1}^M |F(\eta_j)|^2 > 0} \left\{ \frac{\int_\Omega |F(\eta)|^2\, d\omega(\eta)}{\sum_{j=1}^M |F(\eta_j)|^2} \right\}. \qquad (7.2.35)$$

We know that

$$\mathbf{q} = \mathbf{matr}_{X_M}(U_1, ..., U_M), \qquad (7.2.36)$$

and the functions $U_1, ..., U_M$ constitute a basis of $Harm_{0,\ldots,m}$. The reproducing property of $K_{Harm_{0,\ldots,m}}(\cdot,\cdot)$ implies that

$$U_k(\eta) = \int_\Omega K_{Harm_{0,\ldots,m}}(\xi,\eta)\; U_k(\xi)\, d\omega(\xi), \quad \eta \in \Omega. \qquad (7.2.37)$$

Inserting (7.2.27) in (7.2.37) we get

$$U_k(\eta) = \sum_{j=1}^M \delta_j (U_j, U_k)_{\mathcal{L}^2(\Omega)} U_j(\eta), \quad \eta \in \Omega. \qquad (7.2.38)$$

This shows that

$$\delta_j (U_j, U_k)_{\mathcal{L}^2(\Omega)} = \delta_{jk} \tag{7.2.39}$$

for $j, k = 1, ..., M$. Thus, the functions

$$V_j = \sqrt{\delta_j}\, U_j, \quad j = 1, ..., M, \tag{7.2.40}$$

constitute an orthonormal basis in $Harm_{0,\ldots,m}$ such that the matrix

$$\mathbf{v} \; = \; \mathbf{matr}_{X_M}(V_1, ..., V_M) \tag{7.2.41}$$

satisfies the properties

$$\mathbf{v}^T\mathbf{v} = \mathbf{k}\;, \qquad \mathbf{v}\mathbf{v}^T = \mathbf{d}. \tag{7.2.42}$$

Finally the case of equal eigenvalues of $\mathbf{k}$ (i.e. $\delta_{\max} = \delta_{\min} = M/4\pi$) should be discussed in more detail. Under this assumption we obtain

$$K_{Harm_{0,\ldots,m}}(\xi, \eta) \;=\; \frac{M}{4\pi} \sum_{j=1}^{M} L_j(\xi) L_j(\eta), \tag{7.2.43}$$

so that

$$\begin{aligned} L_k(\eta) \;&=\; \int_\Omega K_{Harm_{0,\ldots,m}}(\xi, \eta) L_k(\xi)\, d\omega(\xi) \\ &=\; \frac{M}{4\pi} \sum_{j=1}^{M} (L_j, L_k)_{\mathcal{L}^2(\Omega)} L_j(\eta). \end{aligned} \tag{7.2.44}$$

Therefore it follows that

$$\frac{M}{4\pi}(L_j, L_k)_{\mathcal{L}^2(\Omega)} = \delta_{jk} \tag{7.2.45}$$

for $j, k = 1, ..., M$. Moreover,

$$(L_j, K_{Harm_{0,\ldots,m}}(\eta_k, \cdot))_{\mathcal{L}^2(\Omega)} = L_j(\eta_k) = 0 \tag{7.2.46}$$

for $k \neq j$. This leads us to the following result.

Theorem 7.2.5. *The identity*

$$\sum_{j=1}^{M} |L_j(\xi)|^2 = 1, \quad \xi \in \Omega$$

holds if and only if X_M is a fundamental system relative to $Harm_{0,\ldots,m}$, and the corresponding Lagrangians constitute an orthogonal basis for $Harm_{0,\ldots,m}$, where

$$L_j = \frac{4\pi}{M}\; K_{Harm_{0,\ldots,m}}(\eta_j, \cdot)$$

and

$$(L_j, L_k)_{\mathcal{L}^2(\Omega)} \;=\; \frac{4\pi}{M}\delta_{jk}.$$

Let X_M be a fundamental system relative to $Harm_{0,\ldots,m}$. Consider the approximate formula

$$\int_\Omega Y(\xi)F(\xi)\,d\omega(\xi) = \sum_{j=1}^{M} a_j F(\eta_j) + R^{a_1,\ldots,a_M}_{Harm_{0,\ldots,m}}(F), \tag{7.2.47}$$

where Y is an element of $Harm_{0,\ldots,m}$ and $a_1,\ldots,a_M$ are given by

$$a_j = \int_\Omega Y(\eta)L_j(\eta)\,d\omega(\eta), \quad j=1,\ldots,M. \tag{7.2.48}$$

Then, from Theorem 5.3.3, we obtain for all pseudodifferential operators with summable symbol $\{\Lambda^\wedge(n)\}$ and for all $F \in \mathcal{H}$

$$\begin{aligned} &R^{a_1,\ldots,a_M}_{Harm_{0,\ldots,m}}(F) \\ &= -\sum_{k=1}^{M} \int_\Omega Y(\eta)L_k(\eta)\,d\omega(\eta) \int_\Omega \Lambda_\eta K_{\mathcal{H}^\perp_{0,\ldots,m}(\{\Lambda^\wedge(n)\};\Omega)}(\eta_k,\eta)(\Lambda_\eta F(\eta))\,d\omega(\eta). \end{aligned}$$

Consequently we have

$$R^{a_1,\ldots,a_M}_{Harm_{0,\ldots,m}}(P) = 0 \tag{7.2.49}$$

for all $P \in Harm_{0,\ldots,m}$. The Cauchy–Schwarz inequality yields the following result.

Lemma 7.2.6. *For all $Y \in Harm_{0,\ldots,m}$ and $F \in \mathcal{C}^{(2m+2)}(\Omega)$*

$$\begin{aligned} &\left|R^{a_1,\ldots,a_M}_{Harm_{0,\ldots,m}}(F)\right| \\ &\le \left(\max_{\xi\in\Omega}\sum_{k=1}^{M}|L_k(\xi)|^2\right)^{1/2} \\ &\quad\times\left(\sum_{n=m+1}^{\infty}\frac{2n+1}{4\pi}\frac{1}{|\Lambda^\wedge(n)|^2}\|Y\|_{\mathcal{L}^2(\Omega)}\right)^{1/2}\|\Lambda F\|_{\mathcal{L}^2(\Omega)}. \end{aligned}$$

In particular, for $Y=1$, we obtain

$$\sum_{j=1}^{M} a_j Y_{n,k}(\eta_j) = 4\pi\delta_{n0} \quad \text{for } n=0,\ldots,m, \quad k=1,\ldots,2n+1.$$

Furthermore, we have

$$a_j = \int_\Omega L_j(\eta)\,d\omega(\eta) = \sum_{k=1}^{M}(L_j,L_k)_{\mathcal{L}^2(\Omega)} \tag{7.2.50}$$

for $j = 1, ..., M$, hence,

$$\sum_{j=1}^{M} a_j = \sum_{j=1}^{M}\sum_{k=1}^{M} (L_j, L_k)_{\mathcal{L}^2(\Omega)} = 4\pi. \tag{7.2.51}$$

The weights a_j are the row sums in the matrix $\mathbf{k}^{-1}$. In the special case $\delta_{\max} = \delta_{\min}$ we obtain from Theorem 7.2.5 that

$$a_j = \frac{4\pi}{M} \int_\Omega K_{Harm_{0,\ldots,m}}(\eta_j, \eta)\, d\omega(\eta) = \frac{4\pi}{M} > 0 \tag{7.2.52}$$

for $j = 1, ..., M$.

7.3 Best Approximations

Consider a pseudodifferential operator Λ with summable symbol $\{\Lambda^\wedge(n)\}$. Let us assume that $X_N = \{\eta_1, ..., \eta_N\}$ is an admissible system relative to $Harm_{0,\ldots,m}$. Let $\mathcal{I}$ be a bounded linear functional on $\mathcal{H}(\{\Lambda^\wedge(n)\};\Omega)$. An approximation $\mathcal{L}$ of the form

$$\mathcal{L}F = \sum_{k=1}^{N} a_k F(\eta_k), \tag{7.3.1}$$

exact for the degree m, is called a *best approximation* to $\mathcal{I}$, if, within the class of such approximations, there is one for which the quantity J (cf. Theorem 5.4.2) given by

$$J = (\mathcal{I}_\xi \mathcal{I}_\zeta - 2\mathcal{I}_\xi \mathcal{L}_\zeta + \mathcal{L}_\xi \mathcal{L}_\zeta) K_{\mathcal{H}^\perp_{0,\ldots,m}(\{\Lambda^\wedge(n)\};\Omega)}(\xi, \zeta) \tag{7.3.2}$$

assumes its minimum. The minimum of J can be obtained by solving the uniquely determined quadratic optimization problem

$$\begin{aligned} &\frac{1}{2}\mathcal{I}_\xi \mathcal{I}_\zeta K_{\mathcal{H}^\perp_{0,\ldots,m}(\{\Lambda^\wedge(n)\};\Omega)}(\xi, \zeta) \\ &- \sum_{k=1}^{N} a_k \mathcal{I}_\xi K_{\mathcal{H}^\perp_{0,\ldots,m}(\{\Lambda^\wedge(n)\};\Omega)}(\xi, \eta_k) \\ &+ \frac{1}{2}\sum_{k=1}^{N}\sum_{r=1}^{N} K_{\mathcal{H}^\perp_{0,\ldots,m}(\{\Lambda^\wedge(n)\};\Omega)}(\eta_r, \eta_k) a_k a_r \to \min. \end{aligned} \tag{7.3.3}$$

under the M-constraints

$$\sum_{k=1}^{N} a_k Y_{p,q}(\eta_k) = \mathcal{I}Y_{p,q}, \quad p = 0, ..., m, \quad q = 1, ..., 2p+1. \tag{7.3.4}$$

According to Lagrange's method of multipliers there exists a vector $v \in \mathbb{R}^M$, $v = (v_{0,1}, ..., v_{m,1}, ..., v_{m,2m+1})^T$, such that

$$\begin{aligned} &\frac{\partial}{\partial a_k}\left(\frac{1}{2}\mathcal{I}_\xi \mathcal{I}_\zeta K_{\mathcal{H}^\perp_{0,\dots,m}(\{\Lambda^\wedge(n)\};\Omega)}(\xi,\zeta)\right) \\ &- \frac{\partial}{\partial a_k}\left(\sum_{k=1}^{N} a_k \mathcal{I}_\xi K_{\mathcal{H}^\perp_{0,\dots,m}(\{\Lambda^\wedge(n)\};\Omega)}(\xi,\eta_k)\right) \\ &+ \frac{\partial}{\partial a_k}\left(\frac{1}{2}\sum_{k=1}^{N}\sum_{r=1}^{N} K_{\mathcal{H}^\perp_{0,\dots,m}(\{\Lambda^\wedge(n)\};\Omega)}(\eta_r,\eta_k)\ a_k a_r\right) \\ &+ \frac{\partial}{\partial a_k}\left(\sum_{p=0}^{m}\sum_{q=1}^{2p+1} v_{p,q}\left(\sum_{k=1}^{N} a_k Y_{p,q}(\eta_k) - \mathcal{I}Y_{p,q}\right)\right) = 0, \quad k = 1, ..., N. \end{aligned} \tag{7.3.5}$$

In connection with the constraints (7.3.4) this leads us to the $(N+M)\times(N+M)$ linear system

$$\begin{aligned} \mathbf{g}a + \mathbf{h}^T v &= j, \\ \mathbf{h}a &= i, \end{aligned} \tag{7.3.6}$$

and $\mathbf{g}$, i, j are given as follows:

$$\mathbf{g} = \mathbf{matr}_{X_N}(K_{\mathcal{H}^\perp_{0,\dots,m}(\{\Lambda^\wedge(n)\};\Omega)}(\eta_1,\cdot), \dots, K_{\mathcal{H}^\perp_{0,\dots,m}(\{\Lambda^\wedge(n)\};\Omega)}(\eta_N,\cdot)), \tag{7.3.7}$$

and

$$j = (j_1, ..., j_N)^T, \ j_k = \mathcal{I}_\xi K_{\mathcal{H}^\perp_{0,\dots,m}(\{\Lambda^\wedge(n)\};\Omega)}(\xi,\eta_k), \tag{7.3.8}$$

$$i = (i_{0,1}, ..., i_{m,1}, ..., i_{m,2m+1})^T, \quad i_{p,q} = \mathcal{I}Y_{p,q}. \tag{7.3.9}$$

Observe that

$$\begin{aligned} &K_{\mathcal{H}^\perp_{0,\dots,m}(\{\Lambda^\wedge(n)\};\Omega)}(\eta_r,\eta_s) \\ &= \int_\Omega \Lambda_\eta K_{\mathcal{H}^\perp_{0,\dots,m}(\{\Lambda^\wedge(n)\};\Omega)}(\eta_r,\eta)\Lambda_\eta K_{\mathcal{H}^\perp_{0,\dots,m}(\{\Lambda^\wedge(n)\};\Omega)}(\eta,\eta_s)\,d\omega(\eta) \end{aligned} \tag{7.3.10}$$

for $r, s = 1, ..., N$. Hence, $\mathbf{g}$ is a $N \times N$ matrix of Gram type. Since the functions $K_{\mathcal{H}^\perp_{0,\dots,m}(\{\Lambda^\wedge(n)\};\Omega)}(\eta_r,\cdot)$, $r = 1, ..., N$ are linearly independent, it follows that $\det \mathbf{g} > 0$ (cf. P.J. Davis (1963)), and the matrix $\mathbf{g}$ is positive definite. It is a remarkable fact that the $(N+M) \times (N+M)$ matrix

$$\begin{pmatrix} \mathbf{g} & \mathbf{h}^T \\ \mathbf{h} & \mathbf{0} \end{pmatrix} \tag{7.3.11}$$

has already been considered in spline interpolation with degree of approximation m. We know therefore from Lemma 6.1.10 that (7.3.11) is non-singular. The unique solution of the linear system (7.3.6) is given by

$$a = \mathbf{g}^{-1}(j + \mathbf{h}^T v), \tag{7.3.12}$$

$$v = -(\mathbf{h}\,\mathbf{g}^{-1}\mathbf{h}^T)^{-1}(\mathbf{h}\,\mathbf{g}^{-1}j - i). \tag{7.3.13}$$

The vector $a \in \mathbb{R}^N$, $a = (a_1, ..., a_N)^T$ in (7.3.13) is the only N-tuple for which the quadratic optimization problem (7.3.3) is minimal under the constraints (7.3.4).

Summarizing our results developed above we obtain the following result.

Theorem 7.3.1. *Let $\mathcal{I}$ be a bounded linear functional on $\mathcal{H}(\{\Lambda^\wedge(n)\};\Omega)$. Let $(a_1, ..., a_N)^T$, $(v_{0,1}, ..., v_{m,1}, ..., v_{m,2m+1})^T$ be the solution of the linear system (7.3.6). Then, for every $F \in \mathcal{H}(\{\Lambda^\wedge(n)\};\Omega)$, the linear functional $\mathcal{L}$ of type (7.3.1) represents the best approximation to $\mathcal{I}$. The approximation formula*

$$\mathcal{I}F \simeq \sum_{k=1}^{N} a_k F(\eta_k)$$

has the a posteriori estimate

$$\left| \mathcal{I}F - \sum_{k=1}^{N} a_k F(\eta_k) \right|$$

$$\leq \left(\mathcal{I}_\xi \mathcal{I}_\zeta K_{\mathcal{H}^\perp_{0,\dots,m}(\{\Lambda^\wedge(n)\};\Omega)}(\xi,\zeta) - \sum_{k=1}^{N} a_k \mathcal{I}_\xi K_{\mathcal{H}^\perp_{0,\dots,m}(\{\Lambda^\wedge(n)\};\Omega)}(\xi,\eta_k) \right.$$

$$\left. + \sum_{n=0}^{m} \sum_{j=1}^{2n+1} v_{n,j} \mathcal{I} Y_{n,j} \right)^{1/2} \left(\int_\Omega |\, \Lambda_\eta F(\eta) \,|^2 \, d\omega(\eta) \right)^{1/2}.$$

In particular, for the linear functional $\mathcal{I}$ given by

$$\mathcal{I}F = F^\wedge(n,j) = \int_\Omega F(\eta) Y_{n,j}(\eta)\, d\omega(\eta), \quad (n, j \text{ fixed}, \; n \leq m), \tag{7.3.14}$$

we have the *a posterori estimate*

$$\left| \int_\Omega F(\eta) Y_{n,j}(\eta)\, d\omega(\eta) - \sum_{k=1}^{N} a_k F(\eta_k) \right| \leq \sqrt{v_{n,j}} \sqrt{\int_\Omega |\Lambda_\eta F(\eta)|^2 \, d\omega(\eta)}.$$

For $n = 0$ we obtain

Corollary 7.3.2.

$$\left| \frac{1}{\sqrt{4\pi}} \int_\Omega F(\eta)\, d\omega(\eta) - \sum_{k=1}^{N} a_k F(\eta_k) \right| \leq \sqrt{v_{0,1}} \sqrt{\int_\Omega |\Lambda_\eta F(\eta)|^2 \, d\omega(\eta)}.$$

Remark. An efficient method for solving the linear system (7.3.6) has been proposed by W. FREEDEN, R. REUTER (1982). A technique for preconditioning the system (7.3.6) has been discussed by N. DYN *et al.* (1986) and many others. An alternate method for solving the mentioned linear system by iteration has been presented, e.g. by R.E. BANK *et al.* (1990). Examples have been studied by W. FREEDEN (1979), W. FREEDEN, R. REUTER (1982).

7.4 Interpolatory Spline Integration

Let $X_M = \{\eta_1, ..., \eta_M\}$ be a fundamental system relative to $Harm_{0,...,m}$. Given real numbers $y_1, ..., y_M$ we are then able to interpolate with a unique $Y \in Harm_{0,...,m}$, i.e. $Y(\eta_k) = y_k$, $k = 1, ..., M$. For a given set $X_N = \{\eta_1, ..., \eta_N\}$ consisting of $N > M$ points on Ω, however, this does not hold true in general. Instead we have to replace polynomial interpolation by spline interpolation: we consider a pseudodifferential operator Λ, its symbol $\{\Lambda^\wedge(n)\}$ being summable with respect to $\mathcal{N}$ in the sense of Definition 5.7.1. We write $\mathcal{H} = \mathcal{H}^{\mathcal{N}}(\{\Lambda^\wedge(n)\}; \Omega)$ and assume, in addition, that $\{0, \ldots, m\} \subset \mathcal{N}$, and that $K_{\mathcal{H}}(\cdot, \cdot)$ is strictly positive definite of order m. Within the space $\mathcal{H}$ we look for the solution S_N^F of the spline interpolation problem

$$\|S_N^F\|_{\mathcal{H}^\perp_{0,\ldots,m}} = \inf_{G \in \mathcal{I}_N(y)} \|G\|_{\mathcal{H}^\perp_{0,\ldots,m}} \tag{7.4.1}$$

in the set $\mathcal{I}_N(y)$ of all $\mathcal{H}$-interpolants to the given data:

$$\mathcal{I}_N(y) = \{G \in \mathcal{H} \mid G(\eta_i) = F(\eta_i) = y_i, \ i = 1, ..., N\}. \tag{7.4.2}$$

As is well known, this procedure is reasonable because it produces an interpolating function S_N^F which is as close to being a member of $Harm_{0,...,m}$ as can be achieved for the data. In addition, the spline interpolant S_N^F is the 'smoothest' in the sense of (7.4.1) thereby avoiding that oscillations become too severe for interpolating the data. Moreover, ordinary interpolation by spherical harmonics (i.e. the previous case when $N = M$) may also be included since the uniquely defined interpolant in $Harm_{0,...,m}$ is evidently a solution of the problem (7.4.1). The explicit determination of S_N^F is known from Chapter 6.

For a given bounded linear functional $\mathcal{I}$ on $\mathcal{H}$ and an admissible system $X_N = \{\eta_1, \ldots, \eta_N\}$ relative to $Harm_{0,\ldots,m}$, an approximation $\mathcal{L}$ of the form

$$\mathcal{L}F = \sum_{k=1}^{N} a_k F(\eta_k), \quad F \in \mathcal{H} \tag{7.4.3}$$

is not uniquely determined by the requirement that $\mathcal{L}$ be exact for the degree m provided that $N > M$. However, as we shall see, $\mathcal{L}$ is in fact uniquely determined by the stronger requirement that $\mathcal{I}F = \mathcal{L}F$ whenever $F \in Spline_{0,\ldots,m}(\{\Lambda^\wedge(n)\}; X_N)$.

Theorem 7.4.1. *Let X_N be an admissible system relative to $Harm_{0,\ldots,m}$ and $\mathcal{I}$ be a bounded linear functional on $\mathcal{H}$. Then there is a unique functional $\mathcal{L}$ of the form (7.4.3) satisfying the properties*

(i) $\mathcal{I}Y = \mathcal{L}Y$ whenever $Y \in Harm_{0,\ldots,m}$

(ii) there exists a unique $P \in Harm_{0,\ldots,m}$ such that $(\mathcal{I}_\xi - \mathcal{L}_\xi)K_{\mathcal{H}^\perp_{0,\ldots,m}}(\xi, \eta_k) = P(\eta_k),\ k = 1, \ldots, N$.

For a given function $F \in \mathcal{H}$, let S_N^F denote the unique function of class $Spline_{0,\ldots,m}(\{\Lambda^\wedge(n)\}; X_N)$ that interpolates the data points $(\eta_i, F(\eta_i)) \in \Omega \times \mathbb{R}, i = 1, \ldots, N$. Then $\mathcal{L}$ is also uniquely determined by the property that $\mathcal{L}F = \mathcal{I}S_N^F$ whenever $F \in \mathcal{H}$. Finally, $\mathcal{L}$ is uniquely determined by the requirement that $\mathcal{L}F = \mathcal{I}F$ whenever $F \in Spline_{0,\ldots,m}(\{\Lambda^\wedge(n)\}; X_N)$.

Remark. The proof of Theorem 7.4.1 is similar to one-dimensional results due to I.J. SCHOENBERG (1964A) and T.N.E. GREVILLE (1969). An analogous proof appeared in W. FREEDEN (1981A,B, 1982B).

Proof. If $\mathcal{L}$ is a bounded linear functional on $\mathcal{H}(\{\Lambda^\wedge(n)\}; \Omega)$, the conditions (i), (ii) stated in the first part of Theorem 7.4.1 are equivalent to the system of $(N + M) \times (N + M)$ linear equations

$$\sum_{k=1}^{N} a_k K_{\mathcal{H}^\perp_{0,\ldots,m}(\{\Lambda^\wedge(n)\};\Omega)}(\eta_r, \eta_k) + P(\eta_r) = j_r, \quad r = 1, \ldots, N \tag{7.4.4}$$

$$\sum_{k=1}^{N} a_k Y_{n,j}(\eta_k) = i_{n,j}, \quad n = 0, \ldots, m, \quad j = 1, \ldots, 2n + 1, \tag{7.4.5}$$

for determining the polynomial P and the coefficients $a_1, \ldots, a_N$. Inspection of this linear system shows that the matrix of coefficients is exactly the non-singular matrix known from best approximate integration (cf. Section 7.3). Thus Y and $\mathcal{L}$ are uniquely determined by the equations (7.4.4) and (7.4.5).

We now prove that, for the approximation $\mathcal{L}$ so determined, the assumption $F \in Spline_{0,\ldots,m}(\{\Lambda^\wedge(n)\}; X_N)$ implies $\mathcal{L}F = \mathcal{I}F$.

Indeed, $F \in Spline_{0,\ldots,m}(\{\Lambda^\wedge(n)\}; X_N)$ implies that F is of the form

$$F = Y + \sum_{k=1}^{N} b_k K_{\mathcal{H}^{\perp}_{0,\ldots,m}}(\cdot, \eta_k), Y \in Harm_{0,\ldots,m} \tag{7.4.6}$$

with coefficients $b_1, \ldots, b_N$ satisfying

$$\sum_{k=1}^{N} b_k Y_{n,j}(\eta_k) = 0, \quad n = 0, \ldots, m, \quad j = 1, \ldots, 2n+1. \tag{7.4.7}$$

Hence,

$$\mathcal{I}F = \mathcal{I}Y + \sum_{k=1}^{N} b_k j_k. \tag{7.4.8}$$

Then, by (7.4.4) and (7.4.5), it follows that

$$\mathcal{I}F = \mathcal{L}Y + \sum_{r=1}^{N} b_r \left(\sum_{k=1}^{N} a_k K_{\mathcal{H}^{\perp}_{0,\ldots,m}}(\eta_r, \eta_k) + P(\eta_r) \right). \tag{7.4.9}$$

Reversing the order of summation and observing (7.4.7) we obtain

$$\mathcal{I}F = \mathcal{L}Y + \sum_{k=1}^{N} a_k \sum_{r=1}^{N} b_r K_{\mathcal{H}^{\perp}_{0,\ldots,m}(\{\Lambda^\wedge(n)\};\Omega)}(\eta_r, \eta_k), \tag{7.4.10}$$

i.e.

$$\mathcal{I}F = \mathcal{L}F. \tag{7.4.11}$$

Conversely, let $\mathcal{L}'$ be any functional of the form (7.4.3) having the property that $\mathcal{L}'S = \mathcal{I}S$ whenever $S \in Spline_{0,\ldots,m}(\{\Lambda^\wedge(n)\}; X_N)$. Then for $F \in \mathcal{H}$, $\mathcal{L}'S_N^F = \mathcal{I}S_N^F$ and, from the definition of S_N^F, $\mathcal{L}'S_N^F = \mathcal{L}'F$. Thus it follows that $\mathcal{L}'F = \mathcal{I}S_N^F$.

In particular, for $\mathcal{L}' = \mathcal{L}$, we have $\mathcal{L}F = \mathcal{I}S_N^F$, and therefore $\mathcal{L}'F = \mathcal{I}S_N^F = \mathcal{L}F$ whenever $F \in \mathcal{H}$. But this means that $\mathcal{L}' = \mathcal{L}$ on $\mathcal{H}$.

Thus, $\mathcal{L}$ is uniquely determined either by the assumption that $\mathcal{L}S = \mathcal{I}S$ whenever $S \in Spline_{0,\ldots,m}(\{\Lambda^\wedge(n)\}; X_N)$, or by the property $\mathcal{L}F = \mathcal{I}S_N^F$ whenever $F \in \mathcal{H}$. □

Finally we are able to prove that the best approximation $\mathcal{L}$ of the form (7.4.3), exact for the degree m, to a bounded linear functional $\mathcal{I}$ on $\mathcal{H}$ is precisely the unique approximation given by Theorem 7.4.1 that is exact for functions $F \in Spline_{0,\ldots,m}(\{\Lambda^\wedge(n)\}; X_N)$.

Theorem 7.4.2 *(Schoenberg's theorem). Suppose that $\mathcal{I}$ is a bounded linear functional on $\mathcal{H}$. Let $\mathcal{L}'$ be any approximation to $\mathcal{I}$ of the form (7.4.3), exact for the degree m, and let $\mathcal{L}$ denote the approximation given by Theorem 7.4.1. Let J and J' be the quantities given by (5.4.7) corresponding to $\mathcal{R} = \mathcal{I} - \mathcal{L}$ and $\mathcal{R}' = \mathcal{I} - \mathcal{L}'$, respectively. Then $J \leq J'$ with equality only if $\mathcal{L}' = \mathcal{L}$.*

Proof. Denote by $\mathcal{L}''$ a linear functional defined by $\mathcal{L}'' = \mathcal{L} - \mathcal{L}'$. By virtue of Theorem 7.4.1 and the assumption on $\mathcal{L}'$

$$\mathcal{L}F = \mathcal{I}F = \mathcal{L}'F \tag{7.4.12}$$

holds whenever $F \in Harm_{0,\ldots,m}$. Thus $\mathcal{L}''$ satisfies the assumption of Peano's theorem (Theorem 5.4.1), and for every $F \in \mathcal{H}$ we have

$$\mathcal{L}''F = \int_\Omega (\Lambda_\eta K''(\eta))(\Lambda_\eta F(\eta))\, d\omega(\eta), \tag{7.4.13}$$

where

$$K''(\eta) = \mathcal{L}''_\xi K_{\mathcal{H}^\perp_{0,\ldots,m}}(\xi, \eta), \quad \eta \in \Omega. \tag{7.4.14}$$

As difference of $\mathcal{L}$ and $\mathcal{L}'$ the functional $\mathcal{L}''$ is of the form (7.4.3). Hence, with coefficients $a_1, \ldots, a_N$ satisfying $\sum_{k=1}^N a_k Y_{n,j}(\eta_k) = 0, n = 0, \ldots, m, j = 1, \ldots, 2m+1$ it follows that

$$K''(\eta) = \sum_{k=1}^N a_k K_{\mathcal{H}^\perp_{0,\ldots,m}}(\eta_k, \eta), \quad \eta \in \Omega. \tag{7.4.15}$$

Now we consider the functional $\mathcal{R}$ given by $\mathcal{I} - \mathcal{L}$ and the functional $\mathcal{R}' = \mathcal{I} - \mathcal{L}'$ associated with $\mathcal{L}'$.

Correspondingly, let K and K' be the Peano kernels with respect to $\mathcal{R}$ and $\mathcal{R}'$. Then, because of $\mathcal{L}'' = \mathcal{L} - \mathcal{L}'$, it follows that

$$\mathcal{L}'' = (\mathcal{I} - \mathcal{R}) - (\mathcal{I} - \mathcal{R}') = \mathcal{R}' - \mathcal{R} \tag{7.4.16}$$

and by (7.4.14) and (7.4.16)

$$K'' = K' - K. \tag{7.4.17}$$

Furthermore, in light of (7.4.16) and (7.4.17),

$$\int_\Omega (\Lambda_\eta K(\eta))(\Lambda_\eta K''(\eta))\, d\omega(\eta) = \mathcal{R}_\xi \left(\sum_{k=1}^N a_k K_{\mathcal{H}^\perp_{0,\ldots,m}}(\xi, \eta_k) \right). \tag{7.4.18}$$

Thus, by Theorem 7.4.1 we get

$$\int_{\Omega} (\Lambda_\eta K(\eta))(\Lambda_\eta K''(\eta))\, d\omega(\eta) = \sum_{k=1}^{N} a_k P(\eta_k) = 0. \tag{7.4.19}$$

In view of (7.4.17) and (7.4.19) it is clear that

$$\int_{\Omega} |\Lambda_\eta K'(\eta)|^2\, d\omega(\eta) = \int_{\Omega} |\Lambda_\eta K(\eta)|^2\, d\omega(\eta) + \int_{\Omega} |\Lambda_\eta K''(\eta)|^2\, d\omega(\eta). \tag{7.4.20}$$

Therefore we have $J \leq J'$, where equality holds only if

$$\int_{\Omega} |\Lambda_\eta K''(\eta)|^2\, d\omega(\eta) = \sum_{k=1}^{N}\sum_{r=1}^{N} K_{\mathcal{H}^{\perp}_{0,\ldots,m}}(\eta_k, \eta_r) a_r a_k = 0. \tag{7.4.21}$$

But this means $a_1 = \ldots = a_N = 0$. □

Bibliographical Notes

Based on the integral formulae (cf. W. FREEDEN (1978, 1979, 1980B)) corresponding to iterated Beltrami operators on the unit sphere E. HLAWKA (1981) has presented a comprehensive text on the theory of equidistribution on the sphere. The formulation of the integral formulae corresponding to pseudodifferential operators with summable symbol is due to J. CUI, W. FREEDEN (1997), W. FREEDEN, U. WINDHEUSER (1997). Interpolatory integration on the sphere using spherical harmonics can be found in many textbooks (for example, M. REIMER (1990) and the references therein). Recently, J.R. DRISCOLL, D.H. HEALY (1994) have presented an algorithm (in exact arithmetic) for the computation of Fourier coefficients based on equidistant lattices (i.e. rectangular blocks which do not generate an equidistribution on the sphere in the sense of Weyl). A review paper on well-scattered points on the sphere is E.B. SAFF, A.B.J. KUIJLAARS (1997) (see also the references therein). Best approximation formulae on the sphere have been given in a series of papers by W. FREEDEN (1979, 1980B, 1981A). The role of spherical splines corresponding to Beltrami derivatives in best approximation formulae and the development of cubature formulae exact for spherical splines of given degree can be found in W. FREEDEN (1981A, 1982A, 1982C).

8

SPHERICAL SINGULAR INTEGRALS

We now study some important facts about spherical singular integrals. Since most of the material about singular integrals is rather standard our treatment will be brief. Three types of singular integrals on the sphere will be discussed in more detail, which are of basic interest in numerical applications. The first method is Abel summability (also called Abel–Poisson summability) whose analogue in Euclidean space $\mathbb{R}^n$ is well known (see for example E.M. STEIN, G. WEISS (1971)). A somewhat similar method is Gauß summability (also called Gauß–Weierstraß summability). We shall show, in particular, that Abel and Gauß means of continuous functions F converge in uniform sense to F. It should be pointed out that Abel–Poisson summability is intimately connected with harmonic functions and provides us with very powerful tools in potential theory. In the same way, the Gauß–Weierstraß method is intimately connected with the solution of the diffusion (heat) equation so that it plays an important role in multiscale descriptions of image processing. We also treat some typical methods of summability by means of axisymmetric 'locally supported' kernels.

8.1 Approximate Identities

The concept of convolution (cf. Definition 3.6.6) enables us to introduce *singular integrals* on the unit sphere (cf. H. BERENS *et al.* (1968)).

Definition 8.1.1. Let $\{K_h\}, h \in [h_0, +1), h_0 \in (-1, +1)$, be a subfamily of $\mathcal{L}^1[-1,+1]$ satisfying the condition $(K_h)^\wedge(0) = 1$, $h \in [h_0, 1)$. Then $\{I_h\}, h \in [h_0, 1)$, defined by

$$I_h(F) = K_h * F, \quad F \in \mathcal{X}(\Omega) \tag{8.1.1}$$

is called spherical *singular integral.* The family $\{K_h\}$ is called the *kernel* of the singular integral $\{I_h\}$. A singular integral $\{I_h\}$ is said to be an *approximate identity in* $\mathcal{X}(\Omega)$ if

$$\lim_{\substack{h \to 1 \\ h<1}} \|F - I_h(F)\|_{\mathcal{X}(\Omega)} = 0 \tag{8.1.2}$$

for all $F \in \mathcal{X}(\Omega)$.

Remark. Substituting $h = \mathrm{e}^{-\rho}, \rho \in (0, \infty)$, we may rewrite formula (8.1.2) in the form

$$\lim_{\substack{\rho \to 0 \\ \rho > 0}} \|F - T_\rho(F)\|_{\mathcal{X}(\Omega)} = 0, \quad F \in \mathcal{X}(\Omega),$$

where, by definition, $\Phi_\rho = K_{\mathrm{e}^{-\rho}}, T_\rho(F) = \Phi_\rho * F$ (cf. Chapter 10).

As fundamental result about singular integrals we first mention the following theorem.

Theorem 8.1.2. *Assume that the kernel* $\{K_h\}, h \in [h_0, +1), h_0 \in (-1, 1)$, *is uniformly bounded, i.e. there exists a positive constant* M *(independent of* h*) such that*

$$2\pi \int_{-1}^{+1} |K_h(t)|\, dt \leq M \tag{8.1.3}$$

for all $h \in [h_0, 1)$. *Then the corresponding singular integral* $\{I_h\}$ *is an approximate identity in* $\mathcal{X}(\Omega)$ *if and only if*

$$\lim_{\substack{h \to 1 \\ h < 1}} (K_h)^\wedge(n) = 1 \tag{8.1.4}$$

for all non-negative integers n.

Proof. Given an arbitrary $Y_n \in Harm_n$. Then we know that

$$\int_\Omega (K_h * Y_n)(\eta) Y_k(\eta)\, d\omega(\eta) = (K_h)^\wedge(k) \int_\Omega Y_n(\eta) Y_k(\eta)\, d\omega(\eta) \tag{8.1.5}$$

for all $Y_k \in Harm_k$. The orthogonality of spherical harmonics yields

$$\int_\Omega (K_h * Y_n)(\eta) Y_k(\eta)\, d\omega(\eta) = (K_h)^\wedge(n) \int_\Omega Y_n(\eta) Y_k(\eta)\, d\omega(\eta) \tag{8.1.6}$$

for all $Y_k \in Harm_k$. The completeness of the system of spherical harmonics gives us $K_h * Y_n = (K_h)^\wedge(n) Y_n$. Hence it is easily seen that (8.1.4) is equivalent to

$$\lim_{\substack{h \to 1 \\ h < 1}} \|Y_n - I_h(Y_n)\|_{\mathcal{X}(\Omega)} = 0, \quad Y_n \in Harm_n, \tag{8.1.7}$$

for all non-negative integers n. According to our assumptions the operators I_h are uniformly bounded on Ω. The assertion of theorem 8.1.2, therefore, follows from the Banach–Steinhaus theorem (see for example K. Yosida (1980)). □

If the kernel $\{K_h\}, K_h \in \mathcal{L}^1[-1,+1],\ h \in [h_0, 1), h_0 \in (-1,+1)$, is non-negative, i.e. K_h is non-negative almost everywhere for all $h \in [h_0, 1)$, then (8.1.3) is certainly satisfied (with $M = \sup_{h\in[h_0,1)} K_h^\wedge(0) < \infty$). This leads us to the following result.

Theorem 8.1.3. *Let $\{K_h\}, K_h \in \mathcal{L}^1[-1,+1], h \in [h_0,+1), h_0 \in (-1,+1)$, be a non-negative kernel of a singular integral. Then each of the following assertions is equivalent to (8.1.2) or (8.1.4):*

(i) The limit relation

$$\lim_{\substack{h\to 1\\ h<1}} (K_h)^\wedge(0) = 1 \tag{8.1.8}$$

holds.

(ii) The 'localization property'

$$\lim_{\substack{h\to 1\\ h<1}} \int_{-1}^{\delta} K_h(t)\,dt = 0, \quad \delta \in (-1,+1), \tag{8.1.9}$$

holds.

Proof. It is clear that (8.1.4) implies (i). If (i) holds, then it follows from the non-negativity of the kernel, that

$$\begin{aligned} 0 \le \int_{-1}^{\delta} K_h(t)\,dt &\le \frac{1}{1-\delta}\int_{-1}^{\delta} (1-t)K_h(t)\,dt \\ &\le \frac{1}{1-\delta}\int_{-1}^{+1} (1-t)K_h(t)\,dt \\ &\le \frac{1}{2\pi}\frac{1}{1\ \ \delta}(1-K_h^\wedge(1)). \end{aligned} \tag{8.1.10}$$

This gives (ii) as $h \to 1$. It remains to prove that (ii) implies (8.1.2). This can be done by observing the fact that, for $F \in \mathcal{X}(\Omega)$,

$$\begin{aligned} \|I_h(F) - F\|_{\mathcal{X}(\Omega)} &\le 2\pi \int_{-1}^{+1} \|T_t(F) - F\|_{\mathcal{X}(\Omega)} K_h(t)\,dt \\ &\le 4\pi \|F\|_{\mathcal{X}(\Omega)} \int_{-1}^{\delta} K_h(u)\,du + \sup_{\delta\le t\le 1} \|T_t(F) - F\|_{\mathcal{X}(\Omega)}, \end{aligned} \tag{8.1.11}$$

which via (3.6.16) implies (8.1.2). □

We shall call a non-negative kernel $\{K_h\}, K_h \in \mathcal{L}^1[-1,+1]$, satisfying one of the properties (8.1.2), (8.1.4), (8.1.8), (8.1.9), an *approximate identity generating $\mathcal{L}^1$-kernel.*

From Theorem 8.1.3 we immediately obtain the following.

Theorem 8.1.4. *Assume that $\{K_h\}, K_h \in \mathcal{L}^1[-1,+1], h \in [h_0,+1), h_0 \in (-1,1)$, is a non-negative kernel. Then $\{I_h^{(q)}\}$, $q = 1,2,...$, given by*

$$I_h^{(q)}(F) = K_h^{(q)} * F, \quad F \in \mathcal{X}(\Omega),$$

is an approximate identity in $\mathcal{X}(\Omega)$ if and only if $\{I_h\}$ is an approximate identity in $\mathcal{X}(\Omega)$. In other words, $\{K_h^{(q)}\}$ is an approximate identity generating $\mathcal{L}^1$-kernel for $q \in \mathbb{N}$ if and only if $\{K_h\}$ is an approximate identity generating $\mathcal{L}^1$-kernel.

Proof. According to (3.6.11) we obtain $(K_h^{(q)})^\wedge(n) = ((K_h)^\wedge(n))^q$ for $q = 1,2,\ldots$ and all non-negative integers n. Thus Theorem 8.1.4 follows from (8.1.4). □

Of particular interest for our applications are the first and second iterations (that are the cases $q = 1$ and $q = 2$).

Corollary 8.1.5. *Assume that $\{K_h\}$, $K_h \in \mathcal{L}^1[-1,+1]$, $h \in [h_0,+1), h_0 \in (-1,1)$, is a family of non-negative functions K_h satisfying the property (8.1.9). Then the limit relation (linear case for $q = 1$)*

$$\lim_{\substack{h\to 1\\ h<1}} \| \int_\Omega F(\zeta)K_h(\cdot\zeta)\, d\omega(\zeta) - F(\cdot)\|_{\mathcal{X}(\Omega)} = 0$$

and the limit relation (bilinear case for $q = 2$)

$$\begin{aligned} &\lim_{\substack{h\to 1\\ h<1}} \| \int_\Omega F(\zeta)K_h^{(2)}(\cdot\zeta)\, d\omega(\zeta) - F(\cdot)\|_{\mathcal{X}(\Omega)} \\ &= \lim_{\substack{h\to 1\\ h<1}} \| \int_\Omega \int_\Omega F(\zeta)K_h(\zeta\cdot\eta)\, d\omega(\zeta)K_h(\cdot\eta)\, d\omega(\eta) - F(\cdot)\|_{\mathcal{X}(\Omega)} \\ &= 0 \end{aligned}$$

hold for all $F \in \mathcal{X}(\Omega)$.

From Corollary 8.1.5 we see that $I_h^{(q)}(F)$ is close to F provided that h is close to the value 1. The right choice of the parameter h of an approximation $I_h^{(q)}(F)$, however, depends on a priori information about the function F which is usually not available. Thus there is a practical necessity of developing 'zooming-in techniques', although such an automatic approximation process is of more computational complexity. A first step are summability methods based on the ideas of Gabor and Toeplitz (cf. Chapter 9). Later on spherical wavelet methods come into play (cf. Chapters 10, 11).

8.2 Examples of Spherical Singular Integrals

The results about singular integrals on the sphere will be applied now to study certain important special cases. As a matter of fact we confine ourselves to three types of singular integrals on the sphere which are of basic interest in numerical applications and whose singular integral kernels are already known from Chapter 5. The first method is Abel–Poisson summability whose analogue in Euclidean spaces $\mathbb{R}^n$ is well–known (see for example E. STEIN, G. WEISS (1971)). A somewhat similar method is Gauß–Weierstraß summability. We shall show, in particular, that Abel–Poisson and Gauß–Weierstraß means converge in uniform sense. Both Abel–Poisson singular integral operators $A_{e^{-\rho}}$ and Gauß–Weierstraß singular integral operators G_ρ, $\rho \in (0,\infty)$ define a semigroup of contraction operators of class (C_0) on $\mathcal{X}(\Omega)$. Finally we treat some typical methods of summability on the sphere by means of 'locally supported' kernel functions. These singular integrals satisfy a 'finite contraction property'.

8.2.1 *Abel–Poisson Singular Integral*

The function $A_h(F) : \xi \mapsto A_h(F)(\xi)$, $\xi \in \Omega$, $F \in \mathcal{X}(\Omega)$, defined by

$$A_h(F)(\xi) = \int_\Omega Q_h(\xi\cdot\eta)F(\eta)\,d\omega(\eta), \quad \xi \in \Omega, \tag{8.2.1}$$

is called the *'Abel–Poisson mean'*, where h is a non-negative real number < 1, and $Q_h : [-1,+1] \to \mathbb{R}$ is the *Abel–Poisson kernel* (cf. (5.6.13)). The integral (8.2.1) may be rewritten as follows:

$$A_h(F)(\xi) = \sum_{n=0}^{\infty} \frac{2n+1}{4\pi} h^n \int_\Omega F(\eta)P_n(\xi\cdot\eta)\,d\omega(\eta) \tag{8.2.2}$$

so that

$$\int_\Omega A_h(F)(\xi)Y_n(\xi)\,d\omega(\xi) = h^n \int_\Omega F(\xi)Y_n(\xi)\,d\omega(\xi) \tag{8.2.3}$$

for all $Y_n \in Harm_n$, i.e. $(Q_h)^\wedge(n) = h^n, n = 0,1,\ldots$.

For all $F \in \mathcal{C}(\Omega)$ we have

$$\| A_h(F) \|_{\mathcal{C}(\Omega)} \leq \| F \|_{\mathcal{C}(\Omega)} . \tag{8.2.4}$$

More general,

$$\|A_h(F)\|_{\mathcal{X}(\Omega)} \leq \|F\|_{\mathcal{X}(\Omega)} \tag{8.2.5}$$

for all $F \in \mathcal{X}(\Omega)$. If F is Lipschitz-continuous with Lipschitz constant C_F (i.e. $F \in Lip(\Omega)$), then

$$\|A_h(F) - F\|_{\mathcal{C}(\Omega)} \leq \sqrt{2}(C_F + {h_0}^{-1}\|F\|_{\mathcal{C}(\Omega)})\sqrt{1-h} \tag{8.2.6}$$

for all $h \in (h_0, 1)$, $h_0 \in (0,1)$ fixed. If $F(\xi) > 0$ for all $\xi \in \Omega$, then $A_h(F)(\xi) > 0$ for all $\xi \in \Omega$. Moreover, because of the limit relation,

$$\lim_{h\to 1, h<1} (1-h)^{-1}(h^n - 1) = -n, \tag{8.2.7}$$

it is not difficult to show that the equations

$$-nF^\wedge(n,j) = G^\wedge(n,j), \quad n = 0, 1, \ldots, \quad j = 1, \ldots, 2n+1,$$

are equivalent to

$$\lim_{h\to 1, h<1} \left\| \frac{A_h(F) - F}{1-h} - G \right\|_{\mathcal{C}(\Omega)} = 0$$

provided that $F, G \in \mathcal{C}(\Omega)$. A similar result holds in $(\mathcal{X}(\Omega), \|\cdot\|_{\mathcal{X}(\Omega)})$. Furthermore, H. Berens *et al.* (1968) have shown that, for $F \in \mathcal{L}^p(\Omega), 1 < p < \infty$, the relation

$$\|A_h(F) - F\|_{\mathcal{L}^p(\Omega)} = O(1-h) \tag{8.2.8}$$

is equivalent to the fact there exists a function $G \in \mathcal{L}^p(\Omega)$ such that $G^\wedge(n,j) = -nF^\wedge(n,j), n = 0, 1, \ldots,\ j = 1, \ldots, 2n+1$. In other words, the 'saturation class' of the Abel–Poisson singular integral operators $\{A_{e^{-\rho}}\}$, $\rho \in (0, \infty)$, is given by

$$\mathcal{H}(\mathcal{L}^p(\Omega); -n) = \{F \in \mathcal{L}^p(\Omega) | \exists G \in \mathcal{L}^p(\Omega) : G^\wedge(n,j) = -nF^\wedge(n,j)\},$$

whereas the 'saturation order' of $\{A_{e^{-\rho}}\}, \rho \in (0, \infty)$ is $O(1 - e^{-\rho}), \rho \to 0$.

A classical problem involving the Abel–Poisson mean is the Dirichlet problem of potential theory corresponding to a spherical boundary. More explicitly, for given $F \in \mathcal{C}(\Omega)$, the function $V : \overline{\Omega_{\text{int}}} \to \mathbb{R}$ given by $V(x) = A_r(F)(\xi)$, $x = r\xi$, $\xi \in \Omega$, is the only solution of the interior

Dirichlet problem (i) $V \in \mathcal{C}^{(2)}(\Omega_{\text{int}}) \cap \mathcal{C}^{(0)}(\overline{\Omega_{\text{int}}})$ (ii) $\Delta V = 0$ in Ω_{int}, (iii) $V|\Omega = F$ (cf. Theorem 3.4.1). This is the reason why this kernel is particularly useful in the approximation of harmonic functions. By virtue of the maximum/minimum principle of potential theory we get for all $\xi \in \Omega$

$$\min_{\eta\in\Omega} F(\eta) \leq A_h(F)(\xi) \leq \max_{\eta\in\Omega} F(\eta), \quad F \in \mathcal{C}(\Omega). \tag{8.2.9}$$

Altogether we are able to conclude that $\{A_{e^{-\rho}}\}, \rho \in (0,\infty)$, forms a *semigroup of contraction operators of* (C_0) *on* $\mathcal{X}(\Omega)$, i.e. $\{A_{e^{-\rho}}\}$ satisfies the properties:

- $A_{e^{-\rho}}$ is a linear bounded operator mapping $\mathcal{X}(\Omega)$ into itself, and $A_{e^{-0}} = I$ (identity operator),
- $A_{e^{-(\rho_1+\rho_2)}} = A_{e^{-\rho_1}} A_{e^{-\rho_2}}$ for $0 \leq \rho_1, \rho_2 < \infty$,
- $\lim\limits_{\substack{\rho\to 0\\ \rho>0}} \|A_{e^{-\rho}}(F) - F\|_{\mathcal{X}(\Omega)} = 0$ for $F \in \mathcal{X}(\Omega)$,
- $\|A_{e^{-\rho}}(F)\|_{\mathcal{X}(\Omega)} \leq \|F\|_{\mathcal{X}(\Omega)}$ for $\rho \in (0,\infty)$ and $F \in \mathcal{X}(\Omega)$.

8.2.2 *Gauß–Weierstraß Singular Integral*

Next we discuss the *'Gauß–Weierstraß mean'*

$$G_\rho(F)(\xi) = \int_\Omega W_\rho(\xi\cdot\eta)F(\eta)\, d\omega(\eta), \quad \xi \in \Omega, \quad F \in \mathcal{X}(\Omega), \tag{8.2.10}$$

where ρ is a real positive parameter and $W_\rho : [-1,+1] \to \mathbb{R}, \rho > 0$, is the Gauß–Weierstraß kernel introduced in (5.6.26). $\{G_\rho\}_{\rho>0}$ forms a singular integral of type (8.1.1) by setting $h = 1 - \rho$. Moreover, we have

$$\|G_\rho(F)\|_{\mathcal{X}(\Omega)} \leq \|F\|_{\mathcal{X}(\Omega)}$$

for all $F \in \mathcal{X}(\Omega)$. Thus $G_\rho : \mathcal{X}(\Omega) \to \mathcal{X}(\Omega)$ defines a bounded linear operator for every $\rho \in (0,\infty)$ such that $G_{\rho+\rho'} = G_\rho G_{\rho'}$. That is the set of all operators $G_\rho, 0 < \rho < \infty$, forms a semigroup of contraction operators of class (C_0) on $\mathcal{X}(\Omega)$.

Since the series is absolutely and uniformly convergent, $G_\rho(F)$ can be rewritten in the form

$$G_\rho(F)(\xi) = \sum_{n=0}^{\infty} e^{-n(n+1)\rho}\frac{2n+1}{4\pi}\int_\Omega F(\eta)P_n(\xi\cdot\eta)\, d\omega(\eta). \tag{8.2.11}$$

In what follows we want to show that for sufficiently small $\rho > 0$ the series

$$\frac{1}{\rho}\sum_{n=0}^{\infty}(e^{-n(n+1)\rho} - 1)\frac{2n+1}{4\pi}\int_\Omega F(\eta)P_n(\xi\cdot\eta)\, d\omega(\eta) \tag{8.2.12}$$

represents an approximation for the Beltrami derivative of F at $\xi \in \Omega$.

Theorem 8.2.1. *Suppose that F is twice continuously differentiable on Ω. Then*

$$\lim_{\rho\to 0,\rho>0} \left\| \frac{G_\rho(F)-F}{\rho} - \Delta^* F \right\|_{\mathcal{C}(\Omega)} = 0.$$

Proof. The integral

$$\int_0^\rho G_\tau(\Delta^* F)(\xi)\, d\tau \tag{8.2.13}$$

exists for all $\rho > 0$, $\xi \in \Omega$ and for all $F \in \mathcal{C}^{(2)}(\Omega)$. Moreover it is not difficult to see that

$$\begin{aligned} &\int_\Omega \int_0^\rho G_\tau(\Delta^* F)(\xi)\, d\tau Y_{n,j}(\xi)\, d\omega(\xi) \qquad (8.2.14)\\ &= \int_0^\rho \int_\Omega G_\tau(\Delta^* F)(\xi) Y_{n,j}(\xi)\, d\omega(\xi)\, d\tau \\ &= -\frac{\mathrm{e}^{-n(n+1)\rho}-1}{n(n+1)} \int_\Omega \Delta_\xi^* F(\xi) Y_{n,j}(\xi)\, d\omega(\xi). \end{aligned}$$

Since $(\Delta^*)^\wedge(n) = -n(n+1)$ are the eigenvalues of the Beltrami operator Δ^*, we find for $n = 0, 1, \ldots$

$$\int_\Omega \int_0^\rho G_\tau(\Delta^* F)(\xi)\, d\tau Y_{n,j}(\xi)\, d\omega(\xi) = (\mathrm{e}^{-n(n+1)\rho} - 1) \int_\Omega F(\xi) Y_{n,j}(\xi)\, d\omega(\xi). \tag{8.2.15}$$

On the other hand,

$$\int_\Omega (G_\rho(F)(\xi) - F(\xi)) Y_{n,j}(\xi)\, d\omega(\xi) = (\mathrm{e}^{-n(n+1)\rho} - 1) \int_\Omega F(\xi) Y_{n,j}(\xi)\, d\omega(\xi). \tag{8.2.16}$$

By comparison of (8.2.15) and (8.2.16) we obtain

$$G_\rho(F)(\xi) - F(\xi) = \int_0^\rho G_\tau(\Delta^* F)(\xi)\, d\tau \tag{8.2.17}$$

for all $\rho > 0$ and $\xi \in \Omega$. Therefore it follows that

$$\begin{aligned} \left\| \frac{G_\rho(F)-F}{\rho} - \Delta^* F \right\|_{\mathcal{C}(\Omega)} &= \sup_{\xi\in\Omega} \left| \frac{1}{\rho} \int_0^\rho (G_\tau(\Delta^* F)(\xi) - (\Delta^* F)(\xi))\, d\tau \right| \\ &\le \frac{1}{\rho} \int_0^\rho \sup_{\xi\in\Omega} |G_\tau(\Delta^* F)(\xi) - \Delta_\xi^* F(\xi)|\, d\tau \\ &\le \sup_{0\le\tau\le\rho} \|G_\tau(\Delta^* F) - \Delta^* F\|_{\mathcal{C}(\Omega)}. \end{aligned}$$

Letting ρ tend to 0 we obtain the desired result. □

In the same way we obtain the following corollary.

Corollary 8.2.2. *For $F, H \in \mathcal{X}(\Omega)$ the following statements are equivalent:*

$$-n(n+1)F^{\wedge}(n,j) = H^{\wedge}(n,j), \quad n = 0,1,\ldots, \quad j = 1,\ldots,2n+1,$$

and

$$\lim_{\substack{\rho\to 0\\ \rho>0}} \left\| \frac{G_\rho(F) - F}{\rho} - H \right\|_{\mathcal{X}(\Omega)} = 0$$

If $H = 0$ in $\mathcal{X}(\Omega)$, then $F =$ const.

Moreover, H. BERENS *et al.* (1968) have shown that the 'saturation class' of the Gauß–Weierstraß singular integral operators $\{G_\rho\}, \rho \in (0,\infty)$, is given by

$$\begin{aligned} &\mathcal{H}(\mathcal{L}^p(\Omega); -n(n+1)) \\ &\quad = \{F \in \mathcal{L}^p(\Omega) | \exists G \in \mathcal{L}^p(\Omega) : G^{\wedge}(n,j) = -n(n+1)F^{\wedge}(n,j)\}, \end{aligned}$$

and the 'saturation order' of $\{G_\rho\}, \rho \in (0,\infty)$, is $O(\rho), \rho \to 0$.

A problem involving the Gauß–Weierstraß kernel is the initial-value problem (heat equation)

$$\frac{\partial}{\partial t}U(t;\xi) = \Delta^*_\xi U(t;\xi), \quad t \geq 0, \quad \xi \in \Omega, \tag{8.2.18}$$

$$U(0;\xi) = F(\xi), \quad \xi \in \Omega. \tag{8.2.19}$$

The solution is given by convolution against the Gauß–Weierstraß kernel

$$U(t;\xi) = G_t(F)(\xi) = \int_\Omega W_t(\xi\cdot\eta)F(\eta)\,d\omega(\eta). \tag{8.2.20}$$

Formula (8.2.20) is of fundamental importance in multiscale descriptions of spherical images (confer the comments by W. FREEDEN, M. SCHREINER (1995A)).

8.2.3 *Singular Integrals Corresponding to Locally Supported Kernel Functions*

When a function F defined globally over the unit sphere Ω is analysed by its spherical harmonic expansion, the orthogonal coefficients $F^{\wedge}(n,j)$ are required. For either convenience or experimental design the average of the function over a prescribed spherical cap is often used instead of individual data (see for example H. POLLACK (1973), E.M. GAPOSCHKIN (1980), K.E. KATSAMBALOS (1979)). Of course, the coefficients derived from an averaged function will, in general, be different from those of the original function, the differences depending on the details of the averaging. In addition, aliasing will occur, in the general sense of leakage of information from the higher into the lower spherical harmonics. Many analogies can be made between analysis of one-dimensional time series and two-dimensional analysis on a sphere. (Relevant theoretical details can be found in many publications on numerical analysis, filtering theory, or sampling theory.) Averaging over the sphere may be performed efficiently by axisymmetric kernels with global support and localizing properties (already known are the Abel–Poisson mean and the Gauß–Weierstraß mean), but it is also successful with locally supported kernels.

In what follows we consider averages by 'polynomial means' over spherical caps. Let, therefore, for $k \geq 0$

$$M_h^{(k)}(F)(\xi) = \int_\Omega L_h^{(k)}(\xi \cdot \eta) F(\eta)\, d\omega(\eta), \quad \xi \in \Omega,$$

where

$$L_h^{(k)}(\xi \cdot \eta) = \left((B_h^{(k)})^\wedge(0) \right)^{-1} B_h^{(k)}(\xi \cdot \eta), \quad \xi, \eta \in \Omega,$$

and $B_h^{(k)}$ is the piecewise polynomial function as introduced in Section 5.8. The case $k = 0$ describes an equally weighted average over a spherical cap (cf. L.P. PELLINEN (1966), P.A. MEISSL (1971, 1981)).

For $F \in \mathcal{C}(\Omega)$ we have

$$\begin{aligned} |M_h^{(0)}(F)(\xi) - F(\xi)| &= \|\Omega(\xi;h)\|^{-1} \left| \int_{\Omega(\xi;h)} (F(\eta) - F(\xi))\, d\omega(\eta) \right| \\ &\leq \sup_{h \leq \xi\cdot\eta \leq 1} |F(\eta) - F(\xi)|. \end{aligned} \tag{8.2.21}$$

Hence it is clear that

$$\|M_h^{(0)}(F) - F\|_{\mathcal{C}(\Omega)} \leq \mu(F; 1-h). \tag{8.2.22}$$

Moreover, for $h \in [0,1)$ and $F \in \mathcal{X}(\Omega)$,

$$\|M_h^{(0)}(F)\|_{\mathcal{X}(\Omega)} \leq \|F\|_{\mathcal{X}(\Omega)} \tag{8.2.23}$$

and

$$\lim_{\substack{h\to 1\\ h<1}} \|M_h^{(0)}(F) - F\|_{\mathcal{X}(\Omega)} = 0. \tag{8.2.24}$$

For $F, H \in \mathcal{X}(\Omega)$ the statement $-n(n+1)F^\wedge(n,j) = H^\wedge(n,j)$, $n = 0, 1, \ldots$, $j = 1, \ldots, 2n+1$, is equivalent to the limit relation

$$\lim_{\substack{h\to 1\\ h<1}} \left\| \frac{M_h^{(0)}(F) - F}{h^2} - H \right\|_{\mathcal{X}(\Omega)} = 0. \tag{8.2.25}$$

As a matter of fact, as V.L. Shapiro (1961) has pointed out,

$$\lim_{\substack{h\to 1\\ h<1}} \frac{M_h^{(0)}(F)(\xi) - F(\xi)}{h^2} = \Delta_\xi^* F(\xi), \quad \xi \in \Omega, \tag{8.2.26}$$

provided that F is a member of class $\mathcal{C}^{(2)}(\Omega)$. By similar arguments (cf. H. Berens *et al.* (1968)) we are able to show that, for $F \in \mathcal{L}^2(\Omega)$, the relation

$$\|M_h^{(0)}(F) - F\|_{\mathcal{L}^2(\Omega)} = O((1-h)^2) \tag{8.2.27}$$

is equivalent to the fact that there exists a function $G \in \mathcal{L}^2(\Omega)$ such that

$$G^\wedge(n,j) = -n(n+1)F^\wedge(n,j), \quad n = 0, 1, \ldots, \; j = 1, \ldots, 2n+1.$$

□

Next we discuss higher order averages, i.e. the case $k \geq 1$. For $k = 1, 2, \ldots$, $h \in (-1, +1)$, and $F \in (\Omega)$ it is easy to see that

$$\left\| M_h^{(k)}(F) \right\|_{\mathcal{X}(\Omega)} \leq \|F\|_{\mathcal{X}(\Omega)} \tag{8.2.28}$$

and

$$\lim_{\substack{h\to 1\\ h<1}} \|M_h^{(k)}(F) - F\|_{\mathcal{X}(\Omega)} = 0. \tag{8.2.29}$$

Using the modulus of continuity we obtain for $F \in \mathcal{C}(\Omega)$

$$\| M_h^{(k)}(F) - F \|_{\mathcal{C}(\Omega)} \leq \mu(F; 1-h). \tag{8.2.30}$$

Moreover, it follows that

$$\int_\Omega P_n(\xi\cdot\zeta) M_h^{(k)}(F)(\zeta)\, d\omega(\zeta) = \frac{(B_h^{(k)})^\wedge(n)}{(B_h^{(k)})^\wedge(0)} \int_\Omega P_n(\xi\cdot\zeta) F(\zeta)\, d\omega(\zeta). \tag{8.2.31}$$

Because of $\|L_h^{(k)}\|_{\mathcal{L}^1[-1,+1]} = 1$ we find

$$L_h^{(k)} = \sum_{n=0}^{\infty} \frac{2n+1}{4\pi} (L_h^{(k)})^\wedge(n) P_n \tag{8.2.32}$$

in the sense of $\|\cdot\|_{\mathcal{L}^2[-1,+1]}$. From the results of Section 5.8 it follows for $t \in [-1,+1]$ and $k = 1, 2, \ldots$

$$\left| L_h^{(k)}(t) \right| \leq \frac{1}{2\pi} \frac{k+1}{1-h}. \tag{8.2.33}$$

Moreover, for all $t, t' \in [-1, 1]$ and $k = 1, 2, \ldots$

$$\left| L_h^{(k)}(t) - L_h^{(k)}(t') \right| \leq \frac{1}{2\pi} \frac{2^{k-1}k(k+1)}{(1-h)^{k+1}} |t - t'|. \tag{8.2.34}$$

The values $(B_h^{(k)})^\wedge(n)$ are the eigenvalues of the operator $M_h^{(k)} : \mathcal{C}(\Omega) \to \mathcal{C}(\Omega)$, i.e. $M_h^{(k)} Y_n = (B_h^{(k)})^\wedge(n) Y_n$ for all $Y_n \in Harm_n$.

It formally makes sense to introduce $M_h^{(-1)} : \mathcal{X}(\Omega) \to \mathcal{X}(\Omega)$ as follows:

$$(M_h^{(-1)}(F))(\xi) = T_h(F)(\xi) = \frac{1}{2\pi\sqrt{1-h^2}} \int_{\substack{\xi\cdot\eta=h \\ \eta\in\Omega}} F(\eta)\, d\sigma(\eta). \tag{8.2.35}$$

As we already know from Section 3.6 the translation operator forms an approximate identity in $\mathcal{X}(\Omega)$. However, H. Berens *et al.* (1968) have shown for $F \in \mathcal{L}^p(\Omega), 1 < p < \infty$, that

$$\|M_h^{(k)}(F) - F\|_{\mathcal{L}^p(\Omega)} = O((1-h)^2), \quad k \geq -1. \tag{8.2.36}$$

The second iterations $(L_h^{(k)})^{(2)}(\cdot)$ are given by

$$(L_h^{(k)})^{(2)}(\xi \cdot \eta) = ((B_h^{(k)})^\wedge(0))^{-2} \int_\Omega B_h^{(k)}(\xi \cdot \zeta) B_h^{(k)}(\zeta \cdot \eta)\, d\omega(\zeta).$$

It is obvious from (8.2.33) and (8.2.34) that for all $\xi, \xi', \eta \in \Omega$, and $k = 1, 2, \ldots$

$$\left| (L_h^{(k)})^{(2)}(\xi \cdot \eta) \right| \leq 2 \frac{(k+1)^2}{(1-h)^2}, \tag{8.2.37}$$

and

$$\left| (L_h^{(k)})^{(2)}(\xi \cdot \eta) - (L_h^{(k)})^{(2)}(\xi' \cdot \eta) \right| \leq \frac{1}{2\pi} \frac{2^k k^2 (k+1)^2}{(1-h)^{k+2}} |\xi - \xi'|. \tag{8.2.38}$$

The operators $(M_h^{(k)})^{(2)} : \mathcal{C}(\Omega) \to \mathcal{C}(\Omega)$ are defined by

$$(M_h^{(k)})^{(2)}(F)(\xi) \quad = \quad M_h^{(k)}(M_h^{(k)}(F))(\xi) \tag{8.2.39}$$

$$= \int_\Omega (L_h^{(k)})^{(2)}(\xi\cdot\eta)F(\eta)\,d\omega(\eta), \quad \xi\in\Omega.$$

For all $\xi\in\Omega$ we find that

$$(M_h^{(k)})^{(2)}(F)(\xi)-F(\xi) = \int\limits_{-1+2h^2\le\xi\cdot\eta\le 1} (L_h^{(k)})^{(2)}(\xi\cdot\eta)(F(\eta)-F(\xi))\,d\omega(\eta). \tag{8.2.40}$$

The function $F\in\mathcal{C}(\Omega)$ is uniformly continuous on Ω. Hence,

$$\begin{aligned}&|(M_h^{(k)})^{(2)}(F)(\xi)-F(\xi)| \qquad (8.2.41)\\ &\le \max_{-1+2h^2\le\xi\cdot\eta\le 1}|F(\xi)-F(\eta)| \int\limits_{-1+2h^2\le\xi\cdot\eta\le 1}(L_h^{(k)})^{(2)}(\xi\cdot\eta)\,d\omega(\eta).\end{aligned}$$

But this gives us

$$|(M_h^{(k)})^{(2)}(F)(\xi)-F(\xi)| \le \max_{-1+2h^2\le\xi\cdot\eta\le 1}|F(\xi)-F(\eta)|. \tag{8.2.42}$$

For $h\in(-1,+1)$ and $F\in\mathcal{C}(\Omega)$ we have

$$\|(M_h^{(k)})^{(2)}(F)\|_{\mathcal{C}(\Omega)} \le \|F\|_{\mathcal{C}(\Omega)} \tag{8.2.43}$$

and

$$\|(M_h^{(k)})^{(2)}(F)-F\|_{\mathcal{C}(\Omega)} \le \mu(F;2(1-h^2)). \tag{8.2.44}$$

The function $B_h^{(k)}(\cdot\eta):\xi\mapsto B_h^{(k)}(\xi\cdot\eta), \xi\in\Omega$, admits an expansion in terms of Legendre polynomials as follows:

$$B_h^{(k)}(\xi\cdot\eta) = \sum_{n=0}^{\infty}\frac{2n+1}{4\pi}(B_h^{(k)})^\wedge(n)P_n(\xi\cdot\eta). \tag{8.2.45}$$

Hence it is easy to see that

$$\begin{aligned}&(L_h^{(k)})^{(2)}(\xi,\eta) = (L_h^{(k)})^{(2)}(\xi\cdot\eta)\\ &= \left((B_h^{(k)})^\wedge(0)\right)^{-2}\sum_{n=0}^{\infty}\frac{2n+1}{4\pi}\left((B_h^{(k)})^\wedge(n)\right)^2 P_n(\xi\cdot\eta). \qquad (8.2.46)\end{aligned}$$

Singular integrals with local support turn out to be of particular efficiency in numerical integration of convolution integrals, since numerical effort must be done only on the local support. We know already that

$$\operatorname{supp}(L_h^{(k)} * L_{h'}^{(k)}) = [h \uplus h', 1], \quad 0 < h, h' < 1 \tag{8.2.47}$$

where the operation $\uplus : [-1,1]^2 \to [-1,+1]$ is defined by

$$h \uplus h' = \cos(\min\{\pi, \arccos(h) + \arccos(h')\}) \tag{8.2.48}$$

However, the statement

$$L_h^{(k)} * L_{h'}^{(k)} = L_{h \uplus h'}^{(k)}, \quad 0 < h', h < 1, \tag{8.2.49}$$

does *not* hold true in general. Our singular integrals with local support, therefore, do not form a semigroup of contraction operators (contrary to the Abel–Poisson or Gauß–Weierstraß singular integrals). Nevertheless we are able to find a way to achieve an *approximate contraction procedure* as follows: for $h_1 \in (0,1)$ and $k \geq 0$ let $L_1^{[k]} \in \mathcal{L}^2[-1,+1]$, be given by

$$L_1^{[k]} = L_{h_1}^{(k)}. \tag{8.2.50}$$

Consequently, we have with $(L_1^{[k]})^\wedge(n) = (L_{h_1}^{(k)})^\wedge(n), n = 0, 1, \ldots,$

$$L_1^{[k]} = \sum_{n=0}^{\infty} \frac{2n+1}{4\pi} (L_1^{[k]})^\wedge(n) P_n \tag{8.2.51}$$

and

$$\sum_{n=0}^{\infty} \frac{2n+1}{4\pi} \left(\left(L_1^{[k]} \right)^\wedge (n) \right)^2 < \infty. \tag{8.2.52}$$

Moreover,

$$\sum_{n=0}^{\infty} \frac{2n+1}{4\pi} \left(\left(L_1^{[k]} \right)^\wedge (n) \right)^{2j} < \infty \tag{8.2.53}$$

holds for all integers $j = 1, 2, \ldots$. This makes it possible to introduce kernels $\{L_j^{[k]}\} \subset \mathcal{L}^2[-1,+1], j \in \mathbb{N}$ of the following representation:

$$L_j^{[k]}(t) = \sum_{n=0}^{\infty} \frac{2n+1}{4\pi} \left(\left(L_1^{[k]} \right)^\wedge (n) \right)^j P_n(t), \quad t \in [-1,+1]. \tag{8.2.54}$$

But this yields the property

$$L_j^{[k]} * L_{j'}^{[k]} = L_{j+j'}^{[k]} \tag{8.2.55}$$

for $j, j' \in \mathbb{N}$. According to our construction $L_j^{[k]}$ is the jth iteration of $L_1^{[k]}$ which in turn means that

$$\text{supp} L_j^{[k]} = [h_j, 1] \tag{8.2.56}$$

with

$$h_j = \underbrace{h_1 \uplus \ldots \uplus h_1}_{j\text{times}}. \tag{8.2.57}$$

Hence, our procedure results in a locally supported kernel $\{L_j^{[k]}\}$ with the property (8.2.55). Consequently, the results obtained for the singular integral

$$M_j^{[k]}(F)(\xi) = (L_j^{[k]} * F)(\xi), \xi \in \Omega, F \in \mathcal{X}(\Omega) \tag{8.2.58}$$

can be summarized as follows.

Theorem 8.2.3. $M_j^{[k]}$ *defines for each* $j \in \mathbb{N}, k \geq 1$, *a linear bounded operator from* $\mathcal{X}(\Omega)$ *to* $\mathcal{C}^{(k-1)}(\Omega)$. *For all* $j \in \mathbb{N}$, *and* $F \in \mathcal{X}(\Omega)$

$$\|M_j^{[k]}(F)\|_{\mathcal{X}(\Omega)} \leq \|F\|_{\mathcal{X}(\Omega)},$$

and we have for $j, j' \in \mathbb{N}$

$$M_{j+j'}^{[k]} = M_j^{[k]} M_{j'}^{[k]}.$$

Seen in comparison with a semigroup of contraction operators we have developed a simulated procedure that stops after finite steps for $j = 1$ that is for a fixed 'window size' h_1. Therefore, approximation cannot be performed with arbitrary accuracy. Nevertheless, our approach is of practical importance. In applications only discrete data material is available, so an arbitrarily close approximation doesn't make sense. Instead we are confronted with the problem of reaching in a finite number of steps a numerically relevant approximation (based on a parameter h_1 chosen in close adaptation to the data situation). But, of course, it remains to find the right choice h_1 in practical applications.

8.2.4 *Spherical Up Function*

The disadvantage of a finite contraction process as described above can be overcome by the following approach, which leads us to an *infinite contraction procedure* with locally supported kernels. It is based on a spherical version of the so-called up function, which is in the one-dimensional context described, for example, in V.A. RVACHEV (1990).

Suppose that $k \geq 0$ is fixed. For a given $h \in (-1,1)$ we construct the sequence $\{h_j\}_{j=0,1,\ldots}$ by $h_j = \cos(2^{-j}\arccos(h))$, $j = 1,2,\ldots$. Using this sequence, we define $U_h^{(k)}$, by the infinite convolution

$$U_h^{(k)} = L_{h_0}^{(k)} * L_{h_1}^{(k)} * L_{h_2}^{(k)} * \ldots \tag{8.2.59}$$

Then it is clear, that $U_h^{(k)}$ is an infinitely often differentiable function with support $[h \uplus h, 1]$ and $(U_h^{(k)})^\wedge(0) = 1$. Assume now, that $h \in (-1,1)$ is fixed, and $h_0, h_1, \ldots$ are defined as described above. Then an important property of the introduced function is that

$$U_{h_j}^{(k)} = L_{h_j}^{(k)} * U_{h_{j+1}}^{(k)}, \quad j = 0,1,\ldots. \tag{8.2.60}$$

Hence, we end up with an infinite sequence $U_{h_0}^{(k)}, U_{h_1}^{(k)}, \ldots$, where the supports of the functions are $\operatorname{supp} U_{h_j}^{(k)} = [h_{j-1}, 1]$. Note that $\lim_{j\to\infty} h_j = 1$. It can be easily proved that

$$\lim_{j\to\infty} \|U_{h_j}^{(k)} * F - F\|_{\mathcal{X}(\Omega)} = 0, \quad F \in \mathcal{X}(\Omega).$$

Thus an infinite contraction process with locally supported kernels is found. We can do even better: if we set in addition $h_{-1} = h_{-2} = \ldots = h_0$, and use (8.2.60) as defining equation for $U_{h_j}^{(k)}$, $j = -1, -2, \ldots$, we clearly obtain

$$\lim_{j\to-\infty} \|U_{h_j}^{(k)} * F - F^\wedge(0,1)Y_{0,1}\|_{\mathcal{X}(\Omega)} = 0, \quad F \in \mathcal{X}(\Omega).$$

This gives the fundament for a multiresolution analysis based on locally supported kernels to be described in Chapter 10.

Bibliographical Notes

Singular integrals on the sphere have been studied by H. Berens *et al.* (1968) (see also the references therein). Non-orthogonal expansions in terms of locally supported kernels of the above type ('finite elements on the sphere') have been discussed in more detail by W. Freeden, J. Mason (1990). The 'finite contraction process' for locally supported kernel functions has been presented by W. Freeden, U. Windheuser (1996). The study of singular integrals with locally supported kernels based on the 'up function' closely follows the ideas of V.A. Rvachev (1990).

9

GABOR AND TOEPLITZ TRANSFORMS

In many applications such as image and signal analysis, only spectral information about a function (signal) can be observed. Thus we are confronted with the situation that some incoming information, available for a function F is given as a discrete table of numbers rather than its original form. One can, of course, choose the trial functions in such a way that the countable family is an orthonormal basis of a Hilbert space (for example, the spherical harmonics in $\mathcal{L}^2(\Omega)$). The decomposition of F into these functions is then quite straightforward by taking the Fourier series (that is the orthogonal expansion) of F. However, although Fourier techniques are very powerful, there are some serious deficiencies of the Fourier series approach by means of orthonormal systems: the orthonormalization process and the recovery of the function is in many instances of very undesirable computational complexity, and local changes of the function F will affect the whole table of Fourier coefficients, when localization does not occur in space domain (as in the case of spherical harmonics). This shortcoming was already observed by D. Gabor (1946) who proposed to 'window' the Fourier integral.

In principle, as we know from Section 6.4, (space localizing) kernels generating singular integrals allow us to construct orthonormal bases, which would enable us to keep local changes under control. If the function undergoes a local change, confined to some subdomain of the sphere, only a finite number of Fourier integrals will be seriously affected, reflecting the locality of the change. But there is still a defect in reconstructing functions (signals) from space localizing orthonormalized kernel functions with fixed 'window parameter' h. One problem over and above the computational complexity is that it poorly resolves phenomena shorter than the window.

A 'modulation' of the kernel (i.e. the window function) by means of spherical harmonics of sufficiently high order opens a first perspective to overcome this difficulty. The result is the spherical Gabor transform. It is based on two operations, namely 'rotation' and 'modulation' of the window function. Modulation allows flexibility for reconstructing 'high order frequency parts'. Rotation, of course, localizes the position. Discretization is understood in the sense of using both a discrete grid of 'rotates' and a finite number of appropriately chosen 'modulates', rather than the entire sphere and the whole spectrum. The constituting elements form a system generated from the window function by rotation and modulation. This system is 'overcomplete' (and not necessarily orthogonal). Nevertheless, approximation can be established without orthonormalization by a 'formal Fourier sum expansion' in terms of the non-orthogonal system ('frame') (see I. DAUBECHIES *et al.* (1985) for a more detailed discussion on frames).

9.1 Gabor Transform

Let $K_{h_0} : [-1,+1] \to \mathbb{R}$, $h_0 \in (-1,+1)$, be a piecewise continuous 'cutoff function' (i.e. a member of a non-negative kernel $\{K_h\}, h \in [h_0, 1)$ satisfying the limit relation (8.1.9)). Chopping up the function amounts to multiplying $F \in \mathcal{L}^2(\Omega)$ by the axisymmetric kernel $K_{h_0}(\xi\cdot), \xi \in \Omega$. In other words, we have to compute the '*Gabor coefficients*' of F

$$F^\wedge(n,j;\xi) = (K_{h_0}^{(2)}(1))^{-1/2} \int_\Omega F(\eta) K_{h_0}(\xi \cdot \eta) Y_{n,j}(\eta)\, d\omega(\eta), \quad \xi \in \Omega. \tag{9.1.1}$$

Definition 9.1.1. For arbitrary but fixed $h_0 \in (-1,+1)$, the spherical Gabor transform

$$(GT)_{K_{h_0}} : \mathcal{X}(\Omega) \to \{H : \mathcal{J} \times \Omega \to \mathbb{R}\}$$

is defined by

$$\begin{aligned} ((GT)_{K_{h_0}}(F))(n,j;\xi) &= F^\wedge(n,j;\xi) \qquad\qquad (9.1.2)\\ &= (K_{h_0}^{(2)}(1))^{-1/2} \int_\Omega F(\eta) K_{h_0}(\xi \cdot \eta) Y_{n,j}(\eta)\, d\omega(\eta) \end{aligned}$$

for $n = 0, 1, \ldots, j = 1, \ldots, 2n+1$, and $\xi \in \Omega$.

The spherical Gabor transform is generated by '*rotations*'

$$R_\xi : K_{h_0}(\cdot\eta) \mapsto R_\xi K_{h_0}(\cdot\eta) = K_{h_0}(\xi \cdot \eta)$$

and '*modulations*'

$$M_{n,j} : K_{h_0}(\cdot\eta) \mapsto M_{n,j} K_{h_0}(\cdot\eta) = K_{h_0}(\cdot\eta) Y_{n,j}$$

from the 'window function' K_{h_0}:

$$((GT)_{K_{h_0}}(F))(n,j;\xi) = (K_{h_0}^{(2)}(1))^{-1/2} (M_{n,j} R_\xi K_{h_0}) * F, \quad F \in \mathcal{L}^2(\Omega).$$

Given an arbitrary function $F \in \mathcal{L}^2(\Omega)$ we call

$$(GE)_{K_{h_0}}(F) = (K_{h_0}^{(2)}(1))^{-1/2} \sum_{n=0}^{\infty} \sum_{j=1}^{2n+1} \int_\Omega F^\wedge(n,j;\eta) K_{h_0}(\cdot\eta)\, d\omega(\eta) Y_{n,j} \tag{9.1.3}$$

the *spherical Gabor expansion* relative to *the 'window function'* K_{h_0}.

Denote by $\mathcal{L}^2(\mathcal{J} \times \Omega)$ the space of all functions $H : \mathcal{J} \times \Omega \to \mathbb{R}$ such that

$$\sum_{n=0}^{\infty} \sum_{j=1}^{2n+1} \int_{\Omega} |H(n, j; \xi)|^2 \, d\omega(\xi) < \infty.$$

On the space $\mathcal{L}^2(\mathcal{J} \times \Omega)$ we are able to impose an inner product $(\cdot, \cdot)_{\mathcal{L}^2(\mathcal{J} \times \Omega)}$ corresponding to the norm

$$\|H\|_{\mathcal{L}^2(\mathcal{J} \times \Omega)} = \left(\sum_{n=0}^{\infty} \sum_{j=1}^{2n+1} \int_{\Omega} |H(n, j; \xi)|^2 \, d\omega(\xi) \right)^{1/2}.$$

Theorem 9.1.2. *Suppose that F is of class $\mathcal{L}^2(\Omega)$. Then the restriction of the Gabor transform to the space $\mathcal{L}^2(\Omega)$ forms a mapping into the space $\mathcal{L}^2(\mathcal{J} \times \Omega)$, i.e. $(GT)_{K_{h_0}} : \mathcal{L}^2(\Omega) \to \mathcal{L}^2(\mathcal{J} \times \Omega)$, and we have*

$$\begin{aligned} \|F\|^2_{\mathcal{L}^2(\Omega)} &= \sum_{n=0}^{\infty} \sum_{j=1}^{2n+1} \int_{\Omega} |((GT)_{K_{h_0}}(F))(n, j; \eta)|^2 \, d\omega(\eta) \\ &= \|(F, K_{h_0}(\,\cdot\,)Y_{\cdot,\cdot})_{\mathcal{L}^2(\Omega)}\|^2_{\mathcal{L}^2(\mathcal{J} \times \Omega)}. \end{aligned}$$

Proof. The Parseval identity of the theory of spherical harmonics gives us

$$\begin{aligned} &\int_{\Omega} \sum_{n=0}^{\infty} \sum_{j=1}^{2n+1} |(GT)_{K_{h_0}}(F)(n, j; \eta)|^2 \, d\omega(\eta) \\ &= (K^{(2)}_{h_0}(1))^{-1} \int_{\Omega} \left(\int_{\Omega} |F(\eta)|^2 |K_{h_0}(\xi \cdot \eta)|^2 \, d\omega(\xi) \right) d\omega(\eta). \end{aligned}$$

Now we have

$$\int_{\Omega} |K_{h_0}(\xi \cdot \eta)|^2 \, d\omega(\xi) = K^{(2)}_{h_0}(\eta \cdot \eta) = K^{(2)}_{h_0}(1).$$

This is the desired result. □

Thus it is clear that any function $F \in \mathcal{L}^2(\Omega)$ can be recovered by its Gabor expansion.

Theorem 9.1.3. *Let F be of class $\mathcal{L}^2(\Omega)$. Then*

$$F = (K^{(2)}_{h_0}(1))^{-1/2} \sum_{n=0}^{\infty} \sum_{j=1}^{2n+1} \int\limits_{\Omega} ((GT)_{K_{h_0}}(F))(n, j; \xi) K_{h_0}(\xi \cdot) \, d\omega(\xi) Y_{n,j}$$

in the sense of $\| \cdot \|_{\mathcal{L}^2(\Omega)}$.

By virtue of the Cauchy–Schwarz inequality we obtain for $F \in \mathcal{L}^2(\Omega)$

$$\begin{aligned} |((GT)_{K_{h_0}}(F))(n,j;\xi)| &= (K_{h_0}^{(2)}(1))^{-1/2}|(F, K_{h_0}(\cdot\xi)Y_{n,j})_{\mathcal{L}^2(\Omega)}| \\ &\leq (K_{h_0}^{(2)}(1))^{-1/2}\|F\|_{\mathcal{L}^2(\Omega)}\|K_{h_0}(\cdot\xi)Y_{n,j}\|_{\mathcal{L}^2(\Omega)} \\ &\leq (K_{h_0}^{(2)}(1))^{-1/2}\sup_{\eta\in\Omega}|K_{h_0}(\xi\cdot\eta)|^2\|F\|_{\mathcal{L}^2(\Omega)}. \end{aligned}$$

In other words, $F \in \mathcal{L}^2(\Omega)$ implies that $(GT)_{K_{h_0}}(F) \in \mathcal{L}^2(\mathcal{J}\times\Omega)$ is bounded. But the transform $(GT)_{K_{h_0}}$ is not surjective on $\mathcal{L}^2(\mathcal{J}\times\Omega)$ (note that $\mathcal{L}^2(\mathcal{J}\times\Omega)$ contains unbounded elements). That means that $\mathcal{G} = (GT)_{K_{h_0}}(\mathcal{L}^2(\Omega))$ is a proper subspace of $\mathcal{L}^2(\mathcal{J}\times\Omega)$:

$$\mathcal{G} \subsetneq \mathcal{L}^2(\mathcal{J}\times\Omega).$$

Therefore we are led to the question of how to characterize $\mathcal{G}$ within the framework of $\mathcal{L}^2(\mathcal{J}\times\Omega)$. For that purpose consider the operator P : $\mathcal{L}^2(\mathcal{J}\times\Omega) \to \mathcal{G}$ given by

$$(PH)(n,j;\xi) = \sum_{p=0}^{\infty}\sum_{q=1}^{2n+1}\int_{\Omega} K_{h_0}(n,j;\xi \mid p,q;\eta)H(p,q;\eta)\,d\omega(\eta) \tag{9.1.4}$$

where

$$K_{h_0}(n,j;\xi \mid p,q;\eta) = (K_{h_0}^{(2)}(1))^{-1}\int_{\Omega} K_{h_0}(\xi\cdot\zeta)Y_{n,j}(\zeta)K_{h_0}(\zeta\cdot\eta)Y_{p,q}(\zeta)\,d\omega(\zeta).$$

Our purpose is to show that P defines a projection.

Lemma 9.1.4. *The operator $P : \mathcal{L}^2(\mathcal{J}\times\Omega) \to \mathcal{G}$ defined by (9.1.4) is a projection operator.*

Proof. Assume that $H = \tilde{F} = (GT)_{K_{h_0}}(F) \in \mathcal{G}$. Then we obtain

$$\begin{aligned} (PH)(n,j;\xi) &= (K_{h_0}^{(2)}(1))^{-1}\int_{\Omega} K_{h_0}(\xi\cdot\zeta)Y_{n,j}(\zeta) \\ &\quad \left(\sum_{p=0}^{\infty}\sum_{q=1}^{2n+1}\int_{\Omega} H(p,q;\eta)K_{h_0}(\zeta\cdot\eta)Y_{p,q}(\zeta)\,d\omega(\eta)\right) d\omega(\zeta) \\ &= (K_{h_0}^{(2)}(1))^{-1/2}\int_{\Omega} F(\zeta)K_{h_0}(\xi\cdot\zeta)Y_{n,j}(\zeta)\,d\omega(\zeta) \\ &= \tilde{F}(n,j;\xi) \\ &= H(n,j;\xi), \end{aligned}$$

hence, $PH = H$ for all $H \in \mathcal{G}$.

Next, we show, that for all $H^\perp \in \mathcal{G}^\perp$ we have $PH^\perp = 0$. Assume therefore that $H^\perp \in \mathcal{G}^\perp$, i.e. for all $F \in \mathcal{L}^2(\Omega)$

$$(H^\perp, (GT)_{K_{h_0}}(F))_{\mathcal{L}^2(\mathcal{J}\times\Omega)} = 0. \tag{9.1.5}$$

If $\xi \in \Omega$, $(n,j) \in \mathcal{J}$, we get from (9.1.5) with the special choice $F = [K_{h_0}^{(2)}(1)]^{-1/2} K_{h_0}(\xi\cdot\ )Y_{n,j}(\cdot)$

$$\begin{aligned}
0 &= (H^\perp, (GT)_{K_{h_0}}([K_{h_0}^{(2)}(1)]^{-1/2} K_{h_0}(\xi\cdot\ )Y_{n,j}(\cdot)))_{\mathcal{L}^2(\mathcal{J}\times\Omega)} \\
&= \sum_{p=0}^{\infty}\sum_{q=1}^{2p+1} \int_\Omega H^\perp(p,q;\eta)[K_{h_0}^{(2)}(1)]^{-1/2} \\
&\qquad ((GT)_{K_{h_0}}(K_{h_0}(\xi\cdot\ )Y_{n,j}(\cdot)))(p,q;\eta)\, d\omega(\eta) \\
&= \sum_{p=0}^{\infty}\sum_{q=1}^{2p+1} \int_\Omega H^\perp(p,q;\eta)[K_{h_0}^{(2)}(1)]^{-1} \\
&\qquad \int_\Omega K_{h_0}(\xi\cdot\zeta)Y_{n,j}(\zeta)K_{h_0}(\eta\cdot\zeta)Y_{p,q}(\zeta)\, d\omega(\zeta)\, d\omega(\eta) \\
&= \sum_{p=0}^{\infty}\sum_{q=1}^{2p+1} \int_\Omega H^\perp(p,q;\eta)K_{h_0}(n,j;\xi \mid p,q;\eta)\, d\omega(\eta) \\
&= (PH^\perp)(n,j;\xi).
\end{aligned}$$

Hence, $PH^\perp = 0$ for all $H^\perp \in \mathcal{G}^\perp$. Summarizing our results we therefore obtain $P(\mathcal{L}^2(\mathcal{J}\times\Omega)) = \mathcal{G}$, $P\mathcal{G}^\perp = 0$, $P^2 = P$. □

Finally $\mathcal{G}$ is characterized as follows: $H \in \mathcal{G}$ if and only if

$$H(n,j;\xi) = (K_{h_0}^{(2)}(1))^{-1/2} \sum_{p=0}^{\infty}\sum_{q=1}^{2p+1} \int_\Omega K_{h_0}(n,j;\xi \mid p,q;\eta) H(p,q;\eta)\, d\omega(\eta).$$

Therefore it is clear that $K_{h_0}(n,j;\xi \mid \cdot,\cdot;\cdot) \in \mathcal{G}$ and $K_{h_0}(\cdot,\cdot;\cdot \mid p,q;\eta) \in \mathcal{G}$. The kernel $(n,j;\xi,p,q;\eta) \mapsto K_{h_0}(n,j,\xi \mid p,q;\eta)$ is the reproducing kernel in $\mathcal{G}$.

Theorem 9.1.5. *Let H be an arbitrary element of $\mathcal{L}^2(\mathcal{J}\times\Omega)$. Then the unique function $F_H \in \mathcal{L}^2(\Omega)$ satisfying*

$$\|H - \tilde{F}_H\|_{\mathcal{L}^2(\mathcal{J}\times\Omega)} = \inf_{\tilde{F}\in\mathcal{L}^2(\Omega)} \|H - \tilde{F}\|_{\mathcal{L}^2(\mathcal{J}\times\Omega)}$$

(with $\tilde{F}_H = (GT)_{K_{h_0}}(F_H)$) is given by

$$F_H(\xi) = (K_{h_0}^{(2)}(1))^{-1/2} \sum_{n=0}^{\infty} \sum_{j=1}^{2n+1} \int_{\Omega} H(n,j;\eta) K_{h_0}(\xi \cdot \eta)\, d\omega(\eta) Y_{n,j}(\xi).$$

Proof. We know that $\tilde{F}_H$ is the orthogonal projection of H onto $\mathcal{G}$. This proves Theorem 9.1.5. □

Our considerations have shown that the coefficients in $\mathcal{L}^2(\mathcal{J} \times \Omega)$ for reconstructing a function $F \in \mathcal{L}^2(\Omega)$ are not unique. This can be immediately seen from the identity

$$F = (K_{h_0}^{(2)}(1))^{-1/2} \sum_{n=0}^{\infty} \sum_{j=1}^{2n+1} \int_{\Omega} (\tilde{F}(n,j;\eta) + \tilde{F}^{\perp}(n,j;\eta)) K_{h_0}(\cdot\eta)\, d\omega(\eta) Y_{n,j},$$

where $\tilde{F} = (GT)_{K_{h_0}}(F)$ and $\tilde{F}^{\perp}$ is an arbitrary member of $\mathcal{G}^{\perp}$. But we are able to formulate the following result.

Theorem 9.1.6. *For arbitrary $F \in \mathcal{L}^2(\Omega)$ the coefficient function $\tilde{F} = (GT)_{K_{h_0}}(F) \in \mathcal{G}$ is the unique element in $\mathcal{L}^2(\mathcal{J} \times \Omega)$ which satisfies the minimum norm condition*

$$\|\tilde{F}\|_{\mathcal{L}^2(\mathcal{J}\times\Omega)} = \inf_{\substack{H \in \mathcal{L}^2(\mathcal{J}\times\Omega) \\ (GT)^{-1}_{K_{h_0}}(H) = F}} \|H\|_{\mathcal{L}^2(\mathcal{J}\times\Omega)}.$$

Proof. We know already that $H = \tilde{F} + \tilde{F}^{\perp}$. Thus we are able to deduce that

$$\begin{aligned} \|H\|^2_{\mathcal{L}^2(\mathcal{J}\times\Omega)} &= \|\tilde{F} + \tilde{F}^{\perp}\|^2_{\mathcal{L}^2(\mathcal{J}\times\Omega)} \\ &= \|\tilde{F}\|^2_{\mathcal{L}^2(\mathcal{J}\times\Omega)} + \|\tilde{F}^{\perp}\|^2_{\mathcal{L}^2(\mathcal{J}\times\Omega)} \\ &\geq \|\tilde{F}\|^2_{\mathcal{L}^2(\mathcal{J}\times\Omega)}. \end{aligned}$$

□

As mentioned above, the Gabor transform works by first dividing a 'signal' $F \in \mathcal{L}^2(\Omega)$ into short consecutive segments of fixed size by use of a 'cutoff kernel' (window function) and then computing the Fourier coefficients of each segment. In other words, the Gabor transform maps local changes of the function being represented to local changes of the coefficients in the expansion and thereby also reduces the computational complexity. However, there is still a defect in reconstructing a function using a sole, fixed 'window parameter' $h_0 \in (0,1)$. It poorly resolves phenomena shorter than the window which leads to non-optimal computational costs in many circumstances. This can be remedied by kernels with decreasing window diameters (i.e. $h_n \to 1$ monotonically as $n \to \infty$) exhibiting the so-called 'zooming-in' property (cf. Section 9.3).

Remark. Note that D. GABOR (1946) considered only the 'Gaussian kernel' as 'window function'. Our approach is essentially based on 'approximate identity generating $\mathcal{C}$-kernels' (like the Abel–Poisson kernel, the Gauß–Weierstraß kernel, etc.). All results, however, remain valid even for general kernels $K \in \mathcal{L}^2[-1,+1]$.

9.2 Combined Fourier and Gabor Transform

We now deal with the discussion of spherical harmonic and Gabor expansion.

Corollary 9.2.1. *Let F be of class $\mathcal{L}^2(\Omega)$. Then*

$$
\begin{aligned}
F = & \sum_{n=0}^{m}\sum_{j=1}^{2n+1}\int_\Omega F(\eta)Y_{n,j}(\eta)\,d\omega(\eta)Y_{n,j} \\
& +(K_{h_0}^{(2)}(1))^{-1/2}\sum_{n=m+1}^{\infty}\sum_{j=1}^{2n+1}\int_\Omega \left((GT)_{K_{h_0}}(F)\right)(n,j;\eta)K_{h_0}(\eta\cdot)\,d\omega(\eta)Y_{n,j}
\end{aligned}
$$

in the sense of $\|\cdot\|_{\mathcal{L}^2(\Omega)}$.

Proof. Let

$$F^\perp_{0,\ldots,m} = F - \sum_{n=0}^{m}\sum_{j=1}^{2n+1}\int_\Omega F(\eta)Y_{n,j}(\eta)\,d\omega(\eta)Y_{n,j}.$$

Then it follows easily from the completeness of the spherical harmonics that

$$
\begin{aligned}
F^\perp_{0,\ldots,m} &= (K_{h_0}^{(2)}(1))^{-1}\sum_{n=m+1}^{\infty}\sum_{j=1}^{2n+1}\int_\Omega F^\perp_{0,\ldots,m}(\eta)K_{h_0}^{(2)}(\cdot\eta)Y_{n,j}(\eta)\,d\omega(\eta)Y_{n,j} \\
&= (K_{h_0}^{(2)}(1))^{-1}\sum_{n=m+1}^{\infty}\sum_{j=1}^{2n+1}\int_\Omega F(\eta)K_{h_0}^{(2)}(\cdot\eta)Y_{n,j}(\eta)\,d\omega(\eta)Y_{n,j}.
\end{aligned}
$$

However,

$$\sum_{n=m+1}^{\infty}\sum_{j=1}^{2n+1}\int_\Omega F(\eta)K_{h_0}^{(2)}(\cdot\eta)Y_{n,j}(\eta)\,d\omega(\eta)Y_{n,j}$$

$$= \sum_{n=m+1}^{\infty} \sum_{j=1}^{2n+1} \int_{\Omega} \int_{\Omega} F(\eta) K_{h_0}(\eta \cdot \alpha) Y_{n,j}(\eta) \, d\omega(\eta) K_{h_0}(\cdot \alpha) d\omega(\alpha) Y_{n,j}.$$

But this is the desired result. □

9.3 Toeplitz Transform

In spite of the 'windowing effect caused by a kernel with a fixed window parameter', one needs to have kernels with variable decreasing window diameters achieving the so-called zooming-in capability. This method has its origin in seismic analysis and the first published work was by A. GROSSMANN, J. MORLET (1984). In Euclidean spaces it is based on the simple ideas of translation and dilation of the *'window function'*. The spherical analogue of translation, of course, is 'rotation', but the spherical counterpart of the operation of dilation is more complicated. In fact, as dilation corresponds to narrowing the window, it may be organized analogously even in a spherical approach by increasing parameters h_n (converging to the limit 1) in the kernel representation of singular integrals. In order to be an appropriate window function, the kernel must satisfy the property (8.1.9).

A first attempt to develop an efficient 'zooming-in' technique is the following result.

Theorem 9.3.1. *Assume that* $\{K_h\}$, $K_h \in \mathcal{L}^1[-1,+1]$, $h \in (-1,+1)$, *is an approximate identity generating* $\mathcal{L}^1$*-kernel. Let* $\{h_n\}, h_n \in (-1,+1)$, $n = 1,2,\ldots$, *be a sequence converging to the limit 1 (for example,* $h_n = 1-(1/2)^n, n = 1,2,\ldots$*). Furthermore, suppose that*

$$\begin{pmatrix} t_{11} & & & & & \\ t_{21} & t_{22} & & & & \\ t_{31} & t_{32} & t_{33} & & & \\ \vdots & \vdots & \vdots & \ddots & & \\ t_{Q1} & t_{Q2} & t_{Q3} & & t_{QQ} & \\ \vdots & \vdots & \vdots & & \vdots & \ddots \end{pmatrix}$$

is a Toeplitz-triangular matrix, i.e. the elements $t_{Qj}, 1 \le j \le Q$, *satisfy the following conditions:*

$$t_{Qj} \to 0 \quad \text{for} \quad Q \to \infty \quad (j \text{ fixed}), \tag{9.3.1}$$

$$t_{Qj} \ge 0 \quad \textit{for all} \quad j, \quad 1 \le j \le Q,$$

and

$$\sum_{j=1}^{Q} t_{Qj} = 1 \tag{9.3.2}$$

for all $Q \in \mathbb{N}$. *Then*

$$\lim_{Q\to\infty} \left\| \sum_{n=1}^{Q} t_{Qn} I_{h_n}^{(q)}(F) - F \right\|_{\mathcal{X}(\Omega)} = 0$$

holds for all $F \in \mathcal{X}(\Omega)$.

Proof. By virtue of the triangle inequality we get

$$0 \le \left\| \sum_{n=1}^{Q} t_{Qn} I_{h_n}^{(q)}(F) - F \right\|_{\mathcal{X}(\Omega)} \le \sum_{n=1}^{Q} |t_{Qn}| \; \| I_{h_n}^{(q)}(F) - F \|_{\mathcal{X}(\Omega)}.$$

The required result, therefore, follows from Corollary 8.1.5 in connection with the argument due to Toeplitz. □

The conditions (9.3.1) (9.3.2) can be weakened slightly (cf. G.M. FICHTENHOLZ (1973), U. WINDHEUSER (1995)). Obviously they contain the following examples:

(i) arithmetical summation (Cauchy's example)

$$t_{Qn} = \frac{1}{Q}, \quad n = 1, \ldots, Q$$

(ii) binomial summation

$$(\alpha) \quad t_{Qn} = \frac{\binom{Q-1}{n-1} z^{n}}{(1+z)^{n}}, \quad n = 1, \ldots, Q$$

$$(\beta) \quad t_{Qn} = \frac{\binom{Q-1}{n-1} z^{Q-n}}{(1+z)^{n}}, \quad n = 1, \ldots, Q$$

(iii) Cesàro C_k-summation, $k \ge 1$,

$$t_{Qn} = \frac{\binom{Q-n+k-2}{Q-n-1}}{\binom{Q-1-k}{Q-1}}, \quad n = 1, \ldots, Q.$$

From the literature we know that C_k-summation can be organized recursively. C_1-summation is equivalent to arithmetical summation.

Definition 9.3.2. The Toeplitz transform

$$TT^{(q)} : \mathcal{X}(\Omega) \to \{H : \mathcal{T} \times \Omega \to \mathbb{R}\}$$

with

$$\mathcal{T} = \{(Q,n) \mid Q \in \mathbb{N}, n = 1, \ldots, Q\}$$

is defined by

$$((TT)^{(q)}(F))(Q,n;\xi) = \sqrt{t_{Qn}} \int_\Omega F(\eta) K_{h_n}^{(q)}(\xi \cdot \eta)\, d\omega(\eta), F \in \mathcal{X}(\Omega), \quad \xi \in \Omega.$$

Given an arbitrary function $F \in \mathcal{X}(\Omega)$, we call the limit

$$(TE^{(q)})(F)(\xi) = \lim_{Q \to \infty} \sum_{n=1}^{Q} t_{Qn} \int_\Omega K_{h_n}^{(q)}(\xi \cdot \zeta) F(\zeta)\, d\omega(\zeta) \tag{9.3.3}$$

the *(spherical) Toeplitz (singular integral) expansion of* F at $\xi \in \Omega$ *(relative to the window function* $K_{h_0}^{(q)}$*)*. The Toeplitz singular integral transform acts as a localization operator in the following way. Roughly speaking, if n is large, then $K_{h_n}^{(q)}(\xi \cdot \zeta)$ seen as an axisymmetric function of the variable ζ around the axis ξ is highly concentrated about the point ξ. As h_n approaches the value 1, $K_{h_n}^{(q)}(\xi \cdot \zeta)$ becomes more and more concentrated around ξ. Thus it behaves like a 'small wave', and so the terminology 'wavelet' may be appropriate for the kernel $K_h^{(q)}$. The Toeplitz singular integral transform displays the information of F at various levels of resolution (*multilevel method*). In other words, as h_n approaches 1, the Toeplitz singular integral transform gives sharper and sharper resolution. Obviously, the Toeplitz transform is generated from $K_{h_0}^{(q)}$ by '*dilations*'

$$D_n : K_{h_0}^{(q)}(\xi \cdot) \mapsto D_n K_{h_0}^{(q)}(\xi \cdot) = K_{h_n}^{(q)}(\xi \cdot), \quad n = 1, 2, \ldots$$

and '*rotations*'

$$R_\xi : K_{h_0}^{(q)}(\cdot \eta) \mapsto R_\xi K_{h_0}^{(q)}(\cdot \eta) = K_{h_0}^{(q)}(\xi \cdot \eta), \quad \xi \in \Omega.$$

The function F is characterized by the values $((TT)^{(q)}(F))(Q,n;\xi)$, and F can be recovered by its Toeplitz expansions in the sense of $\|\cdot\|_{\mathcal{X}(\Omega)}$.

Theorem 9.3.3. *Under the above assumptions*

$$\lim_{Q\to\infty} \left\| F - \sum_{n=1}^{Q} \sqrt{t_{Qn}} \int_{\Omega} ((TT)^{(q)}(F))(Q,n;\eta) K_{h_n}^{(q)}(\cdot\eta)\, d\omega(\eta) \right\|_{\mathcal{X}(\Omega)} = 0$$

holds for all $F \in \mathcal{X}(\Omega)$.

Proof. Clearly we have for all $\xi \in \Omega$

$$\sum_{n=1}^{Q} \sqrt{t_{Qn}} \int_{\Omega} ((TT)^{(q)}(F))(Q,n;\eta) K_{h_n}^{(q)}(\xi\cdot\eta)\, d\omega(\eta) = \sum_{n=1}^{Q} t_{Qn} I_{h_n}^{(2q)}(F)(\xi).$$

This proves the assertion. □

Consequently we find the following corollary.

Corollary 9.3.4. *The restriction of the Toeplitz transform to* $\mathcal{L}^2(\Omega)$ *maps* $\mathcal{L}^2(\Omega)$ *into the space*

$$\mathcal{L}^2(\mathcal{T}\times\Omega) = \{H : \mathcal{T}\times\Omega \to \mathbb{R} |\, \lim_{Q\to\infty} \sum_{n=1}^{Q} \left| \int_{\Omega} H(Q,n;\eta) \right|^2 d\omega(\eta) < \infty\},$$

i.e.

$$TT^{(q)} : \mathcal{L}^2(\Omega) \to \mathcal{L}^2(\mathcal{T}\times\Omega),$$

and we have

$$\int_{\Omega} |F(\eta)|^2\, d\omega(\eta) = \lim_{Q\to\infty} \sum_{n=1}^{Q} \int_{\Omega} |((TT)^{(q)}(F))(Q,n;\eta)|^2\, d\omega(\eta).$$

Proof. For all $Q \in \mathbb{N}$

$$\begin{aligned} &\sum_{n=1}^{Q} \int_{\Omega} |((TT)^{(q)}(F))(Q,n;\eta)|^2 d\omega(\eta) \\ &\quad = \sum_{n=1}^{Q} t_{Qn} \sum_{k=0}^{\infty} \sum_{j=1}^{2k+1} |(K_{h_n}^{(q)})^\wedge(k)|^2 |F^\wedge(k,j)|^2 \\ &\quad = \sum_{k=0}^{\infty} \sum_{j=1}^{2n+1} \left(\sum_{n=1}^{Q} t_{Qn} |(K_{h_n}^{(q)})^\wedge(k)|^2 \right) |F^\wedge(k,j)|^2. \end{aligned}$$

Now,

$$\lim_{Q\to\infty} \sum_{n=1}^{Q} t_{Qn} |(K_{h_n}^{(q)})^\wedge(k)|^2 = 1.$$

Therefore the corollary follows from Parseval's identity and the theory of spherical harmonics. □

By virtue of Toeplitz summation information a function is displayed on various levels of resolution (*multilevel approximation*). Of course, the number of levels needed in a particular numerical computation depends on the 'phenotype' of the function under consideration. In most cases, in fact, a 'one-level' computation using an adapted window parameter is of sufficient accuracy when the function is well-behaved. Multilevel computation, however, is indispensible, when 'high frequency phenomena' must be modelled (as, for example, in earth's gravitational field determination). The automatic 'zoom-in process' is stopped if the resolution of the a priori given data is reached and no further refinement of the window parameter is reasonable. Toeplitz summation is intended to improve convergence compared to the original sequence of singular integral values which has been verified by numerical experiments (cf. R. BRAND (1994)). Nevertheless, the above summation process is still not satisfactory as it often suffers from undesirable redundancy. R. BRAND *et al.* (1996), therefore, devised a method which keeps the conceptual advantages of the Topelitz transform but which is much more flexible in modelling local information of a function, eliminating redundancy as far as possible. This can be guaranteed in form of an adaptive hierarchical approximation procedure.

9.4 Combined Fourier and Toeplitz Transform

Of particular importance for future applications is the combination of the Toeplitz transform with Fourier techniques. This concept will be discussed now. The idea is to handle that part of a function which belongs to low order harmonics by Fourier techniques, while the remaining higher frequency part of the function is dealt with the Toeplitz transform. To be more explicit, we want to show that a given function $F \in \mathcal{X}(\Omega)$ admits an expansion of the form

$$F = \sum_{n=0}^{m} \sum_{j=1}^{2n+1} F^{\wedge}(n,j) Y_{n,j} + \sum_{p=1}^{Q} t_{Qp} J_{h_p}^{0,\ldots,m} * F,$$

where $\{J_{h_p}^{0,\ldots,m}\}$ is a kernel with vanishing Legendre coefficients up to a prescribed order m. Since locally supported radial basis functions are perhaps the most important ones from a numerical point of view, we restrict our attention to kernels of this type.

Theorem 9.4.1. *Let k, m be non-negative integers. Suppose that*

$$h_i = h + \frac{i-1}{m+2}(1-h), \quad i = 1, \ldots, m+2, \quad h \in (0, +1),$$

forms an equidistant partition $h = h_1 < \ldots < h_{m+2} < 1$ of the interval $[h, +1]$ (note that the values h_i depend on h, hence, we occasionally write $h_i = h_i(h)$). Then the following statements are valid:

(i) There exist uniquely determined $j_1, \ldots, j_{m+2} \in \mathbb{R}$ with $j_1 + \ldots + j_{m+2} = 1$ such that

$$J_h^{0,\ldots,m}(t) = \sum_{i=1}^{m+2} j_i L_{h_i}^{(k)}(t), \quad t \in [-1, +1],$$

satisfies $(J_h^{0,\ldots,m})^\wedge(n) = 0$ for $n = 0, \ldots, m$.

(ii) The uniquely determined coefficients $j_1, \ldots, j_{m+2}$ are independent of the choice of h.

Proof. Part (i) is just Theorem 5.8.8 with a special choice of $h_i, i = 1, \ldots, m+2$, and with the functions $L_{h_i}^{(k)} = ((B_{h_i}^{(k)})^\wedge(0))^{-1} B_{h_i}^{(k)}$ instead of $B_{h_i}^{(k)}$. So it remains to show the assertion (ii). Let $h \in (0, +1)$ be given, and assume that $j_1, \ldots, j_{m+2}$ are uniquely determined by condition (i). Let $\tilde{h} \in (h, +1)$ be arbitrarily chosen. It suffices to verify that

$$J_{\tilde{h}}^{0,\ldots,m}(t) = \sum_{i=1}^{m+2} j_i L_{\tilde{h}_i}^{(k)}(t), \quad t \in [-1, +1],$$

satisfies

$$\int_{-1}^{1} J_{\tilde{h}}^{0,\ldots,m}(\tilde{t})\tilde{t}^n \, d\tilde{t} = 0, \quad n = 0, \ldots, m. \tag{9.4.1}$$

In order to show this we make a change of variables

$$t(\tilde{t}) = \tilde{t}\frac{1-h}{1-\tilde{h}} + 1 - \frac{1-h}{1-\tilde{h}}, \quad \tilde{t} \in [\tilde{h}, +1],$$

in such a way that $t(h) = h$, $t(1) = 1$. Then we immediately see that

$$B_h^{(k)}(t(\tilde{t})) = B_{\tilde{h}}^{(k)}(\tilde{t}), \quad \tilde{t} \in [\tilde{h}, +1].$$

Even more,

$$B_{h_i(h)}^{(k)}(t(\tilde{t})) = B_{h_i(\tilde{h})}^{(k)}(\tilde{t}), \quad \tilde{t} \in [\tilde{h}, +1].$$

Simple manipulations yield

$$\int_{-1}^{1} J_{\tilde{h}}^{0,\ldots,m}(\tilde{t})\tilde{t}^n \, d\tilde{t}$$

$$
\begin{aligned}
&= \int_{\tilde{h}}^{1} \sum_{i=1}^{m+2} j_i L_{\tilde{h}_i}^{(k)}(\tilde{t})\tilde{t}^n \, d\tilde{t} \\
&= \int_{\tilde{h}}^{1} \sum_{i=1}^{m+2} j_i \left(2\pi \frac{1-\tilde{h}_i}{k+1}\right)^{-1} B_{\tilde{h}_i}^{(k)}(\tilde{t})\tilde{t}^n \, d\tilde{t} \\
&= \int_{h}^{1} \sum_{i=1}^{m+2} j_i \left(2\pi \frac{1-\tilde{h}_i}{k+1}\right)^{-1} B_{h_i}^{(k)}(t)\,(\tilde{t}(t))^n \frac{1-\tilde{h}}{1-h}\, dt \\
&= \int_{h}^{1} \sum_{i=1}^{m+2} j_i \frac{1-h_i}{1-\tilde{h}_i} L_{h_i}^{(k)}(t)\,(\tilde{t}(t))^n \frac{1-\tilde{h}}{1-h}\, dt \\
&= \int_{-1}^{1} \sum_{i=1}^{m+2} j_i L_{h_i}^{(k)}(t)\,(\tilde{t}(t))^n \, dt \\
&= \int_{-1}^{1} J_h^{0,\ldots,m}(t)\,(\tilde{t}(t))^n \, dt \\
&= 0,
\end{aligned}
$$

where we have used the fact that

$$
\frac{1-h_i}{1-\tilde{h}_i} = \frac{1-h}{1-\tilde{h}}, \qquad i = 1, \ldots, m+2.
$$

Consequently, $(\tilde{t}(t))^n$ with $\tilde{t}(t)$ given by

$$
\tilde{t}(t) = t\frac{1-\tilde{h}}{1-h} + 1 - \frac{1-\tilde{h}}{1-h}, \qquad t \in [h, +1],
$$

is a polynomial of degree n so that $(J_h^{0,\ldots,m})^\wedge(n) = 0,\ n = 0, \ldots, m.$ □

Since the values $j_i, i = 1, \ldots, m+2$, are independent of h it follows that

$$
2\pi \int_{-1}^{1} |J_h^{0,\ldots,m}(t)|\, dt \le M
$$

for all h with a suitable constant $M > 0$. Furthermore, because of the fact that $j_1 + \ldots + j_{m+2} = 1$, it follows that

$$
\lim_{\substack{h\to 1\\ h<1}} (J_h^{0,\ldots,m})^\wedge(n) = \sum_{i=1}^{m+2} j_i \lim_{\substack{h\to 1\\ h<1}} (L_{h_i(h)}^{(k)})^\wedge(n) = 1
$$

for all $n \ge m+1$. As a consequence of Theorem 8.1.2 we therefore obtain the following result.

Theorem 9.4.2. *Let $h \geq 0$ and $m \geq 0$ be fixed, and let $J_h^{0,\ldots,m}$ be given as in Theorem 9.4.1. Then, for $F \in \mathcal{X}(\Omega)$,*

$$\lim_{\substack{h \to 1 \\ h<1}} \left\| F - \sum_{n=0}^{m} \sum_{j=1}^{2n+1} F^\wedge(n,j) Y_{n,j} - J_h^{0,\ldots,m} * F \right\|_{\mathcal{X}(\Omega)} = 0.$$

In the same way as described above, the singular integral approach can be generalized to a multilevel approach. To be more specific, we are able to define a Toeplitz transform and a Toeplitz expansion even for the modified kernel $J_h^{0,\ldots,m}$. The formulation of this result is straightforward.

Theorem 9.4.3. *Let k and m be non-negative integers, and let $J_h^{0,\ldots,m}$ be given as in Theorem 9.4.1. Assume that $h_n \in (-1,+1), n = 0,1,\ldots$ is a sequence converging to the limit 1. Furthermore, suppose that (t_{Qj}) is a Toeplitz-triangular matrix. Then*

$$\lim_{Q \to \infty} \left\| F - \sum_{n=0}^{m} \sum_{j=1}^{2n+1} F^\wedge(n,j) Y_{n,j} - \sum_{p=1}^{Q} t_{Qp} J_{h_p}^{0,\ldots,m} * F \right\|_{\mathcal{X}(\Omega)} = 0$$

holds for all $F \in \mathcal{X}(\Omega)$.

9.5 Combined Gabor and Toeplitz Transform

Of more theoretical concern than practical use is a combination of both Gabor transform and Toeplitz singular integral transform: assume that $t_{Qj} (1 \leq j \leq Q)$ is a Toeplitz triangular matrix. Given an arbitrary function $F \in \mathcal{L}^2(\Omega)$ and a sequence $\{h_n\}, h_n \in (-1,+1)$ converging to the limit 1 as $n \to \infty$, we call

$$(GTE)(F) = \lim_{\substack{Q \to \infty \\ N \to \infty}} \sum_{n=1}^{Q} t_{Qn} \tag{9.5.1}$$
$$\times \sum_{p=0}^{N} \sum_{q=1}^{2p+1} \int_\Omega \int_\Omega F(\zeta) K_{h_n}(\eta \cdot \zeta) Y_{p,q}(\zeta) \, d\omega(\zeta) K_{h_n}(\eta \cdot) \, d\omega(\eta) Y_{p,q}$$

the *spherical Gabor–Toeplitz expansion of F* (relative to the window function K_{h_0}).

Definition 9.5.1. The Gabor–Toeplitz transform

$$((GTT)(F))(Q,n,p,q;\xi) : \mathcal{L}^2(\Omega) \to \mathcal{L}^2(\mathcal{T} \times \mathcal{J} \times \Omega)$$

where $\mathcal{L}^2(\mathcal{T} \times \mathcal{J} \times \Omega)$ is the space of all $H : \mathcal{T} \times \mathcal{J} \times \Omega \to \mathbb{R}$ satisfying

$$\lim_{\substack{Q\to\infty \\ N\to\infty}} \sum_{n=1}^{Q} \sum_{p=0}^{N} \sum_{q=1}^{2p+1} \int_\Omega |H(Q,n,p,q;\eta)|^2 \, d\omega(\eta) < \infty,$$

is defined by

$$((GTT)(F))(Q,n,p,q;\xi) = \int_\Omega F(\eta) Y_{p,q}(\eta) K_{h_n}(\xi \cdot \eta) \, d\omega(\eta)$$

for $Q \in \mathbb{N}, n = 1, \ldots, Q, p = 0, 1, \ldots, q = 1, \ldots, 2p+1, \xi \in \Omega$, and $F \in \mathcal{L}^2(\Omega)$.

The Gabor–Toeplitz transform includes '*rotations*'

$$R_\xi : K_{h_0}(\cdot\eta) \mapsto K_{h_0}(\xi \cdot \eta), \quad \eta \in \Omega$$

'*modulations*'

$$M_{p,q} : K_{h_0}(\cdot\eta) \mapsto K_{h_0}(\cdot\eta) Y_{p,q}, \quad \eta \in \Omega$$

and '*dilations*'

$$D_n : K_{h_0}(\cdot\eta) \mapsto K_{h_n}(\cdot\eta) \quad \eta \in \Omega.$$

We mention the following consequence of Theorem 9.1.3.

Theorem 9.5.2. *Suppose that $\{K_h\}$ is an approximate identity generating C-kernel. Let F be of class $\mathcal{C}(\Omega)$. If the series*

$$\sum_{p=0}^{\infty} \sum_{q=1}^{2p+1} \int_\Omega \int_\Omega F(\zeta) K_{h_n}(\eta \cdot \zeta) Y_{p,q}(\zeta) \, d\omega(\zeta) K_{h_n}(\xi \cdot \eta) \, d\omega(\eta) Y_{p,q}(\xi) \quad (9.5.2)$$

exists for all integers $n = 1, 2, \ldots$ and all $\xi \in \Omega$, then

$$F(\xi) = \lim_{\substack{Q\to\infty \\ N\to\infty}} \sum_{n=1}^{Q} t_{Qn} \sum_{p=0}^{N} \sum_{q=1}^{2p+1} \int_\Omega (GTT)(F)(Q,n,p,q;\eta)) K_{h_n}(\xi\cdot\eta) \, d\omega(\eta) Y_{p,q}(\xi) \quad (9.5.3)$$

holds for each $\xi \in \Omega$ (in the ordinary pointwise sense).

Proof. The assertion of Theorem 9.5.2 follows from the representation theorem of the theory of orthogonal expansions and the Toeplitz argument. □

Bibliographical Notes

The Gabor transform is constructed in close orientation to its counterpart in Euclidean spaces which is known as the 'Weyl–Heisenberg' wavelet transform (see for example C. E. HEIL, D.F. WALNUT (1989)). Spherical Toeplitz theory has been developed by W. FREEDEN and M. SCHREINER (1995A,B). Numerical experiments are due to R. BRAND (1994) and R. BRAND *et al.* (1996).

10

CONTINUOUS WAVELET TRANSFORM

Wavelet theory involves breaking up a complicated function (phenomenon) into many simple pieces at different scales and positions. To be more explicit, wavelet analysis allows us to represent a function $F \in \mathcal{L}^2(\Omega)$ as a sum of wavelets $\Psi_{\rho;\eta} = \Psi_\rho(\eta\cdot) = R_\eta D_\rho \Psi$, i.e. dilated and rotated copies of a mother wavelet, Ψ, an integrable function with vanishing integral. The coefficients of the decomposition are nothing but the inner products $(F, R_\eta D_\rho \Psi)_{\mathcal{L}^2(\Omega)}$. This makes wavelet theory comparable with approximation techniques known from Chapter 9. But the *continuous wavelet transform* distinguishes itself from the Gabor or Toeplitz transform in that the zooming-in and zooming-out capability can be organized by decomposing $\mathcal{L}^2(\Omega)$ by a nested sequence of approximating subspaces (*multiresolution analysis*) for a very general class of wavelets. The wavelet approach, therefore, allows greater flexibility with more desirable features such as '*P(acket)-scale discretization*', '*D(aubechies)-scale discretization*' etc., and readiness for better implementation.

The objective of this chapter is to generalize the important subject of the continuous wavelet transform to the spherical case. The spherical continuous wavelet transform arises naturally as a result of an integral reproducing formula that is closely related to the theory of singular integral operators. In the spherical continuous wavelet transform the rotation and dilation parameters vary continuously, i.e. such a reconstruction requires global information at *all* scales ρ and positions η. From the admissibility condition we see that $\Psi^\wedge(0)$ conventionally has to be 0, and, hence, Ψ has to oscillate. This together with a decay property suggested the name wavelet (French: ondelette). In accordance with W. FREEDEN, U. WINDHEUSER (1995, 1996) we present here a general spherical construction of integral wavelet transform formed by rotation and dilation as opposed to translation and dilation on the real line. Two different scale discretized versions of the spherical continuous wavelet transform are considered, namely P(acket)-scale discretized wavelets (also called wavelet packets) and D(auchbechies)-scale discretized wavelets. According to M. DUVAL-DESTIN *et al.* (1993) we mean by P-scale discretized wavelets (wavelet packets) families of functions generated from a single one by simple transformations and whose relative bandwidth is non-constant and can be matched to a given analysed function. Starting from an admissible spherical analysing wavelet Ψ, it is well known that one can associate with it a scaling function, Φ, and then mimic a (continuous) multiresolution construction. The analysis is described in detail. Examples of spherical wavelets are listed. Moreover, the role of the wavelet transform as a mapping from the one-parameter class $\mathcal{L}^2(\Omega)$ into the two-parameter class $\mathcal{L}^2((0,\infty) \times \Omega)$ is characterized by least-squares properties. In an appropriate way, the continuous decomposition over scales can be replaced by discrete ones. The corresponding sequence of scale parameters can be chosen arbitrarily, and the functions appearing in the decomposition (i.e. the wavelet packets) must be matched to that sequence. Another form of discretization is motivated by the ideas of I. DAUBECHIES (1990, 1992), yielding spherical reconstruction formulae comparable with those of the classical Euclidean theory. Fully discretized wavelet transforms are obtained via approximate integration on the sphere. It should be pointed out that the continuous wavelet transform presented here makes use of the rotational invariance on the sphere from which a double simplification is attained. On the one hand, the difficulties originating from the topological structure of the sphere, which is such that there is no coordinate system free of singularities, are under control. On the other hand, invariant pseudodifferential operators representing geophysical observables (such as gravitational or magnetic potential) become comparable via their spherical symbols. This, for example, gives us the possibility to derive a 'pocket guide to physical geodesy' (cf. R. RUMMEL, M. VAN GELDEREN (1995)) by use of spherical wavelets (for more details see W. FREEDEN, U. WINDHEUSER (1995, 1996, 1997)). Furthermore, the principle of developing wavelets with vanishing moments of order m enables us to combine the mth order expansion of the gravitational potential of the earth in terms of spherical harmonics with its expansion in terms of spherical wavelets (of order m). In doing so a triple advantage is indicated: the modelling of the earth's gravitational potential in terms of a spherical harmonic expansion up to an order m takes the low frequency part, the description of the higher order frequency part is performed by wavelet techniques, and an integrated approach via invariant pseudodifferential operator relations is guaranteed for both the spherical harmonic and the spherical wavelet level.

The chapter is organized as follows. In Section 10.1 the linear approach to continuous wavelet transform is presented. Two cases are distinguished, namely wavelets of order $m = -1$ and $m \geq 0$. As representative examples the Abel–Poisson kernel, the Gauß–Weierstraß kernel and 'axisymmetric locally supported kernels' are mentioned. Section 10.2 is concerned with the bilinear theory of spherical wavelets with vanishing moments up to a prescribed order m. Section 10.3 opens perspectives of scale discretization. We show that singular integral operators forming a semigroup of contraction operators of class (C_0) (like Abel–Poisson or Gauß–Weierstraß operators) leads to pyramid algorithms. In Section 10.4 scale and space discretized wavelet transform is established in canonical way. Finally, the promised combined spherical harmonic and wavelet expansion is outlined, the spherical harmonic expansion of a function being responsible for the 'low frequency part' and the wavelet expansion being appropriate for the 'high frequency part'.

It is believed that discretizations of the continuous wavelet transform as discussed here will be a very useful tool in many areas of geomathematics.

10.1 Continuous Wavelet Transform (Linear Theory)

We start with the introduction of *spherical wavelets (of order* $m = -1$*).*

Definition 10.1.1. Assume that $\alpha : [0, \infty) \to \mathbb{R}$ is a positive weight function. Let $\{\Psi_\rho\}, \rho \in (0, \infty)$, be a subfamily of $\mathcal{L}^2[-1, +1]$ such that the following *admissibility conditions* are satisfied:

(i) for $n = 0, 1, \ldots$

$$\int_0^\infty (\Psi_\rho)^\wedge(n)\alpha(\rho)\, d\rho = 1 \tag{10.1.1}$$

(ii) for $R \in (0, \infty)$

$$\sum_{n=0}^\infty \frac{2n+1}{4\pi} \int_R^\infty (\Psi_\rho)^\wedge(n)\alpha(\rho)\, d\rho < \infty \tag{10.1.2}$$

(iii) for all $R \in (0, \infty)$

$$2\pi \int_{-1}^{+1} \left| \int_R^\infty \Psi_\rho(t)\alpha(\rho)\, d\rho \right| dt \leq T, \tag{10.1.3}$$

where T is a positive constant independent of R.

Then $\{\Psi_\rho\}$ is called *spherical (linear) wavelet (of order* $m = -1$*).* The function $\Psi = \Psi_1$ (i.e. $\rho = 1$) is called *spherical mother wavelet.*

Remark. In Euclidean space $\mathbb{R}$ the choice $\alpha(\rho) = \rho^{-1}$, $\rho > 0$, is obligatory for the weight function, i.e. the measure $d\rho/\rho$ is demanded as Haar measure of the affine group. In our spherical approach dilation and rotation are not in relation to each other. As a matter of fact it turns out that any measure of the form $d\mu(\rho) = \alpha(\rho)\, d\rho$, α positive, can be used within our considerations.

Let $\Psi_{\rho;\eta}$ be defined as follows:

$$\Psi_{\rho;\eta} : \xi \mapsto \Psi_{\rho;\eta}(\xi) = \Psi_\rho(\xi \cdot \eta) = R_\eta D_\rho \Psi(\cdot \xi), \quad \xi \in \Omega,$$

where the '*η-rotation operator*' R_η and the '*ρ-dilation operator*' D_ρ, respectively, are given by

$$R_\eta \quad : \quad \Psi(\cdot\xi) \mapsto R_\eta \Psi(\cdot\xi) = \Psi(\eta \cdot \xi), \tag{10.1.4}$$

$$D_\rho \quad : \quad \Psi(\cdot\xi) \mapsto D_\rho \Psi(\cdot\xi) = \Psi_\rho(\cdot\xi). \tag{10.1.5}$$

The associated (linear) *wavelet transform* is defined by

$$\begin{aligned}(WT)(F)(\rho;\eta) &= (\Psi_{\rho;\eta}, F)_{\mathcal{L}^2(\Omega)} \qquad (10.1.6)\\ &= \int_\Omega \Psi_\rho(\xi\cdot\eta) F(\xi)\, d\omega(\xi)\end{aligned}$$

for all $F \in \mathcal{L}^2(\Omega)$.

It is of interest to develop a formula for reconstructing a function F from its wavelet transform.

Theorem 10.1.2 *(reconstruction formula). Let $\{\Psi_\rho\}$ be a wavelet (of order $m = -1$). Then*

$$F = \int_0^\infty (WT)(F)(\rho; \cdot)\alpha(\rho)\, d\rho$$

holds for all $F \in \mathcal{L}^2(\Omega)$ (in $\|\cdot\|_{\mathcal{L}^2(\Omega)}$-sense).

Proof. Let R be a positive number. Then it is not hard to see that

$$\begin{aligned}\int_R^\infty (WT)(F)(\rho;\eta)\alpha(\rho)\, d\rho &= \int_\Omega \Phi_R(\zeta\cdot\eta)F(\zeta)\, d\omega(\zeta)\\ &= (\Phi_R * F)(\eta), \qquad (10.1.7)\end{aligned}$$

where we have used the abbreviation

$$\Phi_R(\zeta\cdot\eta) = \sum_{n=0}^\infty \frac{2n+1}{4\pi} \int_R^\infty (\Psi_\rho)^\wedge(n)\alpha(\rho)\, d\rho P_n(\zeta\cdot\eta), \quad (\zeta,\eta) \in \Omega^2. \tag{10.1.8}$$

According to our construction, Φ_R is square-integrable on $[-1,+1]$ for all $R > 0$. Moreover,

$$\lim_{\substack{R\to 0\\ R>0}} (\Phi_R)^\wedge(n) = 1$$

for all $n \in \mathbb{N}_0$. From (10.1.3) we are able to deduce that the kernel $\{\Phi_R\}$ is uniformly bounded in the sense of (8.1.3). Therefore we find from standard arguments of the theory of singular integrals

$$\lim_{\substack{R\to 0\\ R>0}} \int_\Omega \Phi_R(\cdot\eta)F(\eta)\,d\omega(\eta) = \sum_{n=0}^{\infty} \frac{2n+1}{4\pi} \int_\Omega F(\eta)P_n(\cdot\eta)\,d\omega(\eta)$$

in $\|\cdot\|_{\mathcal{L}^2(\Omega)}$-sense, as required. □

Note that the defining properties of a spherical wavelet (of order -1) do not presume the zero-mean property of Ψ_ρ, or, equivalently, the vanishing of the Legendre transform at 0: $(\Psi_\rho)^\wedge(0) = 0$. The wavelets constructed in this way, therefore, do not satisfy a substantial condition of the classical (Euclidean) wavelet concept on the real line.

In what follows we study spherical wavelets for which the zero-mean property is guaranteed and the usual wavelet terminology is quite appropriate.

Definition 10.1.3. Let $\{\Psi_\rho\}, \rho \in (0,\infty)$, be a subfamily of $\mathcal{L}^2[-1+1]$ such that the following *admissibility conditions* are satisfied:

(i) for $n = 1, 2, \ldots$

$$\int_0^\infty (\Psi_\rho)^\wedge(n)\alpha(\rho)\,d\rho = 1, \tag{10.1.9}$$

(ii) for $\rho \in (0,\infty)$

$$(\Psi_\rho)^\wedge(0) = 0, \tag{10.1.10}$$

(iii) for $R \in (0,\infty)$

$$\sum_{n=1}^{\infty} \frac{2n+1}{4\pi} \int_R^\infty (\Psi_\rho)^\wedge(n)\alpha(\rho)\,d\rho < \infty, \tag{10.1.11}$$

(iv) for all $R \in (0,\infty)$

$$2\pi \int_{-1}^{+1} \left| \int_R^\infty \Psi_\rho(t)\alpha(\rho)\,d\rho \right| dt \le T, \tag{10.1.12}$$

where T is a positive constant independent of R.

Then $\{\Psi_\rho\}$ is called a *spherical (linear) wavelet of order* 0. The function $\Psi = \Psi_1$ (i.e. $\rho = 1$) is called a *spherical mother wavelet.* The associated *(linear) wavelet transform* is defined by

$$(WT)(F)(\rho;\eta) = (\Psi_{\rho;\eta}, F)_{\mathcal{L}^2(\Omega)} \tag{10.1.13}$$

for all $F \in \mathcal{L}^2(\Omega)$.

The condition $(\Psi_\rho)^\wedge(0) = 0$ is satisfied (cf. (10.1.10)), and, in particular, Ψ cannot just be a 'bump function' (like the Abel–Poisson or the Gauß–Weierstraß kernel), but must oscillate. This, together with the 'decay properties' stated in Definition 10.1.3 ensures that Ψ is a wavelet. Furthermore, as necessary condition we immediately obtain from (10.1.9)

$$\lim_{\substack{\rho\to 0\\ \rho>0}} (\Psi_\rho)^\wedge(n) = 0$$

for $n = 1, 2, \ldots$.

We can now formulate a reconstruction formula.

Theorem 10.1.4. *Let $\{\Psi_\rho\}$ be a wavelet of order 0. Suppose that F is of class $\mathcal{L}^2(\Omega)$ satisfying $F^\wedge(0,1) = 0$. Then*

$$F = \int_0^\infty (WT)(F)(\rho;\cdot)\alpha(\rho)\,d\rho$$

(in $\|\cdot\|_{\mathcal{L}^2(\Omega)}$-sense).

Proof. Choose a positive number R. Then we find

$$\begin{aligned}\int_R^\infty (WT)(F))(\rho;\eta)\alpha(\rho)\,d\rho &= \int_\Omega \Phi_R(\zeta\cdot\eta)F(\zeta)\,d\omega(\zeta) \quad (10.1.14)\\ &= (\Phi_R * F)(\eta),\end{aligned}$$

where

$$\Phi_R(\zeta\cdot\eta) = \frac{1}{4\pi} + \sum_{n=1}^{\infty} \frac{2n+1}{4\pi}\int_R^\infty (\Psi_\rho)^\wedge(n)\alpha(\rho)\,d\rho P_n(\zeta\cdot\eta). \quad (10.1.15)$$

Now $\Phi_R \in \mathcal{L}^2[-1,+1]$ for all $R > 0$, and we have

$$\lim_{\substack{R\to 0\\ R>0}} (\Phi_R)^\wedge(n) = 1 \quad (10.1.16)$$

for all $n \in \mathbb{N}$. Using $(\Phi_R)^\wedge(0) = 1$ and observing the uniform boundedness resulting from (10.1.12) the assertion of Theorem 10.1.4 follows from the known arguments of the theory of singular integrals. □

Remark. The linear wavelet analysis discussed here may be formally interpreted as bilinear analysis in taking the Dirac distribution:

$$F(\xi) = \int_0^\infty \int_\Omega (WT)(F)(\rho;\eta)\delta(\eta\cdot\xi)\, d\omega(\eta)\alpha(\rho)\, d\rho,$$

where

$$\delta(\xi\cdot\eta) = \sum_{n=0}^{\infty} \frac{2n+1}{4\pi} P_n(\xi\cdot\eta).$$

Decomposition is done by the wavelet, reconstruction is formally performed by the Dirac distribution.

Definition 10.1.5. Assume that $\{\Psi_\rho\}$ is a wavelet of order 0. Then the associated *scaling function* $\{\Phi_R\}, R \in (0,\infty)$, is defined by

$$(\Phi_R)^\wedge(n) = \begin{cases} \int_R^\infty (\Psi_\rho)^\wedge(n)\alpha(\rho)\, d\rho & \text{for} \quad 1,2,\ldots \\ 1 & \text{for} \quad n=0. \end{cases}$$

It can be readily seen that

$$\int_R^\infty (WT)(F)(\rho;\eta)\alpha(\rho)\, d\rho = (\Phi_{R;\eta}, F)_{\mathcal{L}^2(\Omega)}$$

holds for all $F \in \mathcal{L}^2(\Omega)$ with $F^\wedge(0,1) = 0$. This enables us to extend our results to general topologies.

Corollary 10.1.6. *Let $\{\Psi_\rho\}, \rho \in (0,\infty)$, be a wavelet of order 0. Then, for $F \in \mathcal{X}(\Omega)$ with $\int_\Omega F(\eta)\, d\omega(\eta) = 0$ (note that $(WT)(F)(\rho;\eta) = \Psi_{\rho;\eta} * F$)*

$$\lim_{\substack{R\to 0\\ R>0}} \left\| F - \int_R^\infty (WT)(F)(\rho;\cdot)\alpha(\rho)\, d\rho \right\|_{\mathcal{X}(\Omega)} = 0. \tag{10.1.17}$$

In particular, the relation (10.1.17) holds in uniform sense for $F \in C(\Omega)$ with $F^\wedge(0,1) = 0$.

The construction of the (linear) wavelet may be organized independently of the 'scaling function' Φ_R. This is shown by the following corollary.

Corollary 10.1.7. *Let the non-negative kernel $\{\Phi_R\}$ define an approximate identity. Furthermore, assume that the Legendre coefficients $(\Phi_R)^\wedge(n)$, $n \in \mathbb{N}$, (as functions of R) are differentiable and decreasing. Moreover, suppose that*

$$(\Phi_R)^\wedge(0) = 1,\ R \in (0,\infty); \qquad \lim_{R\to\infty} (\Phi_R)^\wedge(n) = 0, \quad n = 1,2,\ldots .$$

Then the corresponding wavelet $\{\Psi_\rho\}$ of order 0 is given by

$$(\Psi_\rho)^\wedge(n) = -(\alpha(\rho))^{-1} \frac{d}{d\rho}(\Phi_\rho)^\wedge(n)$$

for $\rho \in (0,\infty)$ and $n = 0,1,\ldots$.

The following examples arise quite naturally from the Abel–Poisson and the Gauß–Weierstraß kernel.

Abel–Poisson Wavelet of Order 0. Take as $\{\Phi_R\}$ the Abel–Poisson kernel. Then the Legendre coefficients of the Abel–Poisson wavelets read as follows:

$$(\Psi_\rho)^\wedge(n) = -(\alpha(\rho))^{-1}\frac{d}{d\rho}\mathrm{e}^{-n\rho} = n(\alpha(\rho))^{-1}\mathrm{e}^{-n\rho}, \quad n \in \mathbb{N}_0.$$

Gauß–Weierstraß Wavelet of Order 0. In this case we find

$$(\Psi_\rho)^\wedge(n) = -(\alpha(\rho))^{-1}\frac{d}{d\rho}\mathrm{e}^{-n(n+1)\rho} = n(n+1)(\alpha(\rho))^{-1}\mathrm{e}^{-n(n+1)\rho}, \quad n \in \mathbb{N}_0.$$

The linear wavelet analysis is very often used in Euclidean spaces for signal analysis since it produces a simple reconstruction formula. However, linear theory does not lead to a table of wavelet coefficients as in the bilinear case. Scale discretization of (linear) wavelet theory can be found in W. FREEDEN, U. WINDHEUSER (1996). The introduction of spherical (linear) wavelets of order $m > 0$ is obvious. In the next section we are concerned with bilinear analysis of spherical wavelets of order $m \geq 0$ which provides us with the convenient scheme for the construction of wavelet bases as known from the Eulidean case (see for example C.K. CHUI (1991, 1992), I. DAUBECHIES (1990), Y. MEYER (1993)).

10.2 Continuous Wavelet Transform (Bilinear Theory)

In Euclidean theory (bilinear) continuous wavelet decompositions have been introduced (or reintroduced, since similar tools have been used by mathematicians for a long time in so-called Calderon–Zygmund theory) by A. GROSSMANN, J. MORLET (1984), and many applications have been developed, in particular in signal and image processing. In the meantime a number of monographs has been published (e.g. C.K. CHUI (1991, 1992), I. DAUBECHIES (1990, 1992), A.K. LOUIS *et al.* (1994), Y. MEYER (1993)). Our purpose now is to present the spherical counterpart of continuous wavelet decomposition (as proposed by W. FREEDEN, U. WINDHEUSER (1995, 1996, 1997)).

10.2.1 *Reconstruction Formula*

Let us start from the basic definition of spherical wavelets of order $m \geq 0$.

Definition 10.2.1. Let $\alpha : [0,\infty) \to \mathbb{R}$ be a positive weight function. A subfamily $\{\Psi_\rho\}$, $\rho \in (0,\infty)$, of the space $\mathcal{L}^2[-1,1]$ is called *spherical (bilinear) wavelet of order* m ($m \geq 0$), if $\{\Psi_\rho^\wedge(n)\}, n = 0, 1, \ldots$, satisfies the following *admissibility conditions:*

(i) for $n = m+1, m+2, \ldots$

$$\int_0^\infty (\Psi_\rho^\wedge(n))^2 \alpha(\rho)\, d\rho = 1, \tag{10.2.1}$$

(ii) for $\rho \in (0,\infty)$ and $n = 0, \ldots, m$

$$\Psi_\rho^\wedge(n) = 0, \tag{10.2.2}$$

(iii) for $R \in (0,\infty)$

$$\sum_{n=m+1}^{\infty} \frac{2n+1}{4\pi} \int_R^\infty (\Psi_\rho^\wedge(n))^2 \alpha(\rho)\, d\rho < \infty, \tag{10.2.3}$$

(iv) for all $R \in (0,\infty)$

$$2\pi \int_{-1}^{+1} \left| \int_R^\infty \Psi_\rho^{(2)}(t) \alpha(\rho)\, d\rho \right| dt \le T, \tag{10.2.4}$$

where T is a positive constant independent of R.

As above, the function $\Psi = \Psi_1$ (i.e. $\rho = 1$) is called the *spherical mother wavelet.*

As necessary condition we immediately obtain from (10.2.1)

$$\lim_{\substack{\rho\to 0\\ \rho>0}} \Psi_\rho^\wedge(n) = 0$$

for $n = m+1, m+2, \ldots$. Again, the condition (10.2.2) justifies the notion *wavelet* of order m. The remaining admissibility conditions are of later importance in that they assure the convergence of an associated singular integral procedure in $\|\cdot\|_{\mathcal{X}(\Omega)}$-sense.

If $F, G \in \mathcal{L}^2(\Omega)$ have vanishing moments of order $\le m$, i.e. $F^\wedge(n,j) = G^\wedge(n,j) = 0$ for $n = 0, \ldots, m$, $j = 1, \ldots, 2n+1$, it follows from Parseval's identity and (10.2.1) that

$$\begin{aligned}
&\int_\Omega \int_0^\infty (F, \Psi_{\rho;\eta})_{\mathcal{L}^2(\Omega)} (G, \Psi_{\rho;\eta})_{\mathcal{L}^2(\Omega)} \alpha(\rho)\, d\rho\, d\omega(\eta) \\
&\quad = \int_0^\infty \sum_{n=m+1}^{\infty} \sum_{j=1}^{2n+1} F^\wedge(n,j) G^\wedge(n,j) (\Psi_\rho^\wedge(n))^2 \alpha(\rho)\, d\rho \\
&\quad = \sum_{n=m+1}^{\infty} \sum_{j=1}^{2n+1} F^\wedge(n,j) G^\wedge(n,j)
\end{aligned} \tag{10.2.5}$$

$$= \quad (F, G)_{\mathcal{L}^2(\Omega)}.$$

Denote by $\mathcal{L}^2((0,\infty) \times \Omega)$ the space of all integrable functions $H : (0,\infty) \times \Omega \to \mathbb{R}$ such that

$$\int_\Omega \int_0^\infty |H(\rho;\eta)|^2 \alpha(\rho)\, d\rho\, d\omega(\eta) < \infty. \tag{10.2.6}$$

On the space $\mathcal{L}^2((0,\infty) \times \Omega)$ we are able to impose an inner product $(\cdot,\cdot)_{\mathcal{L}^2((0,\infty)\times\Omega)}$ corresponding to the norm

$$\|H\|_{\mathcal{L}^2((0,\infty)\times\Omega)} = \left(\int_\Omega \int_0^\infty |H(\rho;\eta)|^2 \alpha(\rho)\, d\rho\, d\omega(\eta)\right)^{1/2}. \tag{10.2.7}$$

From (10.2.5) it follows that

$$\left((F, \Psi_{\cdot;\cdot})_{\mathcal{L}^2(\Omega)}, (G, \Psi_{\cdot;\cdot})_{\mathcal{L}^2(\Omega)}\right)_{\mathcal{L}^2((0,\infty)\times\Omega)} = (F, G)_{\mathcal{L}^2(\Omega)} \tag{10.2.8}$$

and

$$\|(F, \Psi_{\cdot;\cdot})_{\mathcal{L}^2(\Omega)}\|^2_{\mathcal{L}^2((0,\infty)\times\Omega)} = \|F\|^2_{\mathcal{L}^2(\Omega)}. \tag{10.2.9}$$

This leads us to introduce the *continuous wavelet transform* as follows.

Definition 10.2.2. Let $\{\Psi_\rho\}, \rho \in (0,\infty)$, be a spherical wavelet of order m. Then the spherical wavelet transform $(WT) : \mathcal{L}^2(\Omega) \to \mathcal{L}^2((0,\infty) \times \Omega)$ is defined by

$$(WT)(F)(\rho;\eta) = (\Psi_{\rho;\eta}, F)_{\mathcal{L}^2(\Omega)} = \int_\Omega \Psi_{\rho;\eta}(\xi) F(\xi)\, d\omega(\xi). \tag{10.2.10}$$

From our above arguments it is clear that the continuous wavelet transform (WT) is invertible on the space of functions $F \in \mathcal{L}^2(\Omega)$ satisfying $F^\wedge(n,j) = 0$ for $n = 0,\ldots,m$, $j = 1,\ldots,2n+1$.

Theorem 10.2.3 *(reconstruction formula). Let $\{\Psi_\rho\}$, $\rho \in (0,\infty)$, be a spherical wavelet of order m. Suppose that $F \in \mathcal{L}^2(\Omega)$ satisfies $F^\wedge(n,j) = 0$ for $n = 0,\ldots,m$, $j = 1,\ldots,2n+1$. Then*

$$F = \int_\Omega \int_0^\infty (WT)(F)(\rho;\eta) \Psi_{\rho;\eta}(\cdot) \alpha(\rho)\, d\rho\, d\omega(\eta)$$

(in the sense of $\|\cdot\|_{\mathcal{L}^2(\Omega)}$).

The reconstruction formula, in fact, is based on the simple idea of dilation and rotation of the mother wavelet.

For reasons of readability we now make some further remarks showing the intimate relations between continuous wavelet transform and singular integral theory, thereby explaining the canonical bridge between spherical wavelet and spherical scaling function. We start with the auxiliary function $H_R : t \mapsto H_R(t)$, $t \in [-1, +1]$, given by

$$H_R(t) = \sum_{n=0}^{m} \frac{2n+1}{4\pi} P_n(t) + \sum_{n=m+1}^{\infty} \frac{2n+1}{4\pi} \left(\int_R^{\infty} (\Psi_\rho^\wedge(n))^2 \alpha(\rho)\, d\rho \right) P_n(t).$$

According to formula (10.2.3) H_R is of class $\mathcal{L}^2[-1,+1]$, and the kernel $\{H_R\}$ is uniformly bounded. Moreover, by virtue of (10.2.1), we find

$$\lim_{\substack{R\to 0\\ R>0}} H_R^\wedge(n) = \int_0^{\infty} (\Psi_\rho^\wedge(n))^2 \alpha(\rho)\, d\rho = 1$$

for $n = m+1, m+2, \ldots$. A simple calculation shows that

$$\begin{aligned} &\int_\Omega \int_R^{\infty} (WT)(F)(\rho;\eta)\Psi_{\rho;\eta}(\xi)\alpha(\rho)\, d\rho\, d\omega(\eta) \\ &\quad = \int_\Omega H_R(\xi\cdot\zeta)F(\zeta)\, d\omega(\zeta) \end{aligned}$$

holds for an arbitrary value $R > 0$ and all $F \in \mathcal{L}^2(\Omega)$ satisfying $F^\wedge(n,j) = 0$, $n = 0,\ldots,m$, $j = 1,\ldots,2n+1$. Thus it is clear from the theory of spherical singular integrals that

$$F = \lim_{\substack{R\to 0\\ R>0}} (H_R * F) = \lim_{\substack{R\to 0\\ R>0}} \int_\Omega H_R(\cdot\zeta)F(\zeta)\, d\omega(\zeta)$$

holds in the sense of $\|\cdot\|_{\mathcal{L}^2(\Omega)}$, which is nothing else than the result of Theorem 10.2.3.

10.2.2 *Scaling Function*

From the arguments given above it becomes obvious that wavelet reconstruction utilizes essential techniques of singular integration theory. We will investigate this aspect more deeply by specifying a 'corresponding spherical scaling function' to the spherical wavelet of order m.

Definition 10.2.4. Let $\{\Psi_\rho\}, \rho \in (0,\infty)$, be a spherical wavelet of order m. Then the corresponding *spherical scaling function* $\{\Phi_R\}, R \in (0,\infty)$, is defined by its table $\{\Phi_R^\wedge(n)\}$ of Legendre coefficients as follows:

$$\Phi_R^\wedge(n) = \begin{cases} 1, & n = 0,\ldots,m \\ \left(\int_R^\infty (\Psi_\rho^\wedge(n))^2 \alpha(\rho)\, d\rho\right)^{1/2}, & n = m+1, m+2, \ldots . \end{cases}$$

From (10.2.3) we see that

$$\sum_{n=0}^{\infty} \frac{2n+1}{4\pi} (\Phi_R^\wedge(n))^2 = \frac{(m+1)^2}{4\pi} + \sum_{n=m+1}^{\infty} \frac{2n+1}{4\pi} \int_R^\infty (\Psi_\rho^\wedge(n))^2 \alpha(\rho)\, d\rho < \infty, \tag{10.2.11}$$

hence, $\Phi_R \in \mathcal{L}^2[-1,+1]$ for each $R \in (0,\infty)$. Furthermore we have for all $(\xi,\zeta) \in \Omega^2$

$$H_R(\xi\cdot\zeta) = \int_\Omega \Phi_R(\xi\cdot\eta)\Phi_R(\zeta\cdot\eta)\, d\omega(\eta) = \Phi_R^{(2)}(\xi\cdot\zeta),$$

that is $H_R = \Phi_R^{(2)} = \Phi_R * \Phi_R$ for each $R \in (0,\infty)$. It is also important to mention that

$$\begin{aligned} &\int_\Omega \int_R^\infty (F, \Psi_{\rho;\eta})_{\mathcal{L}^2(\Omega)}\ \Psi_{\rho;\eta}(\xi)\alpha(\rho)\, d\rho\, d\omega(\eta) \\ &= \int_\Omega (F, \Phi_{R;\eta})_{\mathcal{L}^2(\Omega)}\ \Phi_{R;\eta}(\xi)\, d\omega(\eta) \end{aligned} \tag{10.2.12}$$

holds for all $F \in \mathcal{L}^2(\Omega)$ with $F^\wedge(n,j) = 0$ for $n = 0,\ldots,m$, $j = 1,\ldots,2n+1$ (in the sense of $\|\cdot\|_{\mathcal{L}^2(\Omega)}$), where we have used the usual abbreviations

$$\begin{aligned} \Psi_{\rho;\eta}(\xi) &= \Psi_\rho(\xi\cdot\eta) &= R_\eta D_\rho \Psi(\xi\cdot), \\ \Phi_{R;\eta}(\xi) &= \Phi_R(\xi\cdot\eta) &= R_\eta D_R \Phi(\xi\cdot); \end{aligned} \tag{10.2.13}$$

for $(\xi,\eta) \in \Omega^2, 0 < \rho, R < \infty$. In connection with Theorem 10.2.3 this leads us to the following.

Corollary 10.2.5. *Let $\{\Psi_\rho\}, \rho \in (0,\infty)$, be a spherical wavelet of order m. Assume that $\{\Phi_R\}, R \in (0,\infty)$, is the corresponding spherical scaling function. Then*

$$F = \lim_{\substack{R\to 0\\ R>0}} \int_\Omega (F, \Phi_{R;\eta})_{\mathcal{L}^2(\Omega)} \Phi_{R;\eta}(\cdot)\, d\omega(\eta)$$

holds (in $\|\cdot\|_{\mathcal{L}^2(\Omega)}$-sense) for all $F \in \mathcal{L}^2(\Omega)$ with $F^\wedge(n,j) = 0$ for $n = 0,\ldots,m$, $j = 1,\ldots,2n+1$.

In other words, $\{\Phi_R^{(2)}\}, R \in (0, \infty)$, is a (uniformly bounded) kernel of an approximate identity. But this observation enables us to formulate the reconstruction formula in the general $\mathcal{X}(\Omega)$-context as introduced in Section 8.1.

Corollary 10.2.6. *Under the assumptions of Corollary 10.2.5*

$$\lim_{\substack{R \to 0 \\ R > 0}} \left\| F - \int_\Omega \int_R^\infty (WT)(F)(\rho;\eta) \Psi_{\rho;\eta}(\cdot) \alpha(\rho) \, d\rho \, d\omega(\eta) \right\|_{\mathcal{X}(\Omega)} = 0$$

for all $F \in \mathcal{X}(\Omega)$ with vanishing moments $\leq m$, i.e. $F^\wedge(n,j) = 0$ for $n = 0, \ldots, m$, $j = 1, \ldots, 2n+1$ (note that $(WT)(F)(\rho;\eta) = \Psi_{\rho;\eta} * F$). In particular,

$$\lim_{\substack{R \to 0 \\ R > 0}} \sup_{\xi \in \Omega} \left| F(\xi) - \int_\Omega \int_R^\infty (WT)(F)(\rho;\eta) \Psi_{\rho;\eta}(\xi) \alpha(\rho) \, d\rho \, d\omega(\eta) \right| = 0$$

holds for all functions $F \in \mathcal{C}(\Omega)$ with $F^\wedge(n,j) = 0$ for $n = 0, \ldots, m$, $j = 1, \ldots, 2n+1$.

Conversely, to a suitable kernel $\{\Phi_\rho\}, \rho \in (0, \infty)$, generating an approximate identity in $\mathcal{X}(\Omega)$, we are able to determine in constructive way an (associated) spherical wavelet $\{\Psi_\rho\}, \rho \in (0, \infty)$. To be more specific, we have the following corollary.

Corollary 10.2.7. *Suppose that $\{\Phi_R\}, R \in (0, \infty)$, is the uniformly bounded kernel of an approximate identity. Furthermore, let the coefficients $\Phi_R^\wedge(n), R \in (0, \infty), n = m+1, m+2, \ldots$ as function of R be differentiable and decreasing. Moreover, assume that*

$$\begin{aligned} \Phi_R^\wedge(n) &= 1, \; n = 0, \ldots, m, \\ \lim_{R \to \infty} \Phi_R^\wedge(n) &= 0, \; n = m+1, m+2, \ldots. \end{aligned}$$

Then the associated spherical wavelet $\{\Psi_\rho\}, \rho \in (0, \infty)$, of order m is given by

$$\Psi_\rho^\wedge(n) = (-(\alpha(\rho))^{-1} \frac{d}{d\rho} (\Phi_\rho^\wedge(n))^2)^{1/2} \tag{10.2.14}$$

for $\rho \in (0, \infty)$ and $n = 0, 1, \ldots$.

Proof. It is clear that our assumptions guarantee the non-negativity of $-(\alpha(\rho))^{-1}(d/d\rho)(\Phi_\rho^\wedge(n))^2$ for $\rho \in (0, \infty)$. We have to prove that the admissibility conditions of Definition 10.2.2 are obtainable from (10.2.14). First it follows that

$$\int_R^\infty (\Psi_\rho^\wedge(n))^2 \alpha(\rho)\, d\rho = -\int_R^\infty (\alpha(\rho))^{-1} \frac{d}{d\rho} (\Phi_\rho^\wedge(n))^2 \alpha(\rho)\, d\rho = (\Phi_R^\wedge(n))^2 \tag{10.2.15}$$

for each $R \in (0,\infty)$ and $n = m+1, m+2, \ldots$. Obviously,

$$\left(-(\alpha(\rho))^{-1} \frac{d}{d\rho} (\Phi_\rho^\wedge(n))^2 \right)^{1/2} = 0 \tag{10.2.16}$$

for $n = 0, \ldots, m$. Hence, $\{\Psi_\rho\}$ is derivable from the kernel $\{\Phi_R\}$ as indicated by Definition 10.2.4. Furthermore the following statements are easily verified:

(i) for $n = m+1, m+2, \ldots$

$$\lim_{\substack{R\to 0\\ R>0}} \int_R^\infty (\Psi_\rho^\wedge(n))^2 \alpha(\rho)\, d\rho = \lim_{\substack{R\to 0\\ R>0}} (\Phi_R^\wedge(n))^2 = 1,$$

(ii) for $\rho \in (0,\infty)$ and $n = 0, \ldots, m$

$$\Psi_\rho^\wedge(n) = 0,$$

(iii) for $R \in (0,\infty)$

$$\sum_{n=m+1}^\infty \frac{2n+1}{4\pi} \int_R^\infty (\Psi_\rho^\wedge(n))^2 \alpha(\rho)\, d\rho = \sum_{n=m+1}^\infty \frac{2n+1}{4\pi} (\Phi_R^\wedge(n))^2 < \infty,$$

(iv) for all $R \in (0,\infty)$

$$2\pi \int_{-1}^{+1} \left| \int_R^\infty \Psi_\rho^{(2)}(t) \alpha(\rho)\, d\rho \right| dt = 2\pi \int_{-1}^{+1} |\Phi_R^{(2)}(t)|\, dt \le T.$$

Consequently the admissibility conditions are satisfied, hence, $\{\Psi_\rho\}, \rho \in (0,\infty)$, is a spherical wavelet of order m in the sense of Definition 10.2.1. □

In terms of filtering $\{\Psi_\rho\}$ and $\{\Phi_R\}$, respectively, may be interpreted as *band-pass filter* and *low-pass filter*. Correspondingly, the convolution operators are given by

$$\begin{aligned} R_\rho(F) &= \Psi_\rho * \Psi_\rho * F, \qquad F \in \mathcal{L}^2(\Omega), \\ P_R(F) &= \Phi_R * \Phi_R * F, \quad F \in \mathcal{L}^2(\Omega), \end{aligned}$$

and the spherical harmonic coefficients read as follows:

$$((FT)R_\rho(F))(n,j) = (R_\rho(F))^\wedge(n,j) = F^\wedge(n,j)(\Psi_\rho^\wedge(n))^2,$$

$$((FT)P_R(F))(n,j) \quad = \quad (P_R(F))^\wedge(n,j) = F^\wedge(n,j)(\Phi_R^\wedge(n))^2.$$

Theorem 10.2.3 then yields an approximation of the identity operator by the operators R_ρ and P_R. If we introduce the so-called *scale space*

$$V_R(\Omega) = P_R(\mathcal{L}^2(\Omega)) = \{P_R(F) | F \in \mathcal{L}^2(\Omega)\}, \quad R \in (0,\infty),$$

then it clearly follows that $V_R(\Omega) \subset V_{R'}(\Omega)$ whenever $R \geq R'$. To be more specific, if $G \in V_R(\Omega)$, then there exists a function $F \in \mathcal{L}^2(\Omega)$ such that

$$G = \sum_{n=0}^{\infty}\sum_{j=1}^{2n+1} F^\wedge(n,j)(\Phi_R^\wedge(n))^2 Y_{n,j}. \tag{10.2.17}$$

In other words,

$$\sum_{n=0}^{\infty}\sum_{j=1}^{2n+1} (G^\wedge(n,j))^2(\Phi_R^\wedge(n))^{-4} < \infty \tag{10.2.18}$$

for all $G \in V_R(\Omega)$. Suppose now that $R \geq R'$. Then the monotonicity of $\Phi_R^\wedge(n)$ shows us that

$$\sum_{n=0}^{\infty}\sum_{j=1}^{2n+1} (G^\wedge(n,j))^2(\Phi_{R'}^\wedge(n))^{-4} < \infty, \tag{10.2.19}$$

hence, $G \in V_{R'}(\Omega)$. In addition, we have

$$\lim_{R\to\infty} \sum_{n=0}^{\infty}\sum_{j=1}^{2n+1} F^\wedge(n,j)(\Phi_R^\wedge(n))^2 Y_{n,j} = \sum_{n=0}^{m}\sum_{j=1}^{2n+1} F^\wedge(n,j) Y_{n,j}.$$

Summarizing our results we finally obtain the following properties:

- $Harm_{0,\ldots,m} \subset V_R(\Omega) \subset V_{R'}(\Omega) \subset \mathcal{L}^2(\Omega), \quad 0 < R' \leq R < \infty$
- $\{\lim_{R\to\infty} \Phi_R^{(2)} * F \mid F \in \mathcal{L}^2(\Omega)\} = Harm_{0,\ldots,m}$
- $\overline{\{F \in \mathcal{L}^2(\Omega) \mid F \in V_R(\Omega) \text{ for some } R \in (0,\infty)\}}^{\|\cdot\|_{\mathcal{L}^2(\Omega)}} = \mathcal{L}^2(\Omega)$.

We call such a collection of scale spaces satisfying the last three properties *continuous* (or *infinitesimal*) *multiresolution analysis* of $\mathcal{L}^2(\Omega)$. Note that all $V_R(\Omega)$-spaces are rotation invariant:

$$F \in V_R(\Omega) \text{ if and only if } F(\mathbf{t}\cdot) \in V_R(\Omega), \quad 0 < R < \infty \tag{10.2.20}$$

for all $\mathbf{t} \in O(3)$. This is of particular significance since in geophysical practice we have to deal with invariant pseudodifferential operators. It is also the reason why we think that our wavelet approach is of particular importance seen from geomathematical point of view (note that a large number of observables in geophysics and geodesy are representable by invariant pseudodifferential operators).

Table 10.2.1

ρ	$\Delta_{\Psi_\rho}^{L^*}$	$\Delta_{\Psi_\rho}^{o^{(1)}}$	$\Delta_{\Psi_\rho}^{o^{(1)}} \Delta_{\Psi_\rho}^{L^*}$
0.05	34.418	0.042	1.478
0.1	17.540	0.086	1.517
0.2	8.886	0.188	1.675
0.4	4.576	0.412	1.888
0.8	2.480	0.928	2.302

10.2.3 *Examples*

The results of the last section lead us to the impression that all types of spherical singular integrals listed in Section 8.2 generate in canonical way examples of spherical wavelets (in the sense of Definition 10.2.1). This aspect will be studied now in more detail.

Abel–Poisson–Wavelet of Order m. Consider as spherical scaling function $\{\Phi_R\}, R \in (0, \infty)$, the kernel

$$\Phi_R(t) = \sum_{n=0}^{m} \frac{2n+1}{4\pi} P_n(t) + \sum_{n=m+1}^{\infty} \frac{2n+1}{4\pi} \mathrm{e}^{-nR} P_n(t).$$

The '*Abel–Poisson wavelet*' of order m, $\{\Psi_\rho\}, \rho \in (0, \infty)$,

$$\Psi_\rho(t) = \sum_{n=m+1}^{\infty} \frac{2n+1}{4\pi} \Psi_\rho^\wedge(n) P_n(t), \quad t \in [-1, +1],$$

then arises quite naturally from the Abel–Poisson kernel by

$$\Psi_\rho^\wedge(n) = \begin{cases} (-(\alpha(\rho))^{-1}(d/d\rho)(\mathrm{e}^{-2n\rho}))^{1/2} = \sqrt{2n}(\alpha(\rho))^{-1/2}\mathrm{e}^{-n\rho}, & n > m \\ 0, & \text{else.} \end{cases}$$

Consequently,

$$\Psi_\rho(t) = (\alpha(\rho))^{-1/2} \sum_{n=m+1}^{\infty} \frac{2n+1}{4\pi} \sqrt{2n}\mathrm{e}^{-n\rho} P_n(t), \quad t \in [-1, +1].$$

In Table 10.2.1 we find a list of the space-frequency localization quantities (cf. Section 5.5) corresponding to a selection of parameters ρ.

Table 10.2.2

ρ	$\Delta^{L^*}_{\Psi_\rho}$	$\Delta^{o^{(1)}}_{\Psi_\rho}$	$\Delta^{o^{(1)}}_{\Psi_\rho}\Delta^{L^*}_{\Psi_\rho}$
0.05	4.473	0.359	1.607
0.1	3.167	0.565	1.789
0.2	2.251	0.980	2.208
0.4	1.647	2.211	3.643
0.8	1.425	10.651	15.186

Gauß–Weierstraß–Wavelet of Order m. By virtue of the kernel

$$\Phi_R(t) = \sum_{n=0}^{m} \frac{2n+1}{4\pi} P_n(t) + \sum_{n=m+1}^{\infty} \frac{2n+1}{4\pi} \mathrm{e}^{-n(n+1)R} P_n(t), \qquad (10.2.21)$$

the '*Gauß–Weierstraß–wavelet*' of order m, $\{\Psi_\rho\}, \rho \in (0,\infty)$, is obtainable from

$$\Psi_\rho^\wedge(n) = \begin{cases} \sqrt{2}(\alpha(\rho))^{-1/2}\sqrt{n(n+1)}\mathrm{e}^{-n(n+1)\rho}, & n > m \\ 0, & \text{else.} \end{cases}$$

In other words,

$$\Psi_\rho(t) = \sqrt{2}(\alpha(\rho))^{-1/2} \sum_{n=m+1}^{\infty} \frac{2n+1}{4\pi} \sqrt{n(n+1)}\mathrm{e}^{-n(n+1)\rho} P_n(t).$$

Table 10.2.2 gives information about space–frequency localization for different values of ρ. Figures 10.2.1 and 10.2.2 show a graphical illustration of both kernels for different parameters.

10.2.4 *Least Squares Property*

As we have seen, (WT) is a transformation from the one-parameter space $\mathcal{L}^2(\Omega)$ into the two-parameter space $\mathcal{L}^2((0,\infty)\times\Omega)$. Thus it is clear that (WT) is not surjective on $\mathcal{L}^2((0,\infty)\times\Omega)$. That means that $\mathcal{W} = (WT)(\mathcal{L}^2(\Omega))$ is a proper subspace of $\mathcal{L}^2((0,\infty)\times\Omega)$:

$$\mathcal{W} \subsetneqq \mathcal{L}^2((0,\infty)\times\Omega). \qquad (10.2.22)$$

Therefore one may ask how to characterize $\mathcal{W}$ within the framework of $\mathcal{L}^2((0,\infty)\times\Omega)$. For that purpose consider the operator $P : \mathcal{L}^2((0,\infty)\times\Omega) \to \mathcal{W}$ given by

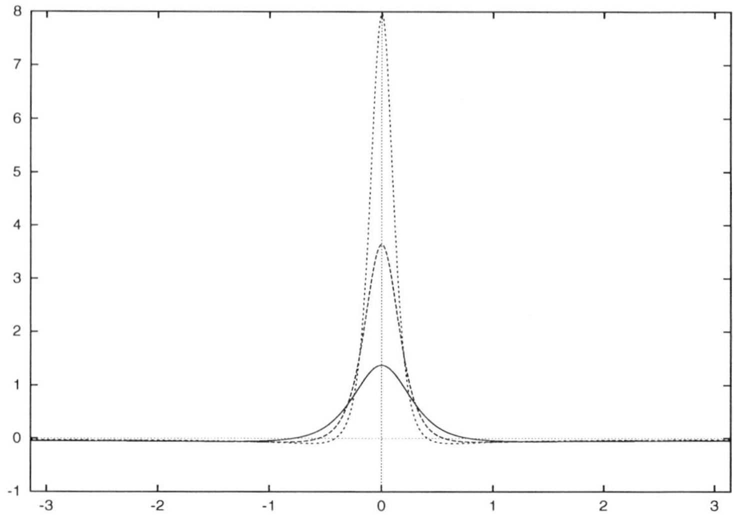

FIG. 10.2.1. Abel–Poisson wavelet of order $m = 0$, $\alpha(\rho) = 1/\rho$, $\rho = 0.2,\ 0.3,\ 0.5$

$$(PH)(\rho';\eta') = \int_0^\infty \int_\Omega K(\rho';\eta' \mid \rho;\eta) H(\rho;\eta)\, d\omega(\eta)\alpha(\rho)\, d\rho, \qquad (10.2.23)$$

where

$$K(\rho';\eta' \mid \rho;\eta) = \int_\Omega \Psi_{\rho';\eta'}(\xi)\Psi_{\rho;\eta}(\xi)\, d\omega(\xi).$$

It can be shown that P defines a projection as in the case of the Gabor transform (see Chapter 9).

Moreover, $\mathcal{W}$ is characterized as follows: $H \in \mathcal{W}$ if and only if

$$H(\rho';\eta') = \int_0^\infty \int_\Omega K(\rho';\eta' \mid \rho;\eta) H(\rho;\eta)\, d\omega(\eta)\alpha(\rho)\, d\rho. \qquad (10.2.24)$$

It can be easily seen, that $K(\rho';\eta' \mid \cdot;\cdot) \in \mathcal{W}$ and $K(\cdot;\cdot \mid \rho;\eta) \in \mathcal{W}$. The kernel $K(\rho';\eta' \mid \rho;\eta)$ is the reproducing kernel in $\mathcal{W}$. The reproducing property (10.2.24) can also be understood in such a way that $H \in \mathcal{W}$ is calculable by superpositions of itself. This shows that there is a kind of linear dependence, which can be interpreted as redundancy. Although it might seem inefficient, such redundancy has certain advantages. Unlike a non-redundant expansion, errors can be detected and corrected. In our context the tendency for correcting errors is expressed in the next theorem.

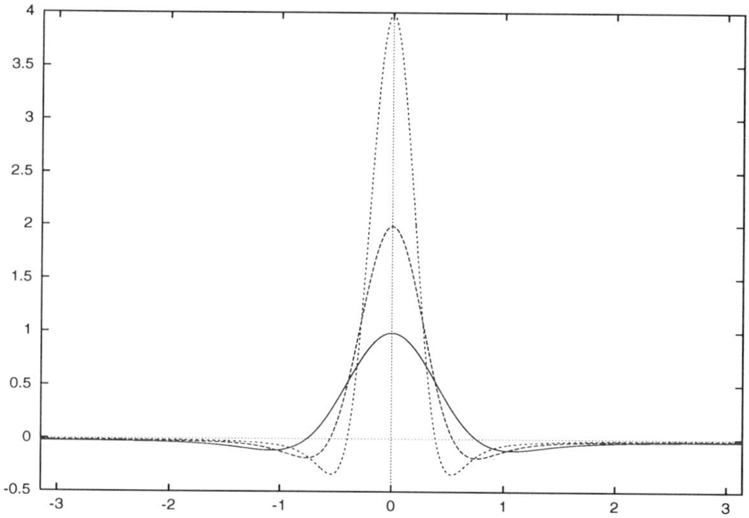

FIG. 10.2.2. Gauß–Weierstraß wavelets of order $m = 0$, $\alpha(\rho) = 1/\rho$, $\rho = 0.025,\ 0.05,\ 0.1$

Theorem 10.2.8. *Let H be an arbitrary element of $\mathcal{L}^2((0,\infty)\times\Omega)$. Then the unique function $F_H \in \mathcal{L}^2(\Omega)$ which satisfies the property*

$$\|H - \tilde{F}_H\|_{\mathcal{L}^2((0,\infty)\times\Omega)} = \inf_{F\in\mathcal{L}^2(\Omega)} \|H - \tilde{F}\|_{\mathcal{L}^2((0,\infty)\times\Omega)}$$

(with $\tilde{F}_H = (WT)(F_H)$) is given by

$$F_H(\xi) = \int_0^\infty \int_\Omega H(\rho;\eta)\Psi_{\rho;\eta}(\xi)\, d\omega(\eta)\alpha(\rho)\, d\rho.$$

Proof. We know that $\tilde{F}_H$ is the orthogonal projection of H onto $\mathcal{W}$. This proves Theorem 10.2.8. □

The linear dependence of $\tilde{F} \in \mathcal{W}$ leads to the effect that the coefficients in $\mathcal{L}^2((0,\infty)\times\Omega)$ for reconstructing a function $F \in \mathcal{L}^2(\Omega)$ are not unique. This can be readily developed from the following identity:

$$F(\xi) = \int_0^\infty \int_\Omega (\tilde{F}(\rho;\eta) + \tilde{F}^\perp(\rho;\eta))\Psi_{\rho;\eta}(\xi)\, d\omega(\eta)\alpha(\rho)\, d\rho$$

where $\tilde{F} = (WT)(F)$ and $\tilde{F}^\perp$ is an arbitrary member of $\mathcal{W}^\perp$. But our considerations above enable us to fomulate the following theorem.

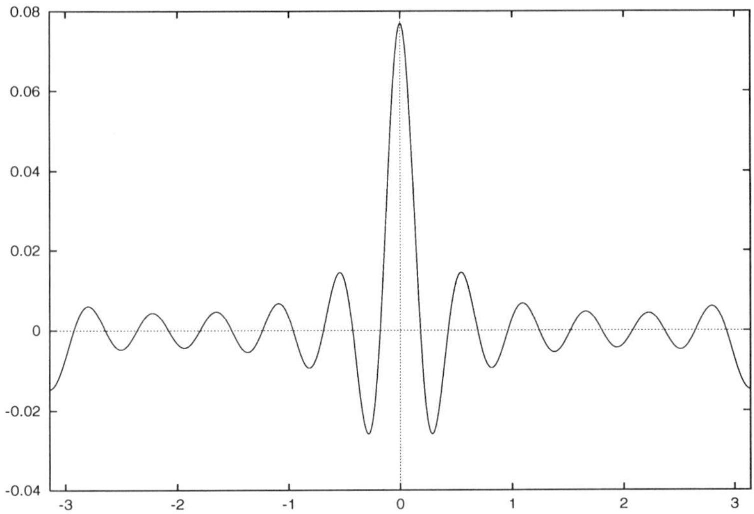

FIG. 10.2.3. Abel–Poisson wavelet of order $m = 10$, $\alpha(\rho) = 1/\rho$, $\rho = 0.5$

Theorem 10.2.9. *For arbitrary $F \in \mathcal{L}^2(\Omega)$ the coefficient function $\tilde{F} = (WT)(F) \in \mathcal{W}$ is the unique element in $\mathcal{L}^2((0,\infty) \times \Omega)$ which satisfies the minimum norm condition*

$$\|\tilde{F}\|_{\mathcal{L}^2((0,\infty)\times\Omega)} = \inf_{\substack{H \in \mathcal{L}^2((0,\infty)\times\Omega) \\ (WT)^{-1}(H)=F}} \|H\|_{\mathcal{L}^2((0,\infty)\times\Omega)}.$$

10.3 Scale Discretized Wavelet Transform

Until now emphasis has been put on the whole scale interval. In what follows, however, discretizations of the wavelet transform will be discussed in form of both wavelet packets and scale discrete Daubechies wavelets. The investigations on wavelet packets are influenced by the work due to M. DUVAL-DESTIN *et al.* (1993) in Euclidean space $\mathbb{R}$. This section ends with the least squares characterization of our scale discrete wavelet transforms.

10.3.1 *P-Scale Discretized Wavelets*

We start from a strictly decreasing sequence $\{\rho_j\}, j \in \mathbb{Z}$, such that

$$\lim_{j\to\infty} \rho_j = 0 \quad \text{and} \quad \lim_{j\to-\infty} \rho_j = \infty$$

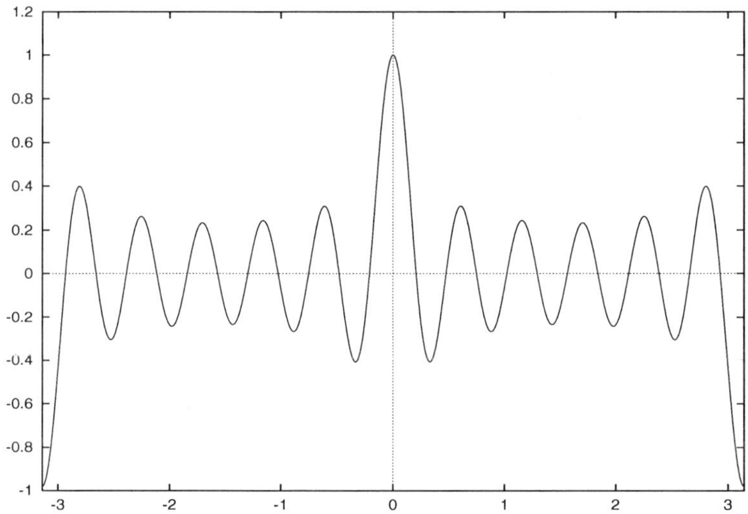

FIG. 10.2.4. Gauß–Weierstraß wavelets of order $m = 10$, $\alpha(\rho) = 1/\rho$, $\rho = 0.2$ (normalized)

(for example, $\rho_j = 2^{-j}\rho,\ j \in \mathbb{Z}, \rho > 0$).

Firstly, we are interested in a way of discretizing the scales such that discretization errors are avoided. Such a procedure canonically produces new types of wavelets that will be called spherical wavelet packets.

Definition 10.3.1. Let $\{\Psi_\rho\}, \rho \in (0,\infty)$, be a spherical wavelet of order m (as defined by Definition 10.2.1). Then the *spherical P(acket)-scale discretized wavelet* (also called *P wavelet packet*) $\{\Psi_j^P\}, j \in \mathbb{Z}$, is defined by

$$(\Psi_j^P)^\wedge(n) = \left(\int_{\rho_{j+1}}^{\rho_j} (\Psi_\rho^\wedge(n))^2 \alpha(\rho)\, d\rho\right)^{1/2}, \quad n = 0, 1, \ldots .$$

The function $\Psi_j^P, j \in \mathbb{Z}$, is called P-wavelet packet member of step j.

Obviously, for $n = m+1, m+2, \ldots$

$$\int_0^\infty (\Psi_\rho^\wedge(n))^2 \alpha(\rho) d\rho = \sum_{j=-\infty}^{\infty} ((\Psi_j^P)^\wedge(n))^2 = 1. \tag{10.3.1}$$

Remark. The sequence $\{\Psi_j^P\}$ is referred to as a spherical P(acket)-scale discretized wavelet, since the functions Ψ_j^P are built up by glueing wavelets together. Such P-scale discretized wavelets are known from Euclidean theory. In M. Duval-Destin *et al.* (1993) they are called wavelet packets, since they are obtained by a segmentation of the reconstruction formula (Theorem 10.2.3) into integrals over domains $\Omega \times [\rho_{j+1}, \rho_j], j \in \mathbb{Z}$. Originally the term 'wavelet packet' refers to a generalization of wavelets by R.R. Coifman *et al.* (1990). Their idea is to increase the frequency resolution of the wavelet transform by further decomposing the wavelet detail space using a full binary tree.

For every function $F \in \mathcal{L}^2(\Omega)$ we find

$$\begin{aligned} &\int_\Omega \int_{\rho_{j+1}}^{\rho_j} (F, \Psi_{\rho;\eta})_{\mathcal{L}^2(\Omega)} \Psi_{\rho;\eta}(\xi) \alpha(\rho)\, d\rho\, d\omega(\eta) \\ &= \sum_{n=0}^{\infty} \frac{2n+1}{4\pi} \left((\Psi_j^P)^\wedge(n)\right)^2 \left(\int_\Omega F(\zeta) P_n(\zeta \cdot \xi)\, d\omega(\zeta)\right) \\ &= \int_\Omega (F, \Psi_{j;\eta}^P)_{\mathcal{L}^2(\Omega)} \Psi_{j;\eta}^P(\xi)\, d\omega(\eta), \end{aligned} \tag{10.3.2}$$

where, by definition,

$$\Psi_{j;\eta}^P(\xi) = \Psi_j^P(\xi \cdot \eta), \quad (\xi, \eta) \in \Omega^2. \tag{10.3.3}$$

Note that

$$\left| \int_\Omega F(\zeta) P_n(\xi \cdot \zeta)\, d\omega(\zeta) \right| \leq \int_\Omega |F(\zeta)|\, d\omega(\zeta) < \infty, \tag{10.3.4}$$

hence, summation and integration in (10.3.2) may be interchanged. Now it follows immediately that

$$\begin{aligned} &\sum_{j=-\infty}^{\infty} \int_\Omega (F, \Psi_{j;\eta}^P)_{\mathcal{L}^2(\Omega)} (G, \Psi_{j;\eta}^P)_{\mathcal{L}^2(\Omega)}\, d\omega(\eta) \\ &= \sum_{n=0}^{\infty} \sum_{k=1}^{2n+1} F^\wedge(n,k) G^\wedge(n,k) \sum_{j=-\infty}^{\infty} \left((\Psi_j^P)^\wedge(n)\right)^2. \end{aligned} \tag{10.3.5}$$

Defining the Hilbert space

$$\mathcal{L}^2(\mathbb{Z} \times \Omega) = \left\{ H : \mathbb{Z} \times \Omega \to \mathbb{R} \mid \sum_{j=-\infty}^{\infty} \int_\Omega |H(j;\eta)|^2\, d\omega(\eta) < \infty \right\}$$

with inner product

$$(H_1, H_2)_{\mathcal{L}^2(\mathbb{Z}\times\Omega)} = \sum_{j=-\infty}^{\infty} \int_{\Omega} H_1(j;\eta) H_2(j;\eta)\, d\omega(\eta)$$

and corresponding norm

$$\|H\|_{\mathcal{L}^2(\mathbb{Z}\times\Omega)} = \sqrt{(H,H)_{\mathcal{L}^2(\mathbb{Z}\times\Omega)}} \tag{10.3.6}$$

we get in connection with Parseval's identity for spherical harmonics and (10.3.1)

$$\left((F, \Psi^P_{.;.})_{\mathcal{L}^2(\Omega)}, (G, \Psi^P_{.;.})_{\mathcal{L}^2(\Omega)}\right)_{\mathcal{L}^2(\mathbb{Z}\times\Omega)} = (F,G)_{\mathcal{L}^2(\Omega)} \tag{10.3.7}$$

for all $F, G \in \mathcal{L}^2(\Omega)$ with $F^\wedge(n,j) = G^\wedge(n,j) = 0, n = 0,\ldots,m,\ j = 1,\ldots,2n+1$. In particular we have

$$\|(F, \Psi^P_{.;.})_{\mathcal{L}^2(\Omega)}\|_{\mathcal{L}^2(\mathbb{Z}\times\Omega)} = \|F\|_{\mathcal{L}^2(\Omega)}. \tag{10.3.8}$$

This gives rise to the following definition.

Definition 10.3.2. Let $\{\Psi^P_j\}, j \in \mathbb{Z}$, be a spherical P-scale discretized wavelet of order m. Then the *P-scale discretized wavelet transform*

$$(WT)^P : \mathcal{L}^2(\Omega) \to \mathcal{L}^2(\mathbb{Z}\times\Omega)$$

is defined by

$$(WT)^P(F)(j;\eta) = (F, \Psi^P_{j;\eta})_{\mathcal{L}^2(\Omega)}, \quad F \in \mathcal{L}^2(\Omega).$$

We now have the following result.

Theorem 10.3.3 *(reconstruction formula). Let $\{\Psi^P_j\}$, $j \in \mathbb{Z}$, be a spherical P-scale discretized wavelet of order m. Then*

$$F = \sum_{j=-\infty}^{\infty} \int_{\Omega} (WT)^P(F)(j;\eta)\Psi^P_{j;\eta}(\,)\, d\omega(\eta)$$

(in the sense of $\|\cdot\|_{\mathcal{L}^2(\Omega)}$) provided that $F \in \mathcal{L}^2(\Omega)$ with $F^\wedge(n,j) = 0, n = 0,\ldots,m,\ j = 1,\ldots,2n+1$.

Proof. It is not hard to verify that

$$\lim_{J\to\infty} \int_{\Omega} \sum_{j=-\infty}^{J-1} (WT)^P(F)(j;\eta)\Psi^P_{j;\eta}(\xi)\, d\omega(\eta)$$
$$-\ \lim_{J\to\infty} \int_{\Omega}\int_{\rho_J}^{\infty} (WT)(F)(\rho;\eta)\Psi_{\rho;\eta}(\zeta)\alpha(\rho)\, d\rho\, d\omega(\eta).$$

This shows the desired result. □

The definition of the corresponding scaling function now follows in canonical way.

Definition 10.3.4. Let $\{\Psi_j^P\}, j \in \mathbb{Z}$, be a spherical P-scale discretized wavelet of order m. Then the corresponding *scaling function* $\{\Phi_J^P\}, J \in \mathbb{Z}$, is defined by its table $\{(\Phi_J^P)^\wedge(n)\}$ of Legendre coefficients

$$(\Phi_J^P)^\wedge(n) = \begin{cases} \left(\sum_{j=-\infty}^{J-1} ((\Psi_j^P)^\wedge(n))^2 \right)^{1/2} & \text{for} \quad n > m \\ 1 & \text{else.} \end{cases}$$

Obviously, for $n = m+1, m+2, \ldots$

$$(\Phi_J^P)^\wedge(n) = \left(\int_{\rho_J}^{\infty} (\Psi_\rho^\wedge(n))^2 \alpha(\rho)\, d\rho \right)^{1/2} = (\Phi_{\rho_J})^\wedge(n) \tag{10.3.9}$$

provided that $\{\Psi_\rho\}$ is the spherical wavelet corresponding to $\{\Psi_J^P\}$. From the reconstruction formula (Theorem 10.3.3) it follows immediately that

$$F = \lim_{J\to\infty} \int_\Omega (F, \Phi_{J;\eta}^P)_{\mathcal{L}^2(\Omega)} \Phi_{J;\eta}^P(\cdot)\, d\omega(\eta) \tag{10.3.10}$$

holds for all $F \in \mathcal{L}^2(\Omega)$ with $F^\wedge(n,j) = 0, n = 0, \ldots, m,\ j = 1, \ldots, 2n+1$ (observe that, as usual, $\Phi_{J;\eta}^P(\xi) = \Phi_J^P(\xi\cdot\eta)$, $(\xi,\eta) \in \Omega^2$). Moreover, for all $F \in \mathcal{L}^2(\Omega)$, we have

$$\begin{aligned} F \;&=\; \int_\Omega \left(F, \Phi_{J;\eta}^P\right)_{\mathcal{L}^2(\Omega)} \Phi_{J;\eta}^P(\cdot)\, d\omega(\eta) \\ &\quad + \sum_{j=J}^{\infty} \int_\Omega (WT)^P(F)(j;\eta)\Psi_{j;\eta}^P(\cdot)\, d\omega(\eta). \end{aligned} \tag{10.3.11}$$

It should be noted that formula (10.3.11), in fact, is valid for all $F \in \mathcal{L}^2(\Omega)$. By construction, the P-scale discretized wavelet leads to a partition of unity as follows:

$$\begin{aligned} \int_0^\infty (\Psi_\rho^\wedge(n))^2 \alpha(\rho)\, d\rho \;&=\; \sum_{j=-\infty}^{+\infty} ((\Psi_j^P)^\wedge(n))^2 \\ &=\; \left(\sum_{j=-\infty}^{J-1} + \sum_{j=J}^{\infty} \right) ((\Psi_j^P)^\wedge(n))^2 \end{aligned}$$

$$
\begin{aligned}
&= ((\Phi_J^P)^\wedge(n))^2 + \sum_{j=J}^{\infty}((\Psi_j^P)^\wedge(n))^2 \\
&= 1
\end{aligned}
$$

for $n = m+1, m+2, \ldots$. P-scale discretized wavelet members may be regarded as differences of two low-pass filters. Indeed, we have

$$
\left((\Psi_j^P)^\wedge(n)\right)^2 = \left((\Phi_{j+1}^P)^\wedge(n)\right)^2 - \left((\Phi_j^P)^\wedge(n)\right)^2, \quad j \in \mathbb{Z}, \tag{10.3.12}
$$

for $n = m+1, m+2, \ldots$ so that in accordance with (10.3.1)

$$
\sum_{j=-\infty}^{\infty} \left(((\Phi_{j+1}^P)^\wedge(n))^2 - ((\Phi_j^P)^\wedge(n))^2\right) = 1.
$$

Moreover,

$$
\begin{aligned}
&\int_\Omega (F, \Psi_{j;\eta}^P)_{\mathcal{L}^2(\Omega)} \Psi_{j;\eta}^P(\xi)\, d\omega(\eta) \qquad\qquad (10.3.13)\\
&= \int_\Omega (F, \Phi_{j+1;\eta}^P)_{\mathcal{L}^2(\Omega)} \Phi_{j+1;\eta}^P(\xi)\, d\omega(\eta) - \int_\Omega (F, \Phi_{j;\eta}^P)_{\mathcal{L}^2(\Omega)} \Phi_{j;\eta}^P(\xi)\, d\omega(\eta).
\end{aligned}
$$

In other words, the jth partial reconstruction of F (i.e. the left-hand side of (10.3.13)) may be expressed as difference of two 'smoothings' of F (i.e. the right-hand side of (10.3.13)) at two consecutive scales $j+1$ and j. Once more, the most important property of the P-scale discretized wavelet transform is (10.3.12). This formula is close to the idea of the 'difference of two smoothings' known in vision theory (see for example D. Marr (1982)).

From already known arguments of the theory of singular integrals about the kernel $\{\Phi_J^P\}$ we obtain the following.

Corollary 10.3.5. *Let $\{\Psi_j^P\}, j \in \mathbb{Z}$, be a spherical P-scale discretized wavelet of order m. Then, with $(WT)^P(F) = \Psi_{j;\eta}^P * F$, $F \in \mathcal{X}(\Omega)$,*

$$
\lim_{J\to\infty} \left\| F - \sum_{j=-\infty}^{J-1} \int_\Omega (WT)^P(F)(j;\eta)\Psi_{j;\eta}^P(\cdot)\, d\omega(\eta) \right\|_{\mathcal{X}(\Omega)} = 0
$$

holds for all functions $F \in \mathcal{X}(\Omega)$ with $F^\wedge(n,j) = 0$, $n = 0, \ldots, m$, $j = 1, \ldots, 2n+1$. In particular,

$$
\lim_{J\to\infty} \sup_{\xi\in\Omega} \left| F(\xi) - \sum_{j=-\infty}^{J-1} \int_\Omega (WT)^P(F)(j;\eta)\Psi_{j;\eta}^P(\xi)\, d\omega(\eta) \right| = 0
$$

holds for all $F \in \mathcal{C}(\Omega)$ with $F^\wedge(n,j) = 0$, $n = 0, \ldots, m$, $j = 1, \ldots, 2n+1$.

Again we discuss the usual examples known from singular integral theory.

Abel–Poisson P-scale Discretized Wavelet of Order m. Remembering the Abel–Poisson kernel we have

$$\Phi_J^P(t) = \sum_{n=0}^{m} \frac{2n+1}{4\pi} P_n(t) + \sum_{n=m+1}^{\infty} \frac{2n+1}{4\pi} \mathrm{e}^{-\rho_J n} P_n(t), \quad t \in [-1,+1],$$

so that we obtain the P-scale discretized wavelet $\{\Psi_j^P\}$ by

$$(\Psi_j^P)^\wedge(n) = \begin{cases} (\mathrm{e}^{-2n\rho_{j+1}} - \mathrm{e}^{-2n\rho_j})^{1/2}, & n > m \\ 0, & \text{else.} \end{cases}$$

Gauß–Weierstraß P-scale Discretized Wavelet of Order m. For the Gauß–Weierstraß kernel

$$\Phi_J^P(t) = \sum_{n=0}^{m} \frac{2n+1}{4\pi} P_n(t) + \sum_{n=m+1}^{\infty} \frac{2n+1}{4\pi} \mathrm{e}^{-n(n+1)\rho_J} P_n(t), \quad t \in [-1,+1], \tag{10.3.14}$$

we get the P-scale discretized wavelet $\{\Psi_j^P\}$ by

$$(\Psi_j^P)^\wedge(n) = \begin{cases} (\mathrm{e}^{-2n(n+1)\rho_{j+1}} - \mathrm{e}^{-2n(n+1)\rho_j})^{1/2}, & n > m \\ 0, & \text{else.} \end{cases}$$

10.3.2 *M-scale Discretized Wavelets.*

We now come to another type of a scale discretized wavelet, namely (*M(odified)-scale discretized wavelet* (also called *M-wavelet packet*)). The idea of establishing an M-scale discretized wavelet (also called *M-wavelet packet*) is based on the observation that the difference and sum of two consecutive members of a scale-discrete scaling function

$$\begin{aligned} \Psi_j^M &= \Phi_{j+1}^P - \Phi_j^P, \quad j \in \mathbb{Z}, \\ \tilde{\Psi}_j^M &= \Phi_{j+1}^P + \Phi_j^P, \quad j \in \mathbb{Z}, \end{aligned}$$

satisfy the property

$$\sum_{j=-\infty}^{\infty} (\Psi_j^M)^\wedge(n)(\tilde{\Psi}_j^M)^\wedge(n) = 1 \tag{10.3.15}$$

for $n = m+1, m+2, \ldots$. Furthermore, we have

$$\int_\Omega (F, \Psi_{j;\eta}^M)_{\mathcal{L}^2(\Omega)} \tilde{\Psi}_{j;\eta}^M(\cdot)\, d\omega(\eta) = \int_\Omega (F, \Psi_{j;\eta}^P)_{\mathcal{L}^2(\Omega)} \Psi_{j;\eta}^P(\cdot)\, d\omega(\eta)$$

for all $F \in \mathcal{L}^2(\Omega)$ with $F^\wedge(n,j) = 0, n = 0, \ldots, m,\ j = 1, \ldots, 2n+1$. Thus it makes sense to introduce the following.

Definition 10.3.6. Let $\{\Phi_J^P\}, J \in \mathbb{Z}$, be a scaling function corresponding to a spherical P-scale discretized wavelet $\{\Psi_j^P\}, j \in \mathbb{Z}$, of order m. Then the spherical M*-scale discretized wavelet* $\{\Psi_j^M\}$ of order m and the corresponding dual M*-scale discretized wavelet* $\{\tilde{\Psi}_j^M\}$, respectively, is defined by

$$(\Psi_j^M)^\wedge(n) = (\Phi_{j+1}^P)^\wedge(n) - (\Phi_j^P)^\wedge(n)$$

and

$$(\tilde{\Psi}_j^M)^\wedge(n) = (\Phi_{j+1}^P)^\wedge(n) + (\Phi_j^P)^\wedge(n).$$

The M-scale discretized wavelet transform is given by

$$(WT)^M(F)(j;\eta) = (\Psi_{j;\eta}^M, F)_{\mathcal{L}^2(\Omega)}, \quad F \in \mathcal{L}^2(\Omega).$$

From equation (10.3.15) we get the following result.

Theorem 10.3.7. *Let* $\{\Psi_j^M\}$ *and* $\{\tilde{\Psi}_j^M\}$ *be an* M*-scale discretized wavelet and corresponding dual* M*-scale discretized wavelet, respectively. Then*

$$F = \sum_{j=-\infty}^{+\infty} \int_\Omega (WT)^M(F)(j;\eta) \tilde{\Psi}_{j;\eta}^M(\cdot)\, d\omega(\eta)$$

holds (in the sense of $\|\cdot\|_{\mathcal{L}^2(\Omega)}$*) for all* $F \in \mathcal{L}^2(\Omega)$ *with* $F^\wedge(n,j) = 0$, $n = 0, \ldots, m,\ j = 1, \ldots, 2n+1$.

In accordance with the construction of $\{\Psi_j^M\}$ and $\{\tilde{\Psi}_j^M\}$ we find for the scaling function $\{\Phi_J^P\}$ the property

$$(\Phi_J^P)^\wedge(n) = \left(\sum_{j=-\infty}^{J-1} (\Psi_j^M)^\wedge(n) (\tilde{\Psi}_j^M)^\wedge(n) \right)^{1/2}$$

for $n = m+1, m+2, \ldots$. This implies the following.

Corollary 10.3.8. *Let $\{\Psi_j^M\}$ and $\{\tilde{\Psi}_j^M\}$ be an M-scale discretized wavelet and corresponding dual M-scale discretized wavelet, respectively. Then*

$$F = \int_\Omega (F, \Phi^P_{J;\eta})_{\mathcal{L}^2(\Omega)} \Phi^P_{J;\eta}(\cdot)\, d\omega(\eta) + \sum_{j=J}^{\infty} \int_\Omega (WT)^M(F)(j;\eta) \tilde{\Psi}^M_{j;\eta}(\cdot)\, d\omega(\eta)$$

holds for all $F \in \mathcal{L}^2(\Omega)$.

From known arguments of the theory of singular integrals we get another corollary.

Corollary 10.3.9. *Under the assumptions of Corollary 10.3.8*

$$\lim_{J\to\infty} \left\| F - \sum_{j=-\infty}^{J-1} \int_\Omega (WT)^M(F)(j;\eta) \tilde{\Psi}^M_{j;\eta}(\cdot)\, d\omega(\eta) \right\|_{\mathcal{X}(\Omega)} = 0$$

holds for all $F \in \mathcal{X}(\Omega)$ with vanishing moments $F^\wedge(n,j)$, $n = 0, \ldots, m$, $j = 1, \ldots, 2n+1$.

Again we consider the standard examples available from singular integral theory.

Abel–Poisson M-scale Discretized Wavelet of Order m.

$$\begin{aligned}
\Psi^M_{j;\eta}(\xi) &= \sum_{n=m+1}^{\infty} \frac{2n+1}{4\pi} (\mathrm{e}^{-n\rho_{j+1}} - \mathrm{e}^{-n\rho_j}) P_n(\xi\cdot\eta), \\
\tilde{\Psi}^M_{j;\eta}(\xi) &= 2\sum_{n=0}^{m} \frac{2n+1}{4\pi} P_n(\xi\cdot\eta) \\
&\quad + \sum_{n=m+1}^{\infty} \frac{2n+1}{4\pi} (\mathrm{e}^{-n\rho_{j+1}} + \mathrm{e}^{-n\rho_j}) P_n(\xi\cdot\eta).
\end{aligned}$$

Gauß–Weierstraß M-scale Discretized Wavelet of Order m.

$$\begin{aligned}
\Psi^M_{j;\eta}(\xi) &= \sum_{n=m+1}^{\infty} \frac{2n+1}{4\pi} (\mathrm{e}^{-n(n+1)\rho_{j+1}} - \mathrm{e}^{-n(n+1)\rho_j}) P_n(\xi\cdot\eta), \\
\tilde{\Psi}^M_{j;\eta}(\xi) &= 2\sum_{n=0}^{m} \frac{2n+1}{4\pi} P_n(\xi\cdot\eta)
\end{aligned}$$

$$+\sum_{n=m+1}^{\infty}\frac{2n+1}{4\pi}(\mathrm{e}^{-n(n+1)\rho_{j+1}}+\mathrm{e}^{-n(n+1)\rho_j})P_n(\xi\cdot\eta).$$

Locally Supported Kernel M-scale Discretized Wavelet of Order 0.

$$\begin{aligned}\Psi^M_{j;\eta}(\xi) &= \sum_{n=0}^{\infty}\frac{2n+1}{4\pi}\left((L^{(k)}_{\mathrm{e}^{-\rho_{j+1}}})^\wedge(n)-(L^{(k)}_{\mathrm{e}^{-\rho_j}})^\wedge(n)\right)P_n(\xi\cdot\eta),\\ \tilde{\Psi}^M_{j;\eta}(\xi) &= \sum_{n=0}^{\infty}\frac{2n+1}{4\pi}\left((L^{(k)}_{\mathrm{e}^{-\rho_{j+1}}})^\wedge(n)+(L^{(k)}_{\mathrm{e}^{-\rho_j}})^\wedge(n)\right)P_n(\xi\cdot\eta).\end{aligned}$$

W. Freeden, U. Windheuser (1996) also discuss S-scale discretized wavelets (S-wavelet packets) which form the bridge between P-scale discretized wavelets and Daubechies-scale discretized wavelets. These considerations are omitted here.

10.3.3 *Multiresolution Analysis by Means of Scale Discretized Wavelets.*

All scale discretized wavelets presented above share the fact that efficient algorithms for reconstruction and decomposition can be given by using the notion of multiresolution analysis. As always, we restrict our attention to the $\mathcal{L}^2(\Omega)$-space. Analogous to arguments in continuous theory we base our multiresolution approach on convolution operators.

Definition 10.3.10. For $F\in\mathcal{L}^2(\Omega)$ denote by R^P_j, P^P_J, the convolution operators given by

$$\begin{aligned}R^P_j(F) &= \Psi^P_j * \Psi^P_j * F = \Psi^M_j * \tilde{\Psi}^M_j * F,\\ P^P_J(F) &= \Phi^P_J * \Phi^P_J * F\end{aligned}$$

or equivalently

$$\begin{aligned}(FT)(R^P_j(F))(n,k) &= F^\wedge(n,k)((\Psi^P_j)^\wedge(n))^2\\ &= F^\wedge(n,k)(\Psi^M_j)^\wedge(n)(\tilde{\Psi}^M_j)^\wedge(n),\\ (FT)(P^P_J(F))(n,k) &= F^\wedge(n,k)((\Phi^P_J)^\wedge(n))^2.\end{aligned}$$

The *'scale spaces'* V^P_J and the *'detail spaces'* W^P_j are defined by

$$V^P_J = P^P_J(\mathcal{L}^2(\Omega)), \tag{10.3.16}$$

$$W^P_j = R^P_j(\mathcal{L}^2(\Omega)), \tag{10.3.17}$$

respectively. The collection $\{V^P_J\}$, of all spaces V^P_J, $J\in\mathbb{Z}$, is called *spherical P-multiresolution analysis* of $\mathcal{L}^2(\Omega)$.

According to our definition the scale spaces V_J^P are the images of $\mathcal{L}^2(\Omega)$ by P_J^P. Loosely spoken, V_J^P contains all J-scale smooth functions of $\mathcal{L}^2(\Omega)$. The notion 'detail space' means that W_j^P contains the 'detail' information needed to go from an approximation at resolution j to an approximation at resolution $j+1$. To be specific, W_J^P denotes the space complementary V_J^P in V_{J+1}^P:

$$V_{J+1}^P = V_J^P + W_J^P = V_{J_0}^P + \sum_{j=J_0}^{J} W_j^P = \sum_{j=-\infty}^{J} W_j^P. \tag{10.3.18}$$

This indicates that the definition of detail spaces is independent of the choice of *P-scale discretized wavelets or M-scale discretized wavelets.*

In analogy to the continuous case we obtain

- $Harm_{0,\ldots,m} \subset V_J^P \subset V_{J'}^P \subset \mathcal{L}^2(\Omega),\ -\infty < J \leq J' < \infty$
- $\{\lim_{J\to-\infty} (\Phi_J^P)^{(2)} * F \mid F \in \mathcal{L}^2(\Omega)\} = Harm_{0,\ldots,m}$
- $\overline{\{F \in \mathcal{L}^2(\Omega) | F \in V_J^P \text{ for some } J \in \mathbb{Z}\}}^{\|\cdot\|_{\mathcal{L}^2(\Omega)}} = \mathcal{L}^2(\Omega).$

Moreover,

$$F \in V_J^P \text{ if and only if } F(\mathbf{t}\cdot) \in V_J^P,\ J \in \mathbb{Z}, \tag{10.3.19}$$

for all $\mathbf{t} \in O(3)$. Furthermore, we have

$$\mathcal{L}^2(\Omega) = V_{J_0}^P + \sum_{j=J_0}^{\infty} W_j^P = \sum_{j=-\infty}^{\infty} W_j^P. \tag{10.3.20}$$

We note that a multiresolution analysis allows us to decompose $\mathcal{L}^2(\Omega)$ by a nested sequence of subspaces which, in turn, illustrates how to approximate a given function $F \in \mathcal{L}^2(\Omega)$ by an approximation in each of the spaces V_J^A. Explicitly written out in terms of the functions

$$\begin{array}{lllll}
\tilde{H}_j^P & = & (WT)^P(F)(j,\cdot), & \tilde{F}_J^P & = & (F, \Phi_J^P)_{\mathcal{L}^2(\Omega)}, \\
\tilde{H}_j^M & = & (WT)^M(F)(j,\cdot), & \tilde{F}_J^M & = & (F, \Phi_J^P)_{\mathcal{L}^2(\Omega)}, \\
H_j^P & = & (\tilde{H}_j^P, \Psi_j^P)_{\mathcal{L}^2(\Omega)}, & F_J^P & = & (\tilde{F}_J^P, \Phi_J^P)_{\mathcal{L}^2(\Omega)}, \\
H_j^M & = & (\tilde{H}_j^M, \tilde{\Psi}_j^M)_{\mathcal{L}^2(\Omega)}, & F_J^M & = & (\tilde{F}_J^M, \Phi_J^P)_{\mathcal{L}^2(\Omega)},
\end{array}$$

with $j, J \in \mathbb{Z}$ and $F \in \mathcal{L}^2(\Omega)$ we have $F_J^A \in V_J^P, A \in \{P, M\}$, and $H_j^A \in W_j^P, A \in \{P, M\}$. A function $F \in \mathcal{L}^2(\Omega)$ with $F^\wedge(n,j) = 0, n = 0,\ldots,m,\ j = 1,\ldots,2n+1$ can be approximated now in the twofold sense

$$F = \lim_{J\to\infty} F_J^A, \quad A \in \{P, M\}$$

or

$$F = \lim_{J \to \infty} \sum_{j=-\infty}^{J-1} H_j^A, \quad A \in \{P, M\}.$$

The figures below show the decomposition and reconstruction scheme.

(i) decomposition scheme

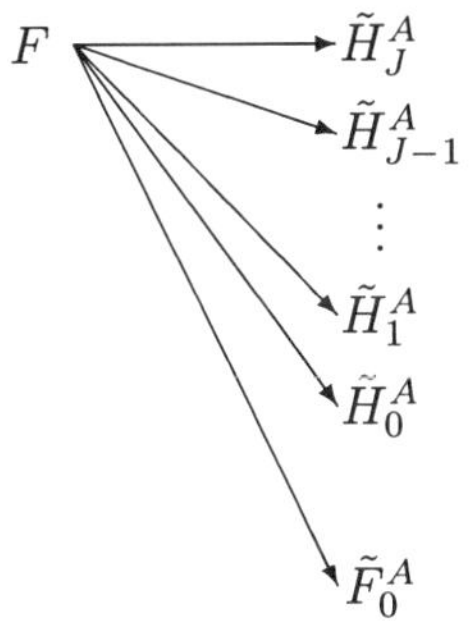

(ii) reconstruction scheme

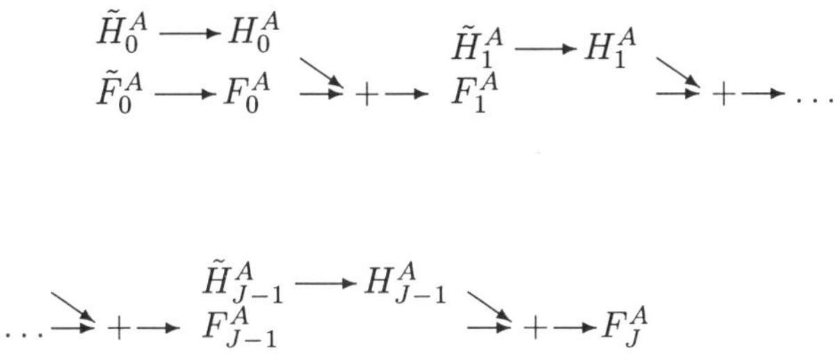

Unfortunately the decomposition formulae do not carry over information from level to level. During the decomposition process we always have to go back to the original function. This difficulty, however, can be overcome by using singular integral operators forming a semigroup of contraction operators of class (C_0) on $\mathcal{L}^2(\Omega)$ (like Abel–Poisson singular integral operators and Gauß–Weierstraß singular integral operators). We shall explain this idea in more detail below.

Definition 10.3.11. A scaling function $\{\Phi_J^P\}$ is called *reproducing scaling function* (briefly: *R-scaling function*), if

$$\Phi_J^P * \Phi_J^P = \Phi_{J-1}^P$$

for all $J \in \mathbb{Z}$.

Again we consider the two standard examples but now for the dyadic scale sequence, namely $\{\rho_j\}, \rho_j = 2^{-j}\rho, j = 0, 1, \ldots, \rho > 0$.

Abel–Poisson R-scaling Function. For all $J \in \mathbb{Z}$ and $(\xi, \eta) \in \Omega^2$ we find

$$\begin{aligned}\sum_{n=0}^{m} \frac{2n+1}{4\pi} P_n(\xi\cdot\eta) + \sum_{n=m+1}^{\infty} \frac{2n+1}{4\pi} \mathrm{e}^{-n2\rho_J} P_n(\xi\cdot\eta) \\ = \sum_{n=0}^{m} \frac{2n+1}{4\pi} P_n(\xi\cdot\eta) + \sum_{n=m+1}^{\infty} \frac{2n+1}{4\pi} \mathrm{e}^{-n\rho_{J-1}} P_n(\xi\cdot\eta).\end{aligned}$$

Gauß–Weierstraß R-scaling Function. For all $J \in \mathbb{Z}$ and $(\xi, \eta) \in \Omega^2$ we get by the same arguments

$$\begin{aligned}\sum_{n=0}^{m} \frac{2n+1}{4\pi} P_n(\xi\cdot\eta) + \sum_{n=m+1}^{\infty} \frac{2n+1}{4\pi} \mathrm{e}^{-n(n+1)2\rho_J} P_n(\xi\cdot\eta) \\ = \sum_{n=0}^{m} \frac{2n+1}{4\pi} P_n(\xi\cdot\eta) + \sum_{n=m+1}^{\infty} \frac{2n+1}{4\pi} \mathrm{e}^{-n(n+1)\rho_{J-1}} P_n(\xi\cdot\eta).\end{aligned}$$

In the following we investigate the scale discretized wavelets associated to R-scaling functions. Firstly we notice that

$$\begin{aligned}(\Psi_{j-1}^P)^\wedge(n) &= \left(((\Phi_j)^\wedge(n))^2 - ((\Phi_{j-1})^\wedge(n))^2\right)^{1/2} \\ &= ((\Phi_j)^\wedge(n) - (\Phi_{j-1})^\wedge(n))^{1/2}((\Phi_j)^\wedge(n) + (\Phi_{j-1})^\wedge(n))^{1/2} \\ &= \left(((\Phi_{j+1})^\wedge(n))^2 - ((\Phi_j)^\wedge(n))^2\right)^{1/2} \\ &\quad \left(((\Phi_{j+1})^\wedge(n))^2 + ((\Phi_j)^\wedge(n))^2\right)^{1/2}.\end{aligned}$$

In connection with

$$(\tilde{\Psi}_j^P)^\wedge(n) = \left(((\Phi_{j+1})^\wedge(n))^2 + ((\Phi_j)^\wedge(n))^2\right)^{1/2}$$

this shows us that

$$(\Psi_{j-1}^P)^\wedge(n) = (\Psi_j^P)^\wedge(n)(\tilde{\Psi}_j^P)^\wedge(n).$$

But this means in the notion of convolutions that

$$\Psi_{j-1}^P = \Psi_j^P * \tilde{\Psi}_j^P \tag{10.3.21}$$

is valid for the associated function $\Psi_j^P \in \mathcal{L}^2[-1, +1]$.

In contrast to (10.3.21) the M-scale discretized wavelets satisfy the equation

$$\begin{aligned}\Psi_j^M * \tilde{\Psi}_j^M &= (\Phi_{j+1}^P - \Phi_j^P) * (\Phi_{j+1}^P + \Phi_j^P) \\ &= (\phi_{j+1}^P)^{(2)} - (\phi_j^P)^{(2)} = \Psi_{j-1}^M.\end{aligned} \tag{10.3.22}$$

This leads us to the following corollary.

Corollary 10.3.12. *Let* $\{\Phi_J^R\}$ *be an R-scaling function. Then*

$$\Psi_{j-1}^P = \Psi_j^P * \tilde{\Psi}_j^P$$

with

$$(\tilde{\Psi}_j^P)^\wedge(n) = \left(((\Phi_{j+1})^\wedge(n))^2 + ((\Phi_j)^\wedge(n))^2\right)^{1/2},$$

for $n = 0, 1, \ldots,$ *while*

$$\Psi_{j-1}^M = \Psi_j^M * \tilde{\Psi}_j^M.$$

At this point, a simplification can be achieved by using the following classification.

Definition 10.3.13. Suppose that $\{\Phi_J^P\}$ is an R-scaling function. Then the associated scale discretized wavelet $\{\Psi_j^P\}$ is called R-scale discretized wavelet , while the M-scale discretized wavelet $\{\Psi_j^M\}$ is called an *RM-scale discretized wavelet.* Correspondingly, the dual M-scale discretized wavelet $\{\tilde{\Psi}_j^M\}$, is called *dual RM-scale discretized wavelet.*

In RA-multiresolution analysis we have

$$\begin{array}{rll} F \in W_j^A & \text{if and only if} & R_j^A(F) \in W_{j-1}^A, \\ F \subset V_j^A & \text{if and only if} & P_j^A(F) \subset V_{j-1}^A. \end{array}$$

In comparison with the aforementioned decomposition algorithms we obtain the following modifications:

$$\begin{array}{ll} \tilde{H}_j^P = (\tilde{H}_{j+1}^P, \tilde{\Psi}_{j+1}^P)_{\mathcal{L}^2(\Omega)}, & \tilde{F}_J^P = (\tilde{F}_{J+1}^P, \Phi_{J+1}^P)_{\mathcal{L}^2(\Omega)}, \\ \tilde{H}_j^M = (\tilde{H}_{j+1}^M, \tilde{\Psi}_{j+1}^M)_{\mathcal{L}^2(\Omega)}, & \tilde{F}_J^M = (\tilde{F}_{J+1}^M, \Phi_{J+1}^P)_{\mathcal{L}^2(\Omega)}. \end{array}$$

The following figure shows the modified decomposition scheme:

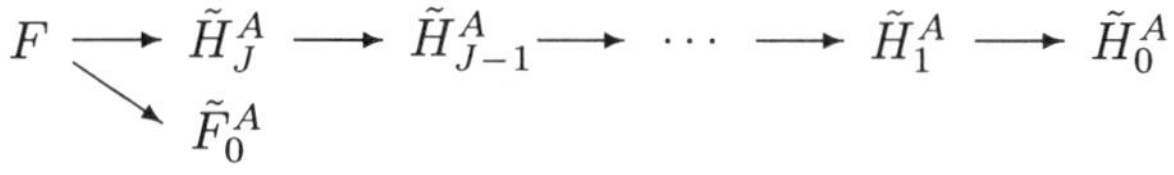

Moreover, RA-scale discretized wavelets, $A = M$, permit us to develop a modified concept which is illustrated below:

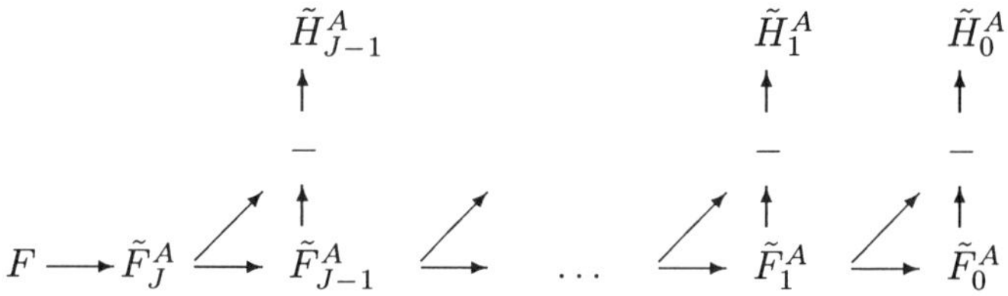

Obviously, decomposition and reconstruction are realized in feasible manner by optically inverse schemata.

Next, our object is to give physical interpretations of the Abel–Poisson and Gauß–Weierstraß wavelets in the context of multiresolution analysis.

Abel–Poisson Multiresolution Analysis. Taking as scaling function the Abel–Poisson kernel may be understood as determination of the harmonic continuation of $F \in \mathcal{L}^2(\Omega)$ to the inner space Ω_{int}. To be specific, choosing $P_J = A_{\mathrm{e}^{-2^{-J}}}$ we are led to the following conclusions: on the one hand side $F_J = P_J(F) \in V_J^A, A \in \{P, M, S\}$, on the other hand $F_J = A_{\mathrm{e}^{-2^{-J}}}(F)$ is an infinitely differentiable function on the sphere Ω_J with radius $r_J = \mathrm{e}^{-2^{-J}}$ around the origin, that is $F \in \mathcal{C}^{(\infty)}(\Omega_J)$. In other words, there is a canonical relationship between scale spaces V_J^A and spheres Ω_J. Multiresolution $\mathcal{L}^2(\Omega) = \overline{\bigcup_{J=-\infty}^{\infty} V_J^A}$ means 'blowing up' the spheres Ω_J to the unit sphere Ω. A graphical illustration of this fact may be found in Figure 10.3.1.

Gauß–Weierstraß Multiresolution Analysis. The physical interpretation of the Gauß–Weierstraß kernel in the context of diffusion equations is well known. Following this approach $F_J = P_J F \in V_J^A$ with $P_J = G_{2^{-J}}$ may be understood as temperature distribution at time $t_J = 2^{-J}$. $F \in \mathcal{L}^2(\Omega)$ is the initial temperature distribution, i.e. the situation at time $t = 0$. Each space V_J^A corresponds to the Jth time scale. The subspaces W_J correspond to elapsed time intervals $[t_{J+1}, t_J]$. See also the graphical illustration in Figure 10.3.2.

Finally, we discuss how locally supported radial basis functions can be used for a mutiresolution analysis. Two approaches to the definition of a multiresolution analysis based on locally supported kernels are discussed

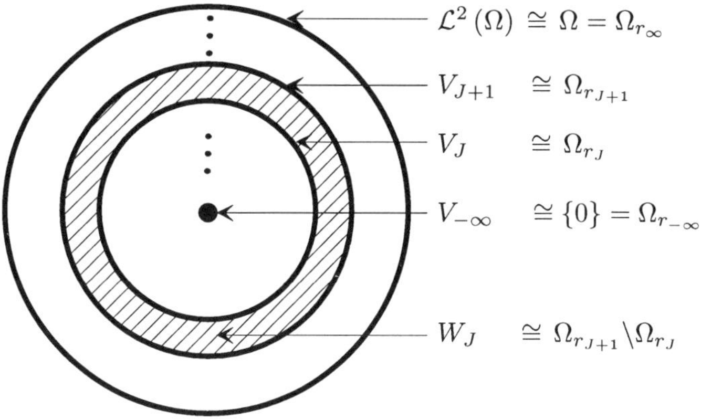

FIG. 10.3.1. Abel–Poisson multiresolution analysis

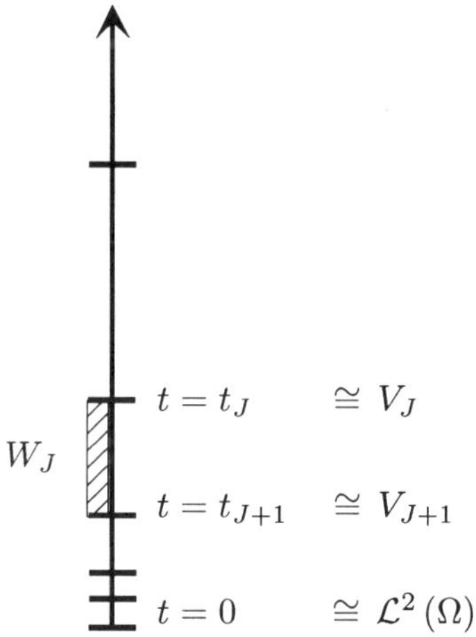

FIG. 10.3.2. Gauß–Weierstraß multiresolution analysis

without using results of a continuous wavelet transform.

Multiresolution Analysis with Locally Supported Kernels. Firstly, using the definition (8.2.54) of $L_J^{[k]}$, $J = 0, 1, \ldots$, we obtain a nested finite sequence

$$Harm_0 \subset \ldots \subset V_{J+1} \subset V_J \subset \ldots \subset V_1 \subset \mathcal{L}^2(\Omega)$$

with

$$V_J = \{L_J^{[k]} * L_J^{[k]} * F \mid F \in \mathcal{L}^2(\Omega)\}.$$

Clearly, we have the properties

- $Harm_0 \subset \ldots \subset V_J \subset V_{J'} \subset \ldots \subset \mathcal{L}^2(\Omega)$, $-\infty < J \leq J' \leq 1$
- $\left\{ \lim_{J \to -\infty} (L_J^{[k]})^{(2)} * F \mid F \in \mathcal{L}^2(\Omega) \right\} = Harm_0.$

Note that this defines only a finite process. But as it is already shown at the end of Chapter 8 we have the additional characteristic that

$$L_J^{(k)} * L_{J'}^{(k)} = L_{J+J'}^{(k)},$$

thus mimiking the notion of an R-scaling function (cf. Definition 10.3.11).

Multiresolution Analysis by Means of the Up Function. An infinite multiresolution analysis with locally supported kernels can be based on the definition (8.2.59) of a function $U_{h_J}^k$, where the values h_J are defined by

$$h_J = \begin{cases} \cos(2^{-J} \arccos(h)) & J = 0, 1, \ldots \\ h & J = -1, -2, \ldots \end{cases},$$

$(h \geq 0)$. Letting

$$V_J = \{U_{h_J}^{(k)} * U_{h_J}^{(k)} * F \mid F \in \mathcal{L}^2(\Omega)\},$$

we obtain an infinite nested sequence

$$Harm_0 \subset \ldots \subset V_J \subset V_{J+1} \subset \ldots \subset \mathcal{L}^2(\Omega)$$

with

$$U_{h_J}^{(k)} = L_{h_J}^{(k)} * U_{h_{J+1}}^{(k)}.$$

Furthermore, the following properties are valid:

- $Harm_0 \subset \ldots \subset V_J \subset V_{J'} \subset \ldots \subset \mathcal{L}^2(\Omega), \quad -\infty < J \leq J' < \infty$
- $\{ \lim_{J \to -\infty} (U_{h_J}^{(k)})^{(2)} * F \mid F \in \mathcal{L}^2(\Omega)\} = Harm_0$
- $\overline{\{F \in \mathcal{L}^2(\Omega) | F \in V_J \text{ for some } J \in \mathbb{Z}\}}^{\|\cdot\|_{\mathcal{L}^2(\Omega)}} = \mathcal{L}^2(\Omega).$

10.3.4 *D-scale Discretized Wavelet Transform*

In the theory of P-scale discretized wavelets the 'scale integral' (cf. Definition 10.3.1) is based on the division of the scale interval into subsequent subintervals. The integral is replaced by a sum of subintegrals, discretization is obtained by a transition from wavelets to 'wavelet packets'. We now continue our discussion with Daubechies' idea of discretization of the continuous formula. This approach follows the concept to discretize the integral over the continuous scale interval $(0, \infty)$ by forming a sum over a discrete scale sequence $\{\rho_j\}, j \in \mathbb{Z}$.

We begin our considerations with the following definition.

Definition 10.3.14. Suppose that $\{\rho_j\}, j \in \mathbb{Z}$, is a scale sequence satisfying

$$\lim_{j\to-\infty} \rho_j = \infty \quad \text{and} \quad \lim_{j\to\infty} \rho_j = 0. \tag{10.3.23}$$

Let $\{\Psi_\rho\}$ be a spherical wavelet of order m with corresponding scale discretized scale function $\{\Phi_j^P\}, \Phi_j^P = \Phi_{\rho_j}, j \in \mathbb{Z}$. The family $\{\Psi_j^D\}, \Psi_j^D = \Psi_{\rho_j}, j \in \mathbb{Z}$, is called *(spherical) Daubechies scale discretized wavelet* (briefly: *D-wavelet*) of order m if for $n = m+1, m+2, \ldots$ the *D-stability conditions*

$$A \leq \sum_{j=-\infty}^{\infty} \left((\Psi_j^D)^\wedge(n)\right)^2 \leq B \tag{10.3.24}$$

with the so-called '*D-stability constants*' A, B with $0 < A \leq B < \infty$,

$$\sum_{n=m+1}^{\infty} \frac{2n+1}{4\pi} \frac{\sum_{j=-\infty}^{J-1} ((\Psi_j^D)^\wedge(n))^2}{\sum_{j=-\infty}^{\infty} ((\Psi_j^D)^\wedge(n))^2} < \infty \tag{10.3.25}$$

for $J \in \mathbb{Z}$ and

$$2\pi \int_{-1}^{+1} \left| \sum_{n=m+1}^{\infty} \frac{2n+1}{4\pi} \frac{\sum_{j=-\infty}^{J-1} \left((\Psi_j^D)^\wedge(n)\right)^2}{\sum_{j\in\mathbb{Z}} \left((\Psi_j^D)^\wedge(n)\right)^2} P_n(t) \right| dt \leq T \tag{10.3.26}$$

for $J \in \mathbb{Z}$ and $T > 0$ independent of J are satisfied.

As usual, we let $\Psi_{j;\eta}^D = \Psi_j^D(\eta\cdot), \eta \in \Omega$. Now the D-wavelet transform is defined as follows.

Definition 10.3.15. Assume that $\{\Psi_j^D\}$ is a spherical D wavelet of order m. Then the spherical *D-wavelet transform* $(WT)^D : \mathcal{L}^2(\Omega) \to \mathcal{L}^2(\mathbb{Z} \times \Omega)$ is defined by

$$(WT)^D(F)(j;\eta) = (F, \Psi_{j;\eta}^D)_{\mathcal{L}^2(\Omega)}, \quad \eta \in \Omega.$$

We are interested in the proof of the following theorem.

Theorem 10.3.16. *Let $\{\Psi_j^D\}$ be a D-wavelet. Suppose that F is of class $\mathcal{L}^2(\Omega)$ satisfying $F^\wedge(n,j) = 0$, $n = 0, \ldots, m$, $j = 1, \ldots, 2n+1$. Then the condition*

$$A\|F\|^2_{\mathcal{L}^2(\Omega)} \leq \|(WT)^D(F)\|^2_{\mathcal{L}^2(\mathbb{Z}\times\Omega)} \leq B\|F\|^2_{\mathcal{L}^2(\Omega)} \tag{10.3.27}$$

is equivalent to the stability condition (10.3.24).

Proof. The inequality (10.3.27) is readily shown to be equivalent to

$$\begin{aligned} A \sum_{n=m+1}^{\infty} \sum_{j=1}^{2n+1} (F^\wedge(n,j))^2 & \\ \leq \sum_{n=m+1}^{\infty} \sum_{j=1}^{2n+1} (F^\wedge(n,j))^2 \left(\sum_{k=-\infty}^{\infty} ((\Psi_k^D)^\wedge(n))^2 \right) & \quad (10.3.28) \\ \leq B \sum_{n=m+1}^{\infty} \sum_{j=1}^{2n+1} (F^\wedge(n,j))^2. & \end{aligned}$$

Assuming that (10.3.24) is valid we obtain (10.3.28) which implies (10.3.27). Substituting F by $Y_{n,j}$ for $n = m+1, m+2, \ldots, j = 1, \ldots, 2n+1$ in (10.3.28) we get the stability condition (10.3.24). This is the required result. □

$(WT)^D$ is a linear, bounded (from below and above), and invertible operator. In order to know more about the representation of this operator we formulate the following definition.

Definition 10.3.17. Let $\{\Psi_j^D\}$ be a spherical D-wavelet of order m. Then the dual spherical D-wavelet $\{\tilde{\Psi}_j^D\}, j \in \mathbb{Z}$, is defined by

$$(\tilde{\Psi}_j^D)^\wedge(n) = \frac{(\Psi_j^D)^\wedge(n)}{\sum\limits_{k=-\infty}^{\infty} ((\Psi_k^D)^\wedge(n))^2}, \quad j \in \mathbb{Z}.$$

Observing the fact that for $n = m+1, m+2, \ldots$

$$\sum_{j=-\infty}^{\infty} ((\tilde{\Psi}_j^D)^\wedge(n))^2 = \left(\sum_{j=-\infty}^{\infty} ((\Psi_j^D)^\wedge(n))^2 \right)^{-1}$$

we find

$$\frac{1}{B} \leq \sum_{j=-\infty}^{\infty} ((\tilde{\Psi}_j^D)^\wedge(n))^2 \leq \frac{1}{A}$$

with A, B being the same constants as above. Moreover,

$$\frac{\sum\limits_{j=-\infty}^{J-1} ((\tilde{\Psi}_j^D)^\wedge(n))^2}{\sum\limits_{j=-\infty}^{\infty} ((\tilde{\Psi}_j^D)^\wedge(n))^2} = \frac{\sum\limits_{j=-\infty}^{J-1} ((\Psi_j^D)^\wedge(n))^2}{\sum\limits_{j=-\infty}^{\infty} ((\Psi_j^D)^\wedge(n))^2},$$

hence, $\{\tilde{\Psi}_j^D\}$ satisfies the stability conditions (10.3.24)–(10.3.26).

Remark. To a dual D-wavelet there generally corresponds no wavelet having continuous scale. An exception is the case $A = B$, when $(\tilde{\Psi}_j^D)^\wedge(n) = (1/A)(\Psi_j^D)^\wedge(n)$.

The construction of dual D-wavelets enables us to verify the following theorem.

Theorem 10.3.18. *Let $\{\Psi_j^D\}$ be a D-wavelet of order m with corresponding dual D-wavelet $\{\tilde{\Psi}_j^D\}$. Then*

$$F = \sum_{j=-\infty}^{\infty} \int_\Omega (WT)^D(F)(j;\eta)\tilde{\Psi}_{j;\eta}^D(\cdot)\, d\omega(\eta)$$

holds for all $F \in \mathcal{L}^2(\Omega)$ with $F^\wedge(n,j) = 0$, $n = 0,\ldots,m$, $j = 1,\ldots,2n+1$ (in $\|\cdot\|_{\mathcal{L}^2(\Omega)}$-sense).

Proof. From Definition 10.3.17 we deduce that

$$\sum_{j=-\infty}^{\infty} (\Psi_j^D)^\wedge(n)(\tilde{\Psi}_j^D)^\wedge(n) = 1.$$

Because of (10.3.25) we get for $n = m+1, m+2, \ldots$ in connection with

$$\begin{aligned}(\Phi_J^D)^\wedge(n) &= \left(\sum_{j=-\infty}^{J-1} (\Psi_j^D)^\wedge(n)(\tilde{\Psi}_j^D)^\wedge(n)\right)^{1/2} \\ &= \left(\frac{\sum_{j=-\infty}^{J-1} ((\Psi_j^D)^\wedge(n))^2}{\sum_{j=-\infty}^{+\infty} ((\Psi_j^D)^\wedge(n))^2}\right)^{1/2}\end{aligned}$$

the inequality

$$\sum_{n=0}^{\infty} \frac{2n+1}{4\pi}((\Phi_J^D)^\wedge(n))^2 < \infty.$$

Furthermore it follows from (10.3.24)

$$\lim_{J\to\infty} (\Phi_J^D)^\wedge(n) = 1$$

for all $n = m+1, m+2, \ldots$. In connection with (10.3.26) this is the desired result. □

Of course, $\{\Phi_J^D\}$ as stated above allows the interpretation as scaling function.

Definition 10.3.19. Let $\{\Psi_j^D\}$ be a spherical D-wavelet of order m with corresponding dual D-wavelet. Then the *D-scaling function* $\{\Phi_J^D\}$ is given by

$$(\Phi_J^D)^\wedge(n) = \begin{cases} \left(\sum\limits_{j=-\infty}^{J-1} (\Psi_j^D)^\wedge(n)(\tilde{\Psi}_j^D)^\wedge(n) \right)^{1/2} & , n > m \\ 1, & \text{else} \end{cases} .$$

It is not hard to verify that

$$\begin{aligned} &\sum_{j=-\infty}^{J-1} \int_\Omega (WT)^D(F)(j;\eta)\tilde{\Psi}_{j;\eta}^D(\xi)\, d\omega(\eta) \\ &= \int_\Omega (F, \Phi_{J;\eta}^D)_{\mathcal{L}^2(\Omega)} \Phi_{J;\eta}^D(\xi)\, d\omega(\eta) \end{aligned}$$

holds for all $F \in \mathcal{L}^2(\Omega)$ with $F^\wedge(n,j) = 0$ for $n = 0,\ldots,m$, $j = 1,\ldots,$ $2n+1$, where $\Phi_{J;\eta}^D(\xi) = \Phi_J^D(\xi\cdot\eta)$, $(\xi,\eta) \in \Omega^2$. This leads us to the following corollary.

Corollary 10.3.20. *Let $\{\Psi_j^D\}$ be a spherical D-wavelet with associated D-scaling function $\{\Phi_J^D\}$, $J \in \mathbb{Z}$. Then*

$$F = \lim_{J\to\infty} \int_\Omega (F, \Phi_{J;\eta}^D)_{\mathcal{L}^2(\Omega)} \Phi_{J;\eta}^D(\cdot)\, d\omega(\eta)$$

holds for all $F \in \mathcal{L}^2(\Omega)$ with $F^\wedge(n,j) = 0$ for $n = 0,\ldots,m$, $j = 1,\ldots,2n+1$ (in $\|\cdot\|_{\mathcal{L}^2(\Omega)}$-sense).

It is not surprising that the approximation can be also established in the framework of $(\mathcal{X}(\Omega), \|\cdot\|_{\mathcal{X}(\Omega)})$.

Corollary 10.3.21. *Let $\{\Psi_j^D\}$ be a D-wavelet of order m with corresponding dual D-wavelet $\{\tilde{\Psi}_j^D\}, j \in \mathbb{Z}$. Then, for $F \in \mathcal{X}(\Omega)$ with $F^\wedge(n,j) = 0$ for $n = 0,\ldots,m$, $j = 1,\ldots,2n+1$ (note that $(WT)^D(F)(j;\eta) = \Psi_{j;\eta}^D * F$),*

$$\lim_{J\to\infty} \left\| F - \sum_{j=-\infty}^{J-1} \int_\Omega (WT)^D(F)(j;\eta)\tilde{\Psi}_{j;\eta}^D(\cdot)\, d\omega(\eta) \right\|_{\mathcal{X}(\Omega)} = 0.$$

In particular, uniform convergence holds for $F \in \mathcal{C}(\Omega)$ with $F^\wedge(n,j) = 0$ for $n = 0,\ldots,m$, $j = 1,\ldots,2n+1$.

Remark. Without going into details (cf. U. WINDHEUSER (1995, 1996)) we mention that the Abel–Poisson wavelet as well as the Gauß–Weierstraß wavelet satisfy the D-stability condition (10.3.24) (with $\rho_j = 2^{-j}, j \in \mathbb{Z}$). Condition (10.3.25) follows from elementary calculations. Moreover, $A \approx B$, hence, $\tilde{\Psi}_j^D \approx (1/A)\Psi_j^D, j \in \mathbb{Z}$.

We end this section with some information about multiresolution analysis caused by the D-wavelet transform.

Definition 10.3.22. Let $\{\Psi_j^D\}$ and $\{\tilde{\Psi}_j^D\}, j \in \mathbb{Z}$, be a spherical D-wavelet of order m and corresponding dual D-wavelet, respectively. The convolution operators R_j^D and P_J^D, respectively, are given by

$$\begin{aligned} R_j^D(F) &= \Psi_j^D * \tilde{\Psi}_j^D * F, \\ P_J^D(F) &= \Phi_J^D * \Phi_J^D * F. \end{aligned}$$

The *'scale spaces'* V_J^D and the *'detail spaces'* W_j^D, respectively, are defined by

$$\begin{aligned} V_J^D &= P_J^D(\mathcal{L}^2(\Omega)), \\ W_j^D &= R_j^D(\mathcal{L}^2(\Omega)). \end{aligned}$$

The collection $\{V_J^D\}$ of all spaces $V_J^D, J \in \mathbb{Z}$, is called *spherical D-multiresolution analysis* of $\mathcal{L}^2(\Omega)$.

We remark that

$$V_{J+1}^D = V_J^D + W_J^D.$$

Furthermore,

- $Harm_{0,\ldots,m} \subset \ldots \subset V_J^D \subset V_{J'}^D \subset \ldots \subset \mathcal{L}^2(\Omega), \quad -\infty < J \le J' < \infty$
- $\{\lim_{J\to-\infty} (\Phi_J^D)^{(2)} * F \mid F \in \mathcal{L}^2(\Omega)\} = Harm_{0,\ldots,m}$
- $\overline{\{F \in \mathcal{L}^2(\Omega) | F \in V_J^D \text{ for some } J \in \mathbb{Z}\}}^{\|\cdot\|_{\mathcal{L}^2(\Omega)}} = \mathcal{L}^2(\Omega)$.

It is clear that

$$F \in V_J^D \text{ if and only if } F(\mathbf{t}\cdot) \in V_J^D$$

for all orthogonal transformations $\mathbf{t}$ and $J \in \mathbb{Z}$.

Letting

$$\tilde{H}_j^D = (WT)^D(F)(j;\cdot),$$

$$\tilde{F}_J^D \quad = \quad (F, \Phi_J^D)_{\mathcal{L}^2(\Omega)},$$

and

$$\begin{aligned} H_j^D &= (\tilde{H}_j^D, \tilde{\Psi}_j^D)_{\mathcal{L}^2(\Omega)}, \\ F_J^D &= (\tilde{F}_J^D, \Phi_J^D)_{\mathcal{L}^2(\Omega)}, \end{aligned} \tag{10.3.29}$$

for $j, J \in \mathbb{Z}$ and $F \in \mathcal{L}^2(\Omega)$ with $F^\wedge(n,j) = 0$ for $n = 0, \ldots, m$, $j = 1, \ldots, 2n+1$ we have $F_J^D \in V_J^D$ and $H_j^D \in W_j^D$. Every function F in $\mathcal{L}^2(\Omega)$ can be approximated as closely as it is desired (in $\|\cdot\|_{\mathcal{L}^2(\Omega)}$- sense) by F_J^D and H_j^D, respectively. That is,

$$F = \lim_{J\to\infty} F_J^D = \lim_{J\to\infty} \sum_{j=-\infty}^{J-1} H_j^D.$$

Decomposition and reconstruction scheme are of the same type as discussed in the previous section. Again it turns out that decomposition becomes particularly efficient if the reproducing property comes into play.

Definition 10.3.23. Let $\{\rho_j\}$ be the dyadic sequence, i.e. $\rho_j = 2^{-j}\rho, j \in \mathbb{Z}, \rho > 0$. Let $\{\Psi_j^D\}$ be a D-wavelet of order m associated to the wavelet $\{\Psi_\rho\}, \rho \in (0,\infty)$ and the corresponding scaling function $\{\Phi_R\}, R \in (0,\infty)$. Then $\{\Psi_j^D\}, j \in \mathbb{Z}$, is called an *RD-wavelet* if the scaling function satisfies

$$\Phi_R * \Phi_R = \Phi_{2R} \tag{10.3.30}$$

for all $R \in (0,\infty)$.

Keeping the convolution (10.3.30) in mind we obtain for integers $n = m+1, m+2, \ldots$

$$\begin{aligned} \frac{\Psi_{2\rho}^\wedge(n)}{\Psi_\rho^\wedge(n)} &= \left(\frac{(\alpha(2\rho))^{-1}(d/d\rho)(\Phi_{2\rho}^\wedge(n))^2}{(\alpha(\rho))^{-1}(d/d\rho)(\Phi_\rho^\wedge(n))^2}\right)^{1/2} \\ &= \left(\frac{(\alpha(2\rho))^{-1}\Phi_{2\rho}^\wedge(n)(d/d\rho)\Phi_{2\rho}^\wedge(n)}{(\alpha(\rho))^{-1}(d/d\rho)\Phi_{2\rho}^\wedge(n)}\right)^{1/2} \\ &= (\frac{\alpha(\rho)}{\alpha(2\rho)}\Phi_{2\rho}^\wedge(n))^{1/2} = \sqrt{\frac{\alpha(\rho)}{\alpha(2\rho)}}\Phi_\rho^\wedge(n). \end{aligned}$$

But this gives us

$$\Psi_{2\rho} = \sqrt{\frac{\alpha(\rho)}{\alpha(2\rho)}}(\Psi_\rho * \Phi_\rho) \tag{10.3.31}$$

for all $\rho \in (0,\infty)$. With $\Phi_j^P = \Phi_{\rho_j}$ we have

$$\Psi_{j-1}^D = \sqrt{\frac{\alpha(\rho)}{\alpha(2\rho)}}(\Psi_j^D * \Phi_{\rho_j}) = \sqrt{\frac{\alpha(\rho)}{\alpha(2\rho)}}(\Psi_j^D * \Phi_j^P). \tag{10.3.32}$$

This results in the following modification of formula (10.3.29)

$$\tilde{H}_j^D = \sqrt{\frac{\alpha(\rho)}{\alpha(2\rho)}}(\tilde{H}_{j+1}^D, \Phi_{j+1}^P)_{\mathcal{L}^2(\Omega)}. \tag{10.3.33}$$

The figures below show the decomposition and reconstruction scheme.

(i) decomposition scheme

$$\begin{array}{ccccccccccc} F & \longrightarrow & \tilde{H}_J^D & \longrightarrow & \tilde{H}_{J-1}^D & \longrightarrow & \cdots & \longrightarrow & \tilde{H}_1^D & \longrightarrow & \tilde{H}_0^D \\ & \searrow & & & & & & & & & \\ & & \tilde{F}_0^D & & & & & & & & \end{array}$$

(ii) reconstruction scheme

$$\begin{array}{lllll} \tilde{H}_0^D \longrightarrow H_0^D & & \tilde{H}_1^D \longrightarrow H_1^D & & \\ & \searrow & & \searrow & \\ \tilde{F}_0^D \longrightarrow F_0^D & \rightarrow + \rightarrow & F_1^D & \rightarrow + \rightarrow & \ldots \end{array}$$

$$\begin{array}{lllll} & & \tilde{H}_{J-1}^D \longrightarrow H_{J-1}^D & & \\ \searrow & & & \searrow & \\ \ldots \rightarrow & + \rightarrow & F_{J-1}^D & \rightarrow + \rightarrow & F_J^D \end{array}$$

10.3.5 *Least Squares Property*

As in the continuous case we can make use of a projection property in the scale discrete case for $A \in \{P, D\}$. We know that

$$(WT)^A : \mathcal{L}^2(\Omega) \to \mathcal{L}^2(\mathbb{Z} \times \Omega),$$

where

$$\mathcal{L}^2(\mathbb{Z} \times \Omega) = \left\{ H : \mathbb{Z} \times \Omega \,\middle|\, \sum_{j \in \mathbb{Z}} \int_\Omega |H(j;\eta)|^2 \, d\omega(\eta) < \infty \right\}$$

and

$$\mathcal{W}^A = (WT)^A(\mathcal{L}^2(\Omega)) \subsetneqq \mathcal{L}^2(\mathbb{Z} \times \Omega).$$

Once more we can define the projection operator $P^A : \mathcal{L}^2(\mathbb{Z} \times \Omega) \to \mathcal{W}^A$ by

$$(P^A H)(j';\eta') = \sum_{j\in\mathbb{Z}} \int_\Omega K^A(j';\eta' \mid j;\eta) H(j;\eta)\, d\omega(\eta), \qquad (10.3.34)$$

where

$$K^A(j';\eta' \mid j;\eta) = \int_\Omega \Psi^A_{j';\eta'}(\xi) \tilde{\Psi}^A_{j;\eta}(\xi)\, d\omega(\xi). \qquad (10.3.35)$$

Then we obtain similarly to results of Chapter 9 that the operator P^A is a projection. In other words, we have the following characterization of $\mathcal{W}^A$ for $A \in \{P, D\}$: $H \in \mathcal{W}^A$ if and only if

$$H(j';\eta') = \sum_{j\in\mathbb{Z}} \int_\Omega K^A(j';\eta' \mid j;\eta) H(j;\eta)\, d\omega(\eta). \qquad (10.3.36)$$

Summarizing our results we obtain the following theorem.

Theorem 10.3.24. *Let $H(j;\eta)$ be an arbitrary element of $\mathcal{L}^2(\mathbb{Z} \times \Omega)$. Then the unique function $F_H \in \mathcal{L}^2(\Omega)$ satisfying the property*

$$\|H - \tilde{F}^A_H\|_{\mathcal{L}^2(\mathbb{Z}\times\Omega)} = \inf_{F\in\mathcal{L}^2(\Omega)} \|H - \tilde{F}^A\|_{L^2(\mathbb{Z}\times\Omega)}$$

is given by

$$F_H(\xi) = \sum_{j\in\mathbb{Z}} \int_\Omega H(j;\eta) \tilde{\Psi}^A_{j;\eta}(\xi)\, d\omega(\eta).$$

Theorem 10.3.25. *For arbitrary $F \in \mathcal{L}^2(\Omega)$ the coefficient function $\tilde{F}^A = (WT)^A(F) \in \mathcal{W}^A$ is the unique element in $\mathcal{L}^2(\mathbb{Z} \times \Omega)$ satisfying the minimum norm property*

$$\|\tilde{F}^A\|_{\mathcal{L}^2(\mathbb{Z}\times\Omega)} = \inf_{\substack{H\in\mathcal{L}^2(\mathbb{Z}\times\Omega) \\ ((WT)^A)^{-1}(H)=F}} \|H\|_{\mathcal{L}^2(\mathbb{Z}\times\Omega)}.$$

10.4 Fully Discretized Wavelet Transform

In previous sections we have outlined several methods of scale discretization thereby preserving the rotational symmetry. So far our wavelet transforms have been continuous in space. For a numerical realization, however, space discretization is necessary. Unfortunately this problem cannot be solved by similar techniques to those known for the real line $\mathbb{R}$. The reason is that there is no canonical (dyadic) hierarchical subdivision of the sphere Ω. Instead we have to remember good candidates of hierarchical (equi-) distributed pointsets on the sphere (cf. Section 7.1).

We first unify our notation. This will help us to discuss the various wavelet types ($A \in \{P, M, D\}$) simultaneously.

Let $\{\Psi_j^A\}$ denote an A-wavelet packet ($A \in \{P, M\}$) or a D-wavelet for $A = D$ as introduced in Section 10.3. Then we designate by $\{\tilde{\Psi}_j^A\}$ the wavelet packet ($A = P$), dual M-wavelet packet ($A = M$), dual D-wavelet ($A = D$), respectively. The reconstruction formula now may be formulated as follows:

$$F = \sum_{j=-\infty}^{\infty} \int_{\Omega} (WT)^A(F)(j;\eta) \tilde{\Psi}_{j;\eta}^A(\cdot)\, d\omega(\eta),$$

provided that $F \in \mathcal{L}^2(\Omega)$ with $F^\wedge(n,j) = 0$ for $n = 0, \ldots, m$, $j = 1, \ldots, 2n+1$. The fully discretized wavelet transform is searched for under the assumption that $\eta \mapsto \Psi_{j;\eta}^A(\xi)$, $\eta \in \Omega$, as well as $\eta \mapsto \tilde{\Psi}_{j;\eta}^A(\xi)$, $\eta \in \Omega$, are (piecewise) continuous on Ω (for all $j \in \mathbb{Z}$ and $\xi \in \Omega$). Note that these conditions are satisfied by most of our wavelet examples.

Under the assumption that $\eta \mapsto (WT)^A(F)(j;\eta)$, $\eta \in \Omega$, is continuous on Ω (for all $j \in \mathbb{Z}$) we obtain the following from Lemma 7.1.6.

Theorem 10.4.1. *Assume that $\{X_N\}$ is an $\{a_k^N\}$-distribution. Then*

$$F = \sum_{j=-\infty}^{+\infty} \left(\lim_{N\to\infty} \frac{1}{4\pi} \sum_{k=1}^{N} a_k^N (WT)^A(F)(j;\eta_k^N) \tilde{\Psi}_{j;\eta_k^N}^A(\cdot) \right)$$

holds for all $F \in \mathcal{L}^2(\Omega)$ satisfying $F^\wedge(n,j) = 0$ for $n = 0, \ldots, m$, $j = 1, \ldots, 2n+1$.

As fully discretized wavelet transform we find the following result.

Theorem 10.4.2. *Assume that $\{X_N\}$ is an $\{a_k^N\}$-distribution. Furthermore, let $\{\Xi_L\}$ be a $\{b_k^L\}$-distribution. Then, under the above assumptions,*

$$F = \sum_{j=-\infty}^{+\infty} \left(\lim_{\substack{N\to\infty \\ L\to\infty}} \sum_{k=1}^{N} \frac{a_k^N}{4\pi} \sum_{r=1}^{L} \frac{b_r^L}{4\pi} F(\xi_r^L) \Psi_{j;\eta_k^N}^A(\xi_r^L)\, \tilde{\Psi}_{j;\eta_k^N}^A(\cdot) \right)$$

holds for all $F \in \mathcal{C}(\Omega)$ with $F^\wedge(n,j) = 0$ for $n = 0, \ldots, m$, $j = 1, \ldots, 2n+1$ (where equality holds in the sense of $\|\cdot\|_{\mathcal{L}^2(\Omega)}$).

Remark. Under the assumption that F and our wavelet and scaling functions are Lipschitz-continuous on Ω, Theorem 10.4.2 can be shown, in addition, to hold in the sense of $\|\cdot\|_{\mathcal{C}(\Omega)}$. The details will be omitted. Furthermore, error estimates are derivable by the same techniques as developed in W. Freeden, M. Schreiner (1995a), R. Brand (1994), and R. Brand *et al.* (1996).

For numerical purposes hierarchical pointsets $\{X_{N_j}\}$, $\{\Xi_{L_j}\}$ (dependent on the scale parameter j) should be used. This finally leads us to the following fully discretized wavelet approximation $F_{J_0}^{\mathrm{appr}}$ of a function $F \in \mathcal{C}(\Omega)$ at $\xi \in \Omega$:

$$\begin{aligned} F_{J_0}^{\mathrm{appr}}(\xi) &= \sum_{k=1}^{N_{J_0}} \frac{a_k^{N_{J_0}}}{4\pi} \sum_{r=1}^{L_{J_0}} \frac{b_r^{L_{J_0}}}{4\pi} F(\xi_r^{L_{J_0}}) \Phi^A_{J_0;\eta_k^{N_{J_0}}}(\xi_r^{L_{J_0}}) \Phi^A_{J_0;\eta_k^{N_{J_0}}}(\xi) \\ &+ \sum_{j=J_0}^{\infty} \sum_{k=1}^{N_j} \frac{a_k^{N_j}}{4\pi} \sum_{r=1}^{L_j} \frac{b_r^{L_j}}{4\pi} F(\xi_r^{L_j}) \Psi^A_{j;\eta_k^{N_j}}(\xi_r^{L_j}) \tilde{\Psi}^A_{j;\eta_k^{N_j}}(\xi). \end{aligned}$$

10.5 Combined Spherical Harmonic and Wavelet Expansion

The principle of developing wavelets with vanishing moments of order m enables us to combine the mth order expansion of a function in terms of spherical harmonics with its expansion in terms of spherical wavelets of order m. To be specific, the combination of both spherical harmonic analysis up to a prescribed order m and wavelet analysis with wavelets of order m leads to the following representation of a function $F \in \mathcal{L}^2(\Omega)$:

$$F(\xi) = \sum_{n=0}^{m} \sum_{j=1}^{2n+1} F^\wedge(n,j) Y_{n,j} + \sum_{j \in \mathbb{Z}} \int_\Omega (WT)^A(F)(j;\eta) \tilde{\Psi}^A_{j;\eta}(\cdot)\, d\omega(\eta)$$

with $A \in \{P, M, D\}$ and $\{\Psi_j^A\}, \{\tilde{\Psi}_j^A\}$ as wavelets and dual wavelets of order m. Observe that if the wavelet coefficients $(WT)^A(F)(j;\eta)$ of a function $F \in \mathcal{L}^2(\Omega)$ are known, the same coefficients can be used to reconstruct $T = \Lambda F$ with Λ (invariant) pseudodifferenital operator (as defined in Section 5.1). In doing so only the dual wavelet has to be modified by convolving K_Λ against $\tilde{\Psi}^A_{j;\eta}$:

$$K_\Lambda(\xi\cdot) * \tilde{\Psi}^A_{j;\eta} = \sum_{n=m+1}^{\infty} \frac{2n+1}{4\pi} (\tilde{\Psi}_j^A)^\wedge(n) \Lambda^\wedge(n) P_n(\xi \cdot \eta), \quad (\xi, \eta) \in \Omega^2,$$

where

$$K_\Lambda(\xi \cdot \eta) = \sum_{n=0}^{\infty} \frac{2n+1}{4\pi} \Lambda^\wedge(n) P_n(\xi \cdot \eta),$$

and the sum is understood in distributional sense.

The combination of spherical harmonic and wavelet expansion is of particular importance for geophysical problems (like the determination of the earth's gravitational or magnetic field), thereby connecting essential geophysical efforts of the past with very promising work in future. As a matter of fact, we are confronted with the situation that in geophysical reality spherical harmonic expansion models are available with accuracy up to a certain value m (see for example R.H. RAPP *et al.* (1991)), but numerical instability and low convergence rates make it difficult to achieve 'longer models'. Therefore we are convinced that the right way is a combined model. For the numerical realization the spherical harmonic part should be subtracted as usual from the geophysical data. Then the calculation of the wavelet coefficients relative to a wavelet of order m can be replaced by the calculation of the coefficients of the corresponding wavelet of order 0 (realize that under this assumption the Legendre coefficients $\Psi_j^\wedge(n)$ for $n = 0, \ldots, m$ may be chosen arbitrarily). In doing so we do not need more sophisticated approximate integration techniques.

Bibliographical Notes

The problem of constructing wavelet transforms on the sphere has seen an enormous increase of activities over the last years. Spherical approaches either use coherent states, homogeneous spaces or function spaces on the tangent bundles. Investigations in these directions have been carried out by B. TORRESANI (1993) and S. DAHLKE, P. MAASS (1996). Other methods to generalize wavelets by convolution kernels or related constructions in the spherical context are due to B. RUBIN (1994), J. GÖTTELMANN (1996), and M. HOLSCHNEIDER (1996) . The papers relevant for our investigations are W. FREEDEN, U. WINDHEUSER (1995, 1996, 1997) and the doctoral thesis by U. WINDHEUSER (1995). Our key idea is to introduce a continuous wavelet transform on the sphere based on approximate convolution identities of spherical singular integral theory and preserving rotational invariance.

11

DISCRETE WAVELET TRANSFORM

In the wavelet reconstruction formulae studied so far, the discretization of the corresponding wavelet transforms must be made directly by computing the relevant integrals for all the necessary values of rotation parameters. In Euclidean theory, however, a multiresolution analysis for performing discrete wavelet analysis and synthesis is known, which is completely recursive and, therefore, ideal for computations. This form of multiresolution analysis is certainly one of the reasons why the wavelet theory has gained such widespread popularity. In what follows we make the attempt to generalize as much as possible from this multiresolution procedure to the spherical context.

We saw in the last chapter that the scale relation between singular integral and wavelet kernel is determined by certain differential equations in terms of their Legendre coefficients. Dilation is understood as an operation acting on the whole scale interval; each scale parameter 'measures' the 'space localizing (i.e. windowing) effect' caused by a wavelet kernel. As the scale parameter continuously approaches the limit value 0, the wavelet kernel as a radially dependent $\mathcal{L}^2$-kernel becomes more and more concentrated around the axis.

The construction of the 'scale discrete wavelets' is based on a new interpretation of dilation (note that we distinguish between discretized wavelets obtained by discretization of the continuous wavelet transform as in the previous chapter and discrete wavelets obtained by the methods of this chapter). Rather than beginning with an operation defined continuously over a scale interval, dilation now is restricted to discrete values. The essential idea is to start from a 'continuous version of a Legendre transform' (i.e. a piecewise continuous function $\gamma_0 : [0, \infty) \to \mathbb{R}$ satisfying an admissibility condition) which acts as a generator of a square-integrable (radially dependent) function Γ_0 by equating the Legendre coefficients of Γ_0 to the functional values of γ_0 at non-negative integers. The jth level dilation operator is defined by letting $\Gamma_j = D_j \Gamma_0$ with Legendre coefficients $\Gamma_j^\wedge(n)$ chosen in such a way that $\Gamma_j^\wedge(n) = \gamma_j(n) = \gamma_0(2^{-j}n)$ for all non-negative integers n. In doing so any scaling function (that is, any 'continuous version of the Legendre transform' γ_0 which is monotonically decreasing on $[0, \infty)$ and continuous at 0 with value 1) provides a scale discrete singular integral. This in turn leads to a scale discrete multiresolution analysis. As a consequence the relation of the scale discrete wavelet associated to a scaling function is expressible by certain difference equations in terms of their Legendre coefficients. The scale index j (associated to the dyadic value 2^{-j} in the discrete formulation of dilation) serves as a 'measure' for decreasing frequency localization of the scale discrete wavelet kernel. To make our investigations more concrete we give some examples. In particular, a spherical counterpart of the Shannon multiresolution analysis is derived as the simplest example satisfying the orthogonality, and de la Vallé Poussin wavelets are presented to smooth out the oscillations which cannot be avoided for the orthogonal (Shannon) wavelets. Other non-band limited examples developed from our new concept of dilation are the P-scale discretized Abel–Poisson wavelets, the P-scale Gauß–Weierstraß wavelets (both known from Section 10.2), polynomial type wavelets, and exponential type wavelets. Finally we are interested in exact fully discrete wavelet transform in case of band-limited wavelets. Essential tools are the interpolatory polynomial integration formulae (known from Chapter 7). The chapter ends with combined spherical harmonic and wavelet reconstruction. We distinguish two cases, namely fixed approximation order and increasing approximation order, in accordance to different choices of dilation operators.

11.1 Scale Discrete Scaling Function

The construction of the 'scale discrete wavelets' as announced above is based on a new definition of dilation which will be explained now. We start by introducing our concept of scale discrete wavelets and associated multiresolution analysis with the definition of a scale discrete scaling function $\{\Phi_j\}$, each member $\Phi_j, j = 0, 1, \ldots$ being a radial basis function on the sphere. The functions Φ_j at different scales j are obtained by a dilation operator D_j applied to the function Φ_0 at scale 0:

$$\Phi_j = D_j \Phi_0, \quad j = 0, 1, \ldots.$$

The definition of D_j is based on the Legendre transform of Φ_0. To be more accurate, we need not only the Legendre transform $\Phi_0^\wedge(n)$, $n = 0, 1, \ldots$, but also a continuous version of it, i.e. a function $\varphi_0 : [0, \infty) \to \mathbb{R}$ with suitable properties. As a matter of fact, we will characterize all the admissible scaling functions by the properties of φ_0.

11.1.1 *Admissibility Condition*

As preparation we introduce the definition of admissibility and the notion of an admissible generator.

Definition 11.1.1. A piecewise continuous function $\gamma_0 : [0, \infty) \to \mathbb{R}$ is said to satisfy an *admissibility condition*, if

$$\sum_{n=0}^{\infty} \frac{2n+1}{4\pi} \left(\sup_{x \in [n,n+1)} |\gamma_0(x)| \right)^2 < \infty. \tag{11.1.1}$$

In this case, the function γ_0 is called an *admissible generator of the function* $\Gamma_0 : [-1, +1] \to \mathbb{R}$ given by

$$\Gamma_0 = \sum_{n=0}^{\infty} \frac{2n+1}{4\pi} \gamma_0(n) P_n. \tag{11.1.2}$$

i.e. $\Gamma_0^\wedge(n) = \gamma_0(n)$, $n = 0, 1, \ldots$.

Remark. In many cases an admissible generator is monotonically decreasing and satisfies, in addition, the estimate $0 \le \gamma_0(x) \le 1$, $x \in [0, \infty)$. Under these assumptions, condition (11.1.1) simply reduces to

$$\sum_{n=0}^{\infty} \frac{2n+1}{4\pi} |\gamma_0(n)|^2 < \infty. \tag{11.1.3}$$

Based on these preliminaries it is not hard to prove the following lemma.

Lemma 11.1.2. *If $\gamma_0 : [0, \infty) \to \mathbb{R}$ is an admissible generator, then Γ_0 defined by (11.1.2) is an element of $\mathcal{L}^2[-1, +1]$, hence, $\Gamma_0(\eta\cdot)$ is an element of $\mathcal{L}^2(\Omega)$ for every $\eta \in \Omega$.*

Proof. Since the Legendre polynomials form an orthogonal basis for the space $\mathcal{L}^2[-1, +1]$ with $(P_n, P_n)_{\mathcal{L}^2[-1,+1]} = 4\pi/(2n + 1)$, the admissibility condition imposed on γ_0 implies the estimate

$$\|\Gamma_0\|^2_{\mathcal{L}^2[-1,+1]} = \sum_{n=0}^{\infty} \frac{2n+1}{4\pi} (\gamma_0(n))^2 \le \sum_{n=0}^{\infty} \frac{2n+1}{4\pi} \left(\sup_{x \in [n,n+1)} |\gamma_0(x)| \right)^2 < \infty.$$

This is the required result. □

For a function γ_0 satisfying the admissibility condition we introduce functions $\gamma_j : [0, \infty) \to \mathbb{R}$ in the following way:

$$\gamma_j(x) = D_j \gamma_0(x) = \gamma_0(2^{-j} x), \quad x \in [0, \infty) \tag{11.1.4}$$

for $j = 0, 1, \ldots$. Then we are able to prove the following result.

Lemma 11.1.3. *Let $\gamma_0 : [0, \infty) \to \mathbb{R}$ satisfy the admissibility condition (11.1.1). Then each function $\gamma_j, j = 0, 1, \ldots,$ defined by (11.1.4) satisfies the admissiblity condition (11.1.1).*

Proof. Assume that γ_j satisfies the admissibility condition (11.1.1) stated in Definition 11.1.1:

$$\sum_{n=0}^{\infty} \frac{2n+1}{4\pi} \left(\sup_{x \in [n,n+1)} |\gamma_j(x)| \right)^2 < \infty.$$

We want to show by induction that γ_{j+1} satisfies the admissibility condition (11.1.1). For that purpose we see that for all $N \in \mathbb{N}$

$$\begin{aligned}
&\sum_{n=0}^{N} \frac{2n+1}{4\pi} \left(\sup_{x \in [n,n+1)} |\gamma_{j+1}(x)| \right)^2 \\
&= \sum_{n=0}^{N} \frac{2n+1}{4\pi} \left(\sup_{x \in [n,n+1)} \left|\gamma_j\left(\tfrac{1}{2}x\right)\right| \right)^2 \\
&= \sum_{n=0}^{N} \frac{2n+1}{4\pi} \left(\sup_{x \in [\frac{n}{2},\frac{n}{2}+\frac{1}{2})} |\gamma_j(x)| \right)^2 \\
&= \sum_{\substack{n=0 \\ n \text{ even}}}^{N} \frac{2n+1}{4\pi} \left(\sup_{x \in [\frac{n}{2},\frac{n+1}{2})} |\gamma_j(x)| \right)^2 + \sum_{\substack{n=1 \\ n \text{ odd}}}^{N} \frac{2n+1}{4\pi} \left(\sup_{x \in [\frac{n}{2},\frac{n+1}{2})} |\gamma_j(x)| \right)^2 \\
&\leq \sum_{n=0}^{[N/2]} \frac{4n+1}{4\pi} \left(\sup_{x \in [n,n+\frac{1}{2})} |\gamma_j(x)| \right)^2 + \sum_{n=0}^{[N/2]} \frac{4n+3}{4\pi} \left(\sup_{x \in [n+\frac{1}{2},n+1)} |\gamma_j(x)| \right)^2 \\
&\leq \sum_{n=0}^{\infty} \frac{2n+1}{4\pi} \left(\sup_{x \in [n,n+1)} |\gamma_j(x)| \right)^2 + 3\sum_{n=0}^{\infty} \frac{2n+1}{4\pi} \left(\sup_{x \in [n,n+1)} |\gamma_j(x)| \right)^2 .
\end{aligned}$$

Since the series on the right-hand side are convergent, Lemma 11.1.3 is proved. □

Lemma 11.1.3 enables us to write

$$\gamma_j = D_1 \gamma_{j-1}, \quad j = 1, 2, \ldots \tag{11.1.5}$$

provided that γ_0 satisfies the admissibility condition (11.1.1). This gives rise to the following definition.

Definition 11.1.4. Suppose that $\gamma_0 : [0, \infty] \to \mathbb{R}$ satisfies (11.1.1). For the generated functions $\Gamma_j \in \mathcal{L}^2[-1,+1]$, $j = 0, 1, \ldots$, given by

$$\Gamma_j = \sum_{n=0}^{\infty} \frac{2n+1}{4\pi} \gamma_j(n) P_n$$

we let

$$\Gamma_j = D_1 \Gamma_{j-1} = D_j \Gamma_0, \quad j = 1, 2, \ldots .$$

D_j is called *jth level dilation operator.*

As an immediate consequence we obtain the following corollary.

Corollary 11.1.5. *If Γ_0 is generated by an admissible generator γ_0, then $\Gamma_j = D_j \Gamma_0 \in \mathcal{L}^2[-1,+1]$ for all $j \in \mathbb{N}_0$.*

Remark. Note that the definition of D_j (Definition 11.1.4) requires both the existence of a generator γ_0, i.e. a 'continuous version of the Legendre transform' and the transition to a radial basis function Γ_0 (generated by γ_0).
We are also able to define the inverse of D_j, denoted by D_{-j}, being applicable to all radial basis functions on the sphere Ω. To be specific,

$$G_{-j} = D_{-j} G = \sum_{n=0}^{\infty} \frac{2n+1}{4\pi} G^{\wedge}(2^j n) P_n$$

whenever $G \in \mathcal{L}^2[-1,+1]$ with

$$G = \sum_{n=0}^{\infty} \frac{2n+1}{4\pi} G^{\wedge}(n) P_n$$

(in the $\|\cdot\|_{\mathcal{L}^2[-1,+1]}$-sense). Indeed, it is not difficult to see that $G_{-j} \in \mathcal{L}^2[-1,+1]$ for all $j \in \mathbb{N}_0$.

11.1.2 *Scale Discrete Scaling Function*

Next, we concentrate on those admissible generators which generate scaling functions.

Definition 11.1.6. Let $\varphi_0 : [0, \infty) \to \mathbb{R}$ satisfy the admissibility condition (11.1.1). φ_0 is called *generator of a scaling function* if it satisfies the following properties:

(i) $\varphi_0(0) = 1$,

(ii) φ_0 is monotonically decreasing,

(iii) φ_0 is continuous at 0.

If φ_0 satisfies the requirements (i), (ii), (iii) of Definition 11.1.6, then φ_0 and its dilates φ_j generate the *scale discrete scaling function* $\{\Phi_j\}, \Phi_j \in \mathcal{L}^2[-1,+1], j = 0, 1, \ldots$, via

$$\Phi_j^{\wedge}(n) = \varphi_j(n), \quad n = 0, 1, \ldots .$$

Remark. Let φ_0 be a generator of a scaling function. Then the admissibility of φ_0 implies the following properties:

(i) $\lim_{x\to\infty} \varphi_0(x) = 0$,

(ii) $\varphi_0(x) \geq 0$ for all $x \in [0, \infty)$.

Now we come to the following result.

Lemma 11.1.7. *Let φ_0 be a generator of a scaling function (in the sense of Definition 11.1.6). Then the sequence $\{\varphi_j(x)\}, j = 0, 1, \ldots$ is monotonically increasing for every $x \in [0, \infty)$, and we have*

$$\lim_{j\to\infty} \varphi_j(x) = 1.$$

Proof. Let $x \in [0, \infty]$. Then it is clear that

$$\lim_{j\to\infty} \varphi_j(x) = \lim_{j\to\infty} \varphi_0(2^{-j}x) = \varphi_0(0) = 1,$$

since φ_0 is continuous at 0. The monotonicity of $\{\varphi_j(x)\}$, $x \in [0, \infty)$, is an immediate consequence of condition (ii) of Definition 11.1.6. □

11.1.3 *Discrete Approximate Identity*

Next we show that the discrete scaling function defines a 'discrete approximate identity' in $\mathcal{L}^2(\Omega)$.

Lemma 11.1.8. *Let φ_0 and its dilates φ_j, $j \in \mathbb{N}$, generate the scale discrete scaling function $\{\Phi_j\}, \Phi_j \in \mathcal{L}^2[-1,+1]$, $j \in \mathbb{N}$. Then, for $F \in \mathcal{L}^2(\Omega)$,*

$$\lim_{j\to\infty} \|F - \Phi_j * F\|_{\mathcal{L}^2(\Omega)} = 0 \tag{11.1.6}$$

and

$$\lim_{j\to\infty} \|F - \Phi_j^{(2)} * F\|_{\mathcal{L}^2(\Omega)} = 0. \tag{11.1.7}$$

Proof. We already know that $\lim_{j\to\infty} \varphi_j(n) = 1$ for every $n \in \mathbb{N}_0$. Thus, $\lim_{j\to\infty} \Phi_j^\wedge(n) = 1$. Hence, the assertion follows from the fact that

$$(\Phi_j * F)^\wedge(n,k) = F^\wedge(n,k)\Phi_j^\wedge(n)$$

and

$$\lim_{j\to\infty} \Phi_j^\wedge(n)F^\wedge(n,k) = F^\wedge(n,k)$$

for every $n \in \mathbb{N}_0$, $k = 1, \ldots, 2n+1$. It is immediately clear that the kernel $\{\Phi_j\}$ in (11.1.6) may be replaced by the iterated kernel $\{\Phi_j^{(2)}\}$, $\Phi_j^{(2)} = \Phi_j * \Phi_j$, $j \in \mathbb{N}_0$. Note that $\Phi_j^{(2)}$ is generated in the sense of Definition 11.1.6 by the function φ_0^2. □

From Lemma 11.1.8 it follows that the $\Phi_j^{(2)} * F, j = 0, 1, \ldots$, provides us with approximations of F at different scales. In terms of filtering $\{\Phi_j^{(2)}\}$ may be interpreted as *low-pass filter*. The corresponding convolution operators $P_j : \mathcal{L}^2(\Omega) \to \mathcal{L}^2(\Omega)$, $j \in \mathbb{N}_0$, are given by

$$P_j(F) = \Phi_j^{(2)} * F, \quad j = 0, 1, \ldots . \tag{11.1.8}$$

Accordingly, we understand the *scale space* V_j to be the image of $\mathcal{L}^2(\Omega)$ under the operator P_j:

$$V_j = P_j(\mathcal{L}^2(\Omega)) = \{\Phi_j^{(2)} * F \mid F \in \mathcal{L}^2(\Omega)\}.$$

The scale spaces V_j define a *(scale discrete) multiresolution analysis* of $\mathcal{L}^2(\Omega)$ in the following sense.

Theorem 11.1.9. *The scale spaces V_j satisfy the following statements:*

(i) $V_0 \subset \ldots \subset V_j \subset V_{j+1} \subset \ldots \subset \mathcal{L}^2(\Omega)$

(ii) $\overline{\bigcup_{j=0}^{\infty} V_j}^{\|\cdot\|_{\mathcal{L}^2(\Omega)}} = \mathcal{L}^2(\Omega)$

(iii) If a function $G \in \mathcal{L}^2[-1,+1]$ satisfies $G(\eta\cdot) \in V_j$, then $D_{-1}G(\eta\cdot) \in V_{j-1}$, $j = 1, 2, \ldots$.

Proof. Part (i) follows from the monotonicity of φ_0. Part (ii) is equivalent to Lemma 11.1.8. Part (iii), finally, is a consequence of the definition of D_j and the construction of $\varphi_j = D_j\varphi_0$. □

The subspaces $V_j \subset \mathcal{L}^2(\Omega)$ can be characterized by spectral information in terms of Fourier coefficients. To be specific, if $\varphi_0 > 0$ for all $x \in [0, \infty)$,

$$V_j = \left\{ F \in \mathcal{L}^2(\Omega) \,\middle|\, \sum_{n=0}^{\infty} \left(\Phi_j^\wedge(n)\right)^{-4} \sum_{j=1}^{2n+1} |F^\wedge(n,j)|^2 < \infty \right\}.$$

Remark. One might expect, that there is some more structure in our (scale discrete) multiresolution, e.g. $V_0 = \{0\}$ or the fact that all scale spaces V_j are finite-dimensional with certain conditions on the dimension, etc. But these properties are, in general, not true. They depend on the special choice of the generator φ_0 of a scaling function and are discussed in more detail later on.

11.2 Scale Discrete Wavelet Transform

The definition of the scale discrete scaling function now allows us to introduce scale discrete wavelets on the sphere. As in the scale discretized case of the last chapter, we represent an $\mathcal{L}^2(\Omega)$-function F by a two-parameter family $(j;\eta), j \in \mathbb{N}_0$, $\eta \in \Omega$, breaking up the function F into 'pieces' at different locations and different levels of resolution. An essential point is the definition of a mother wavelet and its dual wavelet starting from their generators. This definition, of course, has to be done in close relation to a given scaling function. The mother wavelet then is rotated and dilated to establish the discrete version of the wavelet transform of a function. As a matter of fact, we are able to prove a reconstruction formula. Special choices of the mother wavelet and its dual one lead us back to the P- and M-scale discretized wavelets as developed in Chapter 10.

Definition 11.2.1. Let φ_0 be the generator of a scaling function (as defined by Definition 11.1.6). Then the piecewise continuous functions $\psi_0, \tilde{\psi}_0 : [0,\infty) \to \mathbb{R}$ are said to be *generators of the mother wavelet* $\Psi_0 \in \mathcal{L}^2[-1,+1]$ and the *dual mother wavelet* $\tilde{\Psi}_0 \in \mathcal{L}^2[-1,+1]$, respectively, if both of them are admissible generators (in the sense of Definition 11.1.1) and satisfy, in addition, the '*refinement equation*'

$$\psi_0(x)\tilde{\psi}_0(x) = \left(\varphi_0\,(x/2)\right)^2 - \left(\varphi_0(x)\right)^2, \quad x \in [0,\infty).$$

The functions $\Psi_0 \in \mathcal{L}^2[-1,+1]$, $\tilde{\Psi}_0 \in \mathcal{L}^2[-1,+1]$, defined via the Legendre coefficients $\Psi_0^\wedge(n), \tilde{\Psi}_0^\wedge(n)$ given by

$$\begin{aligned} \Psi_0^\wedge(n) &= \psi_0(n), \quad n = 0,1,\ldots, \\ \tilde{\Psi}_0^\wedge(n) &= \tilde{\psi}_0(n), \quad n = 0,1,\ldots, \end{aligned}$$

are called the *mother wavelet* and the *dual mother wavelet*, respectively.

Let us make a couple of simple observations concerning this definition.

Lemma 11.2.2. *The generators* $\psi_0, \tilde{\psi}_0 : [0,\infty) \to \mathbb{R}$ *and their dilates* $\psi_j = D_j\psi_0$, $\tilde{\psi}_j = D_j\tilde{\psi}_0$ *satisfy the following properties:*

(i) $\psi_j(0)\tilde{\psi}_j(0) = 0, \quad j \in \mathbb{N}_0,$

(ii) $\psi_j(x)\tilde{\psi}_j(x) = (\varphi_{j+1}(x))^2 - (\varphi_j(x))^2, \quad j \in \mathbb{N}_0,\ x \in [0,\infty),$

(iii) $(\varphi_0(x))^2 + \sum\limits_{j=0}^{J} \psi_j(x)\tilde{\psi}_j(x) = (\varphi_{J+1}(x))^2, \quad J \in \mathbb{N}_0,\ x \in [0,\infty).$

Proof. Part (i) is clear, since the generator φ_0 of a scaling function satisfies $\varphi_0(0) = \varphi_j(0) = 1$, $j \in \mathbb{N}_0$. Part (ii) follows from the fact that, for $j \in \mathbb{N}_0$, $x \in [0,\infty)$,

$$\begin{aligned} \psi_j(x)\tilde{\psi}_j(x) &= D_j\psi_0(x)D_j\tilde{\psi}_0(x) \\ &= \psi_0(2^{-j}x)\tilde{\psi}_0(2^{-j}x) \\ &= (\varphi_0(2^{-j-1}x))^2 - (\varphi_0(2^{-j}x))^2 \\ &= (\varphi_{j+1}(x))^2 - (\varphi_j(x))^2. \end{aligned}$$

Part (iii), finally, follows from (ii) by summation. □

It is natural, as it was done for the scaling function, to apply the operators D_j directly to the mother wavelet and its dual one. In connection with the rotation operator R_η, this will lead us to the definition of the wavelet $\Psi_{j;\eta}$ and its dual wavelet $\tilde{\Psi}_{j;\eta}$. More explicitly, we have

$$\Psi_j = D_j\Psi_0, \qquad \tilde{\Psi}_j = D_j\tilde{\Psi}_0, \quad j \in \mathbb{N}_0, \tag{11.2.1}$$

and

$$\begin{aligned} (R_\eta\Psi_j)(\xi) &= \Psi_{j;\eta}(\xi) = \Psi_j(\eta\cdot\xi), & \xi \in \Omega, \\ (R_\eta\tilde{\Psi}_j)(\xi) &= \tilde{\Psi}_{j;\eta}(\xi) = \tilde{\Psi}_j(\eta\cdot\xi), & \xi \in \Omega. \end{aligned} \tag{11.2.2}$$

Putting together (11.2.1) and (11.2.2) we therefore obtain for $\xi \in \Omega$

$$\begin{aligned} \Psi_{j;\eta}(\xi) &= (R_\eta D_j\Psi_0)(\xi), \\ \tilde{\Psi}_{j;\eta}(\xi) &= (R_\eta D_j\tilde{\Psi}_0)(\xi). \end{aligned}$$

Definition 11.2.3. Let $\Psi_0 \in \mathcal{L}^2[-1,+1]$ and $\tilde{\Psi}_0 \in \mathcal{L}^2[-1,+1]$ be a mother wavelet and its dual with respect to a scaling function $\Phi_0 \in \mathcal{L}^2[-1,+1]$, respectively. Then, for $F \in \mathcal{L}^2(\Omega)$, the *discrete wavelet transform at scale* $j \in \mathbb{N}_0$ *and position* $\eta \in \Omega$ is defined by

$$(WT)(F)(j;\eta) = (F, \Psi_{j;\eta})_{\mathcal{L}^2(\Omega)}.$$

Our definition, in fact, can be viewed as a discrete counterpart of the continuous wavelet transform.

In analogy to the definition of the operator $P_j : \mathcal{L}^2(\Omega) \to \mathcal{L}^2(\Omega)$ (cf. (11.1.8)) we consider now convolution operators (*band-pass filter*) $R_j : \mathcal{L}^2(\Omega) \to \mathcal{L}^2(\Omega)$, $j \in \mathbb{N}_0$, defined by

$$R_j(F) = (\Psi_j * \tilde{\Psi}_j * F). \tag{11.2.3}$$

$R_j(F)$ can be interpreted as a version of F blurred to the scale j. It describes the 'detail behaviour' of F at scale j. From Lemma 11.2.2 (iii), we can immediately deduce that, for $J \in \mathbb{N}_0$,

$$\Phi_0^{(2)} + \sum_{j=0}^{J} \Psi_j * \tilde{\Psi}_j = \Phi_{J+1}^{(2)}. \tag{11.2.4}$$

Therefore, it follows that the operator P_{J+1} can be decomposed in the following way

$$P_{J+1} = P_0 + \sum_{j=0}^{J} R_j.$$

This gives rise to introduce the *detail space* W_j to be

$$W_j = R_j(\mathcal{L}^2(\Omega)) = \{\Psi_j * \tilde{\Psi}_j * F \mid F \in \mathcal{L}^2(\Omega)\}.$$

The space W_j contains the detail information needed to go from an approximation at resolution j to an approximation at resolution $j+1$. Note that

$$\begin{aligned} V_0 + \sum_{j=0}^{J} W_j &= V_{J+1}, \\ V_j &= V_{j-1} + W_{j-1}. \end{aligned} \tag{11.2.5}$$

It is worth mentioning that the sum decomposition in general is neither direct nor orthogonal. An orthogonal decomposition is described later, when band-limited wavelets come into play.

Any function $F \in \mathcal{L}^2(\Omega)$ can now be decomposed as follows. Starting with $P_{J+1}(F)$ for some J we have $P_{J+1}(F) = P_0(F) + \sum_{j=0}^{J} R_j(F)$. In other words, the partial reconstruction $R_{J+1}(F)$ is nothing other than the difference of two smoothings $P_{J+1}(F) - P_J(F)$ at two consecutive scales: $R_J(F) = P_{J+1}(F) - P_j(F)$ (cf. Lemma 11.2.2).

Furthermore, we have

$$\begin{aligned} (P_j(F))^\wedge(n,k) &= F^\wedge(n,k)(\Phi_j)^\wedge(n)(\Phi_j)^\wedge(n), \\ (R_j(F))^\wedge(n,k) &= F^\wedge(n,k)(\Psi_j)^\wedge(n)(\tilde{\Psi}_j)^\wedge(n). \end{aligned} \tag{11.2.6}$$

The fomulae (11.2.6), therefore, give wavelet decompositions like (11.2.5) an interpretation in terms of Fourier analysis by explaining how the frequency spectrum of a function $F \in V_j$ is divided up between the space V_{j-1} and W_{j-1}, which enhances our understanding of what is meant by 'smoothing' and 'detail'.

Our definition of the discrete wavelet transform developed above enables us to prove a reconstruction formula. In other words, it is possible to reconstruct a function $F \in \mathcal{L}^2(\Omega)$ from its wavelet transform $(WT)(F)$.

Theorem 11.2.4 *(reconstruction formula). Let Ψ_0 $(\tilde{\Psi}_0)$ be the mother wavelet (the dual mother wavelet) with respect to a scale discrete scaling function Φ_0. Then, for $F \in \mathcal{L}^2(\Omega)$,*

$$F = \Phi_0^{(2)} * F + \sum_{j=0}^{\infty} \int_{\Omega} (WT)(F)(j;\eta)\tilde{\Psi}_{j;\eta}(\cdot)\, d\omega(\eta),$$

where the equality is understood in the $\mathcal{L}^2(\Omega)$-sense.

Proof. Let F be of class $\mathcal{L}^2(\Omega)$. From the definition of the spherical convolution and (11.2.4) it follows that

$$\Phi_0^{(2)} * F + \sum_{j=0}^{J} \int_{\Omega} (WT)(F)(j;\eta)\tilde{\Psi}_{j;\eta}(\cdot)\, d\omega(\eta) = \Phi_{J+1}^{(2)} * F.$$

Taking the limit $J \to \infty$ the result follows immediately from Lemma 11.1.8. □

Theorem 11.2.4 can be seen as a discrete version of the inverse continuous wavelet transform developed in the last chapter. For numerical realization the computation of the wavelet coefficients is needed which results in a problem of space discretization.

Up to now, the definition of the mother wavelet Ψ_0 and its dual $\tilde{\Psi}_0$ are quite general. The only condition which has to be satisfied (besides the admissibility) is the 'refinement equation'

$$\psi_0(x)\tilde{\psi}_0(x) = (\varphi_1(x))^2 - (\varphi_0(x))^2, \quad x \in [0,\infty), \tag{11.2.7}$$

where φ_0 is the generator of a scale discrete scaling function.

We are now prepared to derive two construction principles which lead back to results of the last chapter.

P-scale Discrete Wavelet. For the generator $\varphi_0 : [0,\infty) \to \mathbb{R}$ we set

$$\psi_0(x) = \tilde{\psi}_0(x) = \sqrt{(\varphi_1(x))^2 - (\varphi_0(x))^2}, \quad x \in [0,\infty).$$

The monotonicity of φ_0 ensures that the radicand is non-negative. Obviously ψ_0 and $\tilde{\psi}_0$ fulfill (11.2.7). Both of them are piecewise continuous. Furthermore, we have

$$\begin{aligned}
&\sum_{n=0}^{\infty} \frac{2n+1}{4\pi} \left(\sup_{x\in[n,n+1)} |\psi_0(x)| \right)^2 \\
&= \sum_{n=0}^{\infty} \frac{2n+1}{4\pi} \sup_{x\in[n,n+1)} |(\varphi_1(x))^2 - (\varphi_0(x))^2| \\
&\leq \sum_{n=0}^{\infty} \frac{2n+1}{4\pi} \left(\sup_{x\in[n,n+1)} |\varphi_1(x)| \right)^2 + \sum_{n=0}^{\infty} \frac{2n+1}{4\pi} \left(\sup_{x\in[n,n+1)} |\varphi_0(x)| \right)^2 .
\end{aligned}$$

Therefore, the admissibility of φ_0 induces together with Lemma 11.1.3 the admissiblity of ψ_0. But this means that ψ_0 and $\tilde{\psi}_0$ are generators of a mother wavelet. In this case, the dual mother wavelet coincides with the mother wavelet.

M-scale Discrete Wavelets. Starting from $\varphi_0 : [0, \infty) \to \mathbb{R}$ we let

$$\begin{aligned}
\psi_0(x) &= \varphi_1(x) - \varphi_0(x), \quad x \in [0, \infty), \\
\tilde{\psi}_0(x) &= \varphi_1(x) + \varphi_0(x), \quad x \in [0, \infty).
\end{aligned}$$

Hence, ψ_0 and $\tilde{\psi}_0$ are piecewise continuous and satisfy (11.2.7). The admissibility of ψ_0 and $\tilde{\psi}_0$ can be easily seen from the following inequalities:

$$\begin{aligned}
|\psi_0(x)| &= |\varphi_1(x) - \varphi_0(x)| \leq 2\varphi_1(x), \quad x \in [0, \infty), \\
|\tilde{\psi}_0(x)| &= |\varphi_1(x) + \varphi_0(x)| \leq 2\varphi_1(x), \quad x \in [0, \infty),
\end{aligned}$$

where we have used the fact that $\varphi_1(x) \geq \varphi_0(x) \geq 0$ for each $x \in [0, \infty)$.

Remark. The idea of establishing P-scale and M-scale discrete wavelets is based on the observation that the wavelet members may be regarded as differences of two low-pass filters. To be specific,

$$\psi_j(x)\tilde{\psi}_j(x) = (\varphi_{j+1}(x))^2 - (\varphi_j(x))^2, \quad j \in \mathbb{N}_0, \ x \in [0, \infty)$$

so that

$$(\varphi_J(x))^2 + \sum_{j=J}^{\infty} \psi_j(x)\tilde{\psi}_j(x) = 1, \quad x \in [0, \infty).$$

By construction, our wavelets therefore lead to a partition of the unity as follows:

$$\sum_{j=-\infty}^{\infty} \psi_j(x)\tilde{\psi}_j(x) = \sum_{j=-\infty}^{+\infty} ((\varphi_{j+1}(x))^2 - (\varphi_j(x))^2) = 1, \quad x \in [0,\infty).$$

This formula is close to the concept of the 'difference of two smoothings'.

Occasionally, other choices of wavelets (for example, S-scale discrete wavelets) may be helpful but these are not discussed here.

To make the preceeding more concrete, we now give a list of examples. We essentially distinguish two cases, viz. non-band-limited wavelets (Section 11.3) and band-limited wavelets (Section 11.4).

11.3 Non-band-limited Wavelets

In this section we shall take a first look at discrete scaling functions and corresponding wavelets. All wavelets discussed in this section share the property that their generators $\psi_0 : [0,\infty) \to \mathbb{R}$ have a global support, i.e. $\operatorname{supp} \psi_0 = [0,\infty)$. Since there are only a few conditions for a function $\varphi_0 : [0,\infty) \to \mathbb{R}$ to be a generator of a scaling function, there are, of course, various possibilities for its concrete realization. We will concentrate on three types: (i) rational (ii) modified rational (iii) exponential scaling functions.

11.3.1 *Rational Wavelets*

Starting point for the construction of rational wavelets is the generator φ_0 given by

$$\varphi_0(x) = (1+x)^{-s}, \quad x \in [0,\infty).$$

Obviously, $\varphi_0(0) = 1$, φ_0 is monotonously decreasing, and φ_0 is continuous on the interval $[0,\infty)$. Furthermore, (11.1.3) is satisfied, which is easily seen from the fact that

$$\sum_{n=0}^{\infty} \frac{2n+1}{4\pi} |\varphi_0(n)|^2 = \sum_{n=0}^{\infty} \frac{2n+1}{4\pi} \frac{1}{(1+n)^{2s}} < \infty$$

provided that $s > 1$. Therefore, $\varphi_0(x) = (1+x)^{-s}$, defines an admissible generator of a scaling function. The scaling function itself and its dilates $\Phi_j = D_j\Phi_0$ are then given by

$$\begin{aligned}
\Phi_0(t) &= \sum_{n=0}^{\infty} \frac{2n+1}{4\pi} \frac{1}{(1+n)^s} P_n(t), \quad t \in [-1,+1], \\
\Phi_j(t) &= \sum_{n=0}^{\infty} \frac{2n+1}{4\pi} \frac{1}{(1+2^{-j}n)^s} P_n(t), \quad j \in \mathbb{N}_0, t \in [-1,+1].
\end{aligned}$$

Note that the iterated functions $\Phi_j^{(2)}$ are the reproducing kernels of the Sobolev space $\mathcal{H}(\{(1+2^{-j}n)^s\};\Omega)$, respectively.

The P-scale discrete wavelets are given by

$$\begin{aligned}\Psi_j(t) &= \tilde{\Psi}_j(t)\\ &= \sum_{n=0}^{\infty} \frac{2n+1}{4\pi} \sqrt{\frac{1}{(1+2^{-j-1}n)^{2s}} - \frac{1}{(1+2^{-j}n)^{2s}}}\, P_n(t),\ t \in [-1,+1].\end{aligned}$$

The M-wavelets and their duals with respect to φ_0 are of the form

$$\begin{aligned}\Psi_j(t) &= \sum_{n=0}^{\infty} \frac{2n+1}{4\pi} \left(\frac{1}{(1+2^{-j-1}n)^{s}} - \frac{1}{(1+2^{-j}n)^{s}} \right) P_n(t), \quad t \in [-1,+1],\\ \tilde{\Psi}_j(t) &= \sum_{n=0}^{\infty} \frac{2n+1}{4\pi} \left(\frac{1}{(1+2^{-j-1}n)^{s}} + \frac{1}{(1+2^{-j}n)^{s}} \right) P_n(t), \quad t \in [-1,+1].\end{aligned}$$

Figure 11.3.1 gives a graphical impression.

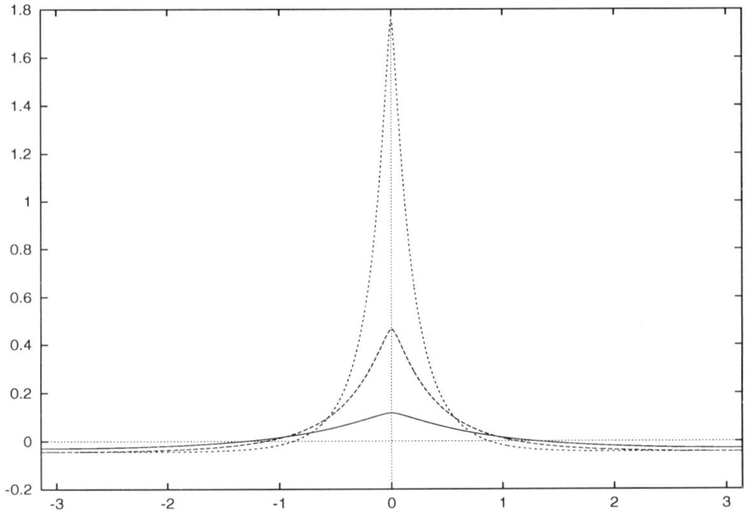

FIG. 11.3.1. Rational P-wavelets $\Psi_j(\cos\vartheta)$, $j = 0, \ldots, 2$, $\varphi_0(x) = (1+x)^{-4}$

11.3.2 *Modified Rational Wavelet*

As a modification of the generator introduced before we consider

$$\varphi_0(x) = (1+x^2)^{-s}, \quad x \in [0, \infty).$$

It is not difficult to verify that φ_0 generates a scale discrete scaling function, provided that $s > 1/2$. Accordingly the functions Φ_j are of the form

$$\Phi_j(t) = \sum_{n=0}^{\infty} \frac{2n+1}{4\pi} \frac{1}{(1+2^{-2j}n^2)^s} P_n(t), \quad t \in [-1, +1].$$

The construction of the corresponding P- and P-scale discrete wavelets are straightforward. We omit the details and give just a graphical impression in Figure 11.3.2.

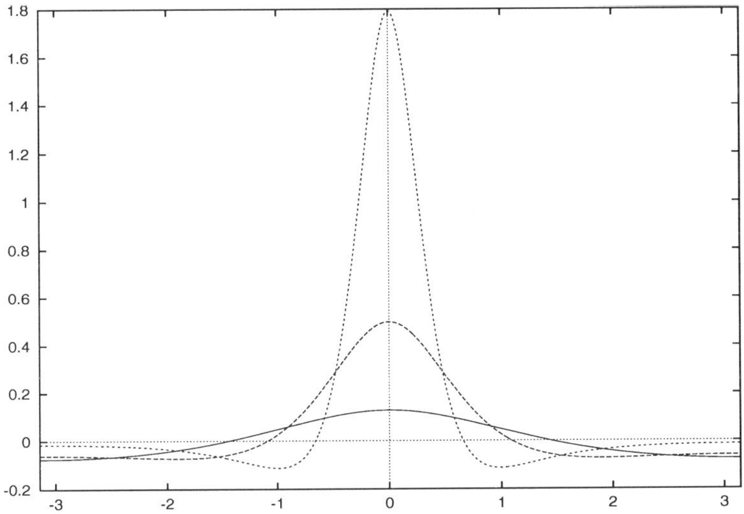

FIG. 11.3.2. Modified rational P-wavelets $\Psi_j(\cos\vartheta)$, $j = 0, \ldots, 2$, $\varphi_0(x) = (1+x^2)^{-4}$

11.3.3 *Exponential Wavelets*

From the Sobolev lemma (cf. Chapter 5) we know that the order of differentiability of the rational scaling function and its modification is finite. A scaling function of class $C^{(\infty)}$ can be obtained via the generator

$$\varphi_0(x) = \mathrm{e}^{-h(x)}, \quad x \in [0, \infty),$$

where $h : [0, \infty) \to \mathbb{R}$ is assumed to satisfy the following conditions:

(i) $h \in C^{(\infty)}[0, \infty)$

(ii) $h(0) = 0,\ h(x) > 0$ for $x > 0$

(iii) $h(x) < h(x')$ whenever $0 < x < x'$

(iv) $\{(e^{h(n)})^{1/2}\}$ is summable (in the sense of Definition 5.2.1).

Then the scaling function and its dilates are of the form

$$\begin{aligned} \Phi_0(t) &= \sum_{n=0}^{\infty} \frac{2n+1}{4\pi} e^{-h(n)} P_n(t), \quad t \in [-1,+1], \\ \Phi_j(t) &= \sum_{n=0}^{\infty} \frac{2n+1}{4\pi} e^{-h(2^{-j}n)} P_n(t), \quad t \in [-1,+1], \end{aligned}$$

respectively. The P-scale discrete wavelets corresponding to this scaling function are given by

$$\Psi_j^{\wedge}(n) = \sqrt{(e^{-h(2^{-j-1}n)})^2 - (e^{-h(2^{-j}n)})^2}, \quad n \in \mathbb{N}_0.$$

Similarly, the M-scale discrete wavelets can be written in terms of the Legendre coefficients as

$$\begin{aligned} \Psi_j^{\wedge}(n) &= e^{-h(2^{-j-1}n)} - e^{-h(2^{-j}n)}, \quad n \in \mathbb{N}_0 \\ \tilde{\Psi}_j^{\wedge}(n) &= e^{-h(2^{-j-1}n)} + e^{-h(2^{-j}n)}, \quad n \in \mathbb{N}_0. \end{aligned}$$

As particularly important examples we mention the *Abel–Poisson wavelet* (obtained from $h(x) = Rx, R > 0$ and the *Gauß–Weierstraß wavelet* obtained from $h(x) = Rx(x+1),\ R > 0$). In both cases the scale discrete wavelets developed here coincide with the scale discretized wavelets introduced in Chapter 10. For a graphical impression, see Figures 11.3.3 and 11.3.4.

11.4 Band-limited Wavelets

If the generator $\varphi_0 : [0,\infty) \to \mathbb{R}$ is chosen in such a way that its support is compact, then the resulting wavelets are band-limited, i.e. only a finite number of Legendre coefficients is different from zero. This allows us to obtain more powerful results such as the construction of orthogonal bases. For example, the multiresolution analysis described in Theorem 11.1.9 is orthogonal in case of the Shannon wavelets, that means $W_i \perp W_j$ whenever $i \neq j$. Furthermore we are able to derive a space discrete exact formulation of the reconstruction formula in case of band-limited wavelets.

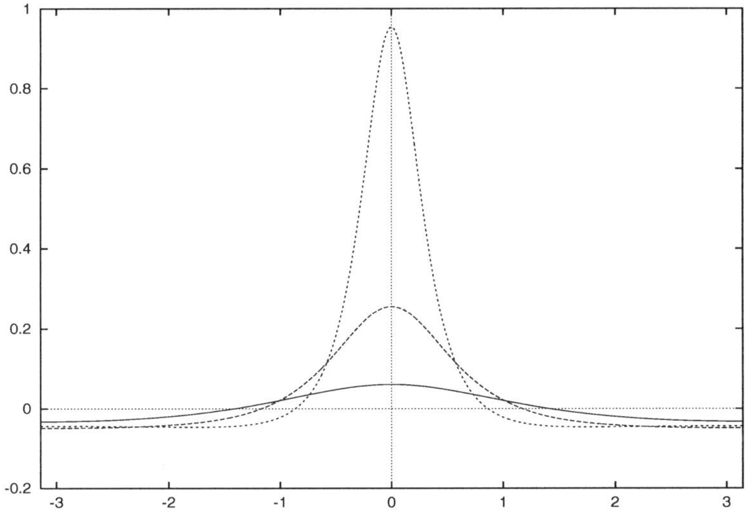

FIG. 11.3.3. Exponential P-wavelets $\Psi_j(\cos\vartheta)$, $j = 0,\ldots,2$, $\varphi_0(x) = \mathrm{e}^{-30x}$

11.4.1 *Shannon Wavelets*

The generator of the Shannon scaling function is given by

$$\varphi_0(x) = \begin{cases} 1 & \text{for} \quad x \in [0,1) \\ 0 & \text{for} \quad x \in [1,\infty), \end{cases} \tag{11.4.1}$$

so that

$$\varphi_j(x) = \begin{cases} 1 & \text{for} \quad x \in [0,2^j) \\ 0 & \text{for} \quad x \in [2^j,\infty). \end{cases} \tag{11.4.2}$$

It is easy to see that all conditions for φ_0 to be a generator of a scaling function are fulfilled. We have

$$\Phi_j(t) = \sum_{n=0}^{2^j-1} \frac{2n+1}{4\pi} P_n(t) \ , \quad t \in [-1,+1]. \tag{11.4.3}$$

A remarkable property is that Φ_j coincides with its iterations

$$\Phi_j^{(k)} = \Phi_j * \Phi_j^{(k-1)} \ , \quad k = 2,3,\ldots,$$

in particular, $\Phi_j^{(2)} = \Phi_j$. Therefore the construction of the P-scale and M-scale wavelets is straightforward. The P-scale discrete wavelets are characterized by

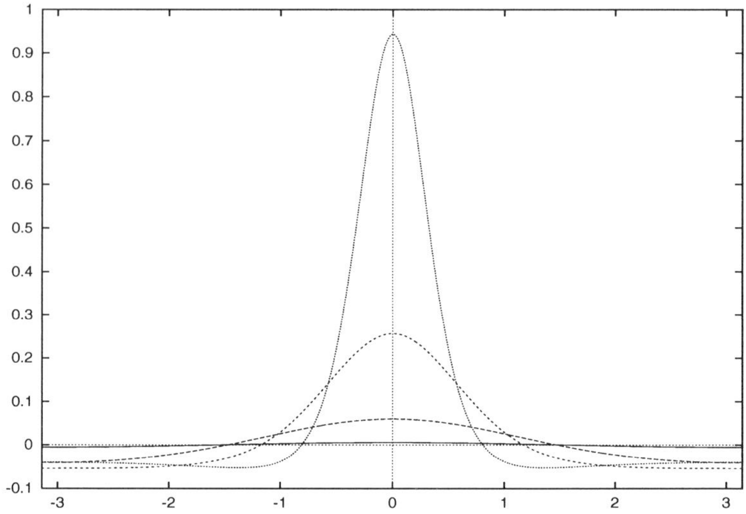

FIG. 11.3.4. Exponential P-wavelets $\Psi_j(\cos\vartheta)$, $j = 0,\ldots,3$, $\varphi_0(x) = \mathrm{e}^{-5x(x+1)}$

$$\tilde{\psi}_j(x) = \psi_j(x) = \begin{cases} 1 & \text{for} \quad x \in [2^j, 2^{j+1}) \\ 0 & \text{elsewhere.} \end{cases}$$

Hence,

$$\tilde{\Psi}_j(t) = \Psi_j(t) = \sum_{n=2^j}^{2^{j+1}-1} \frac{2n+1}{4\pi} P_n(t) \ , \quad t \in [-1,+1]. \tag{11.4.4}$$

For the M-wavelet and its dual we easily obtain

$$\Psi_j(t) = \sum_{n=2^j}^{2^{j+1}-1} \frac{2n+1}{4\pi} P_n(t) \ , \quad t \in [-1,+1], \tag{11.4.5}$$

and

$$\tilde{\Psi}_j(t) = 2\sum_{n=0}^{2^j-1} \frac{2n+1}{4\pi} P_n(t) + \sum_{n=2^j}^{2^{j+1}-1} \frac{2n+1}{4\pi} P_n(t) \ , \quad t \in [-1,+1]. \tag{11.4.6}$$

Since the first sum on the left-hand side of (11.4.6) is orthogonal to Ψ_j with respect to $(\cdot,\cdot)_{\mathcal{L}^2[-1,+1]}$, it does not contribute in calculations, so that

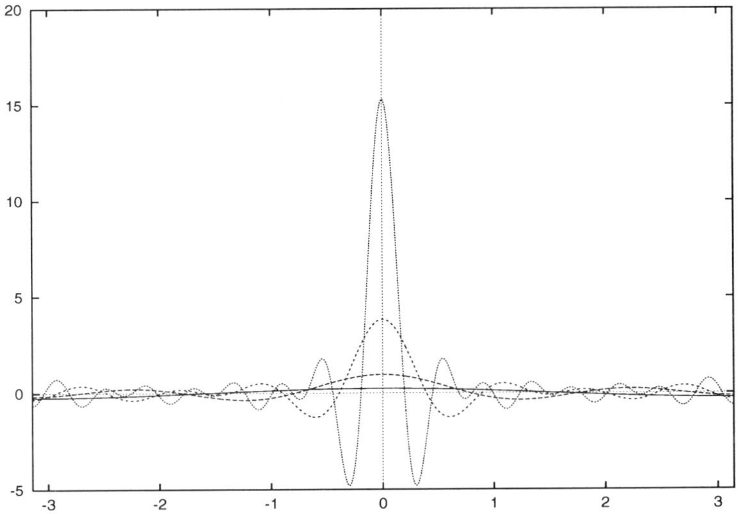

FIG. 11.4.1. Shannon wavelets $\Psi_j(\cos\vartheta)$, $j = 0, \ldots, 3$

Ψ_j, in principle, coincides with Ψ_j. In other words, the M-scale discrete wavelets are not distinguishable from the P-scale discrete wavelets seen from computational point of view. For an illustration, see Figure 11.4.1.

It should be noted that the multiresolution analysis (cf. Section 11.2) has a conceptually more structured form for the particular choice of the Shannon wavelets. This is written down as follows.

Theorem 11.4.1. *Assume that P_j (R_j) are the low pass (band pass) filter convolution operators corresponding to (11.4.2) ((11.4.5), (11.4.6)). Then the scale spaces V_j and the detail spaces W_j given by*

$$\begin{aligned} V_j &- P_j(\mathcal{L}^2(\Omega)), \\ W_j &= R_j(\mathcal{L}^2(\Omega)), \end{aligned}$$

satisfy the properties:

(i) $V_0 \subset \ldots \subset V_j \subset V_{j+1} \subset \ldots \subset \mathcal{L}^2(\Omega)$

(ii) $\overline{\bigcup_{j=0}^{\infty} V_j}^{\|\cdot\|_{\mathcal{L}^2(\Omega)}} = \mathcal{L}^2(\Omega)$

(iii) $\bigcap_{j=0}^{\infty} V_j = Harm_0$

(iv) $V_{j+1} = V_j \oplus W_j$, *where the sum is orthogonal*

(v) If the radial basis function $G \in \mathcal{L}^2[-1,+1]$ *satisfies* $G(\eta\cdot) \in V_j$ *(or* $G(\eta\cdot) \in W_j$*), then* $D_{-1}G(\eta\cdot) \in V_{j-1}$ *(or* $D_{-1}G(\eta\cdot) \in W_{j-1}$*),* $j \geq 1$.

Proof. In comparison to Theorem 11.1.9 we merely have to show the properties (iii) and (iv). But this does not require additional efforts, since we have

$$
\begin{aligned}
V_j &= \bigoplus_{k=0}^{2^j-1} Harm_k, \\
W_j &= \bigoplus_{k=2^j}^{2^{j+1}-1} Harm_k.
\end{aligned}
$$

□

11.4.2 *De la Vallé Poussin Wavelets*

The Shannon wavelets show serious oscillations. If one is willing to give up the orthogonality of the detail spaces W_j, the number of oscillations can be reduced by a modification of the generator φ_0. To this end we consider a somehow 'smoothed' version of the generator of the Shannon wavelets (dependent on a parameter $h \in (0,1)$)

$$
\varphi_0(x) = \begin{cases} 1 & \text{for} \quad x \in [0,h) \\ \frac{1-x}{1-h} & \text{for} \quad x \in [h,1) \\ 0 & \text{for} \quad x \in [1,\infty). \end{cases} \tag{11.4.7}
$$

With the definition (11.4.7) the dilates have the form

$$
\varphi_j(x) = \begin{cases} 1 & \text{for} \quad x \in [0,2^j h) \\ \frac{1-2^{-j}x}{1-h} & \text{for} \quad x \in [2^j h, 2^j) \\ 0 & \text{for} \quad x \in [2^j,\infty) \end{cases}
$$

($j \in \mathbb{N}_0$). For the formulation of the P-wavelets corresponding to the 'de la Vallée Poussin generator' we have to distinguish three cases:

- $h < \frac{1}{2}$

$$\tilde{\Psi}_j^\wedge(n) = \Psi_j^\wedge(n)$$

$$= \begin{cases} 0 & \text{for} \quad n < 2^j h \\ \left(1 - \left(\frac{1-2^{-j}n}{1-h}\right)^2\right)^{1/2} & \text{for} \quad 2^j h \le n < 2^{j+1} h \\ \left(\left(\frac{1-2^{-j-1}n}{1-h}\right)^2 - \left(\frac{1-2^{-j}n}{1-h}\right)^2\right)^{1/2} & \text{for} \quad 2^{j+1} h \le n < 2^j \\ \frac{1-2^{-j-1}n}{1-h} & \text{for} \quad 2^j \le n < 2^{j+1} \\ 0 & \text{for} \quad 2^{j+1} \le n < \infty \end{cases}$$

- $h > \frac{1}{2}$

$$\begin{aligned} \tilde{\Psi}_j^\wedge(n) &= \Psi_j^\wedge(n) \\ &= \begin{cases} 0 & \text{for} \quad 0 \le n < 2^j h \\ \left(1 - \left(\frac{1-2^{-j}n}{1-h}\right)^2\right)^{1/2} & \text{for} \quad 2^j h \le n < 2^j \\ 1 & \text{for} \quad 2^j \le n < 2^{j+1} h \\ \frac{1-2^{-j-1}n}{1-h} & \text{for} \quad 2^{j+1} h \le n < 2^{j+1} \\ 0 & \text{for} \quad 2^{j+1} \le n < \infty \end{cases} \end{aligned}$$

- $h = \frac{1}{2}$

$$\begin{aligned} \tilde{\Psi}_j^\wedge(n) &= \Psi_j^\wedge(n) \\ &= \begin{cases} 0 & \text{for} \quad 0 \le n < 2^{j-1} \\ \left(1 - \left(2 - 2^{j+1}n\right)^2\right)^{1/2} & \text{for} \quad 2^{j-1} \le n < 2^j \\ 2 - 2^{-j}n & \text{for} \quad 2^j \le n < 2^{j+1} \\ 0 & \text{for} \quad 2^{j+1} \le n < \infty. \end{cases} \end{aligned}$$

Compared with the Shannon wavelets there generally are more non-vanishing Legendre coefficients of Ψ_j. This explains the suppressing frequency effect. Note, however, that the Ψ_j's do not satisfy an orthogonality condition. Thus the orthogonal sum decomposition has to be substituted by the sum $V_{j+1} = V_j + W_j$. For the construction of the M-wavelets we again have to distinguish three cases. The details will be omitted. An illustration is given in Figure 11.4.2.

11.4.3 *Cubic Polynomial Wavelets*

Of course, many other suitable choices for φ_0 can be found for practical purposes (cf. M. Schreiner (1996b)), for example

$$\varphi_0(x) = \begin{cases} (1-x)^2(1+2x) & \text{for } x \in [0,1) \\ 0 & \text{for } x \in [1,\infty) \end{cases}$$

(note that $\varphi_0(0) = 1$, $\varphi_0'(0) = 0$, $\varphi_0(1) = 0$, $\varphi_0'(1) = 0$). Numerical tests show that phenomena of oscillations that are still existent for the de la

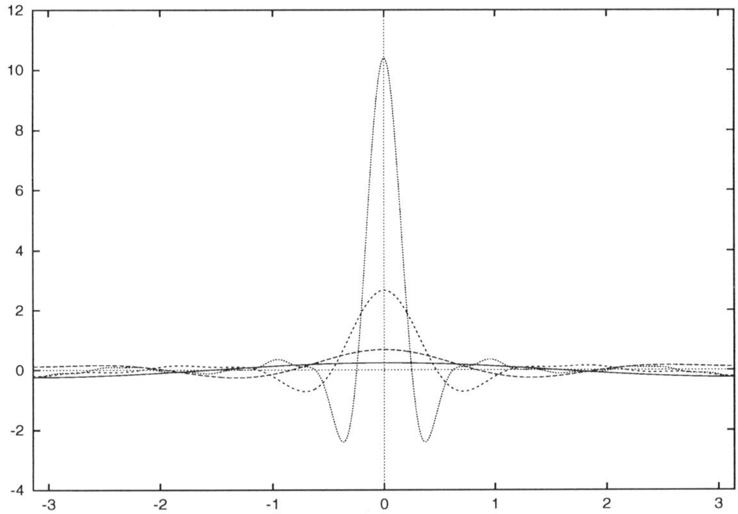

FIG. 11.4.2. De la Vallé Poussin wavelets $\Psi_j(\cos\vartheta)$, $j = 0, \ldots, 3$, $h = 0.5$

Vallé Poussin scaling function, can be suppressed by this choice, cf. Figure 11.4.3.

Remark. It should be remarked that one can think of other ways to introduce to 'smooth' wavelet generators.

11.4.4 *Exact Fully Discrete Wavelet Transform*

We are now interested in fully discrete wavelet approximation. For this purpose we want to show that when using band-limited wavelets, we do not need the wavelet transform $(WT)(F)(j;\eta)$ at all rotates $\eta \in \Omega$. It suffices to know the wavelet transform only at a finite set of rotates η for each scale j. As a matter of fact, the reconstruction can be formulated in terms of simply structured sum representations. More explicitly, the reconstructed function can be expressed in each scale as a linear combination of finitely many dual wavelets $\tilde{\Psi}_j(\eta_k\cdot), \eta_k \in \Omega$.

We start by summarizing those results that are needed for the exact reconstructions.

Lemma 11.4.2. *If $F, G \in Harm_{0,\ldots,m}$, $m \in \mathbb{N}_0$, then the product FG defined by $(FG)(\xi) = F(\xi)G(\xi)$, $\xi \in \Omega$, satisfies $FG \in Harm_{0,\ldots,2m}$.*

Proof. The assertion is an immediate consequence of Corollary 2.2.2. □

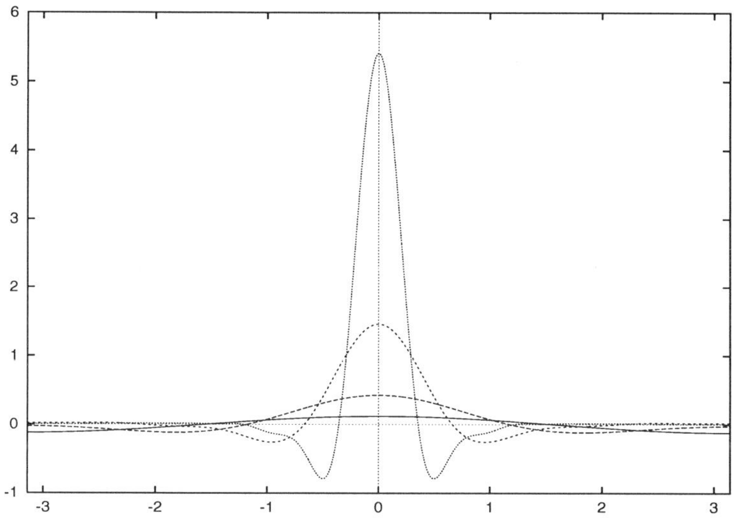

FIG. 11.4.3. Cubic polynomial wavelets $\Psi_j(\cos\vartheta)$, $j = 0, \ldots, 3$

Lemma 11.4.3. *Given a non-negative integer m, then there exist a fundamental system $\eta_1, \ldots, \eta_M \in \Omega$ relative to $Harm_{0,\ldots,m}$, $m \geq 0$, and weights $a_1, \ldots, a_M \in \mathbb{R}$ such that*

$$\int_\Omega F(\xi)\, d\omega(\xi) = \sum_{k=1}^{M} a_k F(\eta_k)$$

holds for all $F \in Harm_{0,\ldots,m}$.

Proof. The interpolatory integration formula follows from Theorem 5.3.3 with $Y \in Harm_0$, $M = (m+1)^2$, and $\Lambda = \partial_{0,\ldots,m}$. □

Next we mention the following result.

Lemma 11.4.4. *Given a non-negative integer m and a function $Y_{n,l} \in Harm_{0,\ldots,m}$. Then there exist a fundamental system $\eta_1, \ldots, \eta_M \in \Omega$ relative to $Harm_{0,\ldots,m}$ and weights $a_1^{n,l}, \ldots, a_M^{n,l} \in \mathbb{R}$ such that*

$$F^\wedge(n,l) = \int_\Omega Y_{n,l}(\xi) F(\xi)\, d\omega(\xi) = \sum_{k=1}^{N} a_k^{n,l} F(\eta_k)$$

and

$$F = \sum_{k=1}^{M} F(\eta_k) L_k$$

holds for all $F \in Harm_{0,\ldots,m}$, *where* $L_1, \ldots, L_M$ *are the Lagrangians in* $Harm_{0,\ldots,m}$ *given by*

$$L_k = \sum_{n=0}^{m} \sum_{l=1}^{2n+1} a_k^{n,l} Y_{n,l}, \quad k = 1, \ldots, M.$$

Proof. Lemma 11.4.4 follows from Corollary 5.3.4 and the spherical Shannon sampling theorem in $Harm_{0,\ldots,m}$ (Corollary 5.3.5). □

We consider now the fully discrete wavelet transform under the following assumptions:

(i) The generator $\varphi_0 : [0, \infty) \to \mathbb{R}$ of a scale discrete scaling function satisfies

$$\text{supp } \varphi_0 \subset [0, C]$$

for some constant $C > 0$.

(ii) The generators $\psi_0, \tilde{\psi}_0 : [0, \infty) \to \mathbb{R}$ of the mother wavelet and the dual mother wavelet, respectively, satisfy

$$\text{supp } \psi_0 \subset [0, C],$$
$$\text{supp } \tilde{\psi}_0 \subset [0, C].$$

Then it follows that

$$\text{supp } \varphi_j \subset [0, 2^j C],$$
$$\text{supp } \psi_j \subset [0, 2^j C],$$
$$\text{supp } \tilde{\psi}_j \subset [0, 2^j C].$$

Hence, there exists a sequence $\{m_j\} \subset \mathbb{N}_0$ such that

$$\Phi_j(\eta \cdot) \in Harm_{0,\ldots,m_j},$$
$$\Psi_j(\eta \cdot) \in Harm_{0,\ldots,m_j},$$
$$\tilde{\Psi}_j(\eta \cdot) \in Harm_{0,\ldots,m_j}$$

for all $j \in \mathbb{N}_0$, $\eta \in \Omega$. This, in turn, results in spaces V_j, W_j such that $V_j \subset Harm_{0,\ldots,m_j}, W_j \subset Harm_{0,\ldots,m_j}$.

Our aim now is to discuss two variants of exact fully discrete wavelet transform.

Variant 1. Let $j \in \mathbb{N}_0$ be fixed. Assume that $\{\eta_1^j, \ldots, \eta_{(2m_j+1)^2}^j\} \subset \Omega$ is a fundamental system relative to $Harm_{0,\ldots,2m_j}$. Furthermore, suppose that the values $a_1^j, \ldots, a_{(2m_j+1)^2}^j$ are the corresponding weights of the exact integration formula (Lemma 11.4.3). Then, for all $\xi \in \Omega$ and $F \in \mathcal{L}^2(\Omega)$,

$$\begin{aligned} R_j(F)(\xi) &= \int_\Omega (WT)(F)(j;\eta)\tilde{\Psi}_{j;\eta}(\xi)\, d\omega(\eta) \\ &= \sum_{k=1}^{(2m_j+1)^2} a_k^j (WT)(F)(j;\eta_k^j)\tilde{\Psi}_{j;\eta_k^j}(\xi). \end{aligned} \tag{11.4.8}$$

The identity (11.4.8) follows from the fact that both $(WT)(F)(j;\cdot)$ and $\tilde{\Psi}_{j;\eta}$ are in $Harm_{0,\ldots,m_j}$ and, by Lemma 11.4.2, their product is in $Harm_{0,\ldots,2m_j}$.

We summarize our results of the first variant of a fully discrete version for the reconstruction formula of Theorem 11.2.4.

Theorem 11.4.5 *(fully discrete reconstruction formula). Let the generators φ_0, ψ_0 and $\tilde{\psi}_0$ of a scale discrete scaling function and the corresponding mother wavelets be compactly supported. Then there exists a sequence $\{m_j\} \subset \mathbb{N}_0$ such that*

$$\begin{aligned} &\Phi_j(\eta\cdot) \in Harm_{0,\ldots,m_j}, \\ &\Psi_j(\eta\cdot) \in Harm_{0,\ldots,m_j}, \\ &\tilde{\Psi}_j(\eta\cdot) \in Harm_{0,\ldots,m_j}, \end{aligned}$$

for all $j \in \mathbb{N}_0$ and all $\eta \in \Omega$. Assume that $\{\eta_1^j, \ldots, \eta_{(2m_j+1)^2}^j\} \subset \Omega$, $j \in \mathbb{N}_0$, is a sequence of fundamental systems relative to $Harm_{0,\ldots,2m_j}$, and let, for every index j, the weights $a_1^j, \ldots, a_{(2m_j+1)^2}^j \in \mathbb{R}$ be obtained from Lemma 11.4.3. Then

$$F(\xi) = (\Phi_0^{(2)} * F)(\xi) + \sum_{j=0}^{\infty} \sum_{k=1}^{(2m_j+1)^2} a_k^j\, (WT)(F)(j;\eta_k^j)\tilde{\Psi}_{j;\eta_k^j}(\xi)$$

holds for all $F \in \mathcal{L}^2(\Omega)$.

Variant 2. Again take a fixed $j \in \mathbb{N}_0$. Let $\{\eta_1^j, \ldots, \eta_{(m_j+1)^2}^j\} \subset \Omega$ be a fundamental system relative to $Harm_{0,\ldots m_j}$. Furthermore, suppose that the values $a_1^{n,l,j}, \ldots, a_{(m_j+1)^2}^{n,l,j}$ are the weights corresponding to $Y_{n,l} \in$

$Harm_{0,\ldots,m_j}$ of the exact integration formula (Lemma 11.4.4). Then, for all $F \in \mathcal{L}^2(\Omega)$,

$$\begin{aligned} R_j(F)(\xi) &= \int_\Omega (WT)(F)(j;\eta)\tilde{\Psi}_{j;\eta}(\xi)d\omega(\eta) \\ &= \sum_{n=0}^{m_j}\sum_{l=1}^{2n+1}\sum_{k=1}^{(m_j+1)^2} a_k^{n,l,j}(WT)(F)(j;\eta_k^j)\tilde{\Psi}_j^\wedge(n)Y_{n,l}(\xi) \end{aligned} \tag{11.4.9}$$

and

$$R_j(F)(\xi) = \sum_{k=1}^{(m_j+1)^2} \tilde{\Psi}_{j;\eta_k^j}(\xi)\int_\Omega (WT)(F)(j;\eta)L_k^j(\eta)d\omega(\eta),$$

where

$$L_k^j = \sum_{n=0}^{m_j}\sum_{l=1}^{2n+1} a_k^{n,l,j}Y_{n,l}.$$

Then (11.4.9) immediately follows from the fact that both expressions $(WT)(F)\ (j;\cdot)$ and $\tilde{\Psi}_{j;\eta}$ are members of $Harm_{0,\ldots,m_j}$. A fully discrete version for the reconstruction formula, therefore, reads as follows.

Theorem 11.4.6 *(fully discrete reconstruction formula). Let the generators φ_0, ψ_0, and $\tilde{\Psi}_0$ of a scale discrete scaling function and the corresponding mother wavelets be compactly supported. Then there exists a sequence $\{m_j\} \subset \mathbb{N}_0$ such that*

$$\begin{aligned} \Phi_j(\eta\cdot) &\in Harm_{0,\ldots,m_j}, \\ \Psi_j(\eta\cdot) &\in Harm_{0,\ldots,m_j}, \\ \tilde{\Psi}_j(\eta\cdot) &\in Harm_{0,\ldots,m_j}, \end{aligned}$$

for all $j \in \mathbb{N}_0$ and all $\eta \in \Omega$. Assume that $\{\eta_1^j, \ldots, \eta_{(m_j+1)^2}^j\} \subset \Omega$, $j \in \mathbb{N}_0$, is a sequence of fundamental systems relative to $Harm_{0,\ldots,m_j}$, and let the weights $a_1^{n,l,j}, \ldots, a_{(m_j+1)^2}^{n,l,j} \in \mathbb{R}$ be obtained from Lemma 11.4.4. Then,

$$\begin{aligned} F(\xi) &= (\Phi_0^{(2)} * F)(\xi) \\ &\quad + \sum_{j=0}^{\infty}\sum_{n=0}^{m_j}\sum_{l=1}^{2n+1} \tilde{\Psi}_j^\wedge(n)Y_{n,l}(\xi)\sum_{k=1}^{(m_j+1)^2} a_k^{n,l,j}(WT)(F)(j;\eta_k^j) \end{aligned} \tag{11.4.10}$$

and

$$F(\xi) = (\Phi_0^{(2)} * F)(\xi) + \sum_{j=0}^{\infty}\sum_{k=1}^{(m_j+1)^2} \tilde{\Psi}_{j;\eta_k^j}(\xi)\int_\Omega (WT)(F)(j;\eta)L_k^j(\eta)\,d\omega(\eta)$$

holds for all $F \in \mathcal{L}^2(\Omega)$.

One cannot expect to get axisymmetric approximations from truncations of the sum on the left-hand side of (11.4.10). But this is the price one has to pay for the fact that less sample points are required in comparison to variant 1.

Remark. Note that the weights in Theorem 11.4.5 and Theorem 11.4.6 have to be determined by the solution of a linear system of equations. However, this has to be done only once, and the coefficients can then be stored elsewhere.

At the end of this section, let us consider our formulae for the special choices of band-limited wavelets studied before. Since both types satisfy $\varphi_0(x) = 0$ for all $x \geq 1$, it follows that

$$(\Phi_0^{(2)} * F)(\xi) = F^\wedge(0,1)Y_{0,1}(\xi) = \frac{1}{\sqrt{4\pi}}F^\wedge(0,1), \quad \xi \in \Omega.$$

Therefore, the term $\Phi_0^{(2)}$ occuring in the formula of Theorem 11.4.5 and Theorem 11.4.6 can be omitted for all $F \in \mathcal{L}^2(\Omega)$ satisfying $F^\wedge(0,1) = 0$. It is also worth mentioning that in the P-wavelet construction ψ_j satisfies $\psi_0(x) = 0$ for all $x \geq 2^{j+1}$. Consequently, $m_j = 2^{j+1} - 1, j \in \mathbb{N}_0$.

11.4.5 *A Pyramid Scheme*

Next we are interested in a pyramid scheme for the recursive determination of the wavelet coefficients, starting from an initial approximation of a given function. For reasons of simplicity we restrict ourselves to the special situation that

$$\operatorname{supp} \varphi_0 = [0,1], \tag{11.4.11}$$
$$\operatorname{supp} \psi_0 = [0,2], \tag{11.4.12}$$
$$\operatorname{supp} \tilde{\psi}_0 = [0,2], \tag{11.4.13}$$

with $\varphi_0(1) = 0$, $\psi_0(2) = \tilde{\psi}_0(2) = 0$. Note that all our generators of band-limited scaling functions satisfy condition (11.4.11). In case of P- or M-wavelets, respectively, the conditions (11.4.12) and (11.4.13) are satisfied as well. An immediate consequence of these assumptions is

$$\begin{aligned} \operatorname{supp} \varphi_j &= [0, 2^j], \\ \operatorname{supp} \psi_j,\ \operatorname{supp} \tilde{\psi}_j &\subset [0, 2^{j+1}], \\ \varphi_j(2^j) = \psi_j(2^{j+1}) = \tilde{\psi}_j(2^{j+1}) &= 0. \end{aligned}$$

Therefore,

$$\Phi_j(\eta\cdot\,) \in Harm_{0,\ldots,2^j-1},$$

$$\Psi_j(\eta\cdot\ ),\ \tilde{\Psi}_j(\eta\cdot\ ) \quad \in \quad Harm_{0,\ldots,2^{j+1}-1}.$$

Furthermore, the scale spaces and detail spaces fulfil

$$\begin{aligned} V_j &= Harm_{0,\ldots,2^j-1}, \\ W_j &\subset Harm_{0,\ldots,2^{j+1}-1}. \end{aligned}$$

Assume now that the system $\{(\eta_i^j, w_i^j)\}_{i=1,\ldots,N_j} \subset \Omega \times \mathbb{R}$ defines an integration formula that is exact of order $2^{j+2}-2$. Then it follows, since $(WT)(F)(j;\ )$ and $\tilde{\Psi}_j(\eta\cdot\ )$ are in $Harm_{0,\ldots,2^{j+1}-1}$, that for all $F \in \mathcal{L}^2(\Omega)$ and all $\xi \in \Omega$

$$\begin{aligned} R_j(F) &= \int_\Omega (WT)(F)(j;\eta)\tilde{\Psi}_{j;\eta}(\xi)\,d\omega(\eta) \\ &= \sum_{i=1}^{N_j} w_i^j (WT)(F)(j;\eta_i^j)\tilde{\Psi}_{j;\eta_i^j}(\xi). \end{aligned}$$

In connection with the results stated in Section 11.4.4 this leads us to the following corollary.

Corollary 11.4.7. *Let the generators* $\varphi_0,\ \psi_0,\ \tilde{\psi}_0 : [0,\infty) \to \mathbb{R}$ *satisfy the assumptions (11.4.11)–(11.4.13). Assume further that there is given a sequence of systems* $\{(\eta_i^j, w_i^j)\}_{i=1,\ldots,N_j},\ j = 0,1,\ldots,$ *which defines an exact integration formula of order* $2^{j+2}-2$ *for each* j*. If* $F \in \mathcal{L}^2(\Omega)$ *with* $F^\wedge(0,1) = 0$*, then*

$$F(\xi) = \sum_{j=0}^{\infty}\sum_{i=1}^{N_j} w_i^j (WT)(F)(j;\eta_i^j)\tilde{\Psi}_{j;\eta_i^j}(\xi), \quad \xi \in \Omega.$$

What we are going to show are the following steps (cf. M. SCHREINER (1996B)). Starting from a level $J \in \mathbb{N}$, there exist vectors $a^j \in \mathbb{R}^{N_j}$, $j = J, J-1, \ldots, 0$ (of course, dependent on the function $F \in \mathcal{L}^2(\Omega)$ under consideration), such that

(i) $(WT)(F)(j;\cdot) = \sum_{i=1}^{N_j} a_i^j \Psi_j(\eta_i^j\cdot\ ), \quad j = J,\ J-1,\ldots,0,$

(ii) the vector a^j is obtainable from a^{j+1},

(iii) the vectors a^j satisfy, in addition,

$$\Phi_{j+1} * F = \sum_{i=1}^{N_j} a_i^j \Phi_{j+1}(\eta_i^j\cdot\ ), \quad j = 0,\ldots,J.$$

Hence, we end up with the following pyramid scheme for the decomposition of an $\mathcal{L}^2(\Omega)$-signal F:

$$\begin{array}{ccccccccc}
F & \longrightarrow & a^J & \longrightarrow & a^{J-1} & \longrightarrow & \cdots & \longrightarrow & a^0 \\
 & & \downarrow & & \downarrow & & & & \downarrow \\
 & & (WT)(F)(J;\cdot) & & (WT)(F)(J-1;\cdot) & & & & (WT)(F)(0;\cdot)
\end{array}$$

The reconstruction from the wavelet coefficients $(WT)(F)(j;\eta_i^j)$ can then be performed as described before:

$$R_j(F) = \tilde{\Psi}_j * (WT)(F)(j;\ ) = \sum_{i=1}^{N_j} w_i^j (WT)(F)(j;\eta_i^j)\tilde{\Psi}_j(\eta_i^j\cdot\ ). \quad (11.4.14)$$

To be more explicit, the reconstruction of the signal F can be carried out in accordance with the following scheme:

$$\begin{array}{ccccccccc}
(WT)(F)(0;\eta_i^0) & & (WT)(F)(1;\eta_i^1) & & (WT)(F)(2;\eta_i^2) & & & & \\
\downarrow & & \downarrow & & \downarrow & & & & \\
R_0(F) & & R_1(F) & & R_2(F) & & & & \\
 & \searrow & & \searrow & & \searrow & & & \\
P_0(F) & \longrightarrow + \longrightarrow & P_1(F) & \longrightarrow + \longrightarrow & P_2(F) & \longrightarrow + \longrightarrow & \cdots
\end{array}$$

We have seen that the wavelet transform $(WT)(F)(j;\cdot)$ is given in terms of the numbers $a_1^j,\ldots,a_{N_j}^j$. Therefore it is also possible to reconstruct the function only in terms of the numbers a_i^j, rather than using the wavelet coefficients of F. Since $\{(\eta_i^j,w_i^j)\}_{i=1,\ldots,N_j}$ defines an integration formula that is exact of order $2^{j+2}-2$, we can rewrite (11.4.14) in the form

$$\begin{aligned}
R_j(F) &= \sum_{i=1}^{N_j} w_i^j (WT)(F)(j;\eta_i^j)\tilde{\Psi}_j(\eta_i^j\cdot\ ) \\
&= \sum_{i=1}^{N_j} w_i^j \sum_{k=1}^{N_j} a_k^j \Psi_j(\eta_k^j\cdot\eta_i^j)\tilde{\Psi}_j(\eta_i^j\cdot\ )
\end{aligned}$$

$$
\begin{aligned}
&= \sum_{k=1}^{N_j} a_k^j \sum_{i=1}^{N_j} w_i^j \Psi_j(\eta_k^j \cdot \eta_i^j) \tilde{\Psi}_j(\eta_i^j \cdot\,) \\
&= \sum_{k=1}^{N_j} a_k^j \int_\Omega \Psi_j(\eta_k^j \cdot \eta) \tilde{\Psi}_j(\eta \cdot\,)\, d\omega(\eta) \\
&= \sum_{k=1}^{N_j} a_k^j (\Psi_j * \tilde{\Psi}_j)(\eta_k^j \cdot\,).
\end{aligned}
$$

In conclusion, the decomposition and reconstruction can thus be simplified in the following fashion:

$$F \longrightarrow a^J \longrightarrow a^{J-1} \longrightarrow \cdots \longrightarrow a^0$$

and

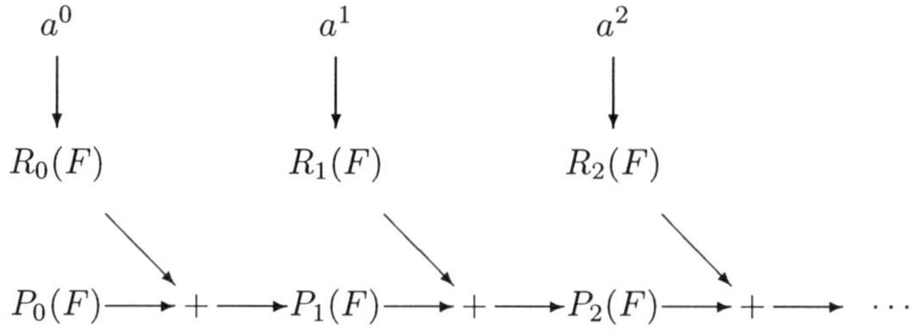

In doing so, the reconstruction of the signal is not performed by means of $\tilde{\Psi}_j$ but in terms of of the convolutions $\Psi_j * \tilde{\Psi}_j$. It turns out later on that the vectors a^j do *not* depend on the special choice of φ_0, ψ_0 or $\tilde{\psi}_0$. In other words, using the second variant one can reconstruct the signal with respect to different wavelets just by the knowledge of the vectors a^j.

The Evaluation of the First Step: The pyramid scheme starts from the vector a^J, where the components a_i^J have to satisfy the equation

$$\Phi_{J+1} * F = \sum_{i=1}^{N_J} a_i^J \Phi_{J+1}(\eta_i^J \cdot\,). \tag{11.4.15}$$

It is not hard to show the following result.

Theorem 11.4.8. *Let $F \in \mathcal{L}^2(\Omega)$, and $J \in \mathbb{N}_0$. Then there exists a vector $a^J \in \mathbb{R}^{N_J}$ such that (11.4.15) is fulfilled.*

Proof. From our assumptions we know that $\Phi_{J+1}^\wedge(n) \neq 0$ for $n = 0, \ldots, 2^{J+1} - 1$ and $\Phi_{J+1}^\wedge(n) = 0$ holds for all $n \geq 2^{J+1}$. Thus, the assertion follows immediately from the fact that $\Phi_{J+1} * F \in Harm_{0,\ldots,2^{J+1}-1}$ and $\{\eta_i^J\}_{i=1,\ldots,N_J}$

is assumed to be an admissible system relative to $Harm_{0,\ldots,2^{J+1}-1}$. □

Next we turn to the question of how the coefficients a_i^J are obtainable. We present two methods of calculation.

If F is band-limited such that

$$F^{\wedge}(n,m) = 0 \text{ for all } n \geq 2^{J+1}, \quad m = 1, \ldots, 2n+1, \tag{11.4.16}$$

then it is easy to find a possible choice of the coefficients a_i^J. Indeed, we have $\Phi_{J+1}(\,\cdot\xi), F \in Harm_{0,\ldots,2^{J+1}-1}$, so that by virtue of the introduced exact integration formula

$$\Phi_{J+1} * F = \int_{\Omega} F(\eta)\Phi_{J+1}(\eta\cdot\,)\,d\omega(\eta) = \sum_{i=1}^{N_J} w_i^J F(\eta_i^J)\Phi_{J+1}(\eta_i^J\cdot\,).$$

In other words, we have

$$a_i^J = w_i^J F(\eta_i^J) \quad \text{for } i = 1, \ldots, N_J. \tag{11.4.17}$$

If F does not satisfy condition (11.4.16), then the numbers a_i^J can still be obtained by observing the fact that (11.4.15) is satisfied if and only if

$$(\Phi_{J+1} * F)(\eta_k^J) = \sum_{i=1}^{N_J} a_i^J \Phi_{J+1}(\eta_i^J \cdot \eta_k^J), \quad k = 1, \ldots, N_J,$$

what follows from the fact that $\{\eta_i^J\}_{i=1,\ldots,N_J}$ is assumed to be admissible relative to $Harm_{0,\ldots,2^{J+1}-1}$. Hence, all possible choices of a_i^J are solutions of the linear system

$$\begin{pmatrix} \Phi_{J+1}(\eta_1^J \cdot \eta_1^J) & \cdots & \Phi_{J+1}(\eta_{N_J}^J \cdot \eta_1^J) \\ \vdots & \ddots & \vdots \\ \Phi_{J+1}(\eta_1^J \cdot \eta_{N_J}^J) & \cdots & \Phi_{J+1}(\eta_{N_J}^J \cdot \eta_{N_J}^J) \end{pmatrix} \begin{pmatrix} a_1^J \\ \vdots \\ a_{N_J}^J \end{pmatrix}$$

$$= \begin{pmatrix} (\Phi_{J+1} * F)(\eta_1^J) \\ \vdots \\ (\Phi_{J+1} * F)(\eta_{N_J}^J) \end{pmatrix}.$$

Notice that the right-hand side of this equation is expressed in terms of convolutions of the scaling function against F, so that there still remains the problem of numerical integration. It should be mentioned, however, that in many applications, it is satisfactory to require satisfy (11.4.15) only approximately. In other words, the solution of a linear system of equations can be avoided, that is to say an integration formula of sufficient exactness is often adequate for the first step.

Next we want to show, that if the vector a^J satisfies (11.4.15) then the wavelet transform $(WT)(F)(J;\cdot)$ is given by

$$(WT)(F)(J;\cdot) = \Psi_J * F = \sum_{i=1}^{N_J} a_i^J \Psi_J(\eta_i^J \cdot\ ). \tag{11.4.18}$$

But this is a consequence of the following result.

Theorem 11.4.9. *Let $F \in \mathcal{L}^2(\Omega)$, $K \in \mathcal{L}^2[-1,+1]$ with $K^\wedge(n) \neq 0$ for $n = 0,\ldots,N$, and $K^\wedge(n) = 0$ for $n \geq N+1$. Assume further that, for $\eta_1,\ldots,\eta_M \in \Omega$,*

$$K * F = \sum_{i=1}^{L} a_i K(\eta_i \cdot\ ). \tag{11.4.19}$$

If $H \in \mathcal{L}^2[-1,+1]$ is another kernel with $H^\wedge(n) = 0$ for all $n \geq N+1$, then

$$H * F = \sum_{i=1}^{L} a_i H(\eta_i \cdot\ ).$$

Proof. For all $n = 0,\ldots,N$, $m = 1,\ldots,2n+1$, we get from (11.4.19) by multiplication with $Y_{n,m}$ and integration over Ω that

$$\int_\Omega (K * F)(\eta) Y_{n,m}(\eta)\, d\omega(\eta) = \sum_{i=1}^{L} a_i \int_\Omega K(\eta_i \cdot \eta) Y_{n,m}(\eta)\, d\omega(\eta),$$

so that via the Funk–Hecke formula

$$K^\wedge(n) F^\wedge(n,m) = K^\wedge(n) \sum_{i=1}^{L} a_i Y_{n,m}(\eta_i).$$

Since $K^\wedge(n) \neq 0$ it follows that

$$F^\wedge(n,m) = \sum_{i=1}^{L} a_i Y_{n,m}(\eta_i).$$

Therefore we get for $\xi \in \Omega$

$$(H * F)(\xi) \quad = \quad \sum_{n=0}^{N} \sum_{m=1}^{2n+1} H^\wedge(n) F^\wedge(n,m) Y_{n,m}(\xi)$$

$$= \sum_{n=0}^{N}\sum_{m=1}^{2n+1} H^\wedge(n) \sum_{i=1}^{L} a_i Y_{n,m}(\eta_i) Y_{n,m}(\xi)$$

$$= \sum_{i=1}^{L} a_i \sum_{n=0}^{N} \frac{2n+1}{4\pi} H^\wedge(n) P_n(\eta_i \cdot \xi)$$

$$= \sum_{i=1}^{L} a_i H(\eta_i \cdot \xi),$$

as required. □

The Pyramid Step. Starting from $a^J \in \mathbb{R}^{N_J}$ with

$$\Phi_{J+1} * F = \sum_{i=1}^{N_J} a_i^J \Phi_{J+1}(\eta_i^J \cdot\ ),$$

we now verify that $a^{J-1} \in \mathbb{R}^{N_{J-1}}$ can be constructed in such a way that

$$\Phi_J * F = \sum_{i=1}^{N_{J-1}} a_i^{J-1} \Phi_J(\eta_i^{J-1} \cdot\ ). \tag{11.4.20}$$

Since $\Phi_J * F \in Harm_{0,\dots,2^J-1}$ and $\{\eta_1^{J-1}, \dots, \eta_{N_{J-1}}^{J-1}\} \subset \Omega$ is admissible relative to $Harm_{0,\dots,2^J-1}$, the existence of a_i^{J-1} is assured. Furthermore, we know from Theorem 11.4.9 that

$$\Phi_J * F = \sum_{i-1}^{N_J} a_i^J \Phi_J(\eta_i^J \cdot\ ).$$

Thus, we finally end up with the equation

$$\sum_{i=1}^{N_{J-1}} a_i^{J-1} \Phi_J(\eta_i^{J-1} \cdot\ ) = \sum_{i=1}^{N_J} a_i^J \Phi_J(\eta_i^J \cdot\ ). \tag{11.4.21}$$

Our purpose is to prove the following.

Theorem 11.4.10. *A solution $a^{J-1} \in \mathbb{R}^{N_{J-1}}$ of equation (11.4.21) is given by*

$$a_k^{J-1} = w_k^{J-1} \sum_{i=1}^{N_J} a_i^J K_{Harm_{0,\dots,2^J-1}}(\eta_i^J \cdot \eta_k^{J-1}). \tag{11.4.22}$$

Provided that the integration weights w_i^J, w_i^{J-1} are all positive, the solution (11.4.22) is characterized to minimize the discrete norm

$$\left(\sum_{i=1}^{N_{J-1}} \frac{1}{w_i^{J-1}} \left(a_i^{J-1}\right)^2 \right)^{1/2} \tag{11.4.23}$$

under all solutions of (11.4.21).

Proof. On $\mathbb{R}^{N_J}$ and $\mathbb{R}^{N_{J-1}}$, respectively, we introduce

$$(\alpha, \beta)_{N_j} = \sum_{i=1}^{N_j} w_i^j \alpha_i \beta_i, \qquad \alpha,\ \beta \in \mathbb{R}^{N_j}, \quad j = J-1, J, \tag{11.4.24}$$

which defines an inner product in $\mathbb{R}^{N_j}$, if all the weights are positive. We let $x^{J-1,n,m} \in \mathbb{R}^{N_{J-1}}$ and $x^{J,n,m} \in \mathbb{R}^{N_J}$ be given by

$$x_i^{j,n,m} = Y_{n,m}(\eta_i^j), \quad i = 1, \ldots, N_j,\ n = 0, \ldots, 2^J - 1,\ m = 1, \ldots, 2n+1,$$

for $j = J-1,\ J$. Then it follows that for $n' = 0, \ldots, 2^J - 1$, $m' = 1, \ldots, 2n'+1$,

$$\begin{aligned}
(x^{J,n,m}, x^{J,n',m'})_{N_J} &= \sum_{i=1}^{N_J} w_i^J x_i^{J,n,m} x_i^{J,n',m'} \\
&= \sum_{i=1}^{N_J} w_i^J Y_{n,m}(\eta_i^J) Y_{n',m'}(\eta_i^J) \\
&= \int_\Omega Y_{n,m}(\eta) Y_{n',m'}(\eta)\, d\omega(\eta) \\
&= \delta_{n,n'} \delta_{m,m'}.
\end{aligned}$$

Analogously, we get

$$(x^{J-1,n,m}, x^{J-1,n',m'})_{N_{J-1}} = \delta_{n,n'} \delta_{m,m'}.$$

The sets

$$\text{span}\ \{x^{j,n,m} \mid n = 0, \ldots, 2^J - 1,\ m = 1, \ldots, 2n+1\}, \quad j = J-1, J, \tag{11.4.25}$$

are subspaces of $\mathbb{R}^{N_J}$ and $\mathbb{R}^{N_{J-1}}$, respectively, which are isomorphic via the canonical basis change $x^{J-1,n,m} \mapsto x^{J,n,m}$.

Equation (11.4.21) is equivalent to

$$\int_\Omega \sum_{i=1}^{N_{J-1}} a_i^{J-1} \Phi_J(\eta_i^{J-1} \cdot \eta) Y_{n,m}(\eta) d\omega(\eta)$$
$$= \int_\Omega \sum_{i=1}^{N_J} a_i^J \Phi_J(\eta_i^J \cdot \eta) Y_{n,m}(\eta) d\omega(\eta)$$

for all $n = 0, \ldots, 2^J - 1$, $m = 1, \ldots, 2n + 1$. Hence, observing the Funk–Hecke formula, we are led to

$$\Phi_J^\wedge(n) \sum_{i=1}^{N_{J-1}} a_i^{J-1} Y_{n,m}(\eta_i^{J-1}) = \Phi_J^\wedge(n) \sum_{i=1}^{N_J} a_i^J Y_{n,m}(\eta_i^J)$$

for all $n = 0, \ldots, 2^J - 1$, $m = 1, \ldots, 2n + 1$. Since $\Phi_J^\wedge(n) \neq 0$ for all $n = 0, \ldots, 2^J - 1$, we may also write

$$\sum_{i=1}^{N_{J-1}} a_i^{J-1} Y_{n,m}(\eta_i^{J-1}) = \sum_{i=1}^{N_J} a_i^J Y_{n,m}(\eta_i^J), \quad n = 0, \ldots, 2^J - 1, \; m = 1, \ldots, 2n+1.$$

This shows that there is no dependence on the scaling function anymore. If we introduce (we may without loss of generality assume that all the weights are different from zero)

$$\tilde{a}_i^{J-1} = \frac{1}{w_i^{J-1}} a_i^{J-1}, \quad \tilde{a}_i^J = \frac{1}{w_i^J} a_i^J,$$

we end up with the equations

$$(\tilde{a}^{J-1}, x^{J-1,n,m})_{N_{J-1}} = (\tilde{a}^J, x^{J,n,m})_{N_J}, \tag{11.4.26}$$

for $n = 0, \ldots, 2^J - 1$, $m = 1, \ldots, 2n + 1$. A solution of (11.4.26) is given by

$$\tilde{a}^{J-1} = \sum_{n=0}^{2^J-1} \sum_{m=1}^{2n+1} (\tilde{a}^J, x^{J,n,m})_{N_J} x^{J-1,n,m},$$

since, for $n' = 0, \ldots, 2^J - 1$, $m' = 1, \ldots, 2n' + 1$,

$$(\tilde{a}^{J-1}, x^{J-1,n',m'})_{N_{J-1}} = \sum_{n=0}^{2^J-1} \sum_{m=1}^{2n+1} (\tilde{a}^J, x^{J,n,m})_{N_J} (x^{J-1,n,m}, x^{J-1,n',m'})_{N_{J-1}}$$

$$= \quad (\tilde{a}^J, x^{J,n',m'})_{N_J}.$$

Therefore we obtain after simple manipulations

$$\begin{aligned}
a_k^{J-1} &= w_k^{J-1}\tilde{a}_k^{J-1} \\
&= w_k^{J-1}\sum_{n=0}^{2^J-1}\sum_{m=1}^{2n+1}(\tilde{a}^J, x^{J,n,m})_{N_j} x_k^{J-1,n,m} \\
&= w_k^{J-1}\sum_{n=0}^{2^J-1}\sum_{m=1}^{2n+1}\sum_{i=1}^{N_J} w_i^J \tilde{a}_i^J Y_{n,m}(\eta_i^J) Y_{n,m}(\eta_k^{J-1}) \\
&= w_k^{J-1}\sum_{i=1}^{N_J} a_i^J \sum_{n=0}^{2^J-1}\sum_{m=1}^{2n+1} Y_{n,m}(\eta_i^J) Y_{n,m}(\eta_k^{J-1}) \\
&= w_k^{J-1}\sum_{i=1}^{N_J} a_i^J K_{Harm_{0,\ldots,2^J-1}}(\eta_i^J \cdot \eta_k^{J-1}),
\end{aligned}$$

which proves the first part of the theorem.

If all the weights w_i^{J-1} and w_i^J are positive, then the bilinear forms (11.4.24) are, in fact, positive definite and the $x^{j,n,m}$, $j = J-1, J$, form orthonormal bases of the subspaces (11.4.25). Standard arguments in linear algebra then show, that a^{J-1} minimizes the norm (11.4.23). □

This theorem opens the perspective to compute a^{J-1} from a^J, so that (11.4.20) is fulfilled. Another application of Theorem 11.4.9 gives us

$$(WT)(F)(J-1;\cdot) = \sum_{i=1}^{N_{J-1}} a_i^{J-1}\Psi_{J-1}(\eta_i^{J-1}\cdot\ ).$$

It is clear that this scheme (Theorem 11.4.10 and Theorem 11.4.9) can be applied recursively to calculate all coefficients $a^J, \ldots, a^0$ and, hence, $(WT)(F)(J;\cdot), \ldots, (WT)(F)(0;\cdot)$.

Let us finally make some comments:

(i) The numerical effort of one pyramid step $j+1 \to j$ is of order $N_j N_{j+1}$. This effort can be drastically reduced by two methods. The first possibility is to use gridded point systems and then to apply FFT methods. To make this statement more concrete we introduce spherical coordinates (ϑ, φ) (cf. (1.2.24). We assume that, for each j, the system $\{(\eta_{i,k}^j, w_{i,k}^j)\}$ is given by $\eta_{i,k}^j = (\vartheta_i^j, \varphi_k^j)$, $i = 1, \ldots, T_j$, $k = 1, \ldots, L_j$, where $0 \le \vartheta_1^j < \ldots < \vartheta_{T_j}^j \le \pi$, $\lambda_k^j = 2\pi(k-1)/L_j$, and L_j is a power of

2. We suppose further that the corresponding integration weights $w_{i,k}^j$ depend only on the latitude, i.e. $w_{i,k}^j = w_i^j$. The pyramid step can then be written as

$$a_{i,k}^j = w_i^j \sum_{i'=1}^{T_{j+1}} \sum_{k'=1}^{L_{j+1}} a_{i',k'}^{j+1} K_{Harm_{0,\ldots,2^{j+1}-1}}(\eta_{i,k}^j \cdot \eta_{i',k'}^{j+1}),$$

where $i = 1, \ldots, T_j$, $k = 1, \ldots, L_j$. Introducing the vectors

$$\begin{aligned} a_i^j &= (a_{i,1}^j, \ldots, a_{i,L_j}^j), \\ S_{i,i'}^{j+1\to j} &= (K_{Harm_{0,\ldots,2^{j+1}-1}}(\eta_{i,1}^j \cdot \eta_{i',1}^{j+1}), \ldots, \\ &\qquad K_{Harm_{0,\ldots,2^{j+1}-1}}(\eta_{i,1}^j \cdot \eta_{i',L_{j+1}}^{j+1})), \end{aligned}$$

we are able to write the pyramid step as follows:

$$a_i^j = w_i^j \sum_{i'=1}^{T_{j+1}} a_{i'}^{j+1} * S_{i,i'}^{j+1\to j},$$

where $*$ means the cyclic discrete convolution and $i = 1, \ldots, T_j$. Thus applying a FFT method for the discrete convolution the numerical effort can be reduced. In J.R. DRISCOLL, R.M. HEALY (1994) there is described a pointsystem with corresponding integration weights associated to an equiangular distribution of the ϑ_i. Using a Gauß quadrature rule in a north–south direction, the numbers T_j can be reduced (cf. M. LANSER (1997)).

(ii) A complete different idea for making the pyramid step more efficient is the use of a panel-clustering method. One takes advantage of the localizing structure of the kernel $K_{Harm_{0,\ldots,2^{j+1}-1}}(\cdot,\cdot)$. The kernel is split into a near field and a far field component. The far field component is then approximated by Legendre polynomials up to a given low degree. For the evaluation one uses for points near at the evaluation position the exact near field of the corresponding kernel. For the remaining points, the approximated far fields are glued together. This is no problem when the addition theorem is applied. A numerical realization of this idea is presented by O. GLOCKNER (1997) in his diploma thesis.

11.5 Scale Discrete Wavelet Transform of Positive Order

In geophysical practice there exist numerous realizations of spherical harmonic models. Numerical instability and the low convergence rates of the spherical harmonics series, however, make it difficult to derive higher order spherical harmonic expansions. In fact, we are confronted with the situation that as the appropriate candidate for the approximation of the low frequency part, the spherical harmonic model should be used. But it is also our opinion that new candidates in modern constructive approximation should come into play when the modelling of the high frequency part is required. This is the reason why we show a strong interest in developing efficient procedures of combined approximation. In the previous chapters we have seen that the Gabor and the Toeplitz transform as well as the continuous wavelet transform admit a combination of these approximation techniques with the (orthogonal) Fourier expansion. In this section we make efforts to find the right way to combine the discrete wavelet transform with an mth truncated spherical harmonic expansion. Starting point of our considerations is the refinement equation

$$\psi_0(x)\tilde{\psi}_0(x) = (\varphi_0(x/2))^2 - (\varphi_0(x))^2, \quad x \in [0, \infty),$$

of Definition 11.2.1. It is clear that $\psi_0(x)\tilde{\psi}_0(x) = 0$ if and only if $(\varphi_0(x/2))^2 = (\varphi_0(x))^2$. In the examples listed above this has been guaranteed for $x = 0$, since $\varphi_0(0) = 1$. Therefore, the mother wavelet or its dual (here we assume both of them) satisfy the mean value condition $\Psi_0^\wedge(0) = 0$, i.e. Ψ_0 has to oscillate. For purposes of combined approximation as announced above we need, however, $(\varphi_0(x/2))^2 = (\varphi_0(x))^2$ for all $x \in [0, m]$. Under these assumptions it may be guaranteed that the wavelets constructed in this way will have more vanishing moments.

11.5.1 *Fixed Order*

Let m be a non-negative integer (note that the case $m = 0$ is exactly what was done before). We start our investigations with a definition.

Definition 11.5.1. A function $\varphi_0^{0,\ldots,m} : [0, \infty) \to \mathbb{R}$ obeying the admissibility condition (11.1.1) is called *generator of a scaling function of order* m, if $\varphi_0^{0,\ldots,m}$ satisfies the following conditions:

(i) $\varphi_0^{0,\ldots,m}(x) = 1$ for all $x \in [0, m]$,

(ii) $\varphi_0^{0,\ldots,m}$ is monotonically decreasing,

(iii) $\varphi_0^{0,\ldots,m}$ is continuous at m.

For such a generator we modify our ordinary dilation operator in such a way that the dilates of $\varphi_0^{0,\ldots,m}$ satisfy the condition $\varphi_0^{0,\ldots,m}(x) = 1$ for all $x \in [0, m]$ but not necessarily for values x in a larger interval.

Definition 11.5.2. Let $\gamma_0 : [0,\infty) \to \mathbb{R}$ be an admissible generator. The *dilation operator of order* m is defined by

$$(D_j^{0,\ldots,m}\gamma_0)(x) = \gamma_0(m + 2^{-j}(x - m)), \quad j \in \mathbb{N}_0.$$

In the same way as introduced in Section 11.1 (Definition 11.1.4) we apply this operator also to the generated kernels.

The meaning of the operator $D_j^{0,\ldots,m}$ is that the dilation is centered around m and not around the origin 0. The significance of this operation becomes clear when we apply the operator to a generator of a scaling function of order m.

Lemma 11.5.3. *Let $\varphi_0^{0,\ldots,m}$ be a generator of a scaling function of order m. Then the following statements are valid:*

(i) For all $j \in \mathbb{N}_0$, $(D_j^{0,\ldots,m}\varphi_0^{0,\ldots,m})(x) = 1$ for all $x \in [0,m]$.

(ii) For all $x \in (m,\infty)$

$$\lim_{j\to\infty} \varphi_j^{0,\ldots,m}(x) = 1,$$

where we have used the canonical notation

$$\varphi_j^{0,\ldots,m} = D_j^{0,\ldots,m}\varphi_0^{0,\ldots,m}.$$

(iii) If $\varphi_j^{0,\ldots,m}$ generates $\Phi_j^{0,\ldots,m} \in \mathcal{L}^2[-1,+1]$, then

$$(\Phi_j^{0,\ldots,m})^\wedge(n) = 1 \quad \text{for all} \quad n = 0,\ldots,m$$

(iv) For all $n = m+1, m+2, \ldots$

$$\lim_{j\to\infty} (\Phi_j^{0,\ldots,m})^\wedge(n) = 1.$$

An immediate consequence of these results is the next statement.

Theorem 11.5.4. *Let $\varphi_0^{0,\ldots,m} : [0,\infty) \to \mathbb{R}$ be a generator of a scaling function of order m:*

$$(\Phi_0^{0,\ldots,m})^\wedge(n) = \varphi_0^{0,\ldots,m}(n), \quad n = 0,1,\ldots.$$

Then, for all $F \in \mathcal{L}^2(\Omega)$, we have

(i) $\Phi_j^{0,\ldots,m} * F = \sum_{n=0}^{m}\sum_{l=1}^{2n+1} F^\wedge(n,l)Y_{n,l} + \sum_{n=m+1}^{\infty}\sum_{l=1}^{2n+1} \varphi_j^{0,\ldots,m}(n)F^\wedge(n,l)Y_{n,l},$

(ii) $\lim_{j\to\infty} \|F - \Phi_j^{0,\ldots,m} * F\|_{\mathcal{L}^2(\Omega)} = 0.$

Part (i) of this theorem shows that the projection of $\Phi_j^{0,\dots,m} * F$ on $Harm_{0,\dots,m}$ is exactly the projection of F on $Harm_{0,\dots,m}$. Roughly speaking, the approximation of $\Phi_j^{0,\dots,m} * F$ is exact for all $F \in Harm_{0,\dots,m}$.

Next we mention the following result.

Corollary 11.5.5. *Under the assumption of Theorem 11.5.4 we have*

$$(i)\ (\Phi_j^{0,\dots,m})^{(2)} * F = \sum_{n=0}^{m} \sum_{l=1}^{2n+1} F^\wedge(n,l) Y_{n,l} + \sum_{n=m+1}^{\infty} \sum_{l=1}^{2n+1} (\varphi_j^{0,\dots,m}(n))^2 F^\wedge(n,l) Y_{n,l}$$

$$(ii)\ \lim_{j\to\infty} \left\| F - (\Phi_j^{0,\dots,m})^{(2)} * F \right\|_{\mathcal{L}^2(\Omega)} = 0.$$

The statement $\varphi_j^{0,\dots,m}(x) = 1$ for all $x \in [0,m]$ has an interesting consequence for the wavelets, as we will now explain. Let $\psi_0^{0,\dots,m}, \tilde{\psi}_0^{0,\dots,m}$ satisfy the assumptions of Definition 11.2.1 so that, in particular,

$$\psi_0^{0,\dots,m}(x)\tilde{\psi}_0^{0,\dots,m}(x) = (\varphi_0^{0,\dots,m}(x/2))^2 - (\varphi_0^{0,\dots,m}(x))^2, \quad x \in [0,m].$$

Then we see that $\psi_0^{0,\dots,m}(x)\tilde{\psi}_0^{0,\dots,m}(x) = 0$ for $x \in [0,m]$. Hence, we may assume that $\psi_0^{0,\dots,m}(x) = \tilde{\psi}_0^{0,\dots,m}(x) = 0$, $x \in [0,m]$. But this shows that the mother wavelet $\Psi_0^{0,\dots,m}$ and its dual $\tilde{\Psi}_0^{0,\dots,m}$ defined as usual by

$$\Psi_0^{0,\dots,m} = \sum_{n=0}^{\infty} \frac{2n+1}{4\pi} \psi_0^{0,\dots,m}(n) P_n,$$

$$\tilde{\Psi}_0^{0,\dots,m} = \sum_{n=0}^{\infty} \frac{2n+1}{4\pi} \tilde{\psi}_0^{0,\dots,m}(n) P_n,$$

can be written as

$$\Psi_0^{0,\dots,m} = \sum_{n=m+1}^{\infty} \frac{2n+1}{4\pi} \psi_0^{0,\dots,m}(n) P_n,$$

$$\tilde{\Psi}_0^{0,\dots,m} = \sum_{n=m+1}^{\infty} \frac{2n+1}{4\pi} \tilde{\psi}_0^{0,\dots,m}(n) P_n.$$

This means that the functions introduced in such a way have vanishing moments up to the prescribed order m. If we now define the wavelets in the form $\Psi_j^{0,\dots,m} = D_j^{0,\dots,m} \Psi_0^{0,\dots,m}$ this has the consequence that

$$(WT)(F)(j;\eta) = 0$$

for all $j \in \mathbb{N}_0$, $\eta \in \Omega$, whenever $F \in Harm_{0,\ldots,m}$. Hence, the corresponding detail spaces

$$W_j^{0,\ldots,m} = \{\Psi_0^{0,\ldots,m} * \tilde{\Psi}_0^{0,\ldots,m} * F \mid F \in \mathcal{L}^2(\Omega)\}$$

show the property that $W_j^{0,\ldots,m} \perp Harm_{0,\ldots,m}$ for all $j \in \mathbb{N}_0$.

Our construction, therefore, enables us to fomulate a reconstruction formula for the wavelets of order m as follows.

Theorem 11.5.6 *(reconstruction formula). Let $\Psi_0^{0,\ldots,m}$ and $\tilde{\Psi}_0^{0,\ldots,m}$ be the mother wavelets of order m with respect to a scale discrete scaling function $\Phi_0^{0,\ldots,m}$ of order m. Suppose that F is of the class $\mathcal{L}^2(\Omega)$. Then*

$$\begin{aligned} F = \sum_{n=0}^{m} \sum_{l=1}^{2n+1} F^\wedge(n,l) Y_{n,l} \quad &+ \quad \sum_{n=m+1}^{\infty} \sum_{l=1}^{2n+1} (\varphi_0^{0,\ldots,m}(n))^2 F^\wedge(n,l) Y_{n,l} \\ &+ \quad \sum_{j=0}^{\infty} \int_\Omega (WT)(F)(j;\eta) \tilde{\Psi}_{j;\eta}^{0,\ldots,m} \, d\omega(\eta), \end{aligned}$$

where the equality is understood in $\|\cdot\|_{\mathcal{L}^2(\Omega)}$-sense

11.5.2 *Increasing Order*

The approach developed above has the appealing property that $W_j^{0,\ldots,m}$ is orthogonal to $Harm_{0,\ldots,m}$ for all j in case of a wavelet transform of order m. Now we modify our concept in such a way that the detail spaces are orthogonal to spherical harmonics of growing order if the scale parameter increases. The idea is simple: we start with a generator of a discrete scaling function of order $m \geq 0$. It is obvious that this function $\varphi_0^{0,\ldots,m}$ also satisfies the properties of a generator of order 0. Therefore, the whole theory developed in the first two sections of this chapter can be applied as well. Consequently the main difference to the approach described before lies in the fact that we now base our considerations on the usual dilation operator D_j instead of $D_j^{0,\ldots,m}$.

Lemma 11.5.7. *Suppose that $\varphi_0^{0,\ldots,m} : [0,\infty) \to \mathbb{R}, m \geq 0$, is a generator of a discrete scaling function of order m. Let φ_j be given by*

$$\varphi_j(x) = (D_j \varphi_0^{0,\ldots,m})(x), \quad x \in [0,\infty)\ ,\ j \in \mathbb{N}_0. \tag{11.5.1}$$

Then

$$\varphi_j(x) = 1, \qquad x \in [0, 2^j m]$$

for all $j \in \mathbb{N}_0$.

Proof. The statement follows immediately by applying the dilation operator to $\varphi_0^{0,\dots,m}$ in connection with the properties stated in Definition 11.5.1. □

Assume now that the functions φ_j defined in Lemma 11.5.7 generate the functions Φ_j in the usual way. Then we obtain the following result.

Theorem 11.5.8. *Assume that $m \geq 1$. Then the discrete scaling function $\{\Phi_j\}$ corresponding to $\{\varphi_j\}$ (as defined by (11.5.1)) has the following property: If $F \in \mathcal{L}^2(\Omega)$ is band-limited (i.e. there exists an integer $N \in \mathbb{N}$ such that $F^\wedge(n,l) = 0$ for all $n > N, l = 1,\dots,2n+1$), then there exists an index $j \in \mathbb{N}_0$ such that*

$$\Phi_j * F = F.$$

Proof. Take j such that $2^j m \geq N$. Then

$$\Phi_j * F = \sum_{n=0}^{2^j m} \sum_{l=1}^{2n+1} F^\wedge(n,l) Y_{n,l} = F.$$

This is the desired result. □

In other words, starting from a generator of order ≥ 1 we find an exact approximation for band-limited functions $F \in \mathcal{L}^2(\Omega)$. The last property can be generalized to the discrete wavelet transform. Assume that $m \geq 1$. Furthermore, let

$$\psi_0(x)\tilde{\psi}_0(x) = (\varphi_0^{0,\dots,m}(x/2))^2 - (\varphi_0^{0,\dots,m}(x))^2, \quad x \in [0,\infty).$$

Moreover, let

$$\Psi_j = D_j \Psi_0, \quad \tilde{\Psi}_j = D_j \tilde{\Psi}_0, \quad j \in \mathbb{N}_0.$$

Then it follows immediately from Lemma 11.5.7 that, for $j \in \mathbb{N}_0$,

$$\psi_j(x)\tilde{\psi}_j(x) = 0\ , \quad x \in [0, 2^j m].$$

But this shows us that $W_j = \{\Psi_j * \widetilde{\Psi_j} * F | F \in \mathcal{L}^2(\Omega)\}$ is orthogonal to $Harm_{0,\dots,2^j m}$, $j \in \mathbb{N}_0$. This fact together with Theorem 11.5.8 establishes the following reconstruction formula for band-limited functions $F \in \mathcal{L}^2(\Omega)$.

Theorem 11.5.9. *Assume that, for $m \geq 1$, $\varphi_0^{0,\dots,m} : [0,\infty) \to \mathbb{R}$ is a generator of a discrete scaling function of order m. Let φ_j be given by $\varphi_j = D_j \varphi_0^{0,\dots,m}$, $j \in \mathbb{N}_0$, where D_j is the (usual) dilation operator. Suppose* *that* $\Psi_0, \tilde{\Psi}_0$ *is a mother wavelet and a dual mother wavelet, respectively. If* $F \in \mathcal{L}^2(\Omega)$ *is band-limited, then there exists an index* $J \in \mathbb{N}$ *such that*

$$F = \Phi_0^{(2)} * F + \sum_{j=0}^{J} \int_{\Omega} (WT)(F)(j;\eta) \tilde{\Psi}_{j;\eta}(\cdot)\, d\omega(\eta),$$

where the equality is understood in the $\|\cdot\|_{\mathcal{L}^2(\Omega)}$*-sense.*

Bibliographical Notes

An essential aim of scale discrete spherical wavelet transform is to provide an easily interpretable representation of functions on the sphere. There are different approaches derived from multiscale constructions on $\mathcal{L}^2(\Omega)$. Examples of this type are P. Schröder, W. Sweldens (1995), D. Potts, M. Tasche (1995) and F.J. Narcowich, J.W. Ward (1996). There is also a paper by S. Dahlke *et al.* (1994) on multiresolution analysis and discrete wavelets on Ω using a latitude-longitude discretization. Even though these schemata often suffer from problems at the poles, they are very helpful to consider as many standard datasets (such as the NASA ETOPO) are sampled on the grid. First techniques of exact space discretization by use of band-limited wavelets have been given by W. Freeden, U. Windheuser (1997). The scale discrete wavelet theory presented in this chapter is due to W. Freeden, M. Schreiner (1997). The pyramid scheme is due to M. Schreiner (1996b). The application of scale discrete wavelet theory as presented here to inverse problems of satellite geodesy and the idea of interpreting regularization as multiresolution are due to W. Freeden *et al.* (1997) and F. Schneider (1997). A method of generalizing wavelet approximations to closed (not necessarily spherical) surfaces and their application to boundary-value problems of potential theory is described in W. Freeden, F. Schneider (1997).

Part II

Vectors

12

VECTOR SPHERICAL HARMONICS

Vector spherical harmonics (as well as tensor spherical harmonics) occur in many areas of theoretical physics and engineering. In particular, they are used in quantum mechanics, or whenever one has to deal with vector fields of any geophysical nature. Important examples are gravitational fields, magnetic fields, electromagnetic fields, or elastic fields (for deformation analysis of the earth's surface using vector spherical harmonics see for example A. BEN-MENAHEM, S.J. SINGH (1981), E.W. GRAFAREND (1982, 1986), J. BRUDOWSKY *et al.* (1992), W. FREEDEN *et al.* (1996), and M. TÜCKS (1996)). Moreover, in all geosciences there is also a growing need for dealing with vector and tensor fields in spherical geometry, as more and more measurements of vectorial and tensorial type become available by modern space techniques such as satellite-to-satellite tracking or satellite gradiometry.

Various applications have led to different ways of definition of vector and tensor spherical harmonics, putting the accent on different issues. So what is our 'philosophy' for their definition? One important aspect is the easy transition from scalar spherical harmonics to the vectorial ones. But in order to generalize our described approximation methods to vectorial data, the transition from scalar to vectorial data should also be applicable to general functions, in order to ensure vectorial wavelet and spline theory. Hence, any technique concerned with the special structure of a specific choice of spherical harmonics (e.g. Wigner or Clebsch–Gordan coefficients) does not suit our purposes well. Furthermore it turns out that a separation into normal and tangential vector (and tensor) fields is an advantage in all applications where there is an underlying differential equation that is separable in spherical coordinates. Thus, the definition of vector spherical harmonics should take into account this separation. Another important feature is that all geosciences are becoming increasingly interested in a consistent concept to handle scalar spherical functions as well as spherical vector and tensor fields in order to combine different types of data derived from various terrestrial and satellite measurement techniques.

All these aspects are guaranteed adequately, if the vector spherical harmonics are defined via certain operators $o^{(i)}, i = 1, 2, 3$, transforming scalar functions into vector fields. In this way we are led to definitions being independent of any particular choice of spherical harmonics as well as any choice of a spherical coordinate system. Further, this approach can be generalized to the tensorial case, thus allowing a unified concept for applications with data of scalar, vectorial, and tensorial type.

This chapter is organized as follows. In Section 12.2 we introduce the vector spherical harmonics in accordance with our notational framework (presented in Chapter 1 and Section 12.1). The interrelations between vector spherical harmonics and homogeneous harmonic vector polynomials are investigated in Section 12.3. This work enables us to deduce closure and completeness properties of the vector spherical harmonics. Section 12.4 is dedicated to a decomposition formula for spherical vector fields (as already announced in Chapter 1). Section 12.5 shows us that the vector spherical harmonics can be regarded as eigenfunctions of a vectorial analogue of the Beltrami operator. Finally, we prove vectorial versions of two keystones in spherical harmonic theory, namely the addition theorem and the Funk–Hecke formula.

12.1 Notation

Assume that $F \in \mathcal{C}^{(1)}(\Omega)$. Adopting the spherical nomenclature presented in Chapter 1 we define operators $o^{(i)} : \mathcal{C}^{(1)}(\Omega) \to c(\Omega)$ as follows:

$$o_\xi^{(1)} F(\xi) = \xi F(\xi), \quad \xi \in \Omega, \tag{12.1.1}$$

$$o_\xi^{(2)} F(\xi) = \nabla_\xi^* F(\xi), \quad \xi \in \Omega, \tag{12.1.2}$$

$$o_\xi^{(3)} F(\xi) = L_\xi^* F(\xi), \quad \xi \in \Omega. \tag{12.1.3}$$

It is clear that $o^{(1)}F$ is a normal vector field, whereas $o^{(2)}F$ and $o^{(3)}F$ are tangential.

It is not difficult to prove the following result.

Lemma 12.1.1. *Let $F, G \in \mathcal{C}^{(1)}(\Omega)$. Then the following statements are valid:*

(i) For all $\xi \in \Omega$, $o_\xi^{(i)} F(\xi) \cdot o_\xi^{(j)} F(\xi) = 0$ whenever $i \neq j$, $\quad i, j \in \{1,2,3\}$.

(ii) If $(F,G)_{\mathcal{L}^2(\Omega)} = 0$, then $(o^{(i)}F, o^{(j)}G)_{l^2(\Omega)} = 0$ for all $i, j \in \{1,2,3\}$.

The integral equations (1.2.51) and (1.2.52) help us to find operators $O^{(i)}$ which are adjoint to $o^{(i)}$, i.e. for $f \in c^{(1)}(\Omega)$ and $G \in \mathcal{C}^{(1)}(\Omega)$ we have

$$(o^{(i)}G, f)_{l^2(\Omega)} = (G, O^{(i)} f)_{\mathcal{L}^2(\Omega)}. \tag{12.1.4}$$

More explicitly,

$$O_\xi^{(1)} f(\xi) = \xi \cdot p_{\mathrm{nor}} f(\xi), \quad \xi \in \Omega, \tag{12.1.5}$$

$$O_\xi^{(2)} f(\xi) = -\nabla_\xi^* \cdot p_{\tan} f(\xi), \quad \xi \in \Omega, \tag{12.1.6}$$

$$O_\xi^{(3)} f(\xi) = -L_\xi^* \cdot p_{\tan} f(\xi), \quad \xi \in \Omega, \tag{12.1.7}$$

provided that $f \in c^{(1)}(\Omega)$. It can be easily seen that these operators satisfy the following relations.

Lemma 12.1.2. *Let $F \in \mathcal{C}^{(2)}(\Omega)$. Then the following statements hold true:*

(i) If $i \neq j$, $i, j \in \{1,2,3\}$, then $O_\xi^{(i)} o_\xi^{(j)} F(\xi) = 0, \quad \xi \in \Omega$.

(ii) For $i = j$

$$O_\xi^{(i)} o_\xi^{(i)} F(\xi) = \begin{cases} F(\xi) & \text{if} \quad i = 1 \\ -\Delta_\xi^* F(\xi) & \text{if} \quad i = 2,3. \end{cases}$$

12.2 Definition of Vector Spherical Harmonics

We start with the definition of vector spherical harmonics. This can be done without using any local coordinate system on the sphere only by aid of the operators $o^{(i)}, O^{(i)}, i = 1,2,3$. The abbreviation

$$0_i = \begin{cases} 0 & \text{if} \quad i = 1 \\ 1 & \text{if} \quad i = 2,3 \end{cases}$$

will simplify the following considerations.

Definition 12.2.1. Any vector field

$$o^{(i)} Y_n, \; n \geq 0_i, \; Y_n \in Harm_n, \quad i = 1,2,3,$$

is called a *vector spherical harmonic of order n and type i.*

$o^{(1)} Y_n$ describes a normal field, while $o^{(2)} Y_n$, $o^{(3)} Y_n$ are tangential fields of order n. Obviously, according to our construction (cf. Lemma 12.1.1),

$$\xi \wedge (o^{(1)} Y_n)(\xi) = 0, \qquad \xi \cdot (o^{(2)} Y_n)(\xi) = 0, \quad \xi \cdot (o^{(3)} Y_n)(\xi) = 0,$$
$$L_\xi^* \cdot (o^{(2)} Y_n)(\xi) = 0, \qquad \nabla_\xi^* \cdot (o^{(3)} Y_n)(\xi) = 0.$$

The set $harm_n^{(i)}$ denotes the set of all vector spherical harmonics of order n and type i. Furthermore, we let

$$harm_0 = harm_0^{(1)},$$

$$harm_n = \bigoplus_{i=1}^{3} harm_n^{(i)}, \quad n \geq 1.$$

It can be easily seen from the orthogonality of (scalar) spherical harmonics of different order and the properties of the $o^{(i)}$-operators that $harm_n$ is orthogonal to $harm_m$ whenever $n \neq m$. If $\{Y_{n,j}\}_{n=0,1,\ldots,j=1,\ldots 2n+1}$ is, as usual, an $\mathcal{L}^2(\Omega)$-orthonormal set of (scalar) spherical harmonics it follows that

$$y_{n,j}^{(i)} = (\mu_n^{(i)})^{-1/2} \quad o^{(i)} Y_{n,j}, \tag{12.2.1}$$

$i \in \{1,2,3\}, n \geq 0_i, j = 1, \ldots, 2n+1$, forms an $l^2(\Omega)$-orthonormal system of vector spherical harmonics, when the values $\mu_n^{(i)}$ are chosen in such a way that

$$\mu_n^{(i)} = \|O^{(i)} o^{(i)} Y_{n,j}\|_{\mathcal{L}^2(\Omega)},$$

i.e.:

$$\mu_n^{(i)} = \begin{cases} 1 & \text{if} \quad i = 1 \\ -(\Delta^*)^\wedge(n) & \text{if} \quad i = 2,3\,. \end{cases}$$

A result of the next section will be that the system of vector spherical harmonics as introduced above is complete in $l^2(\Omega)$ with respect to $(\cdot,\cdot)_{l^2(\Omega)}$ and closed in $c(\Omega)$ with respect to $\|\cdot\|_{c(\Omega)}$. The proof of these results will make use of relations to homogeneous harmonic vector polynomial theory.

12.3 Homogeneous Harmonic Vector Polynomials

The property of scalar spherical harmonics of being restrictions of homogeneous harmonic polynomials to the unit sphere has been of tremendous importance for many results in scalar theory. In this section similar relations between vector spherical harmonics and homogeneous harmonic vector polynomials are developed. Unfortunately, the different nature of the separation into normal/tangential components on the one hand and into cartesian components on the other hand does not provide us with relations of comparable simplicity as in the scalar case. Nevertheless, these interdependencies help us to develop significant results, e.g. the closure and completeness of vector spherical harmonics.

Definition 12.3.1. A vector field $h_n : \mathbb{R}^3 \to \mathbb{R}^3, n \in \mathbb{N}_0$, is called a *homogeneous harmonic vector polynomial of degree* n if $h_n \cdot \varepsilon^i$ is a homogeneous harmonic polynomial of degree n for every $i \in \{1,2,3\}$.

Using the abbreviation

$$Harm_n(\mathbb{R}^3)\varepsilon^i = \{H_n \varepsilon^i \mid H_n \in Harm_n(\mathbb{R}^3)\},$$

the space of all homogeneous harmonic vector polynomials of degree n is characterized by

$$\bigoplus_{i=1}^{3} Harm_n(\mathbb{R}^3)\varepsilon^i.$$

The restriction of a homogeneous harmonic vector polynomial of degree n to the unit sphere Ω does – *in contrast to the scalar case* – in general not yield a spherical harmonic of order n. But we shall see later on that each element of

$$\bigoplus_{i=1}^{3} Harm_n(\Omega)\varepsilon^i$$

is expressible as a linear combination of vector spherical harmonics of different orders.

Suppose that $H_n \in Harm_n(\mathbb{R}^3)$. Then it is immediately clear that ∇H_n is a homogeneous harmonic vector polynomial of degree $n-1$. A simple calculation shows that $x \mapsto x \wedge \nabla_x H_n(x)$, $x \in \mathbb{R}^3$, yields a homogeneous harmonic vector polynomial of degree n. The function $x \mapsto xH_n(x), x \in \mathbb{R}^3$, is, in general, not harmonic, but it follows easily that $x \mapsto ((2n+1)x - |x|^2\nabla_x)H_n(x)$ is a homogeneous harmonic vector polynomial of order $n+1$. This motivates the following.

Definition 12.3.2. For $n \in \mathbb{N}_0, H_n \in Hom_n(\mathbb{R}^3)$, and $i \in \{1,2,3\}$ the operators $k_n^{(i)}, i = 1,2,3$, are defined by

$$\begin{aligned} k_n^{(1)} H_n(x) &= ((2n+1)x - |x|^2\nabla_x)H_n(x), \quad x \in \mathbb{R}^3, \\ k_n^{(2)} H_n(x) &= \nabla_x H_n(x), \quad x \in \mathbb{R}^3, \\ k_n^{(3)} H_n(x) &= x \wedge \nabla_x H_n(x), \quad x \in \mathbb{R}^3. \end{aligned}$$

The aforementioned properties of the operators introduced in Definition 12.3.1 are summarized in the following result.

Lemma 12.3.3. *Let $H_n \in Harm_n(\mathbb{R}^3), n \in \mathbb{N}_0$. Then $k_n^{(i)} H_n$ is a homogeneous harmonic vector polynomial of degree $\deg^{(i)}(n)$, where we have used the abbreviation*

$$\deg^{(i)}(n) = \begin{cases} n+1 & \text{if } i - 1 \\ n-1 & \text{if } i = 2 \\ n & \text{if } i = 3. \end{cases}$$

If $\deg^{(i)}(n) < 0$, then $k_n^{(i)} H_n = 0$.

Proof. Since the other statements are straightforward, we only prove that the cartesian components of $k_n^{(1)} H_n(x) = ((2n+1)x - |x|^2\nabla_x)H_n(x)$ are homogeneous harmonic polynomials of degree $n+1$. For that purpose we observe that the components of $k_n^{(1)} H_n$ are homogeneous of degree $n+1$. Since, for $j \in \{1,2,3\}, \varepsilon^j \cdot \nabla H_n$ is a homogeneous harmonic polynomial of degree $n-1$, we obtain by elementary calculations

$$\Delta_x(k_n^{(1)} H_n(x)) \cdot \varepsilon^j$$

$$
\begin{aligned}
&= \Delta_x \left((2n+1)x_j H_n(x) - |x|^2 \frac{\partial}{\partial x^j} H_n(x) \right) \\
&= 2(2n+1)\varepsilon^j \cdot \nabla H_n(x) \\
&\quad - \left(\Delta_x |x|^2\right) \frac{\partial}{\partial x^j} H_n(x) - \left(2\nabla_x |x|^2\right) \cdot \left(\nabla_x \frac{\partial}{\partial x^j} H_n(x) \right) \\
&= 2(2n+1)\frac{\partial}{\partial x^j} H_n(x) - 6\frac{\partial}{\partial x^j} H_n(x) - 4x \cdot \nabla_x \frac{\partial}{\partial x^j} H_n(x) \\
&= 4(n-1)\frac{\partial}{\partial x^j} H_n(x) - 4(n-1)\frac{\partial}{\partial x^j} H_n(x) \\
&= 0.
\end{aligned}
$$

This yields the desired result. □

With $x = r\xi, r = |x|$, and the representation of the gradient ∇ in polar coordinates (cf. (1.2.31)) it follows that

$$
((2n+1)x - |x|^2 \nabla_x) = \xi \left((2n+1)r - r^2 \frac{\partial}{\partial r} \right) - r\nabla^*_\xi .
$$

Thus, for $Y_n \in Harm_n$ and $x = r\xi, r = |x| > 0$, we see that

$$
\begin{aligned}
k_n^{(1)} r^n Y_n(\xi) &= (n+1)r^{n+1} Y_n(\xi)\xi - r^{n+1}\nabla^*_\xi Y_n(\xi), \\
k_n^{(2)} r^n Y_n(\xi) &= n r^{n-1} Y_n(\xi)\xi + r^{n-1}\nabla^*_\xi Y_n(\xi), \\
k_n^{(3)} r^n Y_n(\xi) &= r^n L^*_\xi Y_n(\xi).
\end{aligned}
$$

But this shows us that the restrictions of $r\xi \mapsto k_n^{(i)} r^n Y_n(\xi)$ to the unit sphere Ω can be written as linear combinations of vector spherical harmonics $o^{(i)}Y_n$. To be specific,

$$
\begin{aligned}
(k_n^{(1)} H_n(x))_{r=1} &= (n+1)o^{(1)}Y_n(\xi) - o^{(2)}Y_n(\xi), \\
(k_n^{(2)} H_n(x))_{r=1} &= n o^{(1)}Y_n(\xi) + o^{(2)}Y_n(\xi), \\
(k_n^{(3)} H_n(x))_{r=1} &= o^{(3)}Y_n(\xi)
\end{aligned}
\tag{12.3.1}
$$

with $H_n(x) = r^n Y_n(\xi), x = r\xi$. By inverting the above equations we find

$$
\begin{aligned}
o^{(1)}Y_n(\xi) &= \frac{1}{2n+1}\left(k_n^{(1)} H_n(x)\right)_{r=1} + \frac{1}{2n+1}\left(k_n^{(2)} H_n(x)\right)_{r=1}, \\
o^{(2)}Y_n(\xi) &= \frac{-n}{2n+1}\left(k_n^{(1)} H_n(x)\right)_{r=1} + \frac{n+1}{2n+1}\left(k_n^{(2)} H_n(x)\right)_{r=1}, \\
o^{(3)}Y_n(\xi) &= \left(k_n^{(3)} H_n(x)\right)_{r=1} .
\end{aligned}
\tag{12.3.2}
$$

In connection with Lemma 12.3.3 it follows that any vector spherical harmonic of order n and type 1 and 2 can be expressed as linear combinations of scalar spherical harmonics of order $n-1$ and $n+1$, whereas the cartesian components of a vector spherical harmonic of order n and type 3 is a linear combination of scalar spherical harmonics of order n.

In conclusion,

$$harm_n^{(i)} \subset \bigoplus_{j=1}^{3} Harm_{n-1}\varepsilon^j \oplus \bigoplus_{j=1}^{3} Harm_{n+1}\varepsilon^j, \quad i=1,2, \tag{12.3.3}$$

$$harm_n^{(3)} \subset \bigoplus_{j=1}^{3} Harm_n\varepsilon^j. \tag{12.3.4}$$

An immediate consequence is the fact that

$$y_n^{(i)}(-\xi) = (-1)^{n+1} y_n^{(i)}(\xi), \quad \xi \in \Omega,\ y_n^{(i)} \in harm_n^{(i)},\ i=1,2,$$

and

$$y_n^{(3)}(-\xi) = (-1)^n y_n^{(3)}(\xi), \quad \xi \in \Omega,\ y_n^{(3)} \in harm_n^{(3)}.$$

Furthermore, the following orthogonality relations are readily obtainable from (12.3.3) and (12.3.4).

Lemma 12.3.4. *Let $y_n^{(i)} \in harm_n^{(i)}$ and $Y_m \in Harm_m$. Then*

$$\int_\Omega y_n^{(i)}(\xi) Y_m(\xi)\, d\omega(\xi) = 0,$$

whenever $i \in \{1,2\}$ and $m \notin \{n-1, n+1\}$ or $i=3$ and $m \neq n$.

Next we are interested in closure and completeness properties of vector spherical harmonics. It is obvious from the corresponding results of (scalar) spherical harmonics that $\bigoplus_{m=0}^{\infty} \bigoplus_{i=1}^{3} Harm_m\varepsilon^i$ is dense in $c(\Omega)$ with respect to $\|\cdot\|_{c(\Omega)}$ and in $l^2(\Omega)$ with respect to $(\cdot,\cdot)_{l^2(\Omega)}$. On the other hand, one can readily show that

$$harm_n \subset \bigoplus_{m=n-1}^{n+1} \bigoplus_{i=1}^{3} Harm_m\varepsilon^i.$$

Summarizing our results we therefore obtain the following theorem.

Theorem 12.3.5. *Let $\{y_{n,j}^{(i)}\}_{\substack{i=1,2,3 \\ n=0_i,\ldots,\ j=1,\ldots,2n+1}}$ be defined as in (12.2.1). Then the following statements are valid:*

(i) The system of vector spherical harmonics is closed in $c(\Omega)$ with respect to $\|\cdot\|_{c(\Omega)}$.

(ii) *The system is complete in* $l^2(\Omega)$ *with respect to* $(\cdot,\cdot)_{l^2(\Omega)}$.

Part (i) of this theorem says that any continuous vector field can be approximated arbitrarily close by finite linear combinations of vector spherical harmonics, while part (ii) is equivalent (cf. Theorem 3.4.7) to the fact that every vector field in $l^2(\Omega)$ can be respresented by its Fourier (orthogonal) series in terms of the system $\{y_{n,j}^{(i)}\}$:

$$f = \sum_{i=1}^{3} \sum_{n=0_i}^{\infty} \sum_{j=1}^{2n+1} (f^{(i)})^\wedge(n,j) y_{n,j}^{(i)}.$$

To be more specific,

$$\lim_{N\to\infty} \left\| f - \sum_{i=1}^{3} \sum_{n=0_i}^{N} \sum_{j=1}^{2n+1} (f^{(i)})^\wedge(n,j) y_{n,j}^{(i)} \right\|_{l^2(\Omega)} = 0, \quad f \in l^2(\Omega),$$

where the *'Fourier coefficients'* are given by

$$(f^{(i)})^\wedge(n,j) = (f, y_{n,j}^{(i)})_{l^2(\Omega)} = \int_\Omega f(\xi) \cdot y_{n,j}^{(i)}(\xi)\, d\omega(\xi).$$

It is obvious that similar properties of the sets $\bigoplus_{m=0}^{\infty} harm_m^{(1)}$ and $\bigoplus_{m=1}^{\infty} (harm_m^{(2)} \oplus harm_m^{(3)})$ may be detected in connection with the spaces $c_{nor}(\Omega)$ and $l^2_{\mathrm{nor}}(\Omega)$, respectively, $c_{\tan}(\Omega)$ and $l^2_{\tan}(\Omega)$.

The approach described above is particularly helpful for the consideration of the vector Laplace equation, since the vector fields $x \mapsto k_n^{(i)} H_n(x)$, $x \in \mathbb{R}^3$, $i = 1,2,3$ satisfy

$$\Delta k_n^{(i)} H_n = 0, \quad i = 1,2,3,$$

where $H_n(x) = |x|^n Y_n(\xi)$, $x = |x|\xi$, $\xi \in \Omega$.

The restrictions of homogeneous vector polynomials can also be used for the definition of vector spherical harmonics, as developed for example by A.R. EDMONDS (1957). In our nomenclature we find the following lemma.

Lemma 12.3.6. *Let* $\{Y_{n,j}\}_{\substack{n=0,1,\ldots \\ j=1,\ldots,2n+1}}$ *be an* $\mathcal{L}^2(\Omega)$*-orthonormal system of spherical harmonics. Then the vector fields*

$$u_{n+1,j}^{(1)} = \frac{1}{\sqrt{(n+1)(2n+1)}} \left((n+1) o^{(1)} Y_{n,j} - o^{(2)} Y_{n,j} \right),$$
$$n = 0,1,\ldots, j = 1,\ldots,2n+1,$$

$$
\begin{aligned}
u_{n-1,j}^{(2)} &= \frac{1}{\sqrt{n(2n+1)}}\left(no^{(1)}Y_{n,j} + o^{(2)}Y_{n,j}\right), \\
& \quad n = 1,2,\ldots, j = 1,\ldots,2n+1, \\
u_{n,j}^{(3)} &= \frac{1}{\sqrt{n(n+1)}}o^{(3)}Y_{n,j}, \quad n = 1,2,\ldots, j = 1,\ldots,2n+1,
\end{aligned}
$$

form an $l^2(\Omega)$-orthonormal set of vector spherical harmonics.

The system of vector spherical harmonics presented in Lemma 12.3.6 does not separate normal and tangential components, since $u_{n+1,j}^{(1)}$ and $u_{n-1,j}^{(2)}$ are of mixed type. Hence, the application of these functions is not well suited for most vectorial problems which are of interest in geomathematics.

12.4 Decomposition of Spherical Vector Fields

As mentioned earlier, the motivation of the above definition of vector spherical harmonics by means of the $o^{(i)}$-operators is based on the fact that any vector field $f \in c^{(1)}(\Omega)$ can be explicitly written out as $f(\xi) = \sum_{i=1}^{3} o^{(i)}F_i(\xi)$, $\xi \in \Omega$, with uniquely determined functions $F_i : \Omega \to \mathbb{R}$. In this section we formulate this result in a rigorous sense. Our particular purpose is to show how the functions F_i can be determined in an explicit way by use of the concept of Green's function with respect to the Beltrami operator (cf. Chapter 4).

Theorem 12.4.1. *Let $f : \Omega \to \mathbb{R}^3$ be a continuously differentiable vector field. Then there exist uniquely determined functions $F_1 \in \mathcal{C}^{(1)}(\Omega)$ and $F_2, F_3 \in \mathcal{C}^{(2)}(\Omega)$ satisfying*

$$\int_\Omega F_i(\xi)\, d\omega(\xi) = 0, \quad i = 2,3, \tag{12.4.1}$$

such that

$$f = \sum_{i=1}^{3} o^{(i)}F_i.$$

The functions F_i are given by

$$
\begin{aligned}
F_1(\xi) &= O_\xi^{(1)} f(\xi), \quad \xi \in \Omega, \\
F_2(\xi) &= -\int_\Omega G(\Delta^*;\xi,\eta) O_\eta^{(2)} f(\eta)\, d\omega(\eta), \quad \xi \in \Omega, \\
F_3(\xi) &= -\int_\Omega G(\Delta^*;\xi,\eta) O_\eta^{(3)} f(\eta)\, d\omega(\eta), \quad \xi \in \Omega.
\end{aligned}
$$

Proof. Any vector field f can be written as $f = f_{\mathrm{nor}} + f_{\mathrm{tan}}$, where $f_{\mathrm{nor}} = p_{\mathrm{nor}} f \in c^{(1)}_{\mathrm{nor}}(\Omega)$, $f_{\mathrm{tan}} = p_{\mathrm{tan}} f \in c^{(1)}_{\mathrm{tan}}(\Omega)$. Clearly, we have $f_{\mathrm{nor}}(\xi) = o^{(1)} F_1(\xi)$ if $F_1(\xi) = O^{(1)} f(\xi), \xi \in \Omega$.

The tangential vector field f_{tan} can be represented in the form $f_{\mathrm{tan}} = \nabla^* F_2 + L^* F_3$. Applying the operators $\nabla^* \cdot$ and $L^* \cdot$, respectively, to f_{tan} we obtain the differential equations $\Delta^* F_2 = \nabla^* \cdot f_{\mathrm{tan}}$, $\Delta^* F_3 = L^* \cdot f_{\mathrm{tan}}$. Since

$$\int_\Omega \nabla^*_\xi \cdot f_{\mathrm{tan}}(\xi)\, d\omega(\xi) = \int_\Omega L^*_\xi \cdot f_{\mathrm{tan}}(\xi)\, d\omega(\xi) = 0,$$

we get the required decomposition from Section 4.3 and our definition of the operators $O^{(i)}$. Furthermore, (12.4.1) is valid.

In order to prove the uniqueness of F_i, $i = 1, 2, 3$, assume that there exists another triple of functions G_i, $i = 1, 2, 3$, such that

$$f = o^{(1)} F_1 + o^{(2)} F_2 + o^{(3)} F_3, \tag{12.4.2}$$

$$f = o^{(1)} G_1 + o^{(2)} G_2 + o^{(3)} G_3. \tag{12.4.3}$$

Then it follows that $F_1 = O^{(1)} f = G_1$, thus F_1 is uniquely defined. Applications of $O^{(2)}$ and $O^{(3)}$ to (12.4.2) and (12.4.3) yields $\Delta^* F_2 = \Delta^* G_2$, $\Delta^* F_3 = \Delta^* G_3$. Hence the normalization conditions (12.4.1) for F_i and G_i, $i = 2, 3$, imply $F_2 = G_2$, $F_3 = G_3$, as required. □

The decomposition theorem also implies an orthogonal decomposition of the space $c^{(\infty)}(\Omega)$. In fact,

$$c^{(\infty)}(\Omega) = c^{(\infty)}_{(1)}(\Omega) \oplus c^{(\infty)}_{(2)}(\Omega) \oplus c^{(\infty)}_{(3)}(\Omega),$$

where

$$\begin{aligned} c^{(\infty)}_{(1)}(\Omega) &= c^{(\infty)}_{\mathrm{nor}}(\Omega), \\ c^{(\infty)}_{(2)}(\Omega) &= \{ f \in c^{(\infty)}(\Omega) \mid O^{(1)} f = O^{(3)} f = 0 \}, \\ c^{(\infty)}_{(3)}(\Omega) &= \{ f \in c^{(\infty)}(\Omega) \mid O^{(1)} f = O^{(2)} f = 0 \}. \end{aligned}$$

Of course, these definitions can be extended as well to the spaces $c^{(k)}(\Omega)$, $0 \le k \le \infty$, or $l^2(\Omega)$. In the case of $l^2(\Omega)$ we are able to write

$$l^2_{(i)}(\Omega) = \overline{\{ o^{(i)} F \mid F \in \mathcal{C}^{(\infty)}(\Omega) \}},$$

where the completion is understood in the $l^2(\Omega)$-sense. Thus, we end up with the orthogonal decompositions

$$l^2(\Omega) = l^2_{\mathrm{nor}}(\Omega) \oplus l^2_{\mathrm{tan}}(\Omega),$$

$$l^2_{\tan}(\Omega) \quad = \quad l^2_{(2)}(\Omega) \oplus l^2_{(3)}(\Omega).$$

Finally, we mention that every $f \in l^2(\Omega)$ can be represented in the form

$$f = \sum_{i=1}^{3} f^{(i)}, \quad f^{(i)} \in l^2_{(i)}(\Omega),$$

where $f^{(i)}$, $i = 1, 2, 3$, is given by

$$f^{(i)} = \sum_{n=0_i}^{\infty} \sum_{j=1}^{2n+1} (f^{(i)})^\wedge(n,j) y^{(i)}_{n,j},$$

with

$$(f^{(i)})^\wedge(n,j) = (f, y^{(i)}_{n,j})_{l^2(\Omega)}$$

(in the $\|\cdot\|_{l^2(\Omega)}$-sense).

12.5 Vectorial Beltrami Operator

In this section we develop a vectorial analogue of the Beltrami operator Δ^*, denoted by $\boldsymbol{\Delta}^*$. In doing so the vector spherical harmonics can be recognized as eigenfunctions of the operator $\boldsymbol{\Delta}^*$. In particular, we notice that the operator $\boldsymbol{\Delta}^*$ corresponds to the orthogonal decomposition with respect to the operators $o^{(i)}, i = 1, 2, 3$, that is to say, $\Delta^* f \in c^{(0)}_{(i)}(\Omega)$ for all $i = 1, 2, 3$ provided that $f \in c^{(2)}_{(i)}(\Omega)$.

Our construction is based on the componentwise application of the (scalar) Beltrami operator Δ^*: if $f \in c^{(2)}(\Omega)$ is of the form

$$f(\xi) = \sum_{i-1}^{3} \varepsilon^i F_i(\xi), \quad \xi \in \Omega,$$

then we let

$$\Delta^* f(\xi) = \sum_{i=1}^{3} \varepsilon^i \Delta^* F_i(\xi), \quad \xi \in \Omega. \tag{12.5.1}$$

Observing this convention we find the following lemma.

Lemma 12.5.1.

(i) Let $F : \Omega \to \mathbb{R}$ be sufficiently smooth. Then we have

$$\begin{aligned} \Delta^* o^{(1)} F &= o^{(1)}(\Delta^* - 2)F + 2o^{(2)}F, \\ \Delta^* o^{(2)} F &= -2o^{(1)}\Delta^* F + o^{(2)}\Delta^* F, \\ \Delta^* o^{(3)} F &= o^{(3)}\Delta^* F. \end{aligned}$$

(ii) Let $f : \Omega \to \mathbb{R}^3$ be a sufficiently smooth vector field. Then

$$
\begin{aligned}
O^{(1)}\Delta^* f &= (\Delta^* - 2)O^{(1)} f + 2O^{(2)} f, \\
O^{(2)}\Delta^* f &= -2\Delta^* O^{(1)} f + \Delta^* O^{(2)} f, \\
O^{(3)}\Delta^* f &= \Delta^* O^{(3)} f.
\end{aligned}
$$

Proof. We prove only the first formula of part (i), since the other assertions follow by similar arguments. Assume firstly that the function under consideration is a spherical harmonic Y_n of class $Harm_n$. Then it follows from (12.3.2) that

$$
\begin{aligned}
\Delta^*_\xi o^{(1)}_\xi Y_n(\xi) &= \Delta^*_\xi \frac{1}{2n+1}\left(k^{(1)}_n r^n Y_n(\xi)|_{r=1} + k^{(2)}_n r^n Y_n(\xi)|_{r=1}\right) \\
&= -\frac{(n+1)(n+2)}{2n+1}(k^{(1)}_n r^n Y_n(\xi))|_{r=1} \\
&\quad -\frac{n(n-1)}{2n+1}(k^{(2)}_n r^n Y_n(\xi))|_{r=1}
\end{aligned}
$$

holds for all $\xi \in \Omega$. Using (12.3.1) we obtain

$$
\begin{aligned}
\Delta^*_\xi o^{(1)}_\xi Y_n(\xi) &= -\frac{(n+1)^2(n+2)}{2n+1} o^{(1)}_\xi Y_n(\xi) + \frac{(n+1)(n+2)}{2n+1} o^{(2)}_\xi Y_n(\xi) \\
&\quad -\frac{(n-1)n^2}{2n+1} o^{(1)}_\xi Y_n(\xi) - \frac{n(n-1)}{2n+1} o^{(2)}_\xi Y_n(\xi) \\
&= (-n(n+1) - 2) o^{(1)}_\xi Y_n(\xi) + 2 o^{(2)}_\xi Y_n(\xi) \\
&= (\Delta^*_\xi - 2) o^{(1)}_\xi Y_n(\xi) + 2 o^{(2)}_\xi Y_n(\xi).
\end{aligned}
$$

Thus our formula is true for every spherical harmonic. The completeness of the spherical harmonics in $l^2(\Omega)$ implies the validity for all sufficiently smooth functions. Part (ii) follows from the adjointness of the operators $o^{(i)}$ and $O^{(i)}$ and the self-adjointness of Δ^*. □

Lemma 12.5.1 helps us to find a vectorial Beltrami operator via the following equations:

$$
\begin{aligned}
(\Delta^* + 2)o^{(1)} Y_n &= -n(n+1)o^{(1)} Y_n + 2o^{(2)} Y_n, \\
\Delta^* o^{(2)} Y_n &= 2n(n+1)o^{(1)} Y_n - n(n+1)o^{(2)} Y_n, \qquad (12.5.2) \\
\Delta^* o^{(3)} Y_n &= -n(n+1)o^{(3)} Y_n.
\end{aligned}
$$

Therefore,

$$
p_{\text{nor}}(\Delta^* + 2)o^{(1)} Y_n = -n(n+1)o^{(1)} Y_n,
$$

$$p_{\tan}\Delta^* o^{(2)} Y_n = -n(n+1)o^{(2)} Y_n, \qquad (12.5.3)$$
$$p_{\tan}\Delta^* o^{(3)} Y_n = -n(n+1)o^{(3)} Y_n.$$

Formula (12.5.3) motivates the introduction of the vectorial Beltrami operator as follows.

Definition 12.5.2. The operator $\mathbf{\Delta}^* : c^{(2)}(\Omega) \to c^{(0)}(\Omega)$ given by

$$\mathbf{\Delta}^* = p_{\mathrm{nor}}(\Delta^* + 2)p_{\mathrm{nor}} + p_{\tan}\Delta^* p_{\tan},$$

is called *vectorial Beltrami operator*, where the application of Δ^* on vector fields is understood in the sense of (12.5.1).

As an immediate consequence of Lemma 12.5.1 we obtain the following.

Lemma 12.5.3. *If $F : \Omega \to \mathbb{R}$ and $f : \Omega \to \mathbb{R}^3$ are sufficiently smooth, then*

$$\mathbf{\Delta}^* o^{(i)} F = o^{(i)} \Delta^* F,$$
$$O^{(i)} \mathbf{\Delta}^* f = \Delta^* O^{(i)} f$$

holds for $i = 1, 2, 3$.

Lemma 12.5.3 helps us to characterize the spectrum of the operator $\mathbf{\Delta}^*$.

Theorem 12.5.4. *Any vector spherical harmonic $y_n \in harm_n$ of order n is an infinitely often differentiable eigenfunction of the operator $\mathbf{\Delta}^*$ with respect to the eigenvalue $(\mathbf{\Delta}^*)^\wedge(n) = (\Delta^*)^\wedge(n)$. Conversely, any infinitely often differentiable eigenfunction of $\mathbf{\Delta}^*$ is a vector spherical harmonic.*

Remark. It is also possible to define $\mathbf{\Delta}^*$ (without projection operators) only by use of differentiation processes in $\mathbb{R}^3$ (cf. T. Gervens(1989)):

$$\mathbf{\Delta}^*_\xi f(\xi) = \Delta^*_\xi f(\xi) - 2(\xi \wedge \nabla_\xi) \wedge f(\xi) - 2f(\xi), \quad f \in c^{(2)}(\Omega).$$

An intrinsic approach within the Riemannian geometry of the sphere is described by M. Schreiner (1994).

12.6 Vectorial Addition Theorem

The previous chapters of this book have shown the central role of the addition theorem of (scalar) spherical harmonics in constructive approximation theory. We shall develop in this section a vectorial analogue of the addition theorem. In what follows this vectorial addition theorem leads us to the definition of vectorial reproducing kernels, which are a basic tool for example in the theory of vector radial basis functions. In addition, it also offers a better insight into some aspects of the theory of vector spherical harmonics.

Let $\{y_{n,j}^{(i)}\}_{\substack{i=1,2,3\\ j=1,\ldots,2n+1}}$ be an $l^2(\Omega)$-orthonormal basis of $harm_n$, as defined in (12.2.1), corresponding to an $\mathcal{L}^2(\Omega)$-orthonormal basis $\{Y_{n,j}\}_{j=1,\ldots,2n+1}$ of $Harm_n$: $y_{n,j}^{(i)} = (\mu_n^{(i)})^{-1/2} o^{(i)} Y_{n,j}$. Then the vectorial addition theorem deals with the question of determining the expression

$$\sum_{j=1}^{2n+1} y_{n,j}^{(i)}(\xi) \otimes y_{n,j}^{(k)}(\eta), \quad \xi, \eta \in \Omega. \tag{12.6.1}$$

Our purpose is to explain how a tensorial version of the Legendre polynomials comes into play. To this end we firstly extend in canonical way the definition of the $o^{(i)}$-operators (cf. (12.1.1)–(12.1.3)) to vector fields. Suppose that $f : \Omega \to \mathbb{R}^3$ is a sufficiently smooth vector field of the representation

$$f(\xi) = \sum_{k=1}^{3} F_k(\xi) \varepsilon^k.$$

Then we set

$$o_\xi^{(i)} f(\xi) = \sum_{k=1}^{3} (o_\xi^{(i)} F_k(\xi)) \otimes \varepsilon^k, \quad i = 1, 2, 3.$$

Thus, $o^{(i)}$ maps functions to vector fields and vector fields to rank-2 tensor fields, respectively. But this means that (12.6.1) can be expressed as follows:

$$\begin{aligned}\sum_{j=1}^{2n+1} y_{n,j}^{(i)}(\xi) \otimes y_{n,j}^{(k)}(\eta) &= (\mu_n^{(i)})^{-1/2} (\mu_n^{(k)})^{-1/2} \sum_{j=1}^{2n+1} o_\xi^{(i)} Y_{n,j}(\xi) \otimes o_\eta^{(k)} Y_{n,j}(\eta) \\ &= (\mu_n^{(i)})^{-1/2} (\mu_n^{(k)})^{-1/2} o_\xi^{(i)} o_\eta^{(k)} \sum_{j=1}^{2n+1} Y_{n,j}(\xi) Y_{n,j}(\eta) \\ &= (\mu_n^{(i)})^{-1/2} (\mu_n^{(k)})^{-1/2} \frac{2n+1}{4\pi} o_\xi^{(i)} o_\eta^{(k)} P_n(\xi \cdot \eta).\end{aligned}$$

Via the definition of the *(i,k)-Legendre-tensor field* of degree n, $\mathbf{p}_n^{(i,k)} : \Omega \times \Omega \to \mathbb{R}^3 \otimes \mathbb{R}^3$ given by

$$\mathbf{p}_n^{(i,k)}(\xi, \eta) = (\mu_n^{(i)})^{-1/2} (\mu_n^{(k)})^{-1/2} o_\xi^{(i)} o_\eta^{(k)} P_n(\xi \cdot \eta), \quad \xi, \eta \in \Omega \tag{12.6.2}$$

this leads us to the following *addition theorem of vector spherical harmonics*.

Theorem 12.6.1. *Let* $\{y_{n,j}^{(i)}\}_{\substack{i=1,2,3\\ j=1,\ldots,2n+1}}$ *be an* $l^2(\Omega)$*-orthonormal basis of* $harm_n$ *as given in (12.2.1). Then*

$$\sum_{j=1}^{2n+1} y_{n,j}^{(i)}(\xi) \otimes y_{n,j}^{(k)}(\eta) = \frac{2n+1}{4\pi}\mathbf{p}_n^{(i,k)}(\xi,\eta), \qquad \xi,\eta \in \Omega,$$

holds for $i,k \in \{1,2,3\}$, *where* $\mathbf{p}_n^{(i,k)}$ *is defined by (12.6.2).*

The actual problem, however, is the evaluation of $\mathbf{p}_n^{(i,k)}$. As a first step we verify the following.

Lemma 12.6.2. *For* $\xi,\eta \in \Omega$,

$$\begin{aligned}
o_\xi^{(2)}(\xi - (\xi\cdot\eta)\eta) &= \mathbf{i}_{\tan}(\xi) - (\eta - (\xi\cdot\eta)\xi)\otimes\eta, \\
o_\xi^{(3)}(\xi - (\xi\cdot\eta)\eta) &= \mathbf{j}_{\tan}(\xi) - (\xi\wedge\eta)\otimes\eta, \\
o_\xi^{(2)}(\eta\wedge\xi) &= -\mathbf{j}_{\tan}(\eta) - \xi\otimes\eta\wedge\xi, \\
o_\xi^{(3)}(\eta\wedge\xi) &= (\xi\cdot\eta)\mathbf{i}_{\tan}(\xi) - (\eta - (\xi\cdot\eta)\xi)\otimes\xi.
\end{aligned}$$

Proof. We prove the first and third formulae. The second and fourth formulae follow by similar arguments, since $L_\xi^* = \xi\wedge\nabla_\xi^*$.

$$\begin{aligned}
o_\xi^{(2)}(\xi - (\xi\cdot\eta)\eta) &= o_\xi^{(2)}\sum_{l=1}^{3}((\xi\cdot\varepsilon^l) - (\xi\cdot\eta)(\eta\cdot\varepsilon^l))\varepsilon^l \\
&= \sum_{l=1}^{3}(\varepsilon^l - (\xi\cdot\varepsilon^l)\xi)\otimes\varepsilon^l - (\eta - (\xi\cdot\eta)\xi)(\eta\cdot\varepsilon^l)\otimes\varepsilon^l \\
&= \mathbf{i}_{\tan}(\xi) - (\eta - (\xi\cdot\eta)\xi)\otimes\eta, \\
o_\xi^{(2)}(\eta\wedge\xi) &= o_\xi^{(2)}\sum_{l=1}^{3}((\eta\wedge\xi)\cdot\varepsilon^l)\varepsilon^l \\
&= o_\xi^{(2)}\sum_{l=1}^{3}((\varepsilon^l\wedge\eta)\cdot\xi)\varepsilon^l \\
&= \sum_{l=1}^{3}(\varepsilon^l\wedge\eta - ((\varepsilon^l\wedge\eta)\cdot\xi)\xi)\otimes\varepsilon^l \\
&= \sum_{l=1}^{3}(-\eta\wedge\varepsilon^l)\otimes\varepsilon^l - ((\eta\wedge\xi)\cdot\varepsilon^l)\xi\otimes\varepsilon^l \\
&= -\mathbf{j}_{\tan}(\eta) - \xi\otimes\eta\wedge\xi.
\end{aligned}$$

□

If $F : [-1,1] \to \mathbb{R}$ is sufficiently smooth, we obtain from Lemma 12.6.2 for $\xi, \eta \in \Omega$:

$$
\begin{aligned}
o_\xi^{(1)} o_\eta^{(1)} F(\xi\cdot\eta) &= F(\xi\cdot\eta)\xi\otimes\eta,\\
o_\xi^{(1)} o_\eta^{(2)} F(\xi\cdot\eta) &= F'(\xi\cdot\eta)\xi\otimes(\xi-(\xi\cdot\eta)\eta),\\
o_\xi^{(1)} o_\eta^{(3)} F(\xi\cdot\eta) &= F'(\xi\cdot\eta)\xi\otimes\eta\wedge\xi.
\end{aligned}
$$

Similar results hold for $o_\xi^{(2)} o_\eta^{(1)} F(\xi\cdot\eta)$ and $o_\xi^{(3)} o_\eta^{(1)} F(\xi\cdot\eta)$. When we treat the tangential operators, we obtain

$$
\begin{aligned}
o_\xi^{(2)} o_\eta^{(2)} F(\xi\cdot\eta) &= \nabla_\xi^*\otimes(F'(\xi\cdot\eta)(\xi-(\xi\cdot\eta)\eta))\\
&= (\nabla_\xi^* F'(\xi\cdot\eta))\otimes(\xi-(\xi\cdot\eta)\eta) + F'(\xi\cdot\eta)\nabla_\xi^*\otimes(\xi-(\xi\cdot\eta)\eta\\
&= F''(\xi\cdot\eta)(\eta-(\xi\cdot\eta)\xi)\otimes(\xi-(\xi\cdot\eta)\eta)\\
&\quad +F'(\xi\cdot\eta)(\mathbf{i}_{\mathrm{tan}}(\xi)-(\eta-(\xi\cdot\eta)\xi)\otimes\eta),
\end{aligned}
$$

$$
\begin{aligned}
o_\xi^{(2)} o_\eta^{(3)} F(\xi\cdot\eta) &= \nabla_\xi^*\otimes(F'(\xi\cdot\eta)\eta\wedge\xi)\\
&= (\nabla_\xi^* F'(\xi\cdot\eta))\otimes\eta\wedge\xi + F'(\xi\cdot\eta)\nabla_\xi^*\otimes(\eta\wedge\xi)\\
&= F''(\xi\cdot\eta)(\eta-(\xi\cdot\eta)\xi)\otimes\eta\wedge\xi\\
&\quad +F'(\xi\cdot\eta)(-\mathbf{j}_{\mathrm{tan}}(\eta)-\xi\otimes\eta\wedge\xi).
\end{aligned}
$$

Similar calculations yield

$$
\begin{aligned}
o_\xi^{(3)} o_\eta^{(2)} F(\xi\cdot\eta) &= F''(\xi\cdot\eta)\xi\wedge\eta\otimes(\xi-(\xi\cdot\eta)\eta)\\
&\quad +F'(\xi\cdot\eta)(\mathbf{j}_{\mathrm{tan}}(\xi)-\xi\wedge\eta\otimes\eta)
\end{aligned}
$$

and

$$
\begin{aligned}
o_\xi^{(3)} o_\eta^{(3)} F(\xi\cdot\eta) &= F''(\xi\cdot\eta)\xi\wedge\eta\otimes\eta\wedge\xi\\
&\quad +F'(\xi\cdot\eta)((\xi\cdot\eta)\mathbf{i}_{\mathrm{tan}}(\xi)-(\eta-(\xi\cdot\eta)\xi)\otimes\xi).
\end{aligned}
$$

Applying these results to P_n we obtain the following identities for the Legendre tensor $\mathbf{p}_n^{(i,k)}$.

Theorem 12.6.3. *Let $n \in \mathbb{N}_0$. Then the identities*

$$
\begin{aligned}
\mathbf{p}_n^{(1,1)}(\xi,\eta) &= P_n(\xi\cdot\eta)\xi\otimes\eta,\\
\mathbf{p}_n^{(1,2)}(\xi,\eta) &= \frac{1}{\sqrt{n(n+1)}}P_n'(\xi\cdot\eta)\xi\otimes(\xi-(\xi\cdot\eta)\eta),
\end{aligned}
$$

$$\begin{aligned}
\mathbf{p}_n^{(1,3)}(\xi,\eta) &= \frac{1}{\sqrt{n(n+1)}} P_n'(\xi\cdot\eta)\xi\otimes\eta\wedge\xi,\\
\mathbf{p}_n^{(2,1)}(\xi,\eta) &= \frac{1}{\sqrt{n(n+1)}} P_n'(\xi\cdot\eta)(\eta-(\xi\cdot\eta)\xi)\otimes\eta,\\
\mathbf{p}_n^{(3,1)}(\xi,\eta) &= \frac{1}{\sqrt{n(n+1)}} P_n'(\xi\cdot\eta)\xi\wedge\eta\otimes\eta,\\
\mathbf{p}_n^{(2,2)}(\xi,\eta) &= \frac{1}{n(n+1)}(P_n''(\xi\cdot\eta)(\eta-(\xi\cdot\eta)\xi)\otimes(\xi-(\xi\cdot\eta)\eta)\\
&\quad + P_n'(\xi\cdot\eta)(\mathbf{i}_{\mathrm{tan}}(\xi)-(\eta-(\xi\cdot\eta)\xi)\otimes\eta)),\\
\mathbf{p}_n^{(2,3)}(\xi,\eta) &= \frac{1}{n(n+1)}(P_n''(\xi\cdot\eta)(\eta-(\xi\cdot\eta)\xi)\otimes\eta\wedge\xi\\
&\quad + P_n'(\xi\cdot\eta)(-\mathbf{j}_{\mathrm{tan}}(\eta)-\xi\otimes\eta\wedge\xi)),\\
\mathbf{p}_n^{(3,2)}(\xi,\eta) &= \frac{1}{n(n+1)}(P_n''(\xi\cdot\eta)\xi\wedge\eta\otimes(\xi-(\xi\cdot\eta)\eta)\\
&\quad + P_n'(\xi\cdot\eta)(\mathbf{j}_{\mathrm{tan}}(\xi)-\xi\wedge\eta\otimes\eta)),\\
\mathbf{p}_n^{(3,3)}(\xi,\eta) &= \frac{1}{n(n+1)}(P_n''(\xi\cdot\eta)\xi\wedge\eta\otimes\eta\wedge\xi\\
&\quad + P_n'(\xi\cdot\eta)((\xi\cdot\eta)\mathbf{i}_{\mathrm{tan}}(\xi)-(\eta-(\xi\cdot\eta)\xi)\otimes\xi))
\end{aligned}$$

hold for all $(\xi,\eta)\in\Omega\times\Omega$.

The case $\xi=\eta$ in the last theorem is of particular interest. Observing $P_n(1)=1, P_n'(1)=\frac{1}{2}n(n+1)$ and $P_n''(1)=\frac{1}{8}n(n+1)(n(n+1)-2)$ (cf. Section 3.2) we obtain the following corollary.

Corollary 12.6.4. *For* $n\in\mathbb{N}_0$ *and all* $\xi\in\Omega$

$$\begin{aligned}
\mathbf{p}_n^{(1,1)}(\xi,\xi) &= \xi\otimes\xi,\\
\mathbf{p}_n^{(1,i)}(\xi,\xi) &= \mathbf{p}_n^{(i,1)}(\xi,\xi)=0, \quad i=2,3,\\
\mathbf{p}_n^{(2,2)}(\xi,\xi) &= \mathbf{p}_n^{(3,3)}(\xi,\xi)=\frac{1}{2}\mathbf{i}_{\mathrm{tan}}(\xi),\\
\mathbf{p}_n^{(2,3)}(\xi,\xi) &= -\mathbf{p}_n^{(3,2)}(\xi,\xi)=-\frac{1}{2}\mathbf{j}_{\mathrm{tan}}(\xi).
\end{aligned}$$

It follows readily that

$$\begin{aligned}
\mathrm{trace}\,((\eta-(\xi\cdot\eta)\xi)\otimes(\xi-(\xi\cdot\eta)\eta)) &= -(\xi\cdot\eta)(1-(\xi\cdot\eta)^2),\\
\mathrm{trace}\,(\xi\wedge\eta\otimes\eta\wedge\xi) &= -(1-(\xi\cdot\eta)^2).
\end{aligned}$$

Hence, from Theorem 12.6.3, we obtain the following result.

Lemma 12.6.5. *For $n \in \mathbb{N}_0$ and all $\xi, \eta \in \Omega$ we have*

$$\begin{aligned}
\text{trace } (\mathbf{p}_n^{(1,1)}(\xi,\eta)) &= P_n(\xi\cdot\eta)(\xi\cdot\eta),\\
\text{trace } (\mathbf{p}_n^{(1,2)}(\xi,\eta)) &= \text{trace } (\mathbf{p}_n^{(2,1)}(\xi,\eta)) = \frac{1}{\sqrt{n(n+1)}}P_n'(\xi\cdot\eta)(1-(\xi\cdot\eta)^2),\\
\text{trace } (\mathbf{p}_n^{(2,2)}(\xi\cdot\eta)) &= \frac{1}{n(n+1)}(P_n''(\xi\cdot\eta)(1-(\xi\cdot\eta)^2)(\xi\cdot\eta)+2P_n'(\xi\cdot\eta)),\\
\text{trace } (\mathbf{p}_n^{(3,3)}(\xi,\eta)) &= P_n(\xi\cdot\eta),\\
\text{trace } (\mathbf{p}_n^{(1,3)}(\xi,\eta)) &= \text{trace } (\mathbf{p}_n^{(3,1)}(\xi,\eta)) = \text{trace } (\mathbf{p}_n^{(2,3)}(\xi,\eta))\\
&= \text{trace } (\mathbf{p}_n^{(3,2)}(\xi,\eta)) = 0.
\end{aligned}$$

Taking $\xi = \eta$ in the last lemma we get in connection with Theorem 12.6.1

Lemma 12.6.6. *Let $n \in \mathbb{N}$ and $i \in \{1,2,3\}$. If $y_{n,j}^{(i)}, j = 1,\ldots,2n+1$, forms an $l^2(\Omega)$-orthonormal basis of $harm_n^{(i)}$, then*

$$\sum_{j=1}^{2n+1} |y_{n,j}^{(i)}(\xi)|^2 = \frac{2n+1}{4\pi}.$$

Every vector spherical harmonic $y_n^{(i)} \in harm_n^{(i)}$ of order n and type i can be represented by its orthogonal expansion

$$y_n^{(i)} = \sum_{j=1}^{2n+1} a_j y_{n,j}^{(i)}$$

with $a_j = (y_n^{(i)}, y_{n,j}^{(i)})_{l^2(\Omega)}$, $j = 1,\ldots,2n+1$. Application of the Cauchy–Schwarz inequality in combination with Lemma 12.6.6 yields the estimate

$$\begin{aligned}
|y_n^{(i)}(\xi)|^2 &\le \left(\sum_{j=1}^{2n+1} a_j^2\right)\left(\sum_{j=1}^{2n+1} |y_{n,j}^{(i)}(\xi)|^2\right) && (12.6.3)\\
&= \frac{2n+1}{4\pi}\sum_{j=1}^{2n+1} a_j^2 && (12.6.4)
\end{aligned}$$

for all $\xi \in \Omega$. Observing $\sum_{j=1}^{2n+1} a_j^2 = \|y_n^{(i)}\|_{l^2(\Omega)}^2$, we finally obtain

Lemma 12.6.7. *Let $y_n^{(i)} \in harm_n^{(i)}$. Then*

$$\|y_n^{(i)}\|_{c(\Omega)} \le \sqrt{\frac{2n+1}{4\pi}}\|y_n^{(i)}\|_{l^2(\Omega)}.$$

Our purpose now is to extend the operators $O^{(i)}$ also to rank-2 tensor fields. For sufficiently smooth $\mathbf{f} : \Omega \to \mathbb{R}^3 \otimes \mathbb{R}^3$ of the form

$$\mathbf{f}(\xi) = \sum_{j=1}^{3} \sum_{k=1}^{3} F_{jk}(\xi) \varepsilon^j \otimes \varepsilon^k$$

we let

$$O^{(i)}\mathbf{f}(\xi) = \sum_{k=1}^{3} O^{(i)} \left(\sum_{j=1}^{3} F_{jk}(\xi)\varepsilon^j \right) \varepsilon^k.$$

According to this definition it is clear that $\mathbf{p}_n^{(i,i)}(\cdot,\cdot)$ is the *reproducing kernel of $harm_n$* in the following sense:

(i) for all $\xi \in \Omega$

$$O_\xi^{(i)} \mathbf{p}_n^{(i,i)}(\xi, \cdot) \in harm_n^{(i)}$$

(ii) for every $f \in harm_n^{(i)}$ and all $\xi \in \Omega$

$$O_\xi^{(i)} f(\xi) = \left(O_\xi^{(i)} \mathbf{p}_n^{(i,i)}(\xi, \cdot), f \right)_{l^2(\Omega)}.$$

Let $a \in \mathbb{R}^3$ and $\eta \in \Omega$ be fixed. Then the vector field

$$\mathbf{p}_n^{(i,k)}(\cdot, \eta) \cdot a = \frac{4\pi}{2n+1} \sum_{j=1}^{2n+1} (y_{n,j}^{(k)}(\eta) \cdot a) y_{n,j}^{(i)}(\cdot)$$

is a vector spherical harmonic of order n and type i. Thus we obtain from (12.6.4) and Lemma 12.6.6

$$|\mathbf{p}_n^{(i,k)}(\xi, \eta) \cdot a|^2 \leq \frac{2n+1}{4\pi} \left(\frac{4\pi}{2n+1} \right)^2 \sum_{j=1}^{2n+1} (y_{n,j}^{(k)}(\xi) \cdot a)^2 \leq |a|^2 \qquad (12.6.5)$$

for all $\xi, \eta \in \Omega$. This leads us to the following.

Lemma 12.6.8. *Let $i, k, l \in \{1, 2, 3\}$. Then, for all $\xi, \eta \in \Omega$,*

$$|\mathbf{p}_n^{(i,k)}(\xi, \eta) \cdot \varepsilon^l| \leq 1.$$

Lemma 12.6.8 generalizes the estimate $|P_n(t)| \leq 1, t \in [-1, +1]$.

12.7 Vectorial Funk–Hecke Formulae

Next we are concerned with generalizations of the Funk–Hecke formula to the vectorial case. Two variants are considered in more detail:

(i) let $g(\cdot,\eta) : \Omega \to \mathbb{R}^3$ be a vector field which is invariant with respect to all orthogonal transformations $\mathbf{t} \in SO(3)$ leaving $\eta \in \Omega$ fixed (this statement will be clarified below). Determine

$$\int_\Omega g(\xi,\eta) \cdot y_n^{(i)}(\xi)\, d\omega(\xi), \tag{12.7.1}$$

where $y_n^{(i)} \in harm_n^{(i)}$.

(ii) let $G \in \mathcal{L}^1[-1,+1], \eta \in \Omega$ fixed. Determine

$$\int_\Omega G(\xi \cdot \eta) y_n^{(i)}(\xi)\, d\omega(\xi), \tag{12.7.2}$$

where $y_n^{(i)} \in harm_n^{(i)}$.

Notice that the integral (12.7.1) is scalar-valued, while (12.7.2) is vector-valued. This difference causes completely different ways of establishing the Funk–Hecke formulae. The first variant uses properties of invariant vector fields, while the second one is based on the cartesian representation of vector spherical harmonics (as developed in Section 12.3).

We start with some topics of representation theory needed for our studies on the Funk–Hecke formula (cf. Section 3.3).

Definition 12.7.1. Let $\mathbf{t} \in SO(3)$. We define the operator $R_\mathbf{t} : l^2(\Omega) \to l^2(\Omega)$ as follows: for $f \in l^2(\Omega)$, $R_\mathbf{t} f(\xi) = \mathbf{t}^T f(\mathbf{t}\xi)$, $\xi \in \Omega$.

Let $S \subset SO(3)$ be a subgroup of $SO(3)$. A subspace $v \subset l^2(\Omega)$ is called *invariant with respect to S* or simply *S-invariant* if $f \in v$ implies that $R_\mathbf{t} f \in v$ for all $\mathbf{t} \in S$ (cf. Section 12.5). If there exists no subspace of v (other than v itself) which is S-invariant, then v is said to be *irreducible.*

Let $f, g \in l^2(\Omega), \mathbf{t} \in SO(3)$. Then it follows that

$$(R_\mathbf{t} f, g)_{l^2(\Omega)} = (f, R_{\mathbf{t}^T} g)_{l^2(\Omega)}. \tag{12.7.3}$$

But this means that the adjoint operator of $R_\mathbf{t}$ is given by $R_{\mathbf{t}^T}$. Let $F \in \mathcal{C}^{(1)}(\Omega)$, $\mathbf{t} \in SO(3)$. Then we find

$$o_\xi^{(1)} R_\mathbf{t} F(\xi) = \xi F(\mathbf{t}\xi) \ = \mathbf{t}^T \mathbf{t}\xi F(\mathbf{t}\xi) \ = R_\mathbf{t} o^{(1)} F(\xi),$$

and

$$o_\xi^{(2)} R_{\mathbf{t}} F(\xi) = \nabla_\xi^* F(\mathbf{t}\xi) \ = \mathbf{t}^T(\nabla^* F)(\mathbf{t}\xi) \ = R_{\mathbf{t}}(o^{(2)} f)(\xi).$$

A similar result holds for $o^{(3)}$. Together with (12.7.3) we get for $F \in \mathcal{C}^{(1)}(\Omega)$ and $f \in c^{(1)}(\Omega)$ and any $\mathbf{t} \in SO(3)$ that for all $i \in \{1,2,3\}$ and $\xi \in \Omega$

$$o_\xi^{(i)} R_{\mathbf{t}} F(\xi) = R_{\mathbf{t}} o^{(i)} F(\xi) \tag{12.7.4}$$

and

$$O_\xi^{(i)} R_{\mathbf{t}} f(\xi) = R_{\mathbf{t}} O^{(i)} f(\xi). \tag{12.7.5}$$

Therefore we obtain in connection with the results of Section 12.5 and Theorem 3.3.2

Lemma 12.7.2.

(i) The space $l^2_{(i)}(\Omega)$ is $SO(3)$-invariant for all $i \in \{1,2,3\}$.

(ii) The set $harm_n^{(i)}$ is an $SO(3)$-invariant subspace of $l^2_{(i)}(\Omega)$ for all $i \in \{1,2,3\}$ and $n \geq 0_i$. Furthermore, $harm_n^{(i)}$ is irreducible.

Assume that $F \in \mathcal{L}^2(\Omega)$ with $R_{\mathbf{t}} F = F$ for all $\mathbf{t} \in SO_\eta(3)$ (cf. Section 3.3). Then we know that there exists a function $\tilde{F} \in \mathcal{L}^2[-1,1]$ such that $F(\xi) = \tilde{F}(\xi \cdot \eta)$, $\xi \in \Omega$. Furthermore we have shown in Theorem 3.3.1 that if F is, in addition, a spherical harmonic of order n, i.e. $F \in Harm_n$, there exists a constant $C \in \mathbb{R}$ such that

$$F(\xi) = C P_n(\xi \cdot \eta), \quad \xi \in \Omega.$$

A generalization of these results to the vectorial case is written down as follows

Lemma 12.7.3. *Let $\eta \in \Omega$ be fixed.*

(i) If $f \in c^{(1)}(\Omega)$ satisfies $R_{\mathbf{t}} f = f$ for all $\mathbf{t} \in SO_\eta(3)$ then there exist functions $F_i \in \mathcal{C}^{(0)}[-1,1]$, $i = 1,2,3$, such that

$$O_\xi^{(i)} f(\xi) = F_i(\xi \cdot \eta), \quad \xi \in \Omega.$$

(ii) Let $i \in \{1,2,3\}$ and $y_n^{(i)} \in harm_n^{(i)}$ with $R_{\mathbf{t}} y_n^{(i)} = y_n^{(i)}$ for all $\mathbf{t} \in SO_\eta(3)$. Then there exists a constant $C \in \mathbb{R}$ such that

$$y_n^{(i)}(\xi) = C o_\xi^{(i)} P_n(\xi \cdot \eta), \quad \xi \in \Omega.$$

Proof. It follows from $R_{\mathbf{t}}f = f$ and (12.7.5) that the (scalar-valued) function $O^{(i)}f$ satisfies $R_{\mathbf{t}}O^{(i)}f = O^{(i)}f$ for all $\mathbf{t} \in SO_\eta(3)$. Therefore, assertion (i) is clear.

In order to prove (ii), we remember that every $y_n^{(i)} \in harm_n^{(i)}$ can be written as $y_n^{(i)} = o^{(i)}Y_n$ with $Y_n \in Harm_n$. Hence, the second part follows from (12.7.4) and Theorem 3.3.1 . □

Let $\eta \in \Omega$ be fixed. Assume that $g(\cdot,\eta) \in c^{(1)}(\Omega)$ is a spherical vector field with $R_{\mathbf{t}}g(\xi,\eta) = g(\xi,\eta)$, $\xi \in \Omega$, for all $\mathbf{t} \in SO_\eta(3)$. Then it follows from the considerations above that for $i \in \{1,2,3\}$ the functions $O_\xi^{(i)}g(\xi,\eta) = G_i(\xi\cdot\eta)$ depend only on the inner product $\xi\cdot\eta$. Thus, we may define in analogy to Theorem 3.6.1

$$(O^{(i)}g)^\wedge(n) = 2\pi\int_{-1}^{1} G_i(t)P_n(t)\,dt. \tag{12.7.6}$$

It follows from (12.1.4) that $y_n^{(i)} = o^{(i)}Y_n \in harm_n^{(i)}$ satisfies

$$\begin{aligned}\int_\Omega g(\xi,\eta)\cdot y_n^{(i)}(\xi)\,d\omega(\xi) &= \int_\Omega O_\xi^{(i)}g(\xi,\eta)Y_n(\xi)\,d\omega(\xi)\\ &= (O^{(i)}g)^\wedge(n)Y_n(\eta).\end{aligned}$$

This proves the first variant of the vectorial Funk–Hecke formula.

Theorem 12.7.4. *Let $\eta \in \Omega$ be fixed. Assume that $g(\cdot,\eta) \in c^{(1)}(\Omega)$ satisfies $R_{\mathbf{t}}g(\xi,\eta) = g(\xi,\eta)$ for all $\mathbf{t} \in SO_\eta(3)$ and all $\xi \in \Omega$. Then, for $i \in \{1,2,3\}$ and $y_n^{(i)} \in harm_n^{(i)}, n \geq 0_i$, it follows that*

$$\int_\Omega g(\xi,\eta)\cdot y_n^{(i)}(\xi)\,d\omega(\xi) = (\mu_n^{(i)})^{-1}(O^{(i)}g)^\wedge(n)O_\eta^{(i)}y_n^{(i)}(\eta),$$

where $(O^{(i)}g)^\wedge(n)$ is defined by (12.7.6).

Using the addition theorem for vector spherical harmonics we obtain immediately the following consequence.

Corollary 12.7.5. *Let $\eta \in \Omega$ be fixed, $g(\cdot,\eta) \in c^{(1)}(\Omega)$. Assume that $R_{\mathbf{t}}g(\xi,\eta) = g(\xi,\eta)$ for all $\mathbf{t} \in SO_\eta(3)$ and all $\xi \in \Omega$. Then it holds for all $\zeta \in \Omega$ and $i \in \{1,2,3\}$*

$$\int_\Omega g(\xi,\eta)^T\mathbf{p}_n^{(i,i)}(\xi,\zeta)\,d\omega(\xi) = (\mu_n^{(i)})^{-1}(O^{(i)}g)^\wedge(n)\;o_\zeta^{(i)}P_n(\zeta\cdot\eta).$$

We come now to the second variant of the vectorial Funk–Hecke formula as announced in (12.7.2). The basic ideas to handle this problem are the representation of vector spherical harmonics by means of restrictions of homogeneous harmonic vector polynomials (cf. Section 12.3) and the componentwise application of the (scalar) Funk–Hecke formula.

Let $Y_n \in Harm_n$ be a spherical harmonic. Then we know from Lemma 12.3.3 that the cartesian components of the spherical vector field

$$\xi \mapsto k_n^{(i)} r^n Y_n(\xi)|_{r=1},\ \xi \in \Omega,$$

($i = 1, 2, 3$) are (scalar) spherical harmonics of order $\deg^{(i)}(n)$. Thus, we get immediately for $G \in \mathcal{L}^1[-1,1]$ and $\eta \in \Omega$

$$\int_\Omega G(\xi \cdot \eta) k_n^{(i)} r^n Y_n(\xi)|_{r=1} d\omega(\xi) = G^\wedge(\deg^{(i)}(n)) k_n^{(i)} r^n Y_n(\eta)|_{r=1}.$$

We know from formulae (12.3.1)–(12.3.2) how a vector spherical harmonic is expressible by restrictions of homogeneous harmonic vector polynomials. Combining these results we get the second variant of the vectorial Funk–Hecke formula:

Theorem 12.7.6. *Let $G \in \mathcal{L}^1[-1,+1]$. Assume that $Y_n \in Harm_n$. Then we have for all $\eta \in \Omega$*

$$\begin{aligned}
\int_\Omega G(\xi \cdot \eta) o_\xi^{(1)} Y_n(\xi)\, d\omega(\xi) &= G^\wedge_{(1,1)}(n) o_\eta^{(1)} Y_n(\eta) + G^\wedge_{(1,2)}(n) o_\eta^{(2)} Y_n(\eta), \\
\int_\Omega G(\xi \cdot \eta) o_\xi^{(2)} Y_n(\xi)\, d\omega(\xi) &= G^\wedge_{(2,1)}(n) o_\eta^{(1)} Y_n(\eta) + G^\wedge_{(2,2)}(n) o_\eta^{(2)} Y_n(\eta), \\
\int_\Omega G(\xi \cdot \eta) o_\xi^{(3)} Y_n(\xi)\, d\omega(\xi) &= G^\wedge_{(3,3)}(n) o_\eta^{(3)} Y_n(\eta),
\end{aligned}$$

where the numbers $G^\wedge_{(i,j)}(n)$ are given by

$$\begin{aligned}
G^\wedge_{(1,1)}(n) &= \frac{1}{2n+1}\left((n+1)G^\wedge(n+1) + nG^\wedge(n-1)\right), \\
G^\wedge_{(1,2)}(n) &= \frac{1}{2n+1}\left(G^\wedge(n-1) - G^\wedge(n+1)\right), \\
G^\wedge_{(2,1)}(n) &= \frac{n(n+1)}{2n+1}\left(G^\wedge(n-1) - G^\wedge(n+1)\right), \\
G^\wedge_{(2,2)}(n) &= \frac{1}{2n+1}\left(nG^\wedge(n+1) + (n+1)G^\wedge(n-1)\right), \\
G^\wedge_{(3,3)}(n) &= G^\wedge(n).
\end{aligned}$$

Notice that the space $l^2_{(3)}(\Omega)$ is invariant with respect to the defined integral operator, while $l^2_{(1)}(\Omega)$ and $l^2_{(2)}(\Omega)$ are not. However, it is clear that $l^2_{(1)}(\Omega) \oplus l^2_{(2)}(\Omega)$ is an invariant subspace of $l^2(\Omega)$.

Bibliographical Notes

Since the fifties a number of researchers have used spherical harmonics for a variety of vectorial problems (e.g. J. BLATT, V. WEISSKOFF (1952), I.M. GEL'FAND, Z.Y. SHAPIRO (1956), A.R. EDMONDS (1957), H.E. MOSES (1974) and many others). Unfortunately, the vector harmonics have not had any standard form. Each major school has invented its own notation and formalism (leaving us today with a very heterogeneous legacy). Moreover, the normalizations used in some of these formalisms are not very rational. An attempt at consolidating and reviewing the literature has been made by K.S. THORNE (1980). It seems to the authors that a number of formulae that previously were derived only under restrictive assumptions have much wider realms of validity by essentially using one system, the vector harmonics of P.M. MORSE, H. FESHBACH (1953). These vector harmonics are intimately related to the 'pure-spin vector harmonics' (see for example A.R. EDMONDS (1957)). Despite the fact that vector spherical harmonics have long been used in physical disciplines, they are only rarely found in the mathematical literature. They have been studied by M. LAGALLY, W. FRANZ (1964). However, basic topics were unknown. In the literature, we have no knowledge about the decomposition theorem (using Green's function with respect to the Beltrami operator), the addition theorem, and the Funk–Hecke formulae developed in Section 12.7. The treatment of these results here is based on T. GERVENS (1989) and W. FREEDEN, T. GERVENS (1991).

13

VECTORIAL APPROXIMATION METHODS

This chapter combines scalar constructive approximation theory with the vectorial concept of spherical harmonics. It turns out that the presented notation and basic settings lead in canonical way to vectorial generalizations of Sobolev spaces, pseudodifferential operators, radial basis functions, splines, singular integrals, wavelets, etc. It should be noted, however, that we have made no attempt to be comprehensive in the coverage of constructive vectorial approximation. Only the cornerstones are mentioned, to enable readers to extract more specialized information from the scalar part of this book for themselves.

13.1 Sobolev Spaces and Pseudodifferential Operators

Let $\{A_n\}_{n=0,1,\ldots}$ be a sequence consisting of non-zero elements A_n. For $i \in \{1,2,3\}$ we set

$$e^{(i)} = \left\{ f \in c^{(\infty)}_{(i)}(\Omega) \,\middle|\, \sum_{n=0_i}^{\infty} \sum_{j=1}^{2n+1} |A_n|^2 (f, y^{(i)}_{n,j})^2_{l^2(\Omega)} < \infty \right\}.$$

Imposing the inner product

$$(f,g)_{h^{(i)}(\{A_n\};\Omega)} = \sum_{n=0_i}^{\infty} \sum_{j=1}^{2n+1} |A_n|^2 (f, y^{(i)}_{n,j})_{l^2(\Omega)} (g, y^{(i)}_{n,j})_{l^2(\Omega)}, \quad f,g \in e^{(i)},$$

on $e^{(i)}$, the space $e^{(i)}$ becomes a pre-Hilbert space. The *Sobolev space* $h^{(i)}(\{A_n\};\Omega)$ is defined to be the completion

$$h^{(i)}(\{A_n\};\Omega) = \overline{e^{(i)}}^{\|\cdot\|_{h^{(i)}(\{A_n\};\Omega)}},$$

where $\|\cdot\|_{h^{(i)}(\{A_n\};\Omega)}$ is, as usual, the norm associated to $(\cdot,\cdot)_{h^{(i)}(\{A_n\};\Omega)}$. Obviously,

$$h^{(i)}(\{1\};\Omega) = l^2_{(i)}(\Omega), \quad i = 1,2,3.$$

If the sequence $\{A_n\}$ is $\left(n+\frac{1}{2}\right)^{0_i}$-summable of order $0_i - 1$, then the space $h^{(i)}(\{A_n\};\Omega)$ possesses a reproducing kernel $\mathbf{k}_{h^{(i)}(\{A_n\};\Omega)} : \Omega\times\Omega \to \mathbb{R}^3 \otimes \mathbb{R}^3$ in the following sense:

(i) for every fixed $\xi \in \Omega$, $O^{(i)}_\xi \mathbf{k}_{h^{(i)}(\{A_n\};\Omega)}(\xi,\cdot) \in h^{(i)}(\{A_n\};\Omega)$

(ii) the reproducing property

$$O^{(i)}_\xi f(\xi) = (O^{(i)}_\xi \mathbf{k}_{h^{(i)}(\{A_n\};\Omega)}(\xi,\cdot), f)_{h^{(i)}(\{A_n\};\Omega)}$$

holds for all $\xi \in \Omega$ and every $f \in h^{(i)}(\{A_n\};\Omega)$.

The addition theorem allows us to represent the reproducing kernel as series expansion of the form

$$\begin{aligned}
\mathbf{k}_{h^{(i)}(\{A_n\};\Omega)}(\xi,\eta) &= \sum_{n=0_i}^{\infty} \sum_{j=1}^{2n+1} \frac{1}{|A_n|^2} y^{(i)}_{n,j}(\xi) \otimes y^{(i)}_{n,j}(\eta) \\
&= \sum_{n=0_i}^{\infty} \frac{1}{|A_n|^2} \frac{2n+1}{4\pi} \mathbf{p}^{(i,i)}_n(\xi,\eta),
\end{aligned}$$

where $(\xi,\eta) \in \Omega \times \Omega$. Examples of sequences $\{A_n\}$ are known from the scalar theory (cf. Chapter 5). For the special case $A_n = (n+\frac{1}{2})^s$ we set

$$\begin{aligned}
h^{(i)}_s(\Omega) &= h^{(i)}\left(\{(n+\tfrac{1}{2})^s\};\Omega\right), \\
h_s(\Omega) &= \bigoplus_{i=1}^{3} h^{(i)}_s(\Omega).
\end{aligned}$$

In analogy to Lemma 5.2.2 it can be deduced that $h^{(i)}_s(\Omega) \subset c(\Omega)$ holds for $s > 1$.

For given $i \in \{1,2,3\}$, let $\{\lambda^{(i)\wedge}(n)\}_{n=0_i,0_i+1\ldots}$ be a sequence of real numbers satisfying

$$\lim_{n\to\infty} \frac{|\lambda^{(i)\wedge}(n)|}{(n+\frac{1}{2})^t} = const. \neq 0$$

for some $t \in \mathbb{R}$. Then we introduce the operator $\lambda^{(i)} : \mathcal{H}_s(\Omega) \to h^{(i)}_{s-t}(\Omega)$ by

$$\lambda^{(i)}(F) = \sum_{n=0_i}^{\infty} \sum_{j=1}^{2n+1} \lambda^{(i)\wedge}(n) F^\wedge(n,j) y^{(i)}_{n,j},$$

where we have assumed that $F \in \mathcal{H}_s(\Omega)$ has the series expansion (converging in $\mathcal{H}_s(\Omega)$-topology) $F = \sum_{n=0}^{\infty}\sum_{j=1}^{2n+1} F^\wedge(n,j) Y_{n,j}$. We call the operator $\lambda^{(i)}$ a *pseudodifferential operator of order t and type i*. Pseudodifferential operators of order $-\infty$ and type i are defined in analogy to the scalar case.

A sequence $\{(\Lambda^{(i)})^\wedge(n)\}_{n=0_i,0_i+1,\dots}$ with

$$\lim_{n\to\infty} \frac{|(\Lambda^{(i)})^\wedge(n)|}{(n+\frac{1}{2})^t} = const. \neq 0$$

defines a pseudodifferential operator $\Lambda^{(i)}$ of order t mapping $h_s^{(i)}(\Omega)$ into $\mathcal{H}_{s-t}(\Omega)$ by

$$\Lambda^{(i)}(f) = \sum_{n=0_i}^{\infty} \sum_{j=1}^{2n+1} (\Lambda^{(i)})^\wedge(n)(f^{(i)})^\wedge(n,j) Y_{n,j},$$

where $f \in h_s^{(i)}(\Omega)$, $i = 1,2,3$, is assumed to fulfil

$$f = \sum_{n=0_i}^{\infty} \sum_{j=1}^{2n+1} (f^{(i)})^\wedge(n,j) y_{n,j}^{(i)}$$

with

$$(f^{(i)})^\wedge(n,j) = (f, y_{n,j}^{(i)})_{l^2(\Omega)}$$

in $h_s^{(i)}(\Omega)$-topology. Again, pseudodifferential operators of order $-\infty$ can be defined in obvious way.

Using this terminology the operators $o^{(i)}$ and $O^{(i)}$ are found to be pseudodifferential operators of order 0 if $i = 1$ and order 1 if $i = 2,3$. In particular, we have

$$\begin{aligned} (O^{(1)})^\wedge(n) &= (o^{(1)})^\wedge(n) = 1, \\ (O^{(i)})^\wedge(n) &= (o^{(i)})^\wedge(n) = \sqrt{n(n+1)}, \quad i = 2,3. \end{aligned}$$

If $\{\Lambda^\wedge(n)\}_{n=0,1\dots}$ is the symbol of a pseudodifferential operator Λ of order t mapping $\mathcal{H}_s(\Omega)$ into $\mathcal{H}_{s-t}(\Omega)$, this operator can also be regarded as a mapping from $h_s(\Omega)$ into $h_{s-t}(\Omega)$ via the definition

$$\Lambda(f) = \sum_{i=1}^{3} \sum_{n=0_i}^{\infty} \sum_{j=1}^{2n+1} \Lambda^\wedge(n)(f^{(i)})^\wedge(n,j) y_{n,j}^{(i)}, \tag{13.1.1}$$

where $f \in h_s(\Omega)$ is given by

$$f = \sum_{i=1}^{3} \sum_{n=0_i}^{\infty} \sum_{j=1}^{2n+1} (f^{(i)})^\wedge(n,j) y_{n,j}^{(i)}.$$

It is clear, that the so defined operator Λ preserves the orthogonal decomposition $h_s(\Omega) = \bigoplus_{i=1}^{3} h_s^{(i)}(\Omega)$, since we have $\Lambda(h_s^{(i)}(\Omega)) \subset h_{s-t}^{(i)}(\Omega)$ for all

$i \in \{1, 2, 3\}$. An example of a pseudodifferential operator of this type is the vectorial Beltrami operator $\boldsymbol{\Delta}^*$. The symbol of $\boldsymbol{\Delta}^*$ is determined by $(\boldsymbol{\Delta}^*)^\wedge(n) = -n(n+1)$, and the order of $\boldsymbol{\Delta}^*$ is equal to 2.

The formalism described above allows the generalization of various approximation methods.

13.2 Vectorial Splines

Let us start our constructive vectorial approximation with a brief description of vector spherical spline theory. Starting point is an $\{(n+\frac{1}{2})^{0_i}\}$-summable sequence $\{A_n\}$, which implies the definition of a Hilbert space $h^{(i)}(\{A_n\};\Omega)$ possessing the reproducing kernel $\mathbf{k}_{h^{(i)}(\{A_n\};\Omega)}(\cdot,\cdot)$, as described above. As usual, we use the shorthand notation $h^{(i)} = h^{(i)}(\{A_n\};\Omega)$ and $\mathbf{k}_{h^{(i)}}(\cdot,\cdot) = \mathbf{k}_{h^{(i)}(\{A_n\};\Omega)}(\cdot,\cdot)$.

For $m \geq 0_i$, we denote by $h^{(i)}_{0_i,\ldots,m}$ the space of all vector spherical harmonics of order less than or equal to m and type i. We equip this space with the inner product

$$(f^{(i)}, g^{(i)})_{h^{(i)}_{0_i,\ldots,m}} = \sum_{n=0_i}^{m} \sum_{j=1}^{2n+1} |A_n|^2 (f^{(i)}, y^{(i)}_{n,j})_{l^2(\Omega)} (g^{(i)}, y^{(i)}_{n,j})_{l^2(\Omega)}.$$

This results canonically in a Hilbert space $h^{(i)}_{0_i,\ldots,m}$ with reproducing kernel

$$\mathbf{k}_{h^{(i)}_{0_i,\ldots,m}}(\xi,\eta) = \sum_{n=0_i}^{m} \sum_{j=1}^{2n+1} |A_n|^{-2} y^{(i)}_{n,j}(\xi) \otimes y^{(i)}_{n,j}(\eta), \quad (\xi,\eta) \in \Omega^2.$$

Let us denote by $h^{(i)\perp}_{0_i,\ldots,m}$ the orthogonal complement of $h^{(i)}_{0_i,\ldots,m}$ in $h^{(i)}$, that is to say $h^{(i)} = h^{(i)}_{0_i,\ldots,m} \oplus h^{(i)\perp}_{0_i,\ldots,m}$. As in the scalar case, the induced topology of $h^{(i)\perp}_{0_i,\ldots,m}$ is described by the inner product

$$(f^{(i)}, g^{(i)})_{h^{(i)\perp}_{0_i,\ldots,m}} = \sum_{n=m+1}^{\infty} \sum_{j=1}^{2n+1} |A_n|^2 (f^{(i)}, y^{(i)}_{n,j})_{l^2(\Omega)} (g^{(i)}, y^{(i)}_{n,j})_{l^2(\Omega)}.$$

It is clear that $\mathbf{k}_{h^{(i)\perp}_{0_i,\ldots,m}}(\cdot,\cdot)$ given by

$$\mathbf{k}_{h^{(i)\perp}_{0_i,\ldots,m}}(\xi,\eta) = \sum_{n=m+1}^{\infty} \sum_{j=1}^{2n+1} |A_n|^{-2} y^{(i)}_{n,j}(\xi) \otimes y^{(i)}_{n,j}(\eta)$$

is the reproducing kernel of $h^{(i)\perp}_{0_i,\ldots,m}$.

Let $X_N = \{\eta_1, \ldots, \eta_N\} \subset \Omega, N \geq \sum_{n=0_i}^{m}(2n+1)$, be an admissible system with respect to $h^{(i)}_{0_i,\ldots,m}$. Any function $s^{(i)}$ of the form

$$s^{(i)} = p^{(i)} + \sum_{k=1}^{N} a_k^{(i)} O_{\eta_k}^{(i)} \mathbf{k}_{h^{(i)\perp}_{0_i,\ldots,m}}(\eta_k, \cdot), \; N \geq \sum_{n=0_i}^{m} (2n+1),$$

is said to be a *vectorial spline of type i in $h^{(i)}$ relative to X_N* if $p^{(i)} \in \bigoplus_{n=0_i}^{m} harm_n^{(i)}$ and the coefficients $a_1^{(i)}, \ldots, a_N^{(i)}$ satisfy the linear system

$$\sum_{k=1}^{N} a_k^{(i)} O_{\eta_k}^{(i)} y_{n,j}^{(i)}(\eta_k) = 0, \quad n = 0_i, \ldots, m, \; j = 1, \ldots, 2n+1.$$

For the purpose of interpolation, assume now that there are given the data points $(\eta_k, y_k) \in \Omega \times \mathbb{R}, \; k = 1, \ldots, N$. We let

$$\mathcal{I}_N^{(i)}(y) = \{f \in h^{(i)} \mid O_{\eta_k}^{(i)} f(\eta_k) = y_k, k = 1, \ldots, N\}.$$

$\mathcal{I}_N^{(i)}(y)$ contains all vector fields in $h^{(i)}$ satisfying the interpolation constraints

$$O_{\eta_k}^{(i)} f(\eta_k) = y_k, \quad k = 1, \ldots, N.$$

The solution of the variational problem of finding the $\|\cdot\|_{h^{(i)\perp}_{0_i,\ldots,m}}$-smallest norm interpolant

$$\|s_N^{(i)}\|_{h^{(i)\perp}_{0_i,\ldots,m}} = \inf_{f \in \mathcal{I}_N^{(i)}(y)} \|f\|_{h^{(i)\perp}_{0_i,\ldots,m}}$$

is given by the spline

$$s_N^{(i)} = \sum_{n=0_i}^{m} \sum_{j=1}^{2n+1} c_{n,j}^{(i)} y_{n,j}^{(i)}(\cdot) + \sum_{k=1}^{N} a_k^{(i)} O_{\eta_k}^{(i)} \mathbf{k}_{h^{(i)\perp}_{0_i,\ldots,m}}(\eta_k, \cdot),$$

where the coefficients $c_{n,j}^{(i)}$ and $a_k^{(i)}$ solve the linear equations

$$\sum_{n=0_i}^{m} \sum_{j=1}^{2n+1} c_{n,j}^{(i)} O_{\eta_k}^{(i)} y_{n,j}^{(i)}(\eta_k) + \sum_{l=1}^{N} a_k^{(i)} O_{\eta_k}^{(i)} O_{\eta_l}^{(i)} \mathbf{k}_{h^{(i)\perp}_{0_i,\ldots,m}}(\eta_l, \eta_k) = y_k,$$
$$k = 1, \ldots, N,$$

and

$$\sum_{k=1}^{N} a_k^{(i)} O_{\eta_k}^{(i)} y_{n,j}^{(i)}(\eta_k) = 0,$$

$n = 0_i, \ldots, m, \; j = 1, \ldots, 2n+1.$

Obviously, this result can be extended to the space $\bigoplus_{i=1}^{3} h^{(i)}_{0_i,\ldots,m}$, where it is worth mentioning that the pointset X_N can be chosen independently of every space $h^{(i)}$, $i \in \{1,2,3\}$.

For results concerning convergence properties or remarks on possible kernel representations, the reader is referred to W. FREEDEN, T. GERVENS (1991).

13.3 Vectorial Wavelets

Let $\{K_h\} \subset \mathcal{L}^2[-1,1], h \in (-1,+1)$, be the kernel of a spherical singular integral $\{I_h\}$ in $\mathcal{L}^2(\Omega)$. Then I_h may be understood to be a pseudodifferential operator with symbol determined by

$$(I_h)^\wedge(n) = K_h^\wedge(n), \quad n = 0, 1, \ldots.$$

If $\{I_h\}$ is an approximate identity, then the limit relation

$$\lim_{\substack{h\to 1\\ h<1}} K_h^\wedge(n) = 1$$

holds for $n = 0,1,2,\ldots$. Thus, the pseudodifferential operator I_h maps $\mathcal{H}_s(\Omega)$ into $\mathcal{H}_s(\Omega)$. By use of the symbol $\{(I_h)^\wedge(n)\}$ we have found a way to interpret this operator as mapping from $h_s(\Omega)$ into $h_s(\Omega)$. Particularly, for $h_0(\Omega) = l^2(\Omega)$, this operator is given (provided that the decay of $(I_h)^\wedge(n)$ to 0 as $n \to \infty$ for fixed h is sufficiently fast) by convolution of f against the kernel

$$\mathbf{k}_h(\xi,\eta) = \sum_{i=1}^{3}\sum_{n=0_i}^{\infty}\sum_{j=1}^{2n+1} K_h^\wedge(n) y^{(i)}_{n,j}(\xi) \otimes y^{(i)}_{n,j}(\eta).$$

More precisely, for $f \in l^2(\Omega)$ we are able to introduce the operator I_h by

$$I_h(f) = \mathbf{k}_h * f = \int_\Omega \mathbf{k}_h(\cdot,\eta) f(\eta)\, d\omega(\eta).$$

$\{I_h\}, h \in [h_0, 1)$ is called a *singular integral.* According to our approach the kernel $\mathbf{k}_h(\cdot,\cdot)$ can be calculated as follows:

$$\mathbf{k}_h(\cdot,\cdot) = \sum_{i=1}^{3} \mathbf{k}_h^{(i,i)}(\cdot,\cdot),$$

where

$$\mathbf{k}_h^{(1,1)}(\xi,\eta) = o_\xi^{(1)} o_\eta^{(1)} K_h(\xi\cdot\eta)$$

and

$$\mathbf{k}_h^{(i,i)}(\xi,\eta) = o_\xi^{(i)} o_\eta^{(i)} \int_\Omega G(\Delta^*;\xi,\zeta) K_h(\zeta\cdot\eta)\, d\omega(\zeta), \quad i = 2,3.$$

For numerical purposes it is of advantage, when all the kernels $\mathbf{k}_h^{(i,i)}$, $i = 1,2,3$, are expressible in terms of elementary functions. Since the application of the $o^{(i)}$-operators can be performed without any difficulty, it is sufficient to require that both kernels

$$K_h(\xi\cdot\eta) = \sum_{n=0}^{\infty} \frac{2n+1}{4\pi} K_h^\wedge(n) P_n(\xi\cdot\eta) \tag{13.3.1}$$

and

$$\int_\Omega G(\Delta^*;\xi,\zeta) K_h(\zeta\cdot\eta)\, d\omega(\zeta) = \sum_{n=1}^{\infty} \frac{2n+1}{4\pi} \frac{1}{-n(n+1)} K_h^\wedge(n) P_n(\xi\cdot\eta) \tag{13.3.2}$$

are available in terms of elementary functions. Different types of kernels can be used. If, in particular, K_h is the Abel–Poisson kernel, then

$$\int_\Omega G(\Delta^*;\xi,\zeta) K_h(\zeta\cdot\eta)\, d\omega(\zeta) = \sum_{n=1}^{\infty} \frac{2n+1}{4\pi} \frac{1}{-n(n+1)} h^n P_n(\xi\cdot\eta).$$

The corresponding kernel can be immediately determined, since we have

$$\frac{1}{n(n+1)} \left(\frac{\partial}{\partial r} r^2 \frac{\partial}{\partial r}\right) r^n P_n(\xi\cdot\eta)|_{r=h} = h^n P_n(\xi\cdot\eta).$$

In other words, a wide variety of elementary kernels is readily at hand.

If K_h is an approximate identity in $\mathcal{L}^2(\Omega)$, it immediately follows that for all $f \in l^2(\Omega)$

$$\lim_{\substack{h\to 1\\ h<1}} \|I_h(f) - f\|_{l^2(\Omega)} = 0.$$

But this means that we have extended the notion of an approximate identity in a canonical way to spherical vector fields. Moreover, approximation techniques based on the property of being an approximate identity easily carry over to the vectorial case. For example, the formulation of a vectorial Toeplitz transform is straightforward. The details are omitted.

The generalization of the spherical Gabor transform to the vectorial case is even easier. Modulation has to be understood in terms of vector spherical harmonics instead of the scalar ones. To be more specific, if

$\{K_h\}$ is a (suitable) C-kernel on the interval $[-1,+1]$, we may define for a fixed number $h_0 \in (-1,1)$ and $f \in l^2(\Omega)$ the *vectorial Gabor transform* $(GT^{(i)})_{K_{h_0}}$ by

$$((GT^{(i)})_{K_{h_0}}(f))(n,j;\xi) = (K_{h_0}^{(2)}(1))^{-1/2} \int_\Omega K_{h_0}(\xi\cdot\eta) f(\eta)\cdot y_{n,j}^{(i)}(\eta)\, d\omega(\eta). \tag{13.3.3}$$

Details are left to the reader.

13.4 Scale Continuous Spherical Wavelets

Using the notion of an approximate identity for spherical vector fields we are also able to generalize the continuous wavelet transform to the vectorial framework. More explicitly, suppose that $\{\Psi_\rho\}, \rho \in (0,\infty)$, is a spherical wavelet of order 0. Replacing $\Psi_\rho(\xi\cdot\eta)$ by

$$\begin{aligned} \mathbf{\Psi}_\rho^{(1,1)}(\xi,\eta) &= o_\xi^{(1)} o_\eta^{(1)} \Psi_\rho(\xi\cdot\eta), \\ \mathbf{\Psi}_\rho^{(i,i)}(\xi,\eta) &= o_\xi^{(i)} o_\eta^{(i)} \int_\Omega G(\Delta^*;\xi,\rho)\Psi_\rho(\zeta\cdot\eta)\, d\omega(\zeta), \quad i=2,3, \end{aligned}$$

we end up with

$$(WT^{(i)})(f)(\rho;\xi) = \int_\Omega \mathbf{\Psi}_\rho^{(i,i)}(\xi,\eta) f(\eta)\, d\omega(\eta), \quad \xi\in\Omega,$$

which is called the *vectorial wavelet transform of order* 0 *and type* i. Proceeding as in Theorem 10.2.3, this transformation admits an inverse in the space of vector fields $f \in l^2_{(i)}(\Omega)$ (note that $\int_\Omega (f(\xi)\cdot\xi) d\omega(\xi) = 0$ must be required for the case $i=1$). In particular, for $f \in l^2_{(i)}(\Omega)$ (with $\int_\Omega (f(\xi)\cdot\xi)\, d\omega(\xi) = 0$ if $i=1$), it follows that

$$f = \int_\Omega \int_0^\infty \mathbf{\Psi}_\rho^{(i,i)}(\eta,\cdot)(WT^{(i)})(f)(\rho;\eta)\alpha(\rho)\, d\rho\, d\omega(\eta)$$

in the sense of $l^2(\Omega)$. Least squares properties may be guaranteed. In analogy to the results of Chapter 10 vectorial multiresolutions can be obtained by obvious arguments.

13.5 Scale Discrete Spherical Wavelets

The construction of a continuous wavelet transform for vector fields is based on the orthogonal decomposition $l^2(\Omega) = \bigoplus_{i=1}^3 l^2_{(i)}(\Omega)$. In a similar fashion, we are able to introduce scale discrete vectorial wavelets.

Let $\varphi_0 : [0,\infty) \to \mathbb{R}$ be a generator of a scaling function. We set

$$\begin{aligned}\boldsymbol{\varphi}_j^{(i,i)}(\xi,\eta) &= \sum_{n=0_i}^{\infty} \frac{2n+1}{4\pi} \sum_{k=1}^{2n+1} \varphi_j(n)\, y_{n,k}^{(i)}(\xi) \otimes y_{n,k}^{(i)}(\eta) \\ &= \sum_{n=0_i}^{\infty} \frac{2n+1}{4\pi} \varphi_j(n) \mathbf{p}_n^{(i,i)}(\xi,\eta), \quad \xi,\eta \in \Omega.\end{aligned}$$

This defines a scale discrete scaling function of type i. Obviously, the identity

$$\lim_{j\to\infty} \int_\Omega \boldsymbol{\varphi}_j^{(i,i)}(\cdot,\eta) f(\eta)\, d\omega(\eta) = f, \quad \xi \in \Omega,$$

holds for all $f \in l^2(\Omega)$, where the equality is understood in $l^2(\Omega)$-sense. Using the projection operators

$$p_j^{(i)}(f) = \int_\Omega \int_\Omega \boldsymbol{\varphi}_j^{(i,i)}(\cdot,\eta) \boldsymbol{\varphi}_j^{(i,i)}(\eta,\zeta) f(\zeta)\, d\omega(\zeta)\, d\omega(\eta), \quad f \in l^2(\Omega), i = 1,2,3,$$

we understand the scale spaces to be the image of $l^2(\Omega)$ under the operator $p_j^{(i)}$:

$$v_j^{(i)} = p_j^{(i)}(l^2(\Omega)) = \{p_j^{(i)}(f) \mid f \in l^2(\Omega)\}$$

yielding a scale discrete multiresolution analysis of $l^2_{(i)}(\Omega)$, as explained in the scalar case (cf. Theorem 11.1.9). The introduction of vectorial wavelets is based on the refinement equation

$$\tilde{\psi}_0(x)\psi_0(x) = (\varphi_0(x/2))^2 - (\varphi_0(x))^2, \quad x \in [0,\infty),$$

where we assume, as before, that both ψ_0 and $\tilde{\psi}_0$ satisfy the admissibility condition. Consequently, we obtain scale discrete vectorial wavelets (and corresponding dual wavelets)

$$\begin{aligned}\boldsymbol{\psi}_j^{(i,i)}(\xi,\eta) &= \sum_{n=0}^{\infty} \frac{2n+1}{4\pi} \psi_j(n) \mathbf{p}_n^{(i,i)}(\xi,\eta), \quad \xi,\eta \in \Omega, \\ \tilde{\boldsymbol{\psi}}_j^{(i,i)}(\xi,\eta) &= \sum_{n=0}^{\infty} \frac{2n+1}{4\pi} \tilde{\psi}_j(n) \mathbf{p}_n^{(i,i)}(\xi,\eta), \quad \xi,\eta \in \Omega.\end{aligned}$$

The vectorial wavelet transform at scale $j \in \mathbb{N}_0$, position $\eta \in \Omega$, and type i is then defined by

$$(WT)^{(i,i)}(f)(j;\xi) = \int_\Omega \boldsymbol{\psi}_j^{(i,i)}(\xi,\eta) f(\eta)\, d\omega(\eta), \quad f \in l^2(\Omega).$$

Introducing now the detail spaces $w_j^{(i)}$ in a canonical way we obtain

$$\begin{aligned} v_o^{(i)} + \sum_{j=0}^{J} w_j^{(i)} &= v_{J+1}^{(i)}, \\ v_J^{(i)} &= v_{J-1}^{(i)} + w_{J-1}^{(i)}, \end{aligned}$$

where, as before, the sum decomposition is in general neither direct nor orthogonal. A key result is the *reconstruction formula.* Let $\varphi_0, \psi_0, \tilde{\psi}_0$ be defined as before. Then, for $f \in l^2(\Omega)$, we have

$$\begin{aligned} f &= \sum_{i=1}^{3} \int_\Omega \int_\Omega \varphi_0^{(i,i)}(\cdot, \eta) \varphi_0^{(i,i)}(\eta, \zeta) f(\zeta)\, d\omega(\zeta)\, d\omega(\eta) \\ &\quad + \sum_{i=1}^{3} \sum_{j=0}^{\infty} \int_\Omega \tilde{\psi}_j(\cdot, \eta) (WT)^{(i,i)}(f)(j; \eta)\, d\omega(\eta), \end{aligned}$$

where the equality is understood in the $l^2(\Omega)$-sense.

There are many ways to construct concrete realizations of vectorial wavelets. As non-band-limited vectorial counterparts we mention rational and exponential wavelets. In the band-limited case vectorial Shannon and de la Vallé Poussin may be derived in obvious way.

Bibliographical Notes

The development of the addition theorem for vector spherical harmonics (T. GERVENS (1989)) and the Funk–Hecke formula (W. FREEDEN, T. GERVENS (1991)) have made possible a variety of constructive vectorial approximation methods. W. FREEDEN, T. GERVENS (1991) have studied vector spherical spline theory. The key ideas enable W. FREEDEN *et al.* (1990), J. BRUDOWSKY (1992), and J. BRUDOWSKY *et al.* (1992) to formulate variational principles for interpolating the displacement field of a homogeneous isotropic elastic body from discrete data. 'Cauchy–Navier splines' as proposed by W. FREEDEN (1990A), W. FREEDEN *et al.* (1996A) turned out to be successful in problems of deformation analysis of the earth's body. The formulation of vector spherical wavelet theory has not yet been published.

Part III

Tensors

14

TENSOR SPHERICAL HARMONICS

Many problems arising in geomathematics are of tensorial nature and spherical geometry. An important tool in dealing with such problems is the theory of tensor spherical harmonics. We present in this book a theory of tensor spherical harmonics generalizing in canonical way our approach to vector spherical harmonics. A striking point is that the tensor spherical harmonics are generated from the scalar ones by use of certain operators mapping scalar functions to tensor fields. As a matter of fact, these formulations (being as always independent of any choice of spherical coordinates) offer the possibility of extending all essential results known for scalar spherical harmonics to the tensorial case, including the definition of a tensorial Beltrami operator, the addition theorem, and tensorial versions of the Funk–Hecke formula. Furthermore, our appoach allows a straightforward transition of scalar approximation methods to the tensorial framework.

Among other areas of application (see for example R. Burridge (1969), R.W. James (1976), M.N. Jones (1980), K.S. Thorne (1980), and F.J. Zerilli (1970)) tensor spherical harmonics play an important role in diverse satellite problems of physical geodesy. Of current interest is the determination of the earth's gravitational field by satellite gradiometry, where tensor valued functionals of the potential are measured at satellite height, (see for example R. Rummel, M. van Gelderen (1992), M. Schreiner (1994), W. Freeden, M. Schreiner (1995c) and W. Freeden *et al.* (1997)).

The outline of the chapter is as follows. After some notation in Section 14.1 we introduce tensor spherical harmonics in Section 14.2. The interrelations between tensorial homogeneous harmonic polynomials and tensor spherical harmonics are discussed in Section 14.3. Based on Green's function with respect to the Beltrami operator, a decomposition theorem for spherical tensor fields is proved in Section 14.4. The tensor spherical harmonics are characterized as eigenfunctions of a tensorial analogue of the Beltrami operator (Section 14.5). Finally tensorial versions of the addition theorem and the Funk–Hecke formula are developed in Section 14.6 and Section 14.7, respectively.

14.1 Notation

As usual, a tensor of rank $k \in \mathbb{N}$ is understood to be an element of $\bigotimes_{l=1}^{k} \mathbb{R}^3$ (cf. M.E. Gurtin (1971)). Using the canonical orthonormal basis $\{\varepsilon^1, \varepsilon^2, \varepsilon^3\}$ of $\mathbb{R}^3$, a tensor $\mathbf{F}$ of rank k can be written as

$$\mathbf{F} = \sum_{i_1,\ldots,i_k=1}^{3} F_{i_1\ldots i_k} \varepsilon^{i_1} \otimes \ldots \otimes \varepsilon^{i_k}, \quad F_{i_1\ldots i_k} \in \mathbb{R}. \tag{14.1.1}$$

The scalar product $\mathbf{F} \cdot \mathbf{G}$ of two rank-k tensors F, G is defined by

$$\mathbf{F} \cdot \mathbf{G} = \sum_{i_1,\ldots,i_k=1}^{3} F_{i_1\ldots i_k}\, G_{i_1\ldots i_k}, \; F_{i_1\ldots i_k}, \quad G_{i_1\ldots i_k} \in \mathbb{R}, \tag{14.1.2}$$

hence, the modulus $|\mathbf{F}|$ of a rank-k tensor is given by

$$|\mathbf{F}| = \left(\sum_{i_1,\ldots,i_k=1}^{3} |F_{i_1\ldots i_k}|^2 \right)^{1/2}. \tag{14.1.3}$$

If

$$\mathbf{F} = \sum_{i_1,\ldots,i_k=1}^{3} F_{i_1\ldots i_k} \varepsilon^{i_1} \otimes \ldots \otimes \varepsilon^{i_k}$$

and

$$\mathbf{G} = \sum_{i_1,\ldots,i_l=1}^{3} G_{i_1\ldots i_l} \varepsilon^{i_1} \otimes \ldots \otimes \varepsilon^{i_l}$$

are a rank-k tensor and a rank-l tensor, respectively, $\mathbf{F} \otimes \mathbf{G}$ is the rank-$(k+l)$ tensor given by

$$\mathbf{F} \otimes \mathbf{G} = \sum_{i_1,\ldots,i_k=1}^{3} \sum_{j_1,\ldots,j_l=1}^{3} F_{i_1\ldots i_k}\, G_{j_1\ldots j_l} \varepsilon^{i_1} \otimes \ldots \otimes \varepsilon^{i_k} \otimes \varepsilon^{j_1} \otimes \ldots \otimes \varepsilon^{j_l}.$$

If $\mathbf{f}$ is a rank-2 tensor given by

$$\mathbf{f} = \sum_{i,k=1}^{3} F_{ik} \varepsilon^{i} \otimes \varepsilon^{k}, \tag{14.1.4}$$

its trace is defined by

$$\operatorname{trace} \mathbf{f} = \sum_{i=1}^{3} F_{ii}. \tag{14.1.5}$$

In what follows we restrict ourselves to tensor fields of rank 2. But it should be noted that most of our considerations carry over to the more general cases in an obvious way.

We define the *transpose* of $\mathbf{f}$ by

$$\mathbf{f}^T = \sum_{i,k=1}^{3} F_{ik}\varepsilon^k \otimes \varepsilon^i \tag{14.1.6}$$

and say $\mathbf{f}$ is *symmetric* if $\mathbf{f} = \mathbf{f}^T$ and *skew-symmetric* if $\mathbf{f} = -\mathbf{f}^T$. As it is well known, any tensor field $\mathbf{f}$ can be decomposed into a symmetric and skew-symmetric part: $\mathbf{f} = \text{sym}\,\mathbf{f} + \text{skew}\,\mathbf{f}$, where $\text{sym}\,\mathbf{f} = \frac{1}{2}(\mathbf{f} + \mathbf{f}^T)$ and $\text{skew}\,\mathbf{f} = \frac{1}{2}(\mathbf{f} - \mathbf{f}^T)$.

Clearly, a spherical rank-2 tensor field $\mathbf{f} : \Omega \to \mathbb{R}^3 \otimes \mathbb{R}^3$ can be represented in terms of its coordinate functions by means of functions: $F_{ik} : \Omega \to \mathbb{R}$ as follows:

$$\mathbf{f}(\xi) = \sum_{i,k=1}^{3} F_{ik}(\xi)\varepsilon^i \otimes \varepsilon^k, \quad \xi \in \Omega. \tag{14.1.7}$$

If $v = \sum_{i=1}^{3} V_i\varepsilon^i$ is a vector field, then the products $v^T\mathbf{f}$ and $\mathbf{f}v$ are defined by

$$v^T\mathbf{f} = \sum_{k=1}^{3}\sum_{i=1}^{3} V_i F_{ik}\varepsilon^k$$

and

$$\mathbf{f}v = \sum_{i=1}^{3}\sum_{k=1}^{3} F_{ik}V_k\varepsilon^i,$$

respectively.

The space of rank-2 tensors on the sphere with continuous coordinate functions is denoted by $\mathbf{c}(\Omega)$. Endowed with the norm

$$\|\mathbf{f}\|_{\mathbf{c}(\Omega)} = \sup_{\xi\in\Omega} |\mathbf{f}(\xi)| \tag{14.1.8}$$

the space $\mathbf{c}(\Omega)$ is a Banach space. Similarly, the space of all p times continuously differentiable tensor fields is denoted by $\mathbf{c}^{(p)}(\Omega)(0 \le p \le \infty)$. We denote by $\mathbf{l}^2(\Omega)$ the space of square-integrable tensor fields on the sphere Ω. Equipped with the inner product

$$(\mathbf{f}, \mathbf{g})_{\mathbf{l}^2(\Omega)} = \int_\Omega \mathbf{f}(\xi)\cdot\mathbf{g}(\xi)\, d\omega(\xi), \quad \mathbf{f}, \mathbf{g} \in \mathbf{l}^2(\Omega), \tag{14.1.9}$$

this space is a Hilbert space.

In analogy to (1.2.16) and (1.2.17) tensor fields can be separated into their tangential and normal parts. We set

$$\begin{aligned}
\mathbf{p}_{*,\mathrm{nor}}\mathbf{f} &= (\mathbf{f}\xi)\otimes\xi && (14.1.10)\\
\mathbf{p}_{*,\mathrm{tan}}\mathbf{f} &= \mathbf{f}-\mathbf{p}_{*,\mathrm{nor}}\mathbf{f} && (14.1.11)\\
\mathbf{p}_{\mathrm{nor},*}\mathbf{f} &= \xi\otimes(\xi^T\mathbf{f}) && (14.1.12)\\
\mathbf{p}_{\mathrm{nor},\mathrm{tan}}\mathbf{f} &= \mathbf{p}_{\mathrm{nor},*}(\mathbf{p}_{*,\mathrm{tan}}\mathbf{f}) = \mathbf{p}_{*,\mathrm{tan}}(\mathbf{p}_{\mathrm{nor},*}\mathbf{f}) && (14.1.13)\\
&= \xi\otimes(\xi^T\mathbf{f})-(\xi^T\mathbf{f}\xi)\xi\otimes\xi.
\end{aligned}$$

The operators $\mathbf{p}_{\mathrm{tan},*}$, $\mathbf{p}_{\mathrm{nor},\mathrm{nor}}$, $\mathbf{p}_{\mathrm{tan},\mathrm{nor}}$, and $\mathbf{p}_{\mathrm{tan},\mathrm{tan}}$ are defined analogously. A vector field $\mathbf{f}\in\mathbf{l}^2(\Omega)$ is called *normal* if $\mathbf{f}=\mathbf{p}_{\mathrm{nor},\mathrm{nor}}\mathbf{f}$ and *tangential* if $\mathbf{f}=\mathbf{p}_{\mathrm{tan},\mathrm{tan}}\mathbf{f}$. It is called *left normal* if $\mathbf{f}=\mathbf{p}_{\mathrm{nor},*}\mathbf{f}$, *left normal/right tangential* if $\mathbf{f}=\mathbf{p}_{\mathrm{nor},\mathrm{tan}}\mathbf{f}$, and so on. These operators admit the definition of the spaces $\mathbf{l}^2_{*,\mathrm{nor}}(\Omega),\mathbf{l}^2_{*,\mathrm{tan}}(\Omega),\ldots$, and $\mathbf{c}^{(p)}_{*,\mathrm{nor}}(\Omega),\mathbf{c}^{(p)}_{*,\mathrm{tan}}(\Omega)$, etc., as described for vector fields above. We end up with the following orthogonal decompositions:

$$\begin{aligned}
\mathbf{l}^2_{*,\mathrm{nor}}(\Omega) &= \mathbf{l}^2_{\mathrm{nor},\mathrm{nor}}(\Omega)\oplus\mathbf{l}^2_{\mathrm{tan},\mathrm{nor}}(\Omega), && (14.1.14)\\
\mathbf{l}^2_{*,\mathrm{tan}}(\Omega) &= \mathbf{l}^2_{\mathrm{nor},\mathrm{tan}}(\Omega)\oplus\mathbf{l}^2_{\mathrm{tan},\mathrm{tan}}(\Omega), && (14.1.15)\\
\mathbf{l}^2_{\mathrm{nor},*}(\Omega) &= \mathbf{l}^2_{\mathrm{nor},\mathrm{nor}}(\Omega)\oplus\mathbf{l}^2_{\mathrm{nor},\mathrm{tan}}(\Omega), && (14.1.16)\\
\mathbf{l}^2_{\mathrm{tan},*}(\Omega) &= \mathbf{l}^2_{\mathrm{tan},\mathrm{nor}}(\Omega)\oplus\mathbf{l}^2_{\mathrm{tan},\mathrm{tan}}(\Omega), && (14.1.17)\\
\mathbf{l}^2(\Omega) &= \mathbf{l}^2_{*,\mathrm{nor}}(\Omega)\oplus\mathbf{l}^2_{*,\mathrm{tan}}(\Omega), && (14.1.18)\\
\mathbf{l}^2(\Omega) &= \mathbf{l}^2_{\mathrm{nor},*}(\Omega)\oplus\mathbf{l}^2_{\mathrm{tan},*}(\Omega). && (14.1.19)
\end{aligned}$$

It is a well known fact (see for example M.E. Gurtin (1971)) that every rank-2 tensor field $\mathbf{f}:\Omega\to\mathbb{R}^3\otimes\mathbb{R}^3$ can be represented as a dyadic sum

$$\mathbf{f}(\xi)=\sum_{i=1}^{p}u_i(\xi)\otimes v_i(\xi),\quad \xi\in\Omega, \tag{14.1.20}$$

in terms of vector fields $u_i,v_i:\Omega\to\mathbb{R}^3$ (for example, take the coordinate expression (14.1.7)). It should be noted that neither the upper index p nor the vector fields u_i,v_i are uniquely defined in (14.1.20).

A left normal tensor $\mathbf{f}$ can be expressed as

$$\mathbf{f}(\xi)=\sum_{i=1}^{3}\xi\otimes v_i(\xi),\quad \xi\in\Omega, \tag{14.1.21}$$

by means of vector fields v_i. For this tensor it follows that

$$\xi^T \mathbf{f}(\xi) = \sum_{i=1}^{3} v_i(\xi), \quad \xi \in \Omega. \tag{14.1.22}$$

Similarly, a multiplication by ξ from the right of a right normal tensor field $\mathbf{f}$, given by

$$\mathbf{f}(\xi) = \sum_{i=1}^{3} u_i(\xi) \otimes \xi, \tag{14.1.23}$$

yields

$$\mathbf{f}(\xi)\xi = \sum_{i=1}^{3} u_i(\xi), \quad \xi \in \Omega. \tag{14.1.24}$$

As in the vectorial case, spherical tensor fields can be discussed in an elegant manner by the use of certain differential processes. Since we are mainly interested in presenting the underlying formalism, we will not spend much effort to formulate our results in their most general setting.

Let $u \in c^{(1)}(\Omega)$ be a vector field given in its coordinate form by

$$u(\xi) = \sum_{i=1}^{3} U_i(\xi)\varepsilon^i, \quad \xi \in \Omega, \quad U_i \in \mathcal{C}^{(1)}(\Omega). \tag{14.1.25}$$

Then we define the operators $\nabla^* \otimes$ and $L^* \otimes$ by

$$\nabla^*_\xi \otimes u(\xi) = \sum_{i=1}^{3} (\nabla^*_\xi U_i(\xi)) \otimes \varepsilon^i, \quad \xi \in \Omega, \tag{14.1.26}$$

$$L^*_\xi \otimes u(\xi) = \sum_{i=1}^{3} (L^*_\xi U_i(\xi)) \otimes \varepsilon^i, \quad \xi \in \Omega. \tag{14.1.27}$$

Clearly, $\nabla^* \otimes u$ and $L^* \otimes u$ are left tangential. But it is an important fact, that even if u is tangential, the tensor fields $\nabla^* \otimes u$ and $L^* \otimes u$ are generally not tangential. It is obvious, that the *product rule* is valid. To be specific, let $F \in \mathcal{C}^{(1)}(\Omega)$ and $u \in c^{(1)}(\Omega)$, then

$$\nabla^*_\xi \otimes (F(\xi)u(\xi)) = \nabla^*_\xi F(\xi) \otimes u(\xi) + F(\xi)\nabla^*_\xi \otimes u(\xi), \quad \xi \in \Omega. \tag{14.1.28}$$

In order to simplify our calculations, we first apply the operators $\nabla^* \otimes$ and $L^* \otimes$ to the local orthonormal triad defined in (1.2.25)–(1.2.27). It follows that for $\varphi \in (0, 2\pi), t \in (-1, 1)$

$$\begin{aligned}
\nabla^* \otimes \varepsilon^r(\varphi, t) &= \varepsilon^\varphi(\varphi, t) \otimes \varepsilon^\varphi(\varphi, t) + \varepsilon^t(\varphi, t) \otimes \varepsilon^t(\varphi, t), \\
\nabla^* \otimes \varepsilon^\varphi(\varphi, t) &= -\varepsilon^\varphi(\varphi, t) \otimes \varepsilon^r(\varphi, t) + \frac{t}{\sqrt{1 - t^2}} \varepsilon^\varphi(\varphi, t) \otimes \varepsilon^t(\varphi, t),
\end{aligned}$$

$$\nabla^* \otimes \varepsilon^t(\varphi,t) = -\varepsilon^t(\varphi,t)\otimes\varepsilon^r(\varphi,t) - \frac{t}{\sqrt{1-t^2}}\varepsilon^\varphi(\varphi,t)\otimes\varepsilon^\varphi(\varphi,t),$$

and

$$\begin{aligned} L^*\otimes\varepsilon^r(\varphi,t) &= \varepsilon^t(\varphi,t)\otimes\varepsilon^\varphi(\varphi,t) - \varepsilon^\varphi(\varphi,t)\otimes\varepsilon^t(\varphi,t), \\ L^*\otimes\varepsilon^\varphi(\varphi,t) &= -\varepsilon^t(\varphi,t)\otimes\varepsilon^r(\varphi,t) + \frac{t}{\sqrt{1-t^2}}\varepsilon^t(\varphi,t)\otimes\varepsilon^t(\varphi,t), \\ L^*\otimes\varepsilon^t(\varphi,t) &= \varepsilon^\varphi(\varphi,t)\otimes\varepsilon^r(\varphi,t) - \frac{t}{\sqrt{1-t^2}}\varepsilon^t(\varphi,t)\otimes\varepsilon^\varphi(\varphi,t). \end{aligned}$$

The product rule (14.1.28) then yields for $u \in c^{(1)}(\Omega)$

$$\nabla^*\otimes u = \mathbf{p}_{\tan,*}\nabla^*\otimes u, \tag{14.1.29}$$
$$L^*\otimes u = \mathbf{p}_{\tan,*}L^*\otimes u. \tag{14.1.30}$$

Furthermore, we have for $u \in c^{(1)}_{\tan}(\Omega)$ and $\xi \in \Omega$

$$\mathbf{p}_{\tan,\mathrm{nor}}\nabla^*_\xi\otimes u(\xi) = -u(\xi)\otimes\xi, \tag{14.1.31}$$
$$\mathbf{p}_{\tan,\mathrm{nor}}L^*_\xi\otimes u(\xi) = -\xi\wedge u(\xi)\otimes\xi. \tag{14.1.32}$$

For the normal vector field $\xi \mapsto F(\xi)\xi$, $\xi\in\Omega$, with $F \in \mathcal{C}^{(1)}(\Omega)$ we obtain

$$\nabla^*_\xi\otimes F(\xi)\xi = F(\xi)\mathbf{i}_{\tan}(\xi) + \nabla^*_\xi F(\xi)\otimes\xi, \tag{14.1.33}$$
$$L^*_\xi\otimes F(\xi)\xi = F(\xi)\mathbf{j}_{\tan}(\xi) + L^*_\xi F(\xi)\otimes\xi. \tag{14.1.34}$$

After these considerations on possible generalizations of the surface gradient operators we shall next investigate analogues to the operators $\nabla^*\cdot$ and $L^*\cdot$ applied to tensor fields.

Let $\mathbf{f}\in\mathbf{c}^{(1)}_{\tan,*}(\Omega)$ be a left tangential rank-2 tensor field given by

$$\mathbf{f}(\xi) = \sum_{i,k=1}^{3} F_{ik}(\xi)\varepsilon^i\otimes\varepsilon^k, \quad F_{ik}\in\mathcal{C}^{(1)}(\Omega),\ i,k=1,2,3. \tag{14.1.35}$$

Then we set

$$\begin{aligned} \nabla^*_\xi\cdot\mathbf{f}(\xi) &= \sum_{k=1}^{3}\left(\nabla^*_\xi\cdot\left(\sum_{i=1}^{3}F_{ik}(\xi)\varepsilon^i\right)\right)\varepsilon^k, \\ L^*_\xi\cdot\mathbf{f}(\xi) &= \sum_{k=1}^{3}\left(L^*_\xi\cdot\left(\sum_{i=1}^{3}F_{ik}(\xi)\varepsilon^i\right)\right)\varepsilon^k. \end{aligned} \tag{14.1.36}$$

Moreover, for $u \in c^{(1)}_{\tan}(\Omega)$ and $v \in c^{(1)}(\Omega)$ with

$$u = \sum_{i=1}^{3} U_i \varepsilon^i, \quad v = \sum_{i=1}^{3} V_i \varepsilon^i,$$

and $U_i, V_i \in \mathcal{C}^{(1)}(\Omega)$ for $i = 1, 2, 3$ we get

$$\begin{aligned}
\nabla^* \cdot (u \otimes v) &= \nabla^* \cdot \left(\sum_{i,k=1}^{3} U_i V_k \varepsilon^i \otimes \varepsilon^k \right) \\
&= \sum_{k=1}^{3} \left(\nabla^* \cdot \left(\sum_{i=1}^{3} U_i V_k \varepsilon^i \right) \varepsilon^k \right) \\
&= \sum_{k=1}^{3} \left(\nabla^* \cdot \left(\sum_{i=1}^{3} U_i \varepsilon^i \right) \right) V_k \varepsilon^k + \sum_{k=1}^{3} (u \cdot \nabla^* V_k) \varepsilon^k \\
&= (\nabla^* \cdot u) v + u \cdot (\nabla^* \otimes v).
\end{aligned}$$

Thus we end up with the *product rule*

$$\nabla^* \cdot (u \otimes v) = (\nabla^* \cdot u) v + u \cdot (\nabla^* \otimes v). \tag{14.1.37}$$

Analogously,

$$L^* \cdot (u \otimes v) = (L^* \cdot u) v + u \cdot (L^* \otimes v). \tag{14.1.38}$$

Furthermore, for $F \in \mathcal{C}^{(1)}(\Omega)$ and $\mathbf{f} \in \mathbf{c}^{(1)}_{\tan,*}(\Omega)$, the result is

$$\nabla^*_\xi \cdot (F(\xi) \mathbf{f}(\xi)) = (\nabla^*_\xi F(\xi)) \cdot \mathbf{f}(\xi) + F(\xi) \nabla^*_\xi \cdot \mathbf{f}(\xi), \ \xi \in \Omega, \tag{14.1.39}$$

and a similar formula is valid for L^*. The surface theorem of Gauß yields the following result.

Lemma 14.1.1. *Let $f \in c^{(1)}(\Omega)$ and $\mathbf{g} \in \mathbf{c}^{(1)}_{\tan,*}(\Omega)$. Then*

$$\begin{aligned}
\int_\Omega (\nabla^*_\xi \otimes f(\xi)) \cdot \mathbf{g}(\xi) \, d\omega(\xi) &= -\int_\Omega f(\xi) \cdot (\nabla^*_\xi \cdot \mathbf{g}(\xi)) \, d\omega(\xi), \\
\int_\Omega (L^*_\xi \otimes f(\xi)) \cdot \mathbf{g}(\xi) \, d\omega(\xi) &= -\int_\Omega f(\xi) \cdot (L^*_\xi \cdot \mathbf{g}(\xi)) \, d\omega(\xi).
\end{aligned}$$

Proof. We only prove the first formula. Let

$$f = \sum_{i=1}^{3} F_i \varepsilon^i, \quad \mathbf{g} = \sum_{i,k=1}^{3} G_{ik} \varepsilon^i \otimes \varepsilon^k.$$

Then

$$(\nabla^* \otimes f) \cdot \mathbf{g} = \left(\sum_{i=1}^{3} \nabla^* F_i \otimes \varepsilon^i \right) \cdot \left(\sum_{i,k=1}^{3} G_{ki} \varepsilon^k \otimes \varepsilon^i \right)$$

$$= \left(\sum_{i=1}^{3} \nabla^* F_i\right) \cdot \left(\sum_{k=1}^{3} G_{ki}\varepsilon^k\right)$$

and

$$\begin{aligned} f \cdot (\nabla^* \cdot \mathbf{g}) &= \left(\sum_{i=1}^{3} F_i \varepsilon^i\right) \cdot \left(\sum_{i,k=1}^{3} \left(\nabla^* \cdot \left(\sum_{k=1}^{3} G_{ki}\varepsilon^k\right)\right) \varepsilon^i\right) \\ &= \left(\sum_{i=1}^{3} F_i \nabla^*\right) \cdot \left(\sum_{k=1}^{3} G_{ki}\varepsilon^k\right). \end{aligned}$$

Thus, the assertion follows from the surface theorem of Gauß. □

Again, we write down the effect of the operators $\nabla^*\cdot$ and $L^*\cdot$ on tensors expressed locally with the help of the triad $\varepsilon^r, \varepsilon^\varphi, \varepsilon^t$. In particular, we get for $\varphi \in (0, 2\pi)$ and $t \in (-1, 1)$

$$\begin{aligned} \nabla^* \cdot (\varepsilon^\varphi(\varphi,t) \otimes \varepsilon^r(\varphi,t)) &= \varepsilon^\varphi(\varphi,t), \\ \nabla^* \cdot (\varepsilon^\varphi(\varphi,t) \otimes \varepsilon^\varphi(\varphi,t)) &= -\varepsilon^r(\varphi,t) + \frac{t}{\sqrt{1-t^2}}\varepsilon^t(\varphi,t), \\ \nabla^* \cdot (\varepsilon^\varphi(\varphi,t) \otimes \varepsilon^t(\varphi,t)) &= -\frac{t}{\sqrt{1-t^2}}\varepsilon^\varphi(\varphi,t), \end{aligned}$$

$$\begin{aligned} \nabla^* \cdot (\varepsilon^t(\varphi,t) \otimes \varepsilon^r(\varphi,t)) &= -\frac{t}{\sqrt{1-t^2}}\varepsilon^r(\varphi,t) + \varepsilon^t(\varphi,t), \\ \nabla^* \cdot (\varepsilon^t(\varphi,t) \otimes \varepsilon^\varphi(\varphi,t)) &= -\frac{t}{\sqrt{1-t^2}}\varepsilon^\varphi(\varphi,t), \\ \nabla^* \cdot (\varepsilon^t(\varphi,t) \otimes \varepsilon^t(\varphi,t)) &= -\varepsilon^r(\varphi,t) - \frac{t}{\sqrt{1-t^2}}\varepsilon^t, \end{aligned}$$

and

$$\begin{aligned} L^* \cdot (\varepsilon^\varphi(\varphi,t) \otimes \varepsilon^r(\varphi,t)) &= -\frac{t}{\sqrt{1-t^2}}\varepsilon^r(\varphi,t) - \varepsilon^t(\varphi,t), \\ L^* \cdot (\varepsilon^\varphi(\varphi,t) \otimes \varepsilon^\varphi(\varphi,t)) &= -\frac{t}{\sqrt{1-t^2}}\varepsilon^\varphi(\varphi,t), \\ L^* \cdot (\varepsilon^\varphi(\varphi,t) \otimes \varepsilon^t(\varphi,t)) &= -\varepsilon^r(\varphi,t) + \frac{t}{\sqrt{1-t^2}}\varepsilon^t(\varphi,t), \end{aligned}$$

$$L^* \cdot (\varepsilon^t(\varphi,t) \otimes \varepsilon^r(\varphi,t)) = -\varepsilon^\varphi(\varphi,t),$$

$$\begin{aligned} L^* \cdot (\varepsilon^t(\varphi,t) \otimes \varepsilon^\varphi(\varphi,t)) &= -\varepsilon^r(\varphi,t) + \frac{t}{\sqrt{1-t^2}}\varepsilon^t(\varphi,t), \\ L^* \cdot (\varepsilon^t(\varphi,t) \otimes \varepsilon^t(\varphi,t)) &= -\frac{t}{\sqrt{1-t^2}}\varepsilon^\varphi(\varphi,t). \end{aligned}$$

Since the tensors $\mathbf{i}_{\tan}$ and $\mathbf{j}_{\tan}$ are locally given by

$$\begin{aligned} \mathbf{i}_{\tan}(\varphi,t) &= \varepsilon^\varphi(\varphi,t) \otimes \varepsilon^\varphi(\varphi,t) + \varepsilon^t(\varphi,t) \otimes \varepsilon^t(\varphi,t), \\ \mathbf{j}_{\tan}(\varphi,t) &= \varepsilon^t(\varphi,t) \otimes \varepsilon^\varphi(\varphi,t) - \varepsilon^\varphi(\varphi,t) \otimes \varepsilon^t(\varphi,t), \end{aligned}$$

we obtain for $\xi \in \Omega$

$$\begin{aligned} \nabla^*_\xi \cdot \mathbf{i}_{\tan}(\xi) &= -2\xi, \\ L^*_\xi \cdot \mathbf{i}_{\tan}(\xi) &= 0, \\ \nabla^*_\xi \cdot \mathbf{j}_{\tan}(\xi) &= 0, \\ L^*_\xi \cdot \mathbf{j}_{\tan}(\xi) &= -2\xi. \end{aligned}$$

Using the product rule (14.1.39) and its counterpart for $L^*\cdot$, the above listed formulae help us to evaluate $\nabla^*\cdot$ and $L^*\cdot$ applied to tensor fields in local representation with respect to $\varepsilon^r, \varepsilon^\varphi$, and ε^t.

Certain combinations of the operators defined before turn out to be of particular importance for the definition of tensor spherical harmonics. In order to motivate this we introduce mappings $\mathbf{q}^{(i,k)} : \mathcal{C}^{(2)}(\Omega) \to \mathbf{c}^{(0)}(\Omega)$ as follows:

$$\mathbf{q}^{(1,1)}_\xi F(\xi) = \xi \otimes \xi F(\xi), \tag{14.1.40}$$

$$\mathbf{q}^{(1,2)}_\xi F(\xi) = \xi \otimes \nabla^*_\xi F(\xi), \tag{14.1.41}$$

$$\mathbf{q}^{(1,3)}_\xi F(\xi) = \xi \otimes L^*_\xi F(\xi), \tag{14.1.42}$$

$$\mathbf{q}^{(2,1)}_\xi F(\xi) = \nabla^*_\xi F(\xi) \otimes \xi, \tag{14.1.43}$$

$$\mathbf{q}^{(3,1)}_\xi F(\xi) = L^*_\xi F(\xi) \otimes \xi, \tag{14.1.44}$$

$$\mathbf{q}^{(2,2)}_\xi F(\xi) = \nabla^*_\xi \otimes \nabla^*_\xi F(\xi), \tag{14.1.45}$$

$$\mathbf{q}^{(2,3)}_\xi F(\xi) = \nabla^*_\xi \otimes L^* F(\xi), \tag{14.1.46}$$

$$\mathbf{q}^{(3,2)}_\xi F(\xi) = L^*_\xi \otimes \nabla^*_\xi F(\xi), \tag{14.1.47}$$

$$\mathbf{q}^{(3,3)}_\xi F(\xi) = L^*_\xi \otimes L^*_\xi F(\xi). \tag{14.1.48}$$

It is clear from the considerations before that $\mathbf{q}^{(1,1)}F$ are left normal/right normal, $\mathbf{q}^{(1,k)}F$ are left normal/right tangential ($k = 2, 3$), and $\mathbf{q}^{(i,1)}F$ are left tangential/right normal ($i = 2, 3$). Furthermore, the tensor fields

$\mathbf{q}^{(i,k)}F, i,k = 2,3$ are left tangential, but in general not right tangential. In particular, (14.1.31) and (14.1.32) show that

$$\begin{aligned}
\mathbf{p}_{*,\mathrm{nor}}\nabla^*_\xi \otimes \nabla^*_\xi F(\xi) &= -\nabla^*_\xi F(\xi) \otimes \xi, &(14.1.49)\\
\mathbf{p}_{*,\mathrm{nor}}\nabla^*_\xi \otimes L^*_\xi F(\xi) &= -L^*_\xi F(\xi) \otimes \xi, &(14.1.50)\\
\mathbf{p}_{*,\mathrm{nor}}L^*_\xi \otimes \nabla^*_\xi F(\xi) &= -L^*_\xi F(\xi) \otimes \xi, &(14.1.51)\\
\mathbf{p}_{*,\mathrm{nor}}L^*_\xi \otimes L^*_\xi F(\xi) &= \nabla^*_\xi F(\xi) \otimes \xi. &(14.1.52)
\end{aligned}$$

Thus, certain combinations of the $\mathbf{q}^{(i,k)}$-operators allow a separation into normal and tangential tensor fields.

The trace of $\mathbf{q}^{(i,k)}F$ is given by

$$\begin{aligned}
\text{trace } \mathbf{q}^{(1,1)}_\xi F(\xi) &= F(\xi),\\
\text{trace } \mathbf{q}^{(2,2)}_\xi F(\xi) &= \text{trace } \mathbf{q}^{(3,3)}_\xi F(\xi) = \Delta^*_\xi F(\xi),
\end{aligned}$$

and

$$\text{trace } \mathbf{q}^{(i,k)}_\xi F(\xi) = 0,$$

if $(i,k) \in \{(1,2),(1,3),(2,1),(3,1),(2,3),(3,2)\}$.

Next we are interested in the adjoint operators of $\mathbf{q}^{(i,k)}$ (denoted by $Q^{(i,k)}$). To be more explicit, we are looking for operators $Q^{(i,k)}$ satisfying

$$(\mathbf{q}^{(i,k)}F, \mathbf{f})_{\mathrm{l}^2(\Omega)} = (F, Q^{(i,k)}\mathbf{f})_{\mathcal{L}^2(\Omega)} \tag{14.1.53}$$

for all $(i,k) \in \{(1,1),\ldots,(3,3)\}$ and all sufficiently smooth functions $F : \Omega \to \mathbb{R}$ and tensor fields $\mathbf{f} : \Omega \to \mathbb{R}^3 \otimes \mathbb{R}^3$. It can be deduced from Lemma 14.1.1 that the operators $Q^{(i,k)}$ may be written down in the following form:

$$\begin{aligned}
Q^{(1,1)}_\xi \mathbf{f}(\xi) &= \xi^T \mathbf{f}(\xi)\xi, &(14.1.54)\\
Q^{(1,2)}_\xi \mathbf{f}(\xi) &= -\nabla^*_\xi \cdot p_{\mathrm{tan}}(\xi^T \mathbf{f}(\xi)), &(14.1.55)\\
Q^{(1,3)}_\xi \mathbf{f}(\xi) &= -L^*_\xi \cdot p_{\mathrm{tan}}(\xi^T \mathbf{f}(\xi)), &(14.1.56)\\
Q^{(2,1)}_\xi \mathbf{f}(\xi) &= -\nabla^*_\xi \cdot p_{\mathrm{tan}}(\mathbf{f}(\xi)\xi), &(14.1.57)\\
Q^{(3,1)}_\xi \mathbf{f}(\xi) &= -L^*_\xi \cdot p_{\mathrm{tan}}(\mathbf{f}(\xi)\xi), &(14.1.58)\\
Q^{(2,2)}_\xi \mathbf{f}(\xi) &= \nabla^*_\xi \cdot p_{\mathrm{tan}}(\nabla^*_\xi \cdot \mathbf{p}_{\mathrm{tan},*}\mathbf{f}(\xi)), &(14.1.59)\\
Q^{(2,3)}_\xi \mathbf{f}(\xi) &= L^*_\xi \cdot p_{\mathrm{tan}}(\nabla^*_\xi \cdot \mathbf{p}_{\mathrm{tan},*}\mathbf{f}(\xi)), &(14.1.60)\\
Q^{(3,2)}_\xi \mathbf{f}(\xi) &= \nabla^*_\xi \cdot p_{\mathrm{tan}}(L^*_\xi \cdot \mathbf{p}_{\mathrm{tan},*}\mathbf{f}(\xi)), &(14.1.61)
\end{aligned}$$

$$Q_\xi^{(3,3)}\mathbf{f}(\xi) = L_\xi^* \cdot p_{\tan}(L_\xi^* \cdot \mathbf{p}_{\tan,*}\mathbf{f}(\xi)), \tag{14.1.62}$$

provided that $\mathbf{f} \in \mathbf{c}^{(2)}(\Omega)$. Let $F : \Omega \to \mathbb{R}$ be sufficiently smooth, then

$$\begin{array}{rcll}
Q^{(1,1)}\mathbf{q}^{(1,1)}F &=& F, & \\
Q^{(i,k)}\mathbf{q}^{(i,k)}F &=& -\Delta^*F & \text{if } (i,k) \in \{(1,2),(1,3),(2,1),(3,1)\}, \\
Q^{(i,k)}\mathbf{q}^{(i,k)}F &=& \Delta^*\Delta^*F & \text{if } (i,k) \in \{(2,2),(2,3),(3,2),(3,3)\}, \\
Q^{(1,1)}\mathbf{q}^{(i,k)}F &=& 0 & \text{if } (i,k) \neq (1,1), \\
Q^{(1,2)}\mathbf{q}^{(i,k)}F &=& 0 & \text{if } (i,k) \neq (1,2), \\
Q^{(1,3)}\mathbf{q}^{(i,k)}F &=& 0 & \text{if } (i,k) \neq (1,3), \\
Q^{(2,1)}\mathbf{q}^{(i,k)}F &=& 0 & \text{if } (i,k) \notin \{(2,1),(2,2),(3,3)\}, \\
Q^{(2,1)}\mathbf{q}^{(2,2)}F &=& \Delta^*F, & \\
Q^{(2,1)}\mathbf{q}^{(3,3)}F &=& -\Delta^*F, & \\
Q^{(3,1)}\mathbf{q}^{(i,k)}F &=& 0 & \text{if } (i,k) \notin \{(3,1),(2,3),(3,2)\}, \\
Q^{(3,1)}\mathbf{q}^{(2,3)}F &=& \Delta^*F, & \\
Q^{(3,1)}\mathbf{q}^{(3,2)}F &=& \Delta^*F, & \\
Q^{(2,2)}\mathbf{q}^{(i,k)}F &=& 0 & \text{if } (i,k) \notin \{(2,1),(2,2)\}, \\
Q^{(2,2)}\mathbf{q}^{(2,1)}F &=& \Delta^*F, & \\
Q^{(2,3)}\mathbf{q}^{(i,k)}F &=& 0 & \text{if } (i,k) \notin \{(3,1),(2,3)\}, \\
Q^{(2,3)}\mathbf{q}^{(3,1)}F &=& \Delta^*F, & \\
Q^{(3,2)}\mathbf{q}^{(i,k)}F &=& 0 & \text{if } (i,k) \notin \{(3,1),(3,2)\}, \\
Q^{(3,2)}\mathbf{q}^{(3,1)}F &=& \Delta^*F, & \\
Q^{(3,3)}\mathbf{q}^{(i,k)}F &=& 0 & \text{if } (i,k) \notin \{(2,1),(3,3)\}, \\
Q^{(3,3)}\mathbf{q}^{(2,1)}F &=& -\Delta^*F. &
\end{array}$$

It is remarkable to mention that, contrary to the vectorial case, 'constant' tangential tensors can be detected, i.e. tensor fields $\mathbf{f} \in \mathbf{c}^{(1)}_{\tan,\tan}(\Omega)$ satisfying $p_{\tan}\nabla^* \cdot \mathbf{f} = 0$ and $p_{\tan}L^* \cdot \mathbf{f} = 0$. Indeed, both tensor fields can be characterized as linear combinations of $\mathbf{i}_{\tan}$ and $\mathbf{j}_{\tan}$.

Important relations for $F \in \mathcal{C}^{(2)}(\Omega)$ are

$$(\nabla^* \otimes \nabla^* + L^* \otimes L^*)F = \mathbf{i}_{\tan}\Delta^*F, \tag{14.1.63}$$

$$(L^* \otimes \nabla^* - \nabla^* \otimes L^*)F = \mathbf{j}_{\tan}\Delta^*F. \tag{14.1.64}$$

In view of the above equations and definitions we finally introduce operators $\mathbf{o}^{(i,k)} : \mathcal{C}^{(2)}(\Omega) \to \mathbf{c}^{(0)}(\Omega)$ by

$$\begin{array}{rcl}
\mathbf{o}_\xi^{(1,1)}F(\xi) &=& \xi \otimes \xi F(\xi), \\
\mathbf{o}_\xi^{(1,2)}F(\xi) &=& \xi \otimes \nabla_\xi^* F(\xi), \\
\mathbf{o}_\xi^{(1,3)}F(\xi) &=& \xi \otimes L_\xi^* F(\xi),
\end{array}$$

$$\begin{aligned}
\mathbf{o}_\xi^{(2,1)} F(\xi) &= \nabla_\xi^* F(\xi) \otimes \xi, \\
\mathbf{o}_\xi^{(3,1)} F(\xi) &= L_\xi^* F(\xi) \otimes \xi, \\
\mathbf{o}_\xi^{(2,2)} F(\xi) &= \mathbf{i}_{\tan}(\xi) F(\xi), \\
\mathbf{o}_\xi^{(2,3)} F(\xi) &= \left(\nabla_\xi^* \otimes \nabla_\xi^* - L_\xi^* \otimes L_\xi^*\right) F(\xi) + 2\nabla_\xi^* F(\xi) \otimes \xi, \\
\mathbf{o}_\xi^{(3,2)} F(\xi) &= \left(\nabla_\xi^* \otimes L_\xi^* + L_\xi^* \otimes \nabla_\xi^*\right) F(\xi) + 2L_\xi^* F(\xi) \otimes \xi, \\
\mathbf{o}_\xi^{(3,3)} F(\xi) &= \mathbf{j}_{\tan}(\xi) F(\xi).
\end{aligned}$$

After our above considerations it is not difficult to prove the following.

Lemma 14.1.2. *Let $F : \Omega \to \mathbb{R}$ be sufficiently smooth. Then the following statements are valid:*

(i) $\mathbf{o}^{(1,1)} F$ *is a normal tensor field.*

(ii) $\mathbf{o}^{(1,2)} F$ *and* $\mathbf{o}^{(1,3)} F$ *are left normal/right tangential.*

(iii) $\mathbf{o}^{(2,1)} F$ *and* $\mathbf{o}^{(3,1)} F$ *are left tangential/right normal.*

(iv) $\mathbf{o}^{(2,2)} F$, $\mathbf{o}^{(2,3)} F$, $\mathbf{o}^{(3,2)} F$ *and* $\mathbf{o}^{(3,3)} F$ *are tangential.*

(v) $\mathbf{o}^{(1,1)} F$, $\mathbf{o}^{(2,2)} F$, $\mathbf{o}^{(2,3)} F$ *and* $\mathbf{o}^{(3,2)} F$ *are symmetric.*

(vi) $\mathbf{o}^{(3,3)} F$ *is skew-symmetric.*

(vii) $\left(\mathbf{o}^{(1,2)} F\right)^T = \mathbf{o}^{(2,1)} F$ *and* $\left(\mathbf{o}^{(1,3)} F\right)^T = \mathbf{o}^{(3,1)} F$.

(viii) For $\xi \in \Omega$

$$\operatorname{trace} \mathbf{o}_\xi^{(i,k)} F(\xi) = \begin{cases} F(\xi) & \text{for} \quad (i,k) = (1,1) \\ 2F(\xi) & \text{for} \quad (i,k) = (2,2) \\ 0 & \text{for} \quad (i,k) \neq (1,1),\ (2,2) \end{cases} .$$

The tangent representation theorem (cf. G.E. Backus (1966, 1967)) asserts that if $\mathbf{p}_{\tan,\tan}\mathbf{f}$ is the tangential part of a tensor field $\mathbf{f} \in \mathbf{c}^{(2)}(\Omega)$, as defined above, there exist unique scalar fields $F_{2,2}, F_{3,3}, F_{2,3}, F_{3,2}$ such that

$$\int_\Omega F_{2,2}(\xi)\, d\omega(\xi) = \int_\Omega F_{3,3}(\xi)\, d\omega(\xi) = 0,$$

$$\int_\Omega F_{3,2}(\xi)(\varepsilon^i \cdot \xi)\, d\omega(\xi) = \int_\Omega F_{2,3}(\xi)(\varepsilon^i \cdot \xi)\, d\omega(\xi) = 0, \ i = 1,2,3,$$

and

$$\mathbf{p}_{\tan,\tan}\mathbf{f} = \mathbf{o}^{(2,2)} F_{2,2} + \mathbf{o}^{(2,3)} F_{2,3} + \mathbf{o}^{(3,2)} F_{3,2} + \mathbf{o}^{(3,3)} F_{3,3}.$$

Furthermore, the following orthogonality relations may be formulated.

Lemma 14.1.3. *Let $F, G : \Omega \to \mathbb{R}$ be sufficiently smooth. Then the following properties hold true:*

(i) *For all $\xi \in \Omega$, $\mathbf{o}_\xi^{(i,k)} F(\xi) \cdot \mathbf{o}_\xi^{(i',k')} F(\xi) = 0$ whenever $(i,k) \neq (i',k')$.*

(ii) *If $(F,G)_{\mathcal{L}^2(\Omega)} = 0$, then $(\mathbf{o}^{(i,k)} F, \mathbf{o}^{(i',k')} G)_{\mathbf{l}^2(\Omega)} = 0$ for all $(i,k), (i',k') \in \{(1,1),(1,2),\ldots,(3,3)\}$.*

As a matter of fact, the analogy to Lemma 12.1.1 justifies our definition of the operators $\mathbf{o}^{(i,k)}$.

The adjoint operators $O^{(i,k)}$ satisfying

$$(\mathbf{o}^{(i,k)} F, \mathbf{f})_{\mathbf{l}^2(\Omega)} = (F, O^{(i,k)} \mathbf{f})_{\mathcal{L}^2(\Omega)}$$

for all sufficiently smooth functions $F : \Omega \to \mathbb{R}$ and tensor fields $\mathbf{f} : \Omega \to \mathbb{R}^3 \otimes \mathbb{R}^3$ can be deduced from the definitions (14.1.54)–(14.1.62). For $\mathbf{f} \in \mathbf{c}^{(2)}(\Omega)$ we have

$$O_\xi^{(1,1)} \mathbf{f}(\xi) = \xi^T \mathbf{f}(\xi) \xi, \tag{14.1.65}$$

$$O_\xi^{(1,2)} \mathbf{f}(\xi) = -\nabla_\xi^* \cdot p_{\tan} \left(\xi^T \mathbf{f}(\xi) \right), \tag{14.1.66}$$

$$O_\xi^{(1,3)} \mathbf{f}(\xi) = -L_\xi^* \cdot p_{\tan} \left(\xi^T \mathbf{f}(\xi) \right), \tag{14.1.67}$$

$$O_\xi^{(2,1)} \mathbf{f}(\xi) = -\nabla_\xi^* \cdot p_{\tan} \left(\mathbf{f}(\xi) \xi \right), \tag{14.1.68}$$

$$O_\xi^{(3,1)} \mathbf{f}(\xi) = -L_\xi^* \cdot p_{\tan} \left(\mathbf{f}(\xi) \xi \right), \tag{14.1.69}$$

$$O_\xi^{(2,2)} \mathbf{f}(\xi) = \mathbf{i}_{\tan}(\xi) \cdot \mathbf{f}(\xi), \tag{14.1.70}$$

$$\begin{aligned} O_\xi^{(2,3)} \mathbf{f}(\xi) = {} & \nabla_\xi^* \cdot p_{\tan} \left(\nabla_\xi^* \cdot \mathbf{p}_{\tan,*} \mathbf{f}(\xi) \right) - L_\xi^* \cdot p_{\tan} \left(L_\xi^* \cdot \mathbf{p}_{\tan,*} \mathbf{f}(\xi) \right) \\ & -2\nabla_\xi^* \cdot p_{\tan} \left(\mathbf{f}(\xi) \xi \right), \end{aligned} \tag{14.1.71}$$

$$\begin{aligned} O_\xi^{(3,2)} \mathbf{f}(\xi) = {} & L_\xi^* \cdot p_{\tan} \left(\nabla_\xi^* \cdot \mathbf{p}_{\tan,*} \mathbf{f}(\xi) \right) + \nabla_\xi^* \cdot p_{\tan} \left(L_\xi^* \cdot \mathbf{p}_{\tan,*} \mathbf{f}(\xi) \right) \\ & -2L_\xi^* \cdot p_{\tan} \left(\mathbf{f}(\xi) \xi \right), \end{aligned} \tag{14.1.72}$$

$$O_\xi^{(3,3)} \mathbf{f}(\xi) = \mathbf{j}_{\tan}(\xi) \cdot \mathbf{f}(\xi). \tag{14.1.73}$$

Provided that $F : \Omega \to \mathbb{R}$ is sufficiently smooth we find

$$O_\xi^{(i',k')} \mathbf{o}_\xi^{(i,k)} F(\xi) = 0 \quad \text{if } (i,k) \neq (i',k'), \tag{14.1.74}$$

while

$$O_\xi^{(i,k)} \mathbf{o}_\xi^{(i,k)} F(\xi) = \begin{cases} F(\xi) & \text{if } (i,k) = (1,1) \\ -\Delta^* F(\xi) & \text{if } (i,k) \in \{(1,2),(1,3) \\ & \qquad (2,1),(3,1)\} \\ 2F(\xi) & \text{if } (i,k) \in \{(2,2),(3,3)\} \\ 2\Delta^*(\Delta^* + 2) F(\xi) & \text{if } (i,k) \in \{(2,3),(3,2)\} \end{cases}. \tag{14.1.75}$$

14.2 Definition of Tensor Spherical Harmonics

Our appoach is based on the notations introduced in Section 14.1. The operators $\mathbf{o}^{(i,k)}$ will help us to construct tensor spherical harmonics which correspond to the orthogonal decomposition of $\mathbf{l}^2(\Omega)$.

It can be easily deduced that for $Y \in Harm_0$

$$\mathbf{o}^{(i,k)}Y = 0 \quad \text{if } (i,k) \in \{(1,2),(1,3),(2,1),(3,1)\}, \tag{14.2.1}$$

while for $Y \in Harm_0 \cup Harm_1$ we find

$$\mathbf{o}^{(i,k)}Y = 0 \quad \text{if } (i,k) \in \{(2,3),(3,2)\}.$$

Thus, for notational convenience, we introduce

$$0_{i,k} = \begin{cases} 0 & \text{for} \quad (i,k) \in \{(1,1),(2,2),(3,3)\} \\ 1 & \text{for} \quad (i,k) \in \{(1,2),(1,3),(2,1),(3,1)\} \\ 2 & \text{for} \quad (i,k) \in \{(2,3),(3,2)\} \end{cases} . \tag{14.2.2}$$

This enables us to define the tensor spherical harmonics as follows.

Definition 14.2.1. Let $(i,k) \in \{(1,1),(1,2),\ldots,(3,3)\}$. We call any tensor field $\mathbf{o}^{(i,k)}Y_n$, $n \geq 0_{i,k}$, with $Y_n \in Harm_n$, *tensor spherical harmonic of order n and type (i,k)*. The space of all tensor spherical harmonics of order n and type (i,k) is denoted by $\mathbf{harm}_n^{(i,k)}$. Furthermore, we set

$$\begin{aligned} \mathbf{harm}_0 &= \mathbf{harm}_0^{(1,1)} \oplus \mathbf{harm}_0^{(2,2)} \oplus \mathbf{harm}_0^{(3,3)}, \\ \mathbf{harm}_1 &= \bigoplus_{\substack{i,k=1 \\ (i,k)\neq(2,3),(3,2)}}^{3} \mathbf{harm}_1^{(i,k)}, \\ \mathbf{harm}_n &= \bigoplus_{i,k=1}^{3} \mathbf{harm}_n^{(i,k)}, \quad \text{for } n \geq 2. \end{aligned}$$

Next we introduce an $\mathbf{l}^2(\Omega)$-orthonormal system of tensor spherical harmonics for $(i,k) \in \{(1,1),(1,2),\ldots,(3,3)\}$, $n \geq 0_{i,k}$, $j = 1,\ldots,2n+1$, by letting

$$\mathbf{y}_{n,j}^{(i,k)} = (\mu_n^{(i,k)})^{-1/2}\mathbf{o}^{(i,k)}Y_{n,j}, \tag{14.2.3}$$

where $\mu_n^{(i,k)}$ is defined by

$$\mu_n^{(i,k)} = \left\| O^{(i,k)}\mathbf{o}^{(i,k)}Y_{n,j} \right\|_{\mathcal{L}^2(\Omega)},$$

that is

$$\mu_n^{(i,k)} = \begin{cases} 1 & \text{if} \quad (i,k) = (1,1) \\ 2 & \text{if} \quad (i,k) \in \{(2,2),(3,3)\} \\ -\Delta^{*\wedge}(n) & \text{if} \quad (i,k) \in \\ & \quad \{(1,2),(1,3),(2,1),(3,1)\} \\ 2\Delta^{*\wedge}(n)\left(\Delta^{*\wedge}(n) - \Delta^{*\wedge}(1)\right) & \text{if} \quad (i,k) \in \{(2,3),(3,2)\} \end{cases}.$$

As a matter of fact, it is not difficult to see that $\left\{\mathbf{y}_{n,j}^{(i,k)}\right\}$ forms an orthonormal system in $\mathbf{l}^2(\Omega)$. We want to show in the next section that this system is complete in $\mathbf{l}^2(\Omega)$ with respect to $(\cdot,\cdot)_{\mathbf{l}^2(\Omega)}$ and closed in $\mathbf{c}(\Omega)$ with respect to $\|\cdot\|_{\mathbf{c}(\Omega)}$.

14.3 Homogeneous Harmonic Tensor Polynomials

We start with the definition of homogeneous harmonic tensor polynomials of rank 2.

Definition 14.3.1. A tensor field $\mathbf{h}_n : \mathbb{R}^3 \to \mathbb{R}^3 \otimes \mathbb{R}^3$, $n \geq 0$, of the form

$$\mathbf{h}_n(x) = \sum_{i,k=1}^{3} H_n^{ik}(x)\varepsilon^i \otimes \varepsilon^k, \quad x \in \mathbb{R}^3,$$

is called *homogeneous harmonic tensor polynomial of degree* n, if every H_n^{ik} is a scalar homogeneous harmonic polynomial of degree n.

The space of all homogeneous harmonic tensor polynomials of degree n, when restricted to the unit sphere Ω, is denoted by $\mathbf{hom}_n$. It is clear that

$$\mathbf{hom}_n = \operatorname{span}\left\{Y_{n,j}\varepsilon^i \otimes \varepsilon^k\right\}_{\substack{i,k=1,2,3 \\ j=1,\ldots,2n+1}},$$

provided that $\{Y_{n,j}\}_{n=0,1,\ldots,\ j=1,\ldots,2n+1}$ is an $\mathcal{L}^2(\Omega)$-orthonormal system.

As immediate consequences of the Corollaries 3.4.3 and 3.4.5 we obtain the following.

Lemma 14.3.2. *The following statements are valid:*

(i) The system $\left\{\varepsilon^i \otimes \varepsilon^k Y_{n,j}\right\}_{\substack{i,k=1,2,3 \\ n=0,1,\ldots,\ j=1,\ldots,2n+1}}$ *is complete in* $\mathbf{l}^2(\Omega)$.

(ii) The system $\{\varepsilon^i \otimes \varepsilon^k Y_{n,j}\}_{\substack{i,k=1,2,3 \\ n=0,1,\ldots,\ j=1,\ldots,2n+1}}$ *is closed in* $\mathbf{c}(\Omega)$ *with respect to* $\|\cdot\|_{\mathbf{c}(\Omega)}$.

In what follows we are interested in the relations between tensor spherical harmonics on the one hand side and homogeneous harmonic tensor polynomials restricted to the unit sphere on the other hand side. To this end we consider operators $\mathbf{k}_n^{(i,k)}$ of the form

$$\begin{aligned}
\mathbf{k}_n^{(1,1)} F(x) &= \left((2n+3)x - |x|^2\nabla_x\right) \otimes \left((2n+1)x - |x|^2\nabla_x\right) F(x), \\
\mathbf{k}_n^{(1,2)} F(x) &= \left((2n-1)x - |x|^2\nabla_x\right) \otimes \nabla_x F(x), \\
\mathbf{k}_n^{(1,3)} F(x) &= \left((2n+1)x - |x|^2\nabla_x\right) \otimes (x \wedge \nabla_x) F(x), \\
\mathbf{k}_n^{(2,1)} F(x) &= \nabla_x \otimes \left((2n+1)x - |x|^2\nabla_x\right) F(x), \\
\mathbf{k}_n^{(2,2)} F(x) &= \nabla_x \otimes \nabla_x F(x), \\
\mathbf{k}_n^{(2,3)} F(x) &= \nabla_x \otimes (x \wedge \nabla_x) F(x), \\
\mathbf{k}_n^{(3,1)} F(x) &= (x \wedge \nabla_x) \otimes \left((2n+1)x - |x|^2\nabla_x\right) F(x), \\
\mathbf{k}_n^{(3,2)} F(x) &= (x \wedge \nabla_x) \otimes \nabla_x F(x), \\
\mathbf{k}_n^{(3,3)} F(x) &= (x \wedge \nabla_x) \otimes (x \wedge \nabla_x) F(x)
\end{aligned}$$

for $x \in \mathbb{R}^3$ and sufficiently smooth function $F : \mathbb{R}^3 \to \mathbb{R}$. Simple calculations in cartesian coordinates lead us in straightforward way to the following result.

Lemma 14.3.3. *Let $H_n, n \in \mathbb{N}_0$, be a homogeneous harmonic polynomial of degree n. Then $\mathbf{k}_n^{(i,k)} H_n$ is a homogeneous harmonic tensor polynomial of degree $\deg^{(i,k)}(n)$, where we use the abbreviation*

$$\deg^{(i,k)}(n) = \begin{cases} n-2 & \textit{for} \quad (i,k) = (2,2) \\ n-1 & \textit{for} \quad (i,k) \in \{(2,3),(3,2)\} \\ n & \textit{for} \quad (i,k) \in \{(1,2),(2,1),(3,3)\} \\ n+1 & \textit{for} \quad (i,k) \in \{(1,3),(3,1)\} \\ n+2 & \textit{for} \quad (i,k) = (1,1) \end{cases} .$$

$\deg^{(i,k)}(n) < 0$ means that $\mathbf{k}_n^{(i,k)} H_n = 0$.

The gradient of a sufficiently smooth function $F : \mathbb{R}^3 \setminus \{0\} \to \mathbb{R}$ is known to be equal to (cf. (1.2.31))

$$\nabla_x F(x) = \xi \frac{\partial}{\partial r} F(r\xi) + \frac{1}{r} \nabla_\xi^* F(r\xi), \qquad x = r\xi,\ r > 0,\ \xi \in \Omega.$$

Similarly, if $f : \mathbb{R}^3 \setminus \{0\} \to \mathbb{R}^3$ is a sufficiently smooth vector field of the form

$$f(x) = \sum_{i=1}^{3} F_i(x)\varepsilon^i, \qquad |x| > 0,$$

we have for $r > 0$, $\xi \in \Omega$,

$$\nabla_x \otimes f(x) \quad = \quad \sum_{i=1}^{3} \nabla_x F_i(x) \otimes \varepsilon^i$$

$$
\begin{aligned}
&= \sum_{i=1}^{3} \left(\xi \frac{\partial}{\partial r} F_i(r\xi) + \frac{1}{r} \nabla_\xi^* F_i(r\xi) \right) \otimes \varepsilon^i \\
&= \xi \otimes \frac{\partial}{\partial r} f(r\xi) + \frac{1}{r} \nabla_\xi^* \otimes f(r\xi).
\end{aligned}
$$

Thus we get for $Y_n \in Harm_n$

$$
\begin{aligned}
&\mathbf{k}_n^{(1,1)} r^n Y_n(\xi) \\
&\quad= \left((2n+3)r\xi - r^2\nabla_x\right) \otimes \left((2n+1)r\xi - r^2\nabla_x\right) r^n Y_n(\xi) \\
&\quad= \left((2n+3)r\xi - r^2\nabla_x\right) \otimes \left((2n+1)r^{n+1}\xi Y_n(\xi) - nr^{n+1}\xi Y_n(\xi)\right. \\
&\qquad\left. -r^{n+1}\nabla_\xi^* Y_n(\xi)\right) \\
&\quad= \left((2n+3)r\xi - r^2\nabla_x\right) \otimes \left((n+1)r^{n+1}\xi Y_n(\xi) - r^{n+1}\nabla_\xi^* Y_n(\xi)\right) \\
&\quad= (2n+3)(n+1)r^{n+2}\xi \otimes \xi Y_n(\xi) - (2n+3)r^{n+2}\xi \otimes \nabla_\xi^* Y_n(\xi) \\
&\qquad -(n+1)r^{n+2}\nabla_\xi^* \otimes \xi Y_n(\xi) + r^{n+2}\nabla_\xi^* \otimes \nabla_\xi^* Y_n(\xi) \\
&\qquad -(n+1)r^2\xi \otimes \frac{\partial}{\partial r} r^{n+1}\xi Y_n(\xi) + r^2\xi \otimes \frac{\partial}{\partial r} r^{n+1}\nabla_\xi^* Y_n(\xi) \\
&\quad= (n+2)(n+1)r^{n+2}\xi \otimes \xi Y_n(\xi) - (n+2)r^{n+2}\xi \otimes \nabla_\xi^* Y_n(\xi) \\
&\qquad -(n+1)r^{n+2}\nabla_\xi^* Y_n(\xi) \otimes \xi - (n+1)r^{n+2}\mathbf{i}_{\tan}(\xi) Y_n(\xi) \\
&\qquad +r^{n+2}\nabla_\xi^* \otimes \nabla_\xi^* Y_n(\xi),
\end{aligned}
$$

where we have used (14.1.33). By restricting this tensor field to the unit sphere Ω we get with (14.1.63)

$$
\begin{aligned}
\mathbf{k}_n^{(1,1)} r^n Y_n(\xi)|_{r=1} &= (n+2)(n+1)\mathbf{o}^{(1,1)} Y_n(\xi) - (n+2)\mathbf{o}^{(1,2)} Y_n(\xi) \\
&\quad -(n+2)\mathbf{o}^{(2,1)} Y_n(\xi) - \frac{1}{2}(n+2)(n+1)\mathbf{o}^{(2,2)} Y_n(\xi) \\
&\quad +\frac{1}{2}\mathbf{o}^{(2,3)} Y_n(\xi).
\end{aligned}
$$

Similar calculations show that all restrictions of $r\xi \mapsto \mathbf{k}_n^{(i,k)} r^n Y_n(\xi)$ to the unit sphere Ω (i.e. $r = 1$) can be written as linear combinations of the tensor spherical harmonics $\mathbf{o}^{(i,k)} Y_n$. In particular it can be shown that for $Y_n \in Harm_n$

$$
\begin{pmatrix}
\mathbf{k}_n^{(1,1)} r^n Y_n|_{r=1} \\
\mathbf{k}_n^{(1,2)} r^n Y_n|_{r=1} \\
\mathbf{k}_n^{(2,1)} r^n Y_n|_{r=1} \\
\mathbf{k}_n^{(2,2)} r^n Y_n|_{r=1} \\
\mathbf{k}_n^{(3,3)} r^n Y_n|_{r=1}
\end{pmatrix}
= A
\begin{pmatrix}
\mathbf{o}^{(1,1)} Y_n \\
\mathbf{o}^{(1,2)} Y_n \\
\mathbf{o}^{(2,1)} Y_n \\
\mathbf{o}^{(2,2)} Y_n \\
\mathbf{o}^{(2,3)} Y_n
\end{pmatrix}
\qquad (14.3.1)
$$

and

$$\begin{pmatrix} \mathbf{k}_n^{(1,3)} r^n Y_n|_{r=1} \\ \mathbf{k}_n^{(2,3)} r^n Y_n|_{r=1} \\ \mathbf{k}_n^{(3,1)} r^n Y_n|_{r=1} \\ \mathbf{k}_n^{(3,2)} r^n Y_n|_{r=1} \end{pmatrix} = B \begin{pmatrix} \mathbf{o}^{(1,3)} Y_n \\ \mathbf{o}^{(3,1)} Y_n \\ \mathbf{o}^{(3,2)} Y_n \\ \mathbf{o}^{(3,3)} Y_n \end{pmatrix}, \tag{14.3.2}$$

where the matrices A and B are defined by

$$A = \begin{pmatrix} (n+1)(n+2) & -(n+2) & -(n+2) & -\frac{1}{2}(n+2)(n+1) & \frac{1}{2} \\ n^2 & n & -(n-1) & -\frac{1}{2}n(n-1) & -\frac{1}{2} \\ (n+1)^2 & -(n+1) & n+2 & \frac{1}{2}(n+2)(n+1) & -\frac{1}{2} \\ n(n-1) & n-1 & n-1 & -\frac{1}{2}n(n-1) & \frac{1}{2} \\ 0 & 0 & 1 & -\frac{1}{2}n(n+1) & -\frac{1}{2} \end{pmatrix}, \tag{14.3.3}$$

$$B = \begin{pmatrix} n+1 & 1 & -\frac{1}{2} & -\frac{1}{2}n(n+1) \\ n & -1 & \frac{1}{2} & \frac{1}{2}n(n+1) \\ 0 & n+2 & -\frac{1}{2} & \frac{1}{2}(n+2)(n+1) \\ 0 & n-1 & \frac{1}{2} & -\frac{1}{2}n(n-1) \end{pmatrix}. \tag{14.3.4}$$

Elementary calculations show that

$$\det(A) = \frac{1}{2} n(2n+3)(2n-1)(n+1)\ (2n+1)^2$$

and

$$\det(B) = -\frac{1}{2} n(n+1)(2n+1)^2.$$

Thus the matrices A and B are regular for $n \geq 1$. The inverse matrices read as follows:

$$A^{-1} = \tag{14.3.5}$$

$$\begin{pmatrix} \frac{1}{4n^2+8n+3} & \frac{1}{4n^2-1} & \frac{1}{4n^2+8n+3} & \frac{1}{4n^2-1} & 0 \\ \frac{-n}{4n^2+8n+3} & \frac{n+1}{4n^2-1} & \frac{-n}{4n^2+8n+3} & \frac{n+1}{4n^2-1} & 0 \\ \frac{-n}{4n^2+8n+3} & \frac{1-n^2}{(4n^2-1)n} & \frac{n(2+n)}{4n^3+12n^2+11n+3} & \frac{n+1}{4n^2-1} & \frac{1}{(n+1)n} \\ \frac{-1}{4n^2+8n+3} & \frac{-1+n}{(4n^2-1)n} & \frac{2+n}{4n^3+12n^2+11n+3} & \frac{-1}{4n^2-1} & \frac{-1}{(n+1)n} \\ \frac{(n-1)n}{4n^2+8n+3} & \frac{n^3+2n^2-n-2}{(1-4n^2)n} & \frac{(2-n-n^2)n}{4n^3+12n^2+11n+3} & \frac{n^2+3n+2}{4n^2-1} & \frac{2-n-n^2}{(n+1)n} \end{pmatrix},$$

$$B^{-1} = \begin{pmatrix} \frac{1}{2n+1} & \frac{1}{2n+1} & 0 & 0 \\ \frac{1}{2n^2+3n+1} & -\frac{1}{n(2n+1)} & \frac{n}{1+3n+2n^2} & \frac{n+1}{n(2n+1)} \\ -\frac{n^2+n-2}{1+3n+2n^2} & \frac{n^2+n-2}{n(2n+1)} & -\frac{(-1+n)n}{1+3n+2n^2} & \frac{3n+2+n^2}{n(2n+1)} \\ -\frac{1}{2n^2+3n+1} & \frac{1}{n(2n+1)} & -\frac{1}{2n^2+3n+1} & -\frac{1}{n(2n+1)} \end{pmatrix}. \tag{14.3.6}$$

For the special case $n = 0$ we see that $Y_0 \in Harm_0$ satisfies

$$\begin{array}{rcrcl} 2\mathbf{o}^{(1,1)}Y_0 & - & \mathbf{o}^{(2,2)}Y_0 & = & \mathbf{k}_0^{(1,1)}Y_0|_\Omega, \\ \mathbf{o}^{(1,1)}Y_0 & + & \mathbf{o}^{(2,2)}Y_0 & = & \mathbf{k}_0^{(2,1)}Y_0|_\Omega, \\ & & \mathbf{o}^{(3,3)}Y_0 & = & \mathbf{k}_0^{(3,1)}Y_0|_\Omega. \end{array}$$

Hence, we obtain the following lemma, in view of Lemma 14.3.3.

Lemma 14.3.4. *Let $\mathbf{y}_n^{(i,k)} \in \mathbf{harm}_n^{(i,k)}$ be a tensor spherical harmonic of order n and type (i,k). Then,*

$$\mathbf{y}_n^{(i,k)} \in \mathbf{hom}_{n-2} \oplus \mathbf{hom}_n \oplus \mathbf{hom}_{n+2},$$

if $(i,k) \in \{(1,1),(1,2),(2,1),(2,2),(2,3)\}$. Moreover,

$$\mathbf{y}_n^{(i,k)} \in \mathbf{hom}_{n-1} \oplus \mathbf{hom}_{n+1},$$

if $(i,k) \in \{(1,3),(3,1),(3,2),(3,3)\}$, (where $\mathbf{hom}_{-1} = \mathbf{hom}_{-2} = \{0\}$).

For $\mathbf{y}_n^{(i,k)} \in \mathbf{harm}_n^{(i,k)}$ and $\xi \in \Omega$ it follows as an immediate consequence that

$$\mathbf{y}_n^{(i,k)}(-\xi) = (-1)^n \mathbf{y}_n^{(i,k)}(\xi) \tag{14.3.7}$$

if $(i,k) \in \{(1,1),(1,2),(2,1),(2,2),(2,3)\}$ and

$$\mathbf{y}_n^{(i,k)}(-\xi) = (-1)^{n+1} \mathbf{y}_n^{(i,k)}(\xi) \tag{14.3.8}$$

if $(i,k) \in \{(1,3),(3,1),(3,2),(3,3)\}$.

Lemma 14.3.4 also yields other orthogonality relations.

Lemma 14.3.5. *For $\mathbf{y}_n^{(i,k)} \in \mathbf{harm}_n^{(i,k)}$ and $Y_m \in Harm_m$*

$$\int_\Omega Y_m(\xi)\mathbf{y}_n^{(i,k)}(\xi)\, d\omega(\xi) = 0$$

if $(i,k) \in \{(1,1),(1,2),(2,1),(2,2),(2,3)\}$ and $m \notin \{n-2,n,n+2\}$ or if $(i,k) \in \{(1,3),(3,1),(3,2),(3,3)\}$ and $m \notin \{n-1,n+1\}$.

It can be readily deduced that

$$\mathbf{hom}_n \subset \bigoplus_{m=-2}^{2} \mathbf{harm}_m$$

for all $n \in \mathbb{N}_0$. Thus, we know that every homogeneous harmonic tensor polynomial restricted to Ω can be expressed as a finite linear combination of tensor spherical harmonics, and vice versa. Hence, the following closure and completeness properties follow directly from the Corollaries 3.4.3 and 3.4.5.

Theorem 14.3.6. *Let* $\left\{\mathbf{y}_{n,j}^{(i,k)}\right\}_{\substack{i,k=1,2,3\\ n=0_{i,k},\ldots,\ j=1,\ldots,2n+1}}$ *be defined as in (14.2.3). Then the following statements are valid:*

(i) The system of tensor spherical harmonics is closed in $\mathbf{c}(\Omega)$ *with respect to* $\|\cdot\|_{\mathbf{c}(\Omega)}$.

(ii) The system is complete in $\mathbf{l}^2(\Omega)$ *with respect to* $\|\cdot\|_{\mathbf{l}^2(\Omega)}$.

Part (i) of this theorem justifies the approximation of continuous tensor fields on the sphere by finite sums of tensor spherical harmonics, while part (ii) is equivalent (see for example P.J. Davis (1963)) to the property, that any tensor field $\mathbf{f} \in \mathbf{l}^2(\Omega)$ can be represented in $\mathbf{l}^2(\Omega)$-sense by its orthogonal expansion in terms of tensor spherical harmonics.

14.4 Decomposition Theorem

As an immediate consequence of the completeness of the tensor spherical harmonics in $\mathbf{l}^2(\Omega)$ and the orthogonality of tensor fields $\mathbf{f}, \mathbf{g} : \Omega \to \mathbb{R}^3 \otimes \mathbb{R}^3$ of the representation $\mathbf{f} = \mathbf{o}^{(i,k)}F$ and $\mathbf{g} = \mathbf{o}^{(i',k')}G$, with $(i,k) \neq (i',k')$, we obtain an orthogonal decompostion of $\mathbf{l}^2(\Omega)$ which generalizes the relations (14.1.14)–(14.1.19) in canonical way. Using the notion

$$\mathbf{l}^2_{(i,k)}(\Omega) = \overline{\operatorname{span}\left\{\mathbf{y}_{n,j}^{(i,k)}\right\}_{n=0_{i,k},\ldots,\ j=1,\ldots,2n+1}}^{\|\cdot\|_{\mathbf{l}^2(\Omega)}}$$

for $(i,k) \in \{(1,1),(1,2),\ldots,(3,3)\}$, we are led to an orthogonal decomposition of $\mathbf{l}^2(\Omega)$ into nine complete subspaces, namely

$$\mathbf{l}^2(\Omega) = \bigoplus_{i,k=1}^{3} \mathbf{l}^2_{(i,k)}(\Omega). \tag{14.4.1}$$

Obviously,

$$\mathbf{l}^2_{(i,k)}(\Omega) = \overline{\left\{\mathbf{o}^{(i,k)}F \;\middle|\; F \in \mathcal{C}^{(\infty)}(\Omega)\right\}}^{\|\cdot\|_{\mathbf{l}^2(\Omega)}}$$

and

$$\mathbf{l}^2_{(i,k)}(\Omega) = \left\{\mathbf{f} \in \mathbf{l}^2(\Omega) \;\middle|\; O^{(i',k')}\mathbf{f} = 0 \text{ for } (i',k') \neq (i,k)\right\},$$

where differentiation is understood in the weak sense. But this means that we are able to define the corresponding projection operators $\mathbf{p}_{(i,k)} : \mathbf{l}^2(\Omega) \longrightarrow \mathbf{l}^2_{(i,k)}(\Omega)$.

Our intend now is to show that, if a tensor field $\mathbf{f} : \Omega \to \mathbb{R}^3 \otimes \mathbb{R}^3$ satisfies some smoothness assumptions, its decomposition according to (14.4.1) can be established by use of the Green functions with respect to the (scalar) Beltrami operator and its first iteration. Our result is based on the tangent representation theorem (cf. G.E. BACKUS (1967)) and the decomposition theorem for spherical vector fields of Section 12.4.

Theorem 14.4.1. *Let* $\mathbf{f} \in \mathbf{c}^{(2)}(\Omega)$. *Then there exist uniquely defined functions* $F_{i,k} \in \mathcal{C}^{(2)}(\Omega)$, $(i,k) \in \{(1,1),(1,2),\ldots,(3,3)\}$ *with* $(F_{i,k}, Y_0)_{\mathcal{L}^2(\Omega)} = 0$ *for all* $Y_0 \in Harm_0$ *if* $(i,k) \in \{(1,2),(1,3),(2,1),(2,3),(3,1),(3,2)\}$ *and* $(F_{i,k}, Y_1)_{\mathcal{L}^2(\Omega)} = 0$ *for all* $Y_1 \in Harm_1$ *if* $(i,k) \in \{(2,3),(3,2)\}$, *in such a way that*

$$\mathbf{f} = \sum_{i,k=1}^{3} \mathbf{o}^{(i,k)} F_{i,k},$$

and the functions $\xi \mapsto F_{i,k}(\xi)$, $\xi \in \Omega$, *are given by*

$$\begin{aligned}
F_{1,1}(\xi) &= O_\xi^{(1,1)}\mathbf{f}(\xi),\\
F_{2,2}(\xi) &= (1/2)O_\xi^{(2,2)}\mathbf{f}(\xi),\\
F_{3,3}(\xi) &= (1/2)O_\xi^{(3,3)}\mathbf{f}(\xi),\\
F_{1,2}(\xi) &= -\int_\Omega G(\Delta^*;\xi,\eta)O_\eta^{(1,2)}\mathbf{f}(\eta)\,d\omega(\eta),\\
F_{1,3}(\xi) &= -\int_\Omega G(\Delta^*;\xi,\eta)O_\eta^{(1,3)}\mathbf{f}(\eta)\,d\omega(\eta),\\
F_{2,1}(\xi) &= -\int_\Omega G(\Delta^*;\xi,\eta)O_\eta^{(2,1)}\mathbf{f}(\eta)\,d\omega(\eta),\\
F_{3,1}(\xi) &= -\int_\Omega G(\Delta^*;\xi,\eta)O_\eta^{(3,1)}\mathbf{f}(\eta)\,d\omega(\eta),\\
F_{2,3}(\xi) &= \int_\Omega G(\Delta^*(\Delta^*+2);\xi,\eta)O_\eta^{(2,3)}\mathbf{f}(\eta)\,d\omega(\eta),\\
F_{3,2}(\xi) &= \int_\Omega G(\Delta^*(\Delta^*+2);\xi,\eta)O_\eta^{(3,2)}\mathbf{f}(\eta)\,d\omega(\eta).
\end{aligned}$$

Proof. With the projection operators defined in Section 14.1 we have

$$\mathbf{f} = \mathbf{p}_{\mathrm{nor,nor}}\mathbf{f} + \mathbf{p}_{\mathrm{nor,tan}}\mathbf{f} + \mathbf{p}_{\mathrm{tan,nor}}\mathbf{f} + \mathbf{p}_{\mathrm{tan,tan}}\mathbf{f}.$$

Each term can be investigated separately. It is clear that $\mathbf{p}_{\mathrm{nor,nor}}\mathbf{f} = \mathbf{o}^{(1,1)}O^{(1,1)}\mathbf{f}$. The left normal/right tangential field $\mathbf{p}_{\mathrm{nor,tan}}\mathbf{f}$ allows a representation of the form $\mathbf{p}_{\mathrm{nor,tan}}\mathbf{f}(\xi) = \xi \otimes u(\xi)$, $\xi \in \Omega$, with $u \in c_{\mathrm{tan}}^{(2)}$. If we write u in the form $u(\xi) = \nabla^* F_{1,2}(\xi) + L^* F_{1,3}(\xi)$ (as proposed in Section 12.4) we find

$$\Delta_\xi^* F_{1,2}(\xi) = -O_\xi^{(1,2)}\mathbf{f}(\xi), \quad \xi \in \Omega,$$

$$\Delta_\xi^* F_{1,3}(\xi) = -O_\xi^{(1,3)}\mathbf{f}(\xi), \quad \xi \in \Omega.$$

The solution is uniquely determined by Theorem 4.3.1 since it can be deduced from (14.2.1) that $(O^{(1,2)}\mathbf{f}, Y_0)_{\mathcal{L}^2(\Omega)} = 0$ and $(O^{(1,3)}\mathbf{f}, Y_0)_{\mathcal{L}^2(\Omega)} = 0$ for all $Y_0 \in Harm_0$. We have for $\xi \in \Omega$

$$\begin{aligned}
F_{1,2}(\xi) &= -\int_\Omega G(\Delta^*;\xi,\eta) O_\eta^{(1,2)}\mathbf{f}(\eta)\, d\omega(\eta), \\
F_{1,3}(\xi) &= -\int_\Omega G(\Delta^*;\xi,\eta) O_\eta^{(1,3)}\mathbf{f}(\eta)\, d\omega(\eta),
\end{aligned}$$

as required. The left tangential/right normal part can be considered in a similar way.

The tensor field

$$\mathbf{p}_{\tan,\tan}\mathbf{f} - \frac{1}{2}\mathbf{o}^{(2,2)}O^{(2,2)}\mathbf{f} - \frac{1}{2}\mathbf{o}^{(3,3)}O^{(3,3)}\mathbf{f}$$

is symmetric and traceless. Based on the tangent representation theorem due to G.E. BACKUS (1966, 1967) we obtain from the formulation

$$\mathbf{p}_{\tan,\tan}\mathbf{f} - \frac{1}{2}\mathbf{o}^{(2,2)}O^{(2,2)}\mathbf{f} - \frac{1}{2}\mathbf{o}^{(3,3)}O^{(3,3)}\mathbf{f} = \mathbf{o}^{(2,3)}F_{2,3} + \mathbf{o}^{(3,2)}F_{3,2}$$

the differential equations

$$\begin{aligned}
\Delta_\xi^*(\Delta_\xi^* - (\Delta^*)^\wedge(1))F_{2,3}(\xi) &= O_\xi^{(2,3)}\mathbf{f}(\xi), \\
\Delta_\xi^*(\Delta_\xi^* - (\Delta^*)^\wedge(1))F_{3,2}(\xi) &= O_\xi^{(3,2)}\mathbf{f}(\xi)
\end{aligned}$$

for $\xi \in \Omega$ (remember that $-(\Delta^*)^\wedge(1) = 2$). By virtue of (14.1.74), (14.1.75) and Theorem 4.3.1 we see that the unique solutions are given by

$$\begin{aligned}
F_{2,3}(\xi) &= \int_\Omega G(\Delta^*(\Delta^*+2);\xi,\eta) O_\eta^{(2,3)}\mathbf{f}(\eta)\, d\omega(\eta), \quad \xi \in \Omega, \\
F_{3,2}(\xi) &= \int_\Omega G(\Delta^*(\Delta^*+2);\xi,\eta) O_\eta^{(3,2)}\mathbf{f}(\eta)\, d\omega(\eta), \quad \xi \in \Omega.
\end{aligned}$$

Consequently, the existence is assured. The uniqueness follows from the completeness of the tensor spherical harmonics. □

As an example for the above developed decomposition procedure for tensor fields we describe how the *Hesse matrix* of a function defined in $\mathbb{R}^3 \setminus \{0\}$ can be decomposed, when restricted to the unit sphere. It turns out that this result is of particular importance in satellite gradiometry, i.e. the determination of the earth's gravitational field from measurements of second order derivatives of the potential at satellite height (see for example R. RUMMEL (1986), W. FREEDEN *et al.* (1994, 1997)). In particular, this decomposition is of advantage for the classification of those types of gradiometry measurements, which ensure existence and uniqueness in a mathematical formulation of the gradiometry problem (cf. M. SCHREINER (1995, 1996A)).

Suppose that the function $H : \mathbb{R}^3 \setminus \{0\} \to \mathbb{R}$ is twice continuously differentiable. We want to show how the Hesse matrix restricted to the unit sphere Ω, i.e.

$$\mathbf{h}(\xi) = \nabla_x \otimes \nabla_x H(x)|_{|x|=1}, \quad \xi \in \Omega,$$

can be decomposed according to the rules of Theorem 14.4.1. In order to evaluate

$$\nabla_x \otimes \nabla_x H(x) = \left(\xi \frac{\partial}{\partial r} + \frac{1}{r}\nabla^*_\xi\right) \otimes \left(\xi \frac{\partial}{\partial r} + \frac{1}{r}\nabla^*_\xi\right) H(r\xi),$$

we firstly see that

$$\begin{aligned}
\xi \frac{\partial}{\partial r} \otimes \xi \frac{\partial}{\partial r} H(r\xi) &= \xi \otimes \xi \left(\frac{\partial}{\partial r}\right)^2 H(r\xi), \\
\xi \frac{\partial}{\partial r} \otimes \frac{1}{r}\nabla^*_\xi H(r\xi) &= -\frac{1}{r^2}\xi \otimes \nabla^*_\xi H(r\xi) + \frac{1}{r}\xi \otimes \nabla^*_\xi \frac{\partial}{\partial r} H(r\xi), \\
\frac{1}{r}\nabla^*_\xi \otimes \xi \frac{\partial}{\partial r} H(r\xi) &= \frac{1}{r}\mathbf{i}_{\tan}(\xi)\frac{\partial}{\partial r} H(r\xi) + \frac{1}{r}\nabla^*_\xi \left(\frac{\partial}{\partial r} H(r\xi)\right) \otimes \xi, \\
\frac{1}{r}\nabla^*_\xi \otimes \frac{1}{r}\nabla^*_\xi H(r\xi) &= \frac{1}{r^2}\nabla^*_\xi \otimes \nabla^*_\xi H(r\xi).
\end{aligned}$$

Summing up these terms we obtain

$$\begin{aligned}
\nabla_x \otimes \nabla_x H(x)|_{|x|=1} &= \xi \otimes \xi \left(\frac{\partial}{\partial r}\right)^2 H(r\xi)|_{r=1} \\
&\quad + \xi \otimes \nabla^*_\xi \left(\frac{\partial}{\partial r} H(r\xi)|_{r=1} - H(\xi)\right) \\
&\quad + \left(\nabla^* \frac{\partial}{\partial r} H(r\xi)|_{r=1}\right) \otimes \xi \\
&\quad + \nabla^*_\xi \otimes \nabla^*_\xi H(\xi)
\end{aligned}$$

$$+ \mathbf{i}_{\tan}(\xi) \frac{\partial}{\partial r} H(r\xi)|_{r=1}.$$

Using (14.1.63) and the definition of the $\mathbf{o}^{(i,k)}$-operators of Section 14.1 we finally arrive at

$$\begin{aligned}
\nabla_x \otimes \nabla_x H(x)|_{|x|=1} &= \mathbf{o}_\xi^{(1,1)} \left(\left(\frac{\partial}{\partial r} \right)^2 H(r\xi)|_{r=1} \right) \\
&\quad + \mathbf{o}_\xi^{(1,2)} \left(\frac{\partial}{\partial r} H(r\xi)|_{r=1} - H(\xi) \right) \\
&\quad + \mathbf{o}_\xi^{(2,1)} \left(\frac{\partial}{\partial r} H(r\xi)|_{r=1} - H(\xi) \right) \\
&\quad + \mathbf{o}_\xi^{(2,2)} \left(\frac{1}{2} \Delta_\xi^* H(\xi) + \frac{\partial}{\partial r} H(r\xi)|_{r=1} \right) \\
&\quad + \mathbf{o}_\xi^{(2,3)} \frac{1}{2} H(\xi).
\end{aligned}$$

In particular, if we consider an outer harmonic $H_{-n-1} : x \mapsto H_{-n-1}(x)$, $H_{-n-1}(r\xi) = r^{-(n+1)} Y_n(\xi)$, $r > 0$, $\xi \in \Omega$, we obtain the following decomposition of the Hesse matrix:

$$\begin{aligned}
\nabla_x \otimes \nabla_x H_{-n-1}(x)|_{|x|=1} &= (n+1)(n+2) \mathbf{o}_\xi^{(1,1)} Y_n(\xi) \\
&\quad - (n+2) \mathbf{o}_\xi^{(1,2)} Y_n(\xi) - (n+2) \mathbf{o}_\xi^{(2,1)} Y_n(\xi) \\
&\quad - \frac{1}{2}(n+1)(n+2) \mathbf{o}_\xi^{(2,2)} Y_n(\xi) \\
&\quad + \frac{1}{2} \mathbf{o}_\xi^{(2,3)} Y_n(\xi).
\end{aligned}$$

14.5 Tensorial Beltrami Operator

The question now is how a tensorial Beltrami operator $\hat{\mathbf{\Delta}}^*$ can be constructed such that the tensor spherical harmonics can be characterized as eigenfunctions of this operator. Our particular aim is to define the operator $\hat{\mathbf{\Delta}}^*$ in such a way that the equations

$$\hat{\mathbf{\Delta}}^* \mathbf{o}^{(i,k)} F = \mathbf{o}^{(i,k)} \Delta^* F, \tag{14.5.1}$$

$$O^{(i,k)} \hat{\mathbf{\Delta}}^* \mathbf{f} = \Delta^* O^{(i,k)} \mathbf{f}, \tag{14.5.2}$$

hold for all sufficiently smooth functions $F : \Omega \to \mathbb{R}$ and tensor fields $\mathbf{f} : \Omega \to \mathbb{R}^3 \otimes \mathbb{R}^3$.

As in the vectorial case (cf. Section 12.5) we start by applying the (scalar) Beltrami operator to the cartesian components of tensor spherical harmonics. If $\mathbf{f} \in \mathbf{c}^{(2)}(\Omega)$ is a spherical tensor field of the form

$$\mathbf{f}(\xi) = \sum_{i,k=1}^{3} F_{ik}(\xi)\varepsilon^i \otimes \varepsilon^k, \quad \xi \in \Omega,$$

we set

$$\Delta_\xi^* \mathbf{f}(\xi) = \sum_{i,k=1}^{3} (\Delta^* F_{ik}(\xi))\, \varepsilon^i \otimes \varepsilon^k, \quad \xi \in \Omega.$$

The application of the Beltrami operator to the cartesian components of a tensor spherical harmonic is easy. Using the results of Section 14.3 this can be done in the following way: express firstly $\mathbf{o}^{(i,k)}Y_n$ as a linear combination of restrictions of homogeneous harmonic tensor polynomials to the unit sphere (cf. (14.3.1) and (14.3.2)). Since the cartesian components of these polynomials are homogeneous harmonic polynomials when restricted to Ω (of degree $\in \{n-2, n-1, \ldots, n+2\}$) the application of Δ^* yields just the eigenvalues of Δ^*. Transforming these results back we obtain the following.

Lemma 14.5.1. *Let $Y_n \in Harm_n$ be a spherical harmonic of order n. Then we have*

$$\begin{aligned}
\Delta^*\mathbf{o}^{(1,1)}Y_n &= (-n(n+1)-4)\mathbf{o}^{(1,1)}Y_n + 2(\mathbf{o}^{(1,2)} + \mathbf{o}^{(2,1)} + \mathbf{o}^{(2,2)})Y_n, \\
\Delta^*\mathbf{o}^{(1,2)}Y_n &= (-n(n+1)-2)\mathbf{o}^{(1,2)}Y_n \\
&\quad + (2n(n+1)\mathbf{o}^{(1,1)} - 2\mathbf{o}^{(2,1)} - n(n+1)\mathbf{o}^{(2,2)} + \mathbf{o}^{(2,3)})Y_n, \\
\Delta^*\mathbf{o}^{(2,1)}Y_n &= (-n(n+1)-2)\mathbf{o}^{(2,1)}Y_n \\
&\quad + (2n(n+1)\mathbf{o}^{(1,1)} - 2\mathbf{o}^{(1,2)} - n(n+1)\mathbf{o}^{(2,2)} + \mathbf{o}^{(2,3)})Y_n, \\
\Delta^*\mathbf{o}^{(2,2)}Y_n &= (-n(n+1)-2)\mathbf{o}^{(2,2)}Y_n \\
&\quad + (4\mathbf{o}^{(1,1)} - 2\mathbf{o}^{(2,1)} - 2\mathbf{o}^{(2,3)})Y_n, \\
\Delta^*\mathbf{o}^{(2,3)}Y_n &= (-n(n+1)+2)\mathbf{o}^{(2,3)}Y_n \\
&\quad + ((2n(n+1)-4)\mathbf{o}^{(1,2)} + (2n(n+1)-4)\mathbf{o}^{(2,1)})Y_n,
\end{aligned}$$

and

$$\begin{aligned}
\Delta^*\mathbf{o}^{(1,3)}Y_n &= (-n(n+1)-2)\mathbf{o}^{(1,3)}Y_n \\
&\quad + (-2\mathbf{o}^{(3,1)} + \mathbf{o}^{(3,2)} + n(n+1)\mathbf{o}^{(3,3)})Y_n, \\
\Delta^*\mathbf{o}^{(3,1)}Y_n &= (-n(n+1)-2)\mathbf{o}^{(3,1)}Y_n \\
&\quad + (-2\mathbf{o}^{(1,3)} + \mathbf{o}^{(3,2)} - n(n+1)\mathbf{o}^{(3,3)})Y_n, \\
\Delta^*\mathbf{o}^{(3,2)}Y_n &= (-n(n+1)+2)\mathbf{o}^{(3,2)}Y_n \\
&\quad + ((2n(n+1)-4)\mathbf{o}^{(1,3)} + (2n(n+1)-4)\mathbf{o}^{(3,1)})Y_n, \\
\Delta^*\mathbf{o}^{(3,3)}Y_n &= (-n(n+1)-2)\mathbf{o}^{(3,3)}Y_n + (-2\mathbf{o}^{(1,3)} - 2\mathbf{o}^{(3,2)})Y_n.
\end{aligned}$$

A direct consequence is that the operator $\hat{\mathbf{\Delta}}^* : \mathbf{c}^{(2)}(\Omega) \longrightarrow \mathbf{c}^{(0)}(\Omega)$ defined by

$$\begin{aligned}\hat{\mathbf{\Delta}}^* &= \mathbf{p}_{\mathrm{nor,nor}}(\Delta^*+4)\mathbf{p}_{\mathrm{nor,nor}} + \mathbf{p}_{\mathrm{nor,tan}}(\Delta^*+2)\mathbf{p}_{\mathrm{nor,tan}} \\ &\quad + \mathbf{p}_{\mathrm{tan,nor}}(\Delta^*+2)\mathbf{p}_{\mathrm{tan,nor}} + \mathbf{p}_{(2,2)}(\Delta^*+2)\mathbf{p}_{(2,2)} \\ &\quad + \mathbf{p}_{(2,3)}(\Delta^*-2)\mathbf{p}_{(2,3)} + \mathbf{p}_{(3,2)}(\Delta^*-2)\mathbf{p}_{(3,2)} \\ &\quad + \mathbf{p}_{(3,3)}(\Delta^*+2)\mathbf{p}_{(3,3)}\end{aligned}$$

fulfils the equations (14.5.1) and (14.5.2). (For the definition of the projection operators, see (14.4.1).)

Undoubtly, the separation of tangential tensor fields with respect to the operators $\mathbf{p}_{(i,k)}$ is an unpleasant task. We shall show, however, that this separation can be avoided. It follows from Lemma 14.5.1 that the spaces $\mathbf{c}^{(2)}_{(i,k)}(\Omega)$, $i,k \in \{2,3\}$, are invariant with respect to the operator $\mathbf{p}_{\mathrm{tan,tan}}\Delta^*$, in the sense that $\mathbf{f}_{(i,k)} \in \mathbf{c}^{(2)}_{(i,k)}(\Omega)$ implies $\mathbf{p}_{\mathrm{tan,tan}}\Delta^*\mathbf{f}_{(i,k)} \in \mathbf{c}^{(0)}_{(i,k)}(\Omega)$.

Definition 14.5.2. Let $\mathbf{f} \in \mathbf{c}(\Omega)$ be a tensor field of the form

$$\mathbf{f}(\xi) = \sum_{i,k=1}^{3} F_{ik}(\xi)\varepsilon^i \otimes \varepsilon^k, \quad \xi \in \Omega.$$

Then we define the operator $\mathbf{J} : \mathbf{c}(\Omega) \longrightarrow \mathbf{c}(\Omega)$ by

$$\mathbf{J}_\xi \mathbf{f}(\xi) = \mathbf{f}(\xi) - \sum_{i,k=1}^{3} F_{i,k}(\xi)(\xi \wedge \varepsilon^i) \otimes (\xi \wedge \varepsilon^k), \quad \xi \in \Omega.$$

After elementary calculations we get for $F \in \mathcal{C}(\Omega)$

$$\begin{aligned}\mathbf{J}\mathbf{o}^{(2,2)}F &= 0, & \mathbf{J}\mathbf{o}^{(2,3)}F &= 2\mathbf{o}^{(2,3)}F, \\ \mathbf{J}\mathbf{o}^{(3,3)}F &= 0, & \mathbf{J}\mathbf{o}^{(3,2)}F &= 2\mathbf{o}^{(3,2)}F,\end{aligned}$$

and

$$\mathbf{J}\mathbf{o}^{(i,k)}F = 0,$$

if $i = 1$ or $k = 1$. Hence we are able to redefine the operator $\hat{\mathbf{\Delta}}^*$ as follows:

$$\begin{aligned}\hat{\mathbf{\Delta}}^*\mathbf{f} &= \mathbf{p}_{\mathrm{nor,nor}}(\Delta^*+4)\mathbf{p}_{\mathrm{nor,nor}}\mathbf{f} + \mathbf{p}_{\mathrm{nor,tan}}(\Delta^*+2)\mathbf{p}_{\mathrm{nor,tan}}\mathbf{f} \\ &\quad + \mathbf{p}_{\mathrm{tan,nor}}(\Delta^*+2)\mathbf{p}_{\mathrm{tan,nor}}\mathbf{f} + \mathbf{p}_{\mathrm{tan,tan}}(\Delta^*+2-2\mathbf{J})\mathbf{p}_{\mathrm{tan,tan}}\mathbf{f},\end{aligned}$$

provided that $\mathbf{f} \in \mathbf{c}^{(2)}(\Omega)$.

As final result we state the following theorem.

Theorem 14.5.3. *The operator* $\hat{\mathbf{\Delta}}^*$ *obeys the following properties:*

(i) The operator $\hat{\mathbf{\Delta}}^* : \mathbf{c}^{(2)}(\Omega) \to \mathbf{c}^{(0)}(\Omega)$ *satisfies for all* $i, k \in \{1, 2, 3\}$

$$\begin{aligned} \hat{\mathbf{\Delta}}^* \mathbf{o}^{(i,k)} &= \mathbf{o}^{(i,k)} \Delta^*, \\ O^{(i,k)} \hat{\mathbf{\Delta}}^* &= \Delta^* O^{(i,k)}. \end{aligned}$$

(ii) Any tensor spherical harmonic $\mathbf{y}_n \in \mathbf{harm}_n$ *satisfies the relation* $\hat{\mathbf{\Delta}}^* \mathbf{y}_n = (\Delta^*)^\wedge(n) \mathbf{y}_n$.

(iii) If $\mathbf{y} \in \mathbf{c}^{(\infty)}(\Omega)$ *satisfies* $\hat{\mathbf{\Delta}}^* \mathbf{y} = \lambda \mathbf{y}$ *for any* $\lambda \in \mathbb{R}$, *then* $\lambda = (\Delta^*)^\wedge(n)$, $n \in \mathbb{N}_0$, *and* $\mathbf{y} \in \mathbf{harm}_n$.

Proof. These statements follow from Lemma 3.1.2, Corollary 3.4.3, and the previous results of this section. □

14.6 Tensorial Addition Theorem

The first chapters of this book have shown that the addition theorem for spherical harmonics plays an important role in spherical approximation. In this section we deal with the generalization of the addition theorem to tensor spherical harmonics.

Let

$$\mathbf{y}_{n,j}^{(i,k)} = \left(\mu_n^{(i,k)}\right)^{-1/2} \mathbf{o}^{(i,k)} Y_{n,j}$$

be an $\mathbf{l}^2(\Omega)$-orthonormal system of tensor spherical harmonics of order n. The problem is to evaluate the rank-4 tensor

$$\sum_{j=1}^{2n+1} \mathbf{y}_{n,j}^{(i,k)}(\xi) \otimes \mathbf{y}_{n,j}^{(l,m)}(\eta), \quad (\xi, \eta) \in \Omega^2, \tag{14.6.1}$$

and to establish rank-4 tensorial versions of the Legendre polynomials for $(i,k), (l,m) \in \{(1,1), (1,2), \ldots, (3,3)\}$.

To this end we start with a generalization of the operators $\mathbf{o}^{(i,k)}$ for sufficiently smooth rank-2 tensor fields. Let $\mathbf{f} : \Omega \to \mathbb{R}^3 \otimes \mathbb{R}^3$ be a smooth tensor field of the form

$$\mathbf{f}(\xi) = \sum_{i,k=1}^{3} F_{ik}(\xi) \varepsilon^i \otimes \varepsilon^k, \quad \xi \in \Omega.$$

For $(l,m) \in \{(1,1), (1,2), \ldots, (3,3)\}$ we set

$$\mathbf{o}_\xi^{(l,m)}\mathbf{f}(\xi) = \sum_{i,k=1}^{3} \left(\mathbf{o}_\xi^{(l,m)} F_{ik}(\xi)\right) \otimes \varepsilon^i \otimes \varepsilon^k. \tag{14.6.2}$$

Hence, the expression (14.6.1) can be reformulated as follows:

$$\begin{aligned}
&\sum_{j=1}^{2n+1} \mathbf{y}_{n,j}^{(i,k)}(\xi) \otimes \mathbf{y}_{n,j}^{(l,m)}(\eta) \\
&= \left(\mu_n^{(i,k)}\right)^{-(1/2)} \left(\mu_n^{(l,m)}\right)^{-(1/2)} \sum_j \mathbf{o}_\xi^{(i,k)} Y_{n,j}(\xi) \otimes \mathbf{o}_\eta^{(l,m)} Y_{n,j}(\eta) \\
&= \left(\mu_n^{(i,k)}\right)^{-(1/2)} \left(\mu_n^{(l,m)}\right)^{-(1/2)} \mathbf{o}_\xi^{(i,k)} \mathbf{o}_\eta^{(l,m)} \sum_{j=1}^{2n+1} Y_{n,j}(\xi) Y_{n,j}(\eta) \\
&= \left(\mu_n^{(i,k)}\right)^{-(1/2)} \left(\mu_n^{(l,m)}\right)^{-(1/2)} \frac{2n+1}{4\pi} \mathbf{o}_\xi^{(i,k)} \mathbf{o}_\eta^{(l,m)} P_n(\xi \cdot \eta).
\end{aligned}$$

Introducing the definition of the *Legendre tensor field of degree* n

$$\mathbf{P}_n^{(i,k,l,m)}(\xi,\eta) : \Omega \times \Omega \to \mathbb{R}^3 \otimes \mathbb{R}^3 \otimes \mathbb{R}^3 \otimes \mathbb{R}^3$$

for $(i,k),(l,m) \in \{(1,1),(1,2),\ldots,(3,3)\}$, by

$$\mathbf{P}_n^{(i,k,l,m)}(\xi,\eta) = (\mu_n^{(i,k)})^{-1/2} (\mu_n^{(l,m)})^{-1/2} \mathbf{o}_\xi^{(i,k)} \mathbf{o}_\eta^{(l,m)} P_n(\xi \cdot \eta), \quad \xi,\eta \in \Omega, \tag{14.6.3}$$

we obtain the following formulation of the *tensorial addition theorem.*

Theorem 14.6.1. *Let* $\left\{\mathbf{y}_{n,j}^{(i,k)}\right\}_{\substack{i,k=1,2,3\\ j=1,\ldots,2n+1}}$ *be an* $\mathbf{l}^2(\Omega)$*-orthonormal system of tensor spherical harmonics of order* n*. For index pairs* $(i,k),(l,m) \in \{(1,1),(1,2),\ldots,(3,3)\}$ *and* $\xi,\eta \in \Omega$*, we have*

$$\sum_{j=1}^{2n+1} \mathbf{y}_{n,j}^{(i,k)}(\xi) \otimes \mathbf{y}_{n,j}^{(l,m)}(\eta) = \frac{2n+1}{4\pi} \mathbf{P}_n^{(i,k,l,m)}(\xi,\eta),$$

where the Legendre tensor is given by (14.6.3).

Of course, it remains to express the rank-4 tensor $\mathbf{P}_n^{(i,k,l,m)}(\xi,\eta)$ in explicit form. Starting point is the next lemma which can be verified by use of local coordinates introduced in (1.2.4).

Lemma 14.6.2. *Suppose that* $F \in \mathcal{C}^{(2)}[-1,1]$*. Let* $\eta \in \Omega$ *fixed. Then, for* $\xi \in \Omega$*, we have*

$$\mathbf{o}_\xi^{(1,1)} F(\xi \cdot \eta) \quad = \quad F(\xi \cdot \eta) \xi \otimes \xi,$$

$$
\begin{aligned}
\mathbf{o}_\xi^{(1,2)} F(\xi\cdot\eta) &= F'(\xi\cdot\eta)\xi\otimes(\eta-(\xi\cdot\eta)\xi),\\
\mathbf{o}_\xi^{(1,3)} F(\xi\cdot\eta) &= F'(\xi\cdot\eta)\xi\otimes(\xi\wedge\eta),\\
\mathbf{o}_\xi^{(2,1)} F(\xi\cdot\eta) &= F'(\xi\cdot\eta)(\eta-(\xi\cdot\eta)\xi)\otimes\xi,\\
\mathbf{o}_\xi^{(2,2)} F(\xi\cdot\eta) &= F(\xi\cdot\eta)\mathbf{i}_{\mathrm{tan}}(\xi),\\
\mathbf{o}_\xi^{(2,3)} F(\xi\cdot\eta) &= F''(\xi\cdot\eta)\left[(\eta-(\xi\cdot\eta)\xi)\otimes\right.\\
&\qquad \left.(\eta-(\xi\cdot\eta)\xi)-(\xi\wedge\eta)\otimes(\xi\wedge\eta)\right],\\
\mathbf{o}_\xi^{(3,1)} F(\xi\cdot\eta) &= F'(\xi\cdot\eta)(\xi\wedge\eta)\otimes\xi,\\
\mathbf{o}_\xi^{(3,2)} F(\xi\cdot\eta) &= F''(\xi\cdot\eta)\left[(\eta-(\xi\cdot\eta)\xi)\otimes\right.\\
&\qquad \left.(\xi\wedge\eta)+(\xi\wedge\eta)\otimes(\eta-(\xi\cdot\eta)\xi)\right],\\
\mathbf{o}_\xi^{(3,3)} F(\xi\cdot\eta) &= F(\xi\cdot\eta)\mathbf{j}_{\mathrm{tan}}(\xi).
\end{aligned}
$$

Combining Lemma 14.6.2 with (14.6.2) and (14.6.3) we obtain the following result after some lengthy calculations.

Theorem 14.6.3. *Assume that $\Phi_n^1,\ldots,\Phi_n^9$ for $n\in\mathbb{N}_0$ and $t\in(-1,1)$ are defined by*

$$
\begin{aligned}
\Phi_n^1(t) &= P_n(t),\\
\Phi_n^2(t) &= P_n'(t),\\
\Phi_n^3(t) &= \sqrt{1-t^2}P_n'(t),\\
\Phi_n^4(t) &= -n(n+1)P_n(t)+2tP_n'(t),\\
\Phi_n^5(t) &= -\frac{n(n+1)}{1-t^2}P_n(t)+\frac{2t}{1-t^2}P_n'(t),\\
\Phi_n^6(t) &= -n(n+1)P_n(t)+tP_n'(t),\\
\Phi_n^7(t) &= 2\frac{n(n+1)t}{\sqrt{1-t^2}}P_n(t)-\frac{2t^2+n^2t^2+nt^2-n^2-n+2}{\sqrt{1-t^2}}P_n'(t),\\
\Phi_n^8(t) &= -\frac{n(n+1)(n^2t^2+nt^2+4t^2+8-n-n^2)}{1-t^2}P_n(t)\\
&\quad +4\frac{t(t^2+n^2t^2+nt^2+5-n^2-n)}{1-t^2}P_n'(t),\\
\Phi_n^9(t) &= 3\frac{n(n+1)t}{1-t^2}P_n(t)-\frac{4t^2+n^2t^2+nt^2+2-n^2-n}{1-t^2}P_n'(t).
\end{aligned}
$$

Let $\xi,\eta\in\Omega$ with $\xi\neq\pm\eta$, or equivalently $(\xi\cdot\eta)^2\neq 1$. We define the orthonormal sets of vectors $\left\{\varepsilon_\xi^1,\varepsilon_\xi^2,\varepsilon_\xi^3\right\}$ and $\left\{\varepsilon_\eta^1,\varepsilon_\eta^2,\varepsilon_\eta^3\right\}$ by

$$\begin{aligned} \varepsilon_\xi^1 &= \xi & \varepsilon_\eta^1 &= \eta \\ \varepsilon_\xi^2 &= \frac{1}{\sqrt{1-(\xi\cdot\eta)^2}}(\eta - (\xi\cdot\eta)\xi) & \varepsilon_\eta^2 &= \frac{1}{\sqrt{1-(\xi\cdot\eta)^2}}(\xi - (\xi\cdot\eta)\eta) \\ \varepsilon_\xi^3 &= \frac{1}{\sqrt{1-(\xi\cdot\eta)^2}}\xi\wedge\eta & \varepsilon_\eta^3 &= \frac{1}{\sqrt{1-(\xi\cdot\eta)^2}}\eta\wedge\xi. \end{aligned}$$

Then, using the abbreviation

$$\tilde{\mathbf{P}}_n^{(i,k,l,m)}(\xi,\eta) = \sqrt{\mu_n^{(i,k)}}\sqrt{\mu_n^{(l,m)}}\mathbf{P}_n^{(i,k,l,m)}(\xi,\eta),$$

the following relations hold true:

$$\begin{aligned}
\tilde{\mathbf{P}}_n^{(1,1,1,1)}(\xi,\eta) &= \Phi_n^1(\xi\cdot\eta)\varepsilon_\xi^1\otimes\varepsilon_\xi^1\otimes\varepsilon_\eta^1\otimes\varepsilon_\eta^1, \\
\tilde{\mathbf{P}}_n^{(1,1,1,2)}(\xi,\eta) &= \Phi_n^3(\xi\cdot\eta)\varepsilon_\xi^1\otimes\varepsilon_\xi^1\otimes\varepsilon_\eta^1\otimes\varepsilon_\eta^2, \\
\tilde{\mathbf{P}}_n^{(1,1,1,3)}(\xi,\eta) &= -\Phi_n^3(\xi\cdot\eta)\varepsilon_\xi^1\otimes\varepsilon_\xi^1\otimes\varepsilon_\eta^1\otimes\varepsilon_\eta^3, \\
\tilde{\mathbf{P}}_n^{(1,1,2,1)}(\xi,\eta) &= \Phi_n^3(\xi\cdot\eta)\varepsilon_\xi^1\otimes\varepsilon_\xi^1\otimes\varepsilon_\eta^2\otimes\varepsilon_\eta^1, \\
\tilde{\mathbf{P}}_n^{(1,1,2,2)}(\xi,\eta) &= \Phi_n^1(\xi\cdot\eta)\varepsilon_\xi^1\otimes\varepsilon_\xi^1\otimes\varepsilon_\eta^2\otimes\varepsilon_\eta^2 + \Phi_n^1(\xi\cdot\eta)\varepsilon_\xi^1\otimes\varepsilon_\xi^1\otimes\varepsilon_\eta^3\otimes\varepsilon_\eta^3, \\
\tilde{\mathbf{P}}_n^{(1,1,2,3)}(\xi,\eta) &= \Phi_n^4(\xi\cdot\eta)\varepsilon_\xi^1\otimes\varepsilon_\xi^1\otimes\varepsilon_\eta^2\otimes\varepsilon_\eta^2 - \Phi_n^4(\xi\cdot\eta)\varepsilon_\xi^1\otimes\varepsilon_\xi^1\otimes\varepsilon_\eta^3\otimes\varepsilon_\eta^3, \\
\tilde{\mathbf{P}}_n^{(1,1,3,1)}(\xi,\eta) &= -\Phi_n^3(\xi\cdot\eta)\varepsilon_\xi^1\otimes\varepsilon_\xi^1\otimes\varepsilon_\eta^3\otimes\varepsilon_\eta^1, \\
\tilde{\mathbf{P}}_n^{(1,1,3,2)}(\xi,\eta) &= -\Phi_n^4(\xi\cdot\eta)\varepsilon_\xi^1\otimes\varepsilon_\xi^1\otimes\varepsilon_\eta^2\otimes\varepsilon_\eta^3 - \Phi_n^4(\xi\cdot\eta)\varepsilon_\xi^1\otimes\varepsilon_\xi^1\otimes\varepsilon_\eta^3\otimes\varepsilon_\eta^2, \\
\tilde{\mathbf{P}}_n^{(1,1,3,3)}(\xi,\eta) &= \Phi_n^1(\xi\cdot\eta)\varepsilon_\xi^1\otimes\varepsilon_\xi^1\otimes\varepsilon_\eta^2\otimes\varepsilon_\eta^3 - \Phi_n^1(\xi\cdot\eta)\varepsilon_\xi^1\otimes\varepsilon_\xi^1\otimes\varepsilon_\eta^3\otimes\varepsilon_\eta^2, \\
\tilde{\mathbf{P}}_n^{(1,2,1,2)}(\xi,\eta) &= \Phi_n^6(\xi\cdot\eta)\varepsilon_\xi^1\otimes\varepsilon_\xi^2\otimes\varepsilon_\eta^1\otimes\varepsilon_\eta^2 - \Phi_n^2(\xi\cdot\eta)\varepsilon_\xi^1\otimes\varepsilon_\xi^3\otimes\varepsilon_\eta^1\otimes\varepsilon_\eta^3, \\
\tilde{\mathbf{P}}_n^{(1,2,1,3)}(\xi,\eta) &= -\Phi_n^6(\xi\cdot\eta)\varepsilon_\xi^1\otimes\varepsilon_\xi^2\otimes\varepsilon_\eta^1\otimes\varepsilon_\eta^3 - \Phi_n^2(\xi\cdot\eta)\varepsilon_\xi^1\otimes\varepsilon_\xi^3\otimes\varepsilon_\eta^1\otimes\varepsilon_\eta^2, \\
\tilde{\mathbf{P}}_n^{(1,2,2,1)}(\xi,\eta) &= \Phi_n^6(\xi\cdot\eta)\varepsilon_\xi^1\otimes\varepsilon_\xi^2\otimes\varepsilon_\eta^2\otimes\varepsilon_\eta^1 - \Phi_n^2(\xi\cdot\eta)\varepsilon_\xi^1\otimes\varepsilon_\xi^3\otimes\varepsilon_\eta^3\otimes\varepsilon_\eta^1, \\
\tilde{\mathbf{P}}_n^{(1,2,2,2)}(\xi,\eta) &= \Phi_n^3(\xi\cdot\eta)\varepsilon_\xi^1\otimes\varepsilon_\xi^2\otimes\varepsilon_\eta^2\otimes\varepsilon_\eta^2 + \Phi_n^3(\xi\cdot\eta)\varepsilon_\xi^1\otimes\varepsilon_\xi^2\otimes\varepsilon_\eta^3\otimes\varepsilon_\eta^3, \\
\tilde{\mathbf{P}}_n^{(1,2,2,3)}(\xi,\eta) &= \Phi_n^7(\xi\cdot\eta)\varepsilon_\xi^1\otimes\varepsilon_\xi^2\otimes\varepsilon_\eta^2\otimes\varepsilon_\eta^2 - \Phi_n^7(\xi\cdot\eta)\varepsilon_\xi^1\otimes\varepsilon_\xi^2\otimes\varepsilon_\eta^3\otimes\varepsilon_\eta^3 \\
&\quad -2\,\Phi_n^5(\xi\cdot\eta)(\varepsilon_\xi^1\otimes\varepsilon_\xi^3\otimes\varepsilon_\eta^2\otimes\varepsilon_\eta^3 + \varepsilon_\xi^1\otimes\varepsilon_\xi^3\otimes\varepsilon_\eta^3\otimes\varepsilon_\eta^2), \\
\tilde{\mathbf{P}}_n^{(1,2,3,1)}(\xi,\eta) &= -\Phi_n^6(\xi\cdot\eta)\varepsilon_\xi^1\otimes\varepsilon_\xi^2\otimes\varepsilon_\eta^3\otimes\varepsilon_\eta^1 - \Phi_n^2(\xi\cdot\eta)\varepsilon_\xi^1\otimes\varepsilon_\xi^3\otimes\varepsilon_\eta^2\otimes\varepsilon_\eta^1, \\
\tilde{\mathbf{P}}_n^{(1,2,3,2)}(\xi,\eta) &= -\Phi_n^7(\xi\cdot\eta)\varepsilon_\xi^1\otimes\varepsilon_\xi^2\otimes\varepsilon_\eta^2\otimes\varepsilon_\eta^3 - \Phi_n^7(\xi\cdot\eta)\varepsilon_\xi^1\otimes\varepsilon_\xi^2\otimes\varepsilon_\eta^3\otimes\varepsilon_\eta^2 \\
&\quad -2\,\Phi_n^5(\xi\cdot\eta)(\varepsilon_\xi^1\otimes\varepsilon_\xi^3\otimes\varepsilon_\eta^2\otimes\varepsilon_\eta^2 - \varepsilon_\xi^1\otimes\varepsilon_\xi^3\otimes\varepsilon_\eta^3\otimes\varepsilon_\eta^3), \\
\tilde{\mathbf{P}}_n^{(1,2,3,3)}(\xi,\eta) &= \Phi_n^3(\xi\cdot\eta)\varepsilon_\xi^1\otimes\varepsilon_\xi^2\otimes\varepsilon_\eta^2\otimes\varepsilon_\eta^3 - \Phi_n^3(\xi\cdot\eta)\varepsilon_\xi^1\otimes\varepsilon_\xi^2\otimes\varepsilon_\eta^3\otimes\varepsilon_\eta^2, \\
\tilde{\mathbf{P}}_n^{(1,3,1,3)}(\xi,\eta) &= -\Phi_n^2(\xi\cdot\eta)\varepsilon_\xi^1\otimes\varepsilon_\xi^2\otimes\varepsilon_\eta^1\otimes\varepsilon_\eta^2 + \Phi_n^6(\xi\cdot\eta)\varepsilon_\xi^1\otimes\varepsilon_\xi^3\otimes\varepsilon_\eta^1\otimes\varepsilon_\eta^3, \\
\tilde{\mathbf{P}}_n^{(1,3,2,1)}(\xi,\eta) &= -\Phi_n^2(\xi\cdot\eta)\varepsilon_\xi^1\otimes\varepsilon_\xi^2\otimes\varepsilon_\eta^3\otimes\varepsilon_\eta^1 - \Phi_n^6(\xi\cdot\eta)\varepsilon_\xi^1\otimes\varepsilon_\xi^3\otimes\varepsilon_\eta^2\otimes\varepsilon_\eta^1, \\
\tilde{\mathbf{P}}_n^{(1,3,2,2)}(\xi,\eta) &= -\Phi_n^3(\xi\cdot\eta)\varepsilon_\xi^1\otimes\varepsilon_\xi^3\otimes\varepsilon_\eta^2\otimes\varepsilon_\eta^2 - \Phi_n^3(\xi\cdot\eta)\varepsilon_\xi^1\otimes\varepsilon_\xi^3\otimes\varepsilon_\eta^3\otimes\varepsilon_\eta^3, \\
\tilde{\mathbf{P}}_n^{(1,3,2,3)}(\xi,\eta) &= -2\,\Phi_n^5(\xi\cdot\eta)(\varepsilon_\xi^1\otimes\varepsilon_\xi^2\otimes\varepsilon_\eta^2\otimes\varepsilon_\eta^3 + \varepsilon_\xi^1\otimes\varepsilon_\xi^2\otimes\varepsilon_\eta^3\otimes\varepsilon_\eta^2)
\end{aligned}$$

$$
\begin{aligned}
& -\Phi_n^7(\xi\cdot\eta)\varepsilon_\xi^1\otimes\varepsilon_\xi^3\otimes\varepsilon_\eta^2\otimes\varepsilon_\eta^2+\Phi_n^7(\xi\cdot\eta)\varepsilon_\xi^1\otimes\varepsilon_\xi^3\otimes\varepsilon_\eta^3\otimes\varepsilon_\eta^3,\\
\tilde{\mathbf{P}}_n^{(1,3,3,1)}(\xi,\eta) &= -\Phi_n^2(\xi\cdot\eta)\varepsilon_\xi^1\otimes\varepsilon_\xi^2\otimes\varepsilon_\eta^2\otimes\varepsilon_\eta^1+\Phi_n^6(\xi\cdot\eta)\varepsilon_\xi^1\otimes\varepsilon_\xi^3\otimes\varepsilon_\eta^3\otimes\varepsilon_\eta^1,\\
\tilde{\mathbf{P}}_n^{(1,3,3,2)}(\xi,\eta) &= 2\,\Phi_n^5(\xi\cdot\eta)(\varepsilon_\xi^1\otimes\varepsilon_\xi^2\otimes\varepsilon_\eta^3\otimes\varepsilon_\eta^3-\varepsilon_\xi^1\otimes\varepsilon_\xi^2\otimes\varepsilon_\eta^2\otimes\varepsilon_\eta^2)\\
&\quad+\Phi_n^7(\xi\cdot\eta)\varepsilon_\xi^1\otimes\varepsilon_\xi^3\otimes\varepsilon_\eta^2\otimes\varepsilon_\eta^3+\Phi_n^7(\xi\cdot\eta)\varepsilon_\xi^1\otimes\varepsilon_\xi^3\otimes\varepsilon_\eta^3\otimes\varepsilon_\eta^2,\\
\tilde{\mathbf{P}}_n^{(1,3,3,3)}(\xi,\eta) &= -\Phi_n^3(\xi\cdot\eta)\varepsilon_\xi^1\otimes\varepsilon_\xi^3\otimes\varepsilon_\eta^2\otimes\varepsilon_\eta^3+\Phi_n^3(\xi\cdot\eta)\varepsilon_\xi^1\otimes\varepsilon_\xi^3\otimes\varepsilon_\eta^3\otimes\varepsilon_\eta^2,\\
\tilde{\mathbf{P}}_n^{(2,1,2,1)}(\xi,\eta) &= \Phi_n^6(\xi\cdot\eta)\varepsilon_\xi^2\otimes\varepsilon_\xi^1\otimes\varepsilon_\eta^2\otimes\varepsilon_\eta^1-\Phi_n^2(\xi\cdot\eta)\varepsilon_\xi^3\otimes\varepsilon_\xi^1\otimes\varepsilon_\eta^3\otimes\varepsilon_\eta^1,\\
\tilde{\mathbf{P}}_n^{(2,1,2,2)}(\xi,\eta) &= \Phi_n^3(\xi\cdot\eta)\varepsilon_\xi^2\otimes\varepsilon_\xi^1\otimes\varepsilon_\eta^2\otimes\varepsilon_\eta^2+\Phi_n^3(\xi\cdot\eta)\varepsilon_\xi^2\otimes\varepsilon_\xi^1\otimes\varepsilon_\eta^3\otimes\varepsilon_\eta^3,\\
\tilde{\mathbf{P}}_n^{(2,1,2,3)}(\xi,\eta) &= \Phi_n^7(\xi\cdot\eta)\varepsilon_\xi^2\otimes\varepsilon_\xi^1\otimes\varepsilon_\eta^2\otimes\varepsilon_\eta^2-\Phi_n^7(\xi\cdot\eta)\varepsilon_\xi^2\otimes\varepsilon_\xi^1\otimes\varepsilon_\eta^3\otimes\varepsilon_\eta^3\\
&\quad-2\,\Phi_n^5(\xi\cdot\eta)(\varepsilon_\xi^3\otimes\varepsilon_\xi^1\otimes\varepsilon_\eta^2\otimes\varepsilon_\eta^3+\varepsilon_\xi^3\otimes\varepsilon_\xi^1\otimes\varepsilon_\eta^3\otimes\varepsilon_\eta^2),\\
\tilde{\mathbf{P}}_n^{(2,1,3,1)}(\xi,\eta) &= -\Phi_n^6(\xi\cdot\eta)\varepsilon_\xi^2\otimes\varepsilon_\xi^1\otimes\varepsilon_\eta^3\otimes\varepsilon_\eta^1-\Phi_n^2(\xi\cdot\eta)\varepsilon_\xi^3\otimes\varepsilon_\xi^1\otimes\varepsilon_\eta^2\otimes\varepsilon_\eta^1,\\
\tilde{\mathbf{P}}_n^{(2,1,3,2)}(\xi,\eta) &= -\Phi_n^7(\xi\cdot\eta)\varepsilon_\xi^2\otimes\varepsilon_\xi^1\otimes\varepsilon_\eta^2\otimes\varepsilon_\eta^3-\Phi_n^7(\xi\cdot\eta)\varepsilon_\xi^2\otimes\varepsilon_\xi^1\otimes\varepsilon_\eta^3\otimes\varepsilon_\eta^2\\
&\quad-2\,\Phi_n^5(\xi\cdot\eta)(\varepsilon_\xi^3\otimes\varepsilon_\xi^1\otimes\varepsilon_\eta^2\otimes\varepsilon_\eta^2-\varepsilon_\xi^3\otimes\varepsilon_\xi^1\otimes\varepsilon_\eta^3\otimes\varepsilon_\eta^3),\\
\tilde{\mathbf{P}}_n^{(2,1,3,3)}(\xi,\eta) &= \Phi_n^3(\xi\cdot\eta)\varepsilon_\xi^2\otimes\varepsilon_\xi^1\otimes\varepsilon_\eta^2\otimes\varepsilon_\eta^3-\Phi_n^3(\xi\cdot\eta)\varepsilon_\xi^2\otimes\varepsilon_\xi^1\otimes\varepsilon_\eta^3\otimes\varepsilon_\eta^2,\\
\tilde{\mathbf{P}}_n^{(2,2,2,2)}(\xi,\eta) &= \Phi_n^1(\xi\cdot\eta)\varepsilon_\xi^2\otimes\varepsilon_\xi^2\otimes\varepsilon_\eta^2\otimes\varepsilon_\eta^2+\Phi_n^1(\xi\cdot\eta)\varepsilon_\xi^2\otimes\varepsilon_\xi^2\otimes\varepsilon_\eta^3\otimes\varepsilon_\eta^3\\
&\quad+\Phi_n^1(\xi\cdot\eta)\varepsilon_\xi^3\otimes\varepsilon_\xi^3\otimes\varepsilon_\eta^2\otimes\varepsilon_\eta^2+\Phi_n^1(\xi\cdot\eta)\varepsilon_\xi^3\otimes\varepsilon_\xi^3\otimes\varepsilon_\eta^3\otimes\varepsilon_\eta^3,\\
\tilde{\mathbf{P}}_n^{(2,2,2,3)}(\xi,\eta) &= \Phi_n^4(\xi\cdot\eta)\varepsilon_\xi^2\otimes\varepsilon_\xi^2\otimes\varepsilon_\eta^2\otimes\varepsilon_\eta^2-\Phi_n^4(\xi\cdot\eta)\varepsilon_\xi^2\otimes\varepsilon_\xi^2\otimes\varepsilon_\eta^3\otimes\varepsilon_\eta^3\\
&\quad+\Phi_n^4(\xi\cdot\eta)\varepsilon_\xi^3\otimes\varepsilon_\xi^3\otimes\varepsilon_\eta^2\otimes\varepsilon_\eta^2-\Phi_n^4(\xi\cdot\eta)\varepsilon_\xi^3\otimes\varepsilon_\xi^3\otimes\varepsilon_\eta^3\otimes\varepsilon_\eta^3,\\
\tilde{\mathbf{P}}_n^{(2,2,3,1)}(\xi,\eta) &= -\Phi_n^3(\xi\cdot\eta)\varepsilon_\xi^2\otimes\varepsilon_\xi^2\otimes\varepsilon_\eta^3\otimes\varepsilon_\eta^1-\Phi_n^3(\xi\cdot\eta)\varepsilon_\xi^3\otimes\varepsilon_\xi^3\otimes\varepsilon_\eta^3\otimes\varepsilon_\eta^1,\\
\tilde{\mathbf{P}}_n^{(2,2,3,2)}(\xi,\eta) &= -\Phi_n^4(\xi\cdot\eta)\varepsilon_\xi^2\otimes\varepsilon_\xi^2\otimes\varepsilon_\eta^2\otimes\varepsilon_\eta^3-\Phi_n^4(\xi\cdot\eta)\varepsilon_\xi^2\otimes\varepsilon_\xi^2\otimes\varepsilon_\eta^3\otimes\varepsilon_\eta^2\\
&\quad-\Phi_n^4(\xi\cdot\eta)\varepsilon_\xi^3\otimes\varepsilon_\xi^3\otimes\varepsilon_\eta^2\otimes\varepsilon_\eta^3-\Phi_n^4(\xi\cdot\eta)\varepsilon_\xi^3\otimes\varepsilon_\xi^3\otimes\varepsilon_\eta^3\otimes\varepsilon_\eta^2,\\
\tilde{\mathbf{P}}_n^{(2,2,3,3)}(\xi,\eta) &= \Phi_n^1(\xi\cdot\eta)\varepsilon_\xi^2\otimes\varepsilon_\xi^2\otimes\varepsilon_\eta^2\otimes\varepsilon_\eta^3-\Phi_n^1(\xi\cdot\eta)\varepsilon_\xi^2\otimes\varepsilon_\xi^2\otimes\varepsilon_\eta^3\otimes\varepsilon_\eta^2\\
&\quad+\Phi_n^1(\xi\cdot\eta)\varepsilon_\xi^3\otimes\varepsilon_\xi^3\otimes\varepsilon_\eta^2\otimes\varepsilon_\eta^3-\Phi_n^1(\xi\cdot\eta)\varepsilon_\xi^3\otimes\varepsilon_\xi^3\otimes\varepsilon_\eta^3\otimes\varepsilon_\eta^2,\\
\tilde{\mathbf{P}}_n^{(2,3,2,3)}(\xi,\eta) &= \Phi_n^8(\xi\cdot\eta)\varepsilon_\xi^2\otimes\varepsilon_\xi^2\otimes\varepsilon_\eta^2\otimes\varepsilon_\eta^2-\Phi_n^8(\xi\cdot\eta)\varepsilon_\xi^2\otimes\varepsilon_\xi^2\otimes\varepsilon_\eta^3\otimes\varepsilon_\eta^3\\
&\quad+4\,\Phi_n^9(\xi\cdot\eta)(\varepsilon_\xi^2\otimes\varepsilon_\xi^3\otimes\varepsilon_\eta^2\otimes\varepsilon_\eta^3+\varepsilon_\xi^2\otimes\varepsilon_\xi^3\otimes\varepsilon_\eta^3\otimes\varepsilon_\eta^2)\\
&\quad+4\,\Phi_n^9(\xi\cdot\eta)(\varepsilon_\xi^3\otimes\varepsilon_\xi^2\otimes\varepsilon_\eta^2\otimes\varepsilon_\eta^3+\varepsilon_\xi^3\otimes\varepsilon_\xi^2\otimes\varepsilon_\eta^3\otimes\varepsilon_\eta^2)\\
&\quad-\Phi_n^8(\xi\cdot\eta)\varepsilon_\xi^3\otimes\varepsilon_\xi^3\otimes\varepsilon_\eta^2\otimes\varepsilon_\eta^2+\Phi_n^8(\xi\cdot\eta)\varepsilon_\xi^3\otimes\varepsilon_\xi^3\otimes\varepsilon_\eta^3\otimes\varepsilon_\eta^3,\\
\tilde{\mathbf{P}}_n^{(2,3,3,1)}(\xi,\eta) &= -\Phi_n^7(\xi\cdot\eta)\varepsilon_\xi^2\otimes\varepsilon_\xi^2\otimes\varepsilon_\eta^3\otimes\varepsilon_\eta^1-2\,\Phi_n^5(\xi\cdot\eta)\varepsilon_\xi^2\otimes\varepsilon_\xi^3\otimes\varepsilon_\eta^2\otimes\varepsilon_\eta^1\\
&\quad-2\,\Phi_n^5(\xi\cdot\eta)\varepsilon_\xi^3\otimes\varepsilon_\xi^2\otimes\varepsilon_\eta^2\otimes\varepsilon_\eta^1+\Phi_n^7(\xi\cdot\eta)\varepsilon_\xi^3\otimes\varepsilon_\xi^3\otimes\varepsilon_\eta^3\otimes\varepsilon_\eta^1,\\
\tilde{\mathbf{P}}_n^{(2,3,3,2)}(\xi,\eta) &= -\Phi_n^8(\xi\cdot\eta)\varepsilon_\xi^2\otimes\varepsilon_\xi^2\otimes\varepsilon_\eta^2\otimes\varepsilon_\eta^3-\Phi_n^8(\xi\cdot\eta)\varepsilon_\xi^2\otimes\varepsilon_\xi^2\otimes\varepsilon_\eta^3\otimes\varepsilon_\eta^2\\
&\quad+4\,\Phi_n^9(\xi\cdot\eta)(\varepsilon_\xi^2\otimes\varepsilon_\xi^3\otimes\varepsilon_\eta^2\otimes\varepsilon_\eta^2-\varepsilon_\xi^2\otimes\varepsilon_\xi^3\otimes\varepsilon_\eta^3\otimes\varepsilon_\eta^3)
\end{aligned}
$$

$$
\begin{aligned}
& +4\,\Phi_n^9(\xi\cdot\eta)(\varepsilon_\xi^3\otimes\varepsilon_\xi^2\otimes\varepsilon_\eta^2\otimes\varepsilon_\eta^2-\varepsilon_\xi^3\otimes\varepsilon_\xi^2\otimes\varepsilon_\eta^3\otimes\varepsilon_\eta^3)\\
& +\Phi_n^8(\xi\cdot\eta)\varepsilon_\xi^3\otimes\varepsilon_\xi^3\otimes\varepsilon_\eta^2\otimes\varepsilon_\eta^3+\Phi_n^8(\xi\cdot\eta)\varepsilon_\xi^3\otimes\varepsilon_\xi^3\otimes\varepsilon_\eta^3\otimes\varepsilon_\eta^2,\\
\tilde{\mathbf{P}}_n^{(2,3,3,3)}(\xi,\eta) &= \Phi_n^4(\xi\cdot\eta)\varepsilon_\xi^2\otimes\varepsilon_\xi^2\otimes\varepsilon_\eta^2\otimes\varepsilon_\eta^3-\Phi_n^4(\xi\cdot\eta)\varepsilon_\xi^2\otimes\varepsilon_\xi^2\otimes\varepsilon_\eta^3\otimes\varepsilon_\eta^2\\
& -\Phi_n^4(\xi\cdot\eta)\varepsilon_\xi^3\otimes\varepsilon_\xi^3\otimes\varepsilon_\eta^2\otimes\varepsilon_\eta^3+\Phi_n^4(\xi\cdot\eta)\varepsilon_\xi^3\otimes\varepsilon_\xi^3\otimes\varepsilon_\eta^3\otimes\varepsilon_\eta^2,\\
\tilde{\mathbf{P}}_n^{(3,1,3,1)}(\xi,\eta) &= -\Phi_n^2(\xi\cdot\eta)\varepsilon_\xi^2\otimes\varepsilon_\xi^1\otimes\varepsilon_\eta^2\otimes\varepsilon_\eta^1+\Phi_n^6(\xi\cdot\eta)\varepsilon_\xi^3\otimes\varepsilon_\xi^1\otimes\varepsilon_\eta^3\otimes\varepsilon_\eta^1,\\
\tilde{\mathbf{P}}_n^{(3,1,3,2)}(\xi,\eta) &= 2\,\Phi_n^5(\xi\cdot\eta)(\varepsilon_\xi^2\otimes\varepsilon_\xi^1\otimes\varepsilon_\eta^3\otimes\varepsilon_\eta^3-\varepsilon_\xi^2\otimes\varepsilon_\xi^1\otimes\varepsilon_\eta^2\otimes\varepsilon_\eta^2)\\
& +\Phi_n^7(\xi\cdot\eta)\varepsilon_\xi^3\otimes\varepsilon_\xi^1\otimes\varepsilon_\eta^2\otimes\varepsilon_\eta^3+\Phi_n^7(\xi\cdot\eta)\varepsilon_\xi^3\otimes\varepsilon_\xi^1\otimes\varepsilon_\eta^3\otimes\varepsilon_\eta^2,\\
\tilde{\mathbf{P}}_n^{(3,1,3,3)}(\xi,\eta) &= -\Phi_n^3(\xi\cdot\eta)\varepsilon_\xi^3\otimes\varepsilon_\xi^1\otimes\varepsilon_\eta^2\otimes\varepsilon_\eta^3+\Phi_n^3(\xi\cdot\eta)\varepsilon_\xi^3\otimes\varepsilon_\xi^1\otimes\varepsilon_\eta^3\otimes\varepsilon_\eta^2,\\
\tilde{\mathbf{P}}_n^{(3,2,3,2)}(\xi,\eta) &= 4\,\Phi_n^9(\xi\cdot\eta)(\varepsilon_\xi^2\otimes\varepsilon_\xi^2\otimes\varepsilon_\eta^2\otimes\varepsilon_\eta^2-\varepsilon_\xi^2\otimes\varepsilon_\xi^2\otimes\varepsilon_\eta^3\otimes\varepsilon_\eta^3)\\
& +\Phi_n^8(\xi\cdot\eta)\varepsilon_\xi^2\otimes\varepsilon_\xi^3\otimes\varepsilon_\eta^2\otimes\varepsilon_\eta^3+\Phi_n^8(\xi\cdot\eta)\varepsilon_\xi^2\otimes\varepsilon_\xi^3\otimes\varepsilon_\eta^3\otimes\varepsilon_\eta^2\\
& +\Phi_n^8(\xi\cdot\eta)\varepsilon_\xi^3\otimes\varepsilon_\xi^2\otimes\varepsilon_\eta^2\otimes\varepsilon_\eta^3+\Phi_n^8(\xi\cdot\eta)\varepsilon_\xi^3\otimes\varepsilon_\xi^2\otimes\varepsilon_\eta^3\otimes\varepsilon_\eta^2\\
& -4\,\Phi_n^9(\xi\cdot\eta)(\varepsilon_\xi^3\otimes\varepsilon_\xi^3\otimes\varepsilon_\eta^2\otimes\varepsilon_\eta^2-\varepsilon_\xi^3\otimes\varepsilon_\xi^3\otimes\varepsilon_\eta^3\otimes\varepsilon_\eta^3),\\
\tilde{\mathbf{P}}_n^{(3,2,3,3)}(\xi,\eta) &= -\Phi_n^4(\xi\cdot\eta)\varepsilon_\xi^2\otimes\varepsilon_\xi^3\otimes\varepsilon_\eta^2\otimes\varepsilon_\eta^3+\Phi_n^4(\xi\cdot\eta)\varepsilon_\xi^2\otimes\varepsilon_\xi^3\otimes\varepsilon_\eta^3\otimes\varepsilon_\eta^2\\
& -\Phi_n^4(\xi\cdot\eta)\varepsilon_\xi^3\otimes\varepsilon_\xi^2\otimes\varepsilon_\eta^2\otimes\varepsilon_\eta^3+\Phi_n^4(\xi\cdot\eta)\varepsilon_\xi^3\otimes\varepsilon_\xi^2\otimes\varepsilon_\eta^3\otimes\varepsilon_\eta^2,\\
\tilde{\mathbf{P}}_n^{(3,3,3,3)}(\xi,\eta) &= \Phi_n^1(\xi\cdot\eta)\varepsilon_\xi^2\otimes\varepsilon_\xi^3\otimes\varepsilon_\eta^2\otimes\varepsilon_\eta^3-\Phi_n^1(\xi\cdot\eta)\varepsilon_\xi^2\otimes\varepsilon_\xi^3\otimes\varepsilon_\eta^3\otimes\varepsilon_\eta^2\\
& +\Phi_n^1(\xi\cdot\eta)(\varepsilon_\xi^3\otimes\varepsilon_\xi^2\otimes\varepsilon_\eta^3\otimes\varepsilon_\eta^2-\varepsilon_\xi^3\otimes\varepsilon_\xi^2\otimes\varepsilon_\eta^2\otimes\varepsilon_\eta^3).
\end{aligned}
$$

Remark. (i) The cases not listed in Theorem 14.6.3 can be easily deduced from symmetry arguments. (ii) If $\xi = \pm\eta$ the value of $\mathbf{P}_n^{(i,k,l,m)}(\xi, \pm\xi)$ can be evaluated by taking the limit $\eta \to \xi$ and (if necessary) via the relations (14.3.7) and (14.3.8).

Since the cases $(i, k) = (l, m)$ are of particular importance (e.g. for the formulation of reproducing kernels in spline interpolation (cf. Chapter 15) we list the explicit representation of $\mathbf{P}_n^{(i,k,i,k)}(\xi, \xi)$ in the next theorem. The proof follows by taking the limit $\eta \to \xi$, keeping also in mind that $P_n(1) = 1$ and $P_n'(1) = n(n+1)/2$.

In doing so, we introduce two abbreviations for special tensors which turn out to be useful in the formulation of the next theorem, namely:

$$
\begin{aligned}
\mathbf{A} &= \sum_{i=1}^{3}\varepsilon^i\otimes\varepsilon^i\otimes\varepsilon^i\otimes\varepsilon^i-\sum_{\substack{i,k=1\\ i\neq k}}^{3}\varepsilon^i\otimes\varepsilon^i\otimes\varepsilon^k\otimes\varepsilon^k,\\
\mathbf{B} &= \sum_{\substack{i,k=1\\ i\neq k}}^{3}\left(\varepsilon^i\otimes\varepsilon^k\otimes\varepsilon^i\otimes\varepsilon^k+\varepsilon^i\otimes\varepsilon^k\otimes\varepsilon^k\otimes\varepsilon^i\right),
\end{aligned}
$$

where $\{\varepsilon^1, \varepsilon^2, \varepsilon^3\}$ is the canonical orthonormal basis in $\mathbb{R}^3$. It can be easily seen, that these definitions do not depend on the special choice of the orthonormal basis. In analogy to the definition of $\mathbf{p}_{\tan,\tan}$ in Section 14.1 we are able to introduce a similar operator $\mathbf{P}_{\tan}$, projecting a rank-4 tensor on its tangential part in $\mathbb{R}^3 \otimes \mathbb{R}^3 \otimes \mathbb{R}^3 \otimes \mathbb{R}^3$. We finally define

$$\begin{aligned}
\mathbf{A}_{\tan}(\xi) &= \mathbf{P}_{\tan}\mathbf{A}(\xi), \ \xi \in \Omega, \\
\mathbf{B}_{\tan}(\xi) &= \mathbf{P}_{\tan}\mathbf{B}(\xi), \ \xi \in \Omega.
\end{aligned}$$

Now we are prepared to formulate the following theorem.

Theorem 14.6.4. *Let $n \in \mathbb{N}_0$, $\xi \in \Omega$. Then the following statements are valid:*

$$\begin{aligned}
\mathbf{P}_n^{(1,1,1,1)}(\xi,\xi) &= \xi \otimes \xi \otimes \xi \otimes \xi, \\
\mathbf{P}_n^{(1,2,1,2)}(\xi,\xi) &= \frac{1}{2}\left(\sum_{i=1}^{3} \xi \otimes \varepsilon^i \otimes \xi \otimes \varepsilon^i - \xi \otimes \xi \otimes \xi \otimes \xi\right), \\
\mathbf{P}_n^{(1,3,1,3)}(\xi,\xi) &= \frac{1}{2}\left(\sum_{i=1}^{3} \xi \otimes \varepsilon^i \otimes \xi \otimes \varepsilon^i - \xi \otimes \xi \otimes \xi \otimes \xi\right), \\
\mathbf{P}_n^{(2,1,2,1)}(\xi,\xi) &= \frac{1}{2}\left(\sum_{i=1}^{3} \varepsilon^i \otimes \xi \otimes \varepsilon^i \otimes \xi - \xi \otimes \xi \otimes \xi \otimes \xi\right), \\
\mathbf{P}_n^{(2,2,2,2)}(\xi,\xi) &= \frac{1}{2}\mathbf{i}_{\tan}(\xi) \otimes \mathbf{i}_{\tan}(\xi), \\
\mathbf{P}_n^{(2,3,2,3)}(\xi,\xi) &= -\frac{1}{2}\frac{n(n+1)+4}{n(n+1)-2}\mathbf{A}_{\tan}(\xi) + \frac{n(n+1)+1}{n(n+1)-2}\mathbf{B}_{\tan}(\xi), \\
\mathbf{P}_n^{(3,1,3,1)}(\xi,\xi) &= \frac{1}{2}n(n+1)\left(\sum_{i=1}^{3} \varepsilon^i \otimes \xi \otimes \varepsilon^i \otimes \xi - \xi \otimes \xi \otimes \xi \otimes \xi\right), \\
\mathbf{P}_n^{(3,2,3,2)}(\xi,\xi) &= \frac{n(n+1)+1}{n(n+1)-2}\mathbf{A}_{\tan}(\xi) - \frac{1}{2}\frac{n(n+1)+4}{n(n+1)-2}\mathbf{B}_{\tan}(\xi), \\
\mathbf{P}_n^{(3,3,3,3)}(\xi,\xi) &= \frac{1}{2}\mathbf{j}_{\tan}(\xi) \otimes \mathbf{j}_{\tan}(\xi).
\end{aligned}$$

Observing (14.3.7) and (14.3.8), the values of $\mathbf{P}_n^{(i,k,i,k)}(\xi, -\xi)$ can be derived from this theorem.

Consider a rank-4 tensor

$$\mathbf{T} = \sum_{i=1}^{3}\sum_{j=1}^{3}\sum_{k=1}^{3}\sum_{l=1}^{3} T_{ijkl}\varepsilon^i \otimes \varepsilon^j \otimes \varepsilon^k \otimes \varepsilon^l.$$

Then its trace is defined by (cf. M.E. GURTIN (1971))

$$\text{trace } \mathbf{T} = \sum_{i=1}^{3} \sum_{k=1}^{3} T_{ikik}.$$

Using this definition we easily see that the trace of $\mathbf{P}_n^{(i,k,i,k)}(\xi, \xi)$ is given by

$$\text{trace } \mathbf{P}_n^{(i,k,i,k)}(\xi, \xi) = 1.$$

Hence, the next result is a direct consequence of Theorem 14.6.1.

Lemma 14.6.5. *For the orthonormal system* $\left\{\mathbf{y}_{n,j}^{(i,k)}\right\}_{j=1,\ldots,2n+1}$ *of tensor spherical harmonics of order* n *we have for* $(i,k) \in \{(1,1),(1,2),\ldots,(3,3)\}$, $n \geq 0_{i,k}$ *and* $\xi \in \Omega$

$$\sum_{j=1}^{2n+1} \mathbf{y}_{n,j}^{(i,k)}(\xi) \cdot \mathbf{y}_{n,j}^{(i,k)}(\xi) = \frac{2n+1}{4\pi}. \tag{14.6.4}$$

Every tensor spherical harmonic $\mathbf{y}_n^{(i,k)} \in \mathbf{harm}_n^{(i,k)}$ can be written as linear combination $\mathbf{y}_n^{(i,k)} = \sum_{j=1}^{2n+1} c_j \mathbf{y}_{n,j}^{(i,k)}$ with $c_j = (\mathbf{y}_n^{(i,k)}, \mathbf{y}_{n,j}^{(i,k)})_{\mathbf{l}^2(\Omega)}$. By the Cauchy–Schwarz inequality we obtain for $\xi \in \Omega$

$$\left|\mathbf{y}_n^{(i,k)}(\xi)\right|^2 \leq \left(\sum_{j=1}^{2n+1} c_j^2\right) \left(\sum_{j=1}^{2n+1} \left|\mathbf{y}_{n,j}^{(i,k)}(\xi)\right|^2\right).$$

By Parseval's identity we have

$$\sum_{j=1}^{2n+1} c_j^2 = \sum_{j=1}^{2n+1} (\mathbf{y}_n^{(i,k)}, \mathbf{y}_{n,j}^{(i,k)})_{\mathbf{l}^2(\Omega)}^2 = \left\|\mathbf{y}_n^{(i,k)}\right\|_{\mathbf{l}^2(\Omega)}^2.$$

Therefore we find in connection with Lemma 14.6.5

Lemma 14.6.6. *For any tensor spherical harmonic* $\mathbf{y}_n^{(i,k)} \in \mathbf{harm}_n^{(i,k)}$

$$\sup_{\xi \in \Omega} \left|\mathbf{y}_n^{(i,k)}(\xi)\right| \leq \sqrt{\frac{2n+1}{4\pi}} \left\|\mathbf{y}_n^{(i,k)}\right\|_{\mathbf{l}^2(\Omega)}.$$

Similar to the vectorial case we are able to generalize the definition of the operators $O^{(i,k)}$ to tensor fields of fourth rank. If

$$\mathbf{F}(\xi) = \sum_{p=1}^{3}\sum_{q=1}^{3}\sum_{r=1}^{3}\sum_{s=1}^{3} F_{pqrs}(\xi)\varepsilon^p \otimes \varepsilon^q \otimes \varepsilon^r \otimes \varepsilon^s, \quad \xi \in \Omega,$$

is sufficiently smooth, then we set

$$O_\xi^{(i,k)}\mathbf{F}(\xi) = \sum_{r=1}^{3}\sum_{s=1}^{3} O_\xi^{(i,k)} \left(\sum_{p=1}^{3}\sum_{q=1}^{3} F_{pqrs}(\xi)\varepsilon^p \otimes \varepsilon^q\right) \varepsilon^r \otimes \varepsilon^s.$$

It follows from the previous investigations that $\mathbf{P}_n^{(i,k,i,k)}(\cdot,\cdot)$ is the reproducing kernel of $\mathbf{harm}_n^{(i,k)}$ in the sense that

(i) for all $\xi \in \Omega$

$$O_\xi^{(i,k)}\mathbf{P}_n^{(i,k,i,k)}(\xi,\cdot) \in \mathbf{harm}_n^{(i,k)}$$

(ii) for every $\mathbf{f} \in \mathbf{harm}_n^{(i,k)}$ and all $\xi \in \Omega$

$$O_\xi^{(i,k)}\mathbf{f}(\xi) = (O_\xi^{(i,k)}\mathbf{P}_n^{(i,k,i,k)}(\xi,\cdot),\mathbf{f})_{\mathbf{l}^2(\Omega)}$$

At the end of this section we want to write down the analogue of the estimate $|P_n(t)| \leq 1$, $t \in [-1,+1]$ for the Legendre tensor $\mathbf{P}_n^{(i,k,l,m)}(\xi,\eta)$. Let $\mathbf{a} \in \mathbb{R}^3 \otimes \mathbb{R}^3$ and $\eta \in \Omega$ be fixed. Then

$$\mathbf{P}_n^{(i,k,l,m)}(\cdot,\eta)\cdot\mathbf{a} = \frac{4\pi}{2n+1}\sum_{j=1}^{2n+1}\left(\mathbf{y}_{n,j}^{(l,m)}(\eta)\cdot\mathbf{a}\right)\mathbf{y}_{n,j}^{(i,k)}(\cdot) \tag{14.6.5}$$

is a tensor spherical harmonic of order n and type (i,k). Hence, we obtain from Lemma 14.6.6 and (14.6.4) that

$$\begin{aligned}
\left|\mathbf{P}_n^{(i,k,l,m)}(\xi,\eta)\cdot\mathbf{a}\right|^2 &\leq \frac{2n+1}{4\pi}\left\|\frac{4\pi}{2n+1}\sum_{j-1}^{2n+1}\left(\mathbf{y}_{n,j}^{(l,m)}(\eta)\cdot\mathbf{a}\right)\right\|_{\mathbf{l}^2(\Omega)}^2 \\
&= \frac{4\pi}{2n+1}\sum_{j=1}^{2n+1}\left(\mathbf{y}_{n,j}^{(l,m)}(\eta)\cdot\mathbf{a}\right)^2 \\
&\leq |\mathbf{a}|^2.
\end{aligned}$$

But this leads us to the following result.

Lemma 14.6.7. *Let $i,k,l,m,p,q \in \{1,2,3\}$. Then for all $\xi,\eta \in \Omega$*

$$\left|\mathbf{P}_n^{(i,k,l,m)}(\xi,\eta)\cdot(\varepsilon^p\otimes\varepsilon^q)\right| \leq 1.$$

14.7 Tensorial Funk–Hecke Formula

As in the vectorial case, our purpose is to prove two different generalizations of the Funk–Hecke formula known from the scalar theory. In order to formulate these variants we have to solve the following problems:

(i) Let $\eta \in \Omega$ be fixed and $\mathbf{h}(\cdot, \eta) \in \mathbf{c}^{(2)}(\Omega)$ be invariant with respect to rotations $\mathbf{t} \in SO(3)$ satisfying $\mathbf{t}\eta = \eta$. What is the value of

$$\int_\Omega \mathbf{h}(\xi, \eta) \cdot \mathbf{y}_n^{(i,k)}(\xi)\, d\omega(\xi) \tag{14.7.1}$$

for $\mathbf{y}_n^{(i,k)} \in \mathbf{harm}_n^{(i,k)}$?

(ii) Let $H \in \mathcal{L}^1[-1,1]$. Determine

$$\int_\Omega H(\xi \cdot \eta)\mathbf{y}_n^{(i,k)}(\xi)\, d\omega(\xi) \tag{14.7.2}$$

for $\mathbf{y}_n^{(i,k)} \in \mathbf{harm}_n^{(i,k)}$.

Notice that (14.7.1) is scalar-valued, while the value of (14.7.2) is a tensor. This difference goes along with the different methods of investigating the two formulae. For the first version of the Funk–Hecke formula we have to investigate rotational invariant tensor fields (i.e. we have to consider representations of the group $SO(3)$); the second one can be established using the relations between tensor spherical harmonics and homogeneous harmonic tensor polynomials as described in Section 14.3.

Let us first generalize the definition of the operator $R_\mathbf{t}$ (cf. Definition 12.7.1) to the tensorial situation:

Definition 14.7.1. Let $\mathbf{f}$ be a tensor field of class $\mathbf{l}^2(\Omega)$ and let $\mathbf{t} \in SO(3)$. Then we set

$$R_\mathbf{t}\mathbf{f}(\xi) = \mathbf{t}^T\mathbf{f}(\mathbf{t}\xi)\mathbf{t}, \quad \xi \in \Omega.$$

Let $G \subset SO(3)$ be a subgroup of $SO(3)$. A subspace $\mathbf{v} \subset \mathbf{l}^2(\Omega)$ is called *invariant with respect to* G or simply G*-invariant* if $\mathbf{f} \in \mathbf{v}$ implies that $R_\mathbf{t}\mathbf{f} \in \mathbf{v}$ for all $\mathbf{t} \in G$. If a G-invariant subspace $\mathbf{v}$ does not contain a subspace which is also G-invariant (besides $\mathbf{v}$ itself) then $\mathbf{v}$ is called *irreducible*.

It is obvious that for $\mathbf{f}, \mathbf{g} \in \mathbf{l}^2(\Omega)$ we have

$$(R_\mathbf{t}\mathbf{f}, \mathbf{g})_{\mathbf{l}^2(\Omega)} = (\mathbf{f}, R_{\mathbf{t}^T}\mathbf{g})_{\mathbf{l}^2(\Omega)},$$

i.e. $R_{\mathbf{t}^T}$ is the adjoint operator of $R_\mathbf{t}$. For $F \in \mathcal{C}^{(2)}(\Omega)$ and $\mathbf{f} \in \mathbf{c}^{(2)}(\Omega)$ it can be easily verified that

$$\mathbf{o}_\xi^{(i,k)} R_\mathbf{t} F(\xi) = R_\mathbf{t} \mathbf{o}_\xi^{(i,k)} F(\xi), \tag{14.7.3}$$

$$O_\xi^{(i,k)} R_{\mathbf{t}} \mathbf{f}(\xi) = R_{\mathbf{t}} O_\xi^{(i,k)} \mathbf{f}(\xi) \tag{14.7.4}$$

hold for all $\xi \in \Omega$ and $(i,k) \in \{(1,1),(1,2),\ldots,(3,3)\}$.

Therefore we find, in connection with the results of Section 3.3, the following fundamental properties (cf. also Lemma 12.7.2).

Lemma 14.7.2.

(i) *For all* $(i,k) \in \{(1,1),(1,2),\ldots,(3,3)\}$ *the space* $\mathbf{l}^2_{(i,k)}(\Omega) \subset \mathbf{l}^2(\Omega)$ *is* $SO(3)$*-invariant.*

(ii) *The set* $\mathbf{harm}_n^{(i,k)}$ *is an irreducible invariant subspace of* $\mathbf{l}^2(\Omega)$ *with respect to* $SO(3)$ *for all* $(i,k) \in \{(1,1),(1,2),\ldots,(3,3)\}$ *and all* $n \geq 0_{i,k}$.

In analogy to the proof of Lemma 12.7.3 we find the following.

Lemma 14.7.3. *Let* $\eta \in \Omega$ *be fixed.*

(i) *If* $\mathbf{f} \in \mathbf{c}^{(2)}(\Omega)$ *with* $R_{\mathbf{t}}\mathbf{f} = \mathbf{f}$ *for all* $\mathbf{t} \in SO_\eta(3)$*, then there exist* $F_{i,k} \in C^{(0)}[-1,1]$ *for* $i,k \in \{1,2,3\}$ *such that*

$$O^{(i,k)}\mathbf{f}(\xi) = F_{i,k}(\xi \cdot \eta), \quad \xi \in \Omega.$$

(ii) *Let* $(i,k) \in \{(1,1),(1,2),\ldots,(3,3)\}$ *and* $\mathbf{y}_n^{(i,k)} \in \mathbf{harm}_n^{(i,k)}$ *such that* $R_{\mathbf{t}}\mathbf{y}_n^{(i,k)} = \mathbf{y}_n^{(i,k)}$ *for all* $\mathbf{t} \in SO_\eta(3)$*. Then there exists a constant* $C \in \mathbb{R}$ *such that*

$$\mathbf{y}_n^{(i,k)}(\xi) = C\, \mathbf{o}_\xi^{(i,k)} P_n(\xi \cdot \eta), \quad \xi \in \Omega.$$

Suppose that $\eta \in \Omega$ is fixed. Assume that $\mathbf{h}(\cdot,\eta) \in \mathbf{c}^{(2)}(\Omega)$ with $R_{\mathbf{t}}\mathbf{h}(\xi,\eta) = \mathbf{h}(\xi,\eta)$ for all $\mathbf{t} \in SO_\eta(3)$. Then we know that, for $(i,k) \in \{(1,1),(1,2),\ldots,(3,3)\}$, the function $O_\xi^{(i,k)}\mathbf{h}(\xi,\eta) = H_{i,k}(\xi \cdot \eta)$ depends only on the inner product $\xi \cdot \eta$. Thus, we may define

$$\left(O^{(i,k)}\mathbf{h}\right)^\wedge(n) = 2\pi \int_{-1}^{1} H_{i,k}(t) P_n(t)\, dt. \tag{14.7.5}$$

We obtain, for $\mathbf{y}_n^{(i,k)} = \mathbf{o}^{(i,k)} Y_n \in \mathbf{harm}_n^{(i,k)}$, that

$$\begin{aligned}\int_\Omega \mathbf{h}(\xi,\eta) \cdot \mathbf{y}_n^{(i,k)}(\xi)\, d\omega(\xi) &= \int_\Omega O_\xi^{(i,k)} \mathbf{h}(\xi,\eta) Y_n(\xi)\, d\omega(\xi) \\ &= \left(O^{(i,k)}\mathbf{h}\right)^\wedge(n) Y_n(\eta).\end{aligned}$$

This proves the *first version of the tensorial Funk–Hecke formulae.*

Theorem 14.7.4. *Let $\eta \in \Omega$ be fixed. Assume that $\mathbf{h}(\cdot, \eta) \in \mathbf{c}^{(2)}(\Omega)$ satisfies $R_{\mathbf{t}}\mathbf{h}(\xi, \eta) = \mathbf{h}(\xi, \eta)$ for all $\mathbf{t} \in SO_\eta(3)$ and $\xi \in \Omega$. Then, for $(i,k) \in \{(1,1),(1,2),\ldots,(3,3)\}$ and $\mathbf{y}_n^{(i,k)} \in \mathbf{harm}_n^{(i,k)}$, $n \geq 0_{i,k}$,*

$$\int_\Omega \mathbf{h}(\xi, \eta) \cdot \mathbf{y}_n^{(i,k)}(\xi)\, d\omega(\xi) = (\mu_n^{(i,k)})^{-1} \left(O^{(i,k)}\mathbf{h}\right)^\wedge (n)\, O_\eta^{(i,k)} \mathbf{y}_n^{(i,k)}(\eta),$$

where $\left(O^{(i,k)}\mathbf{h}\right)^\wedge (n)$ is defined in (14.7.5).

Next, we come to a second tensorial version of the Funk–Hecke formula, as already anounced in (14.7.2). Let $(i,k) \in \{(1,1),(1,2),\ldots,(3,3)\}$. Consider $\mathbf{k}_n^{(i,k)} r^n Y_n|_{r=1}$, $n \in \mathbb{N}_0$, $Y_n \in Harm_n$, which are the restrictions of homogeneous harmonic tensor polynomials to the unit sphere. The cartesian components of $\mathbf{k}_n^{(i,k)} r^n Y_n|_{r=1}$ are spherical harmonics of order $\deg^{(i,k)}(n)$ (cf. Lemma 14.3.3). Hence, it follows for $H \in \mathcal{L}^1[-1,1]$ and all $\eta \in \Omega$ that

$$\int_\Omega H(\xi \cdot \eta) \mathbf{k}_n^{(i,k)} r^n Y_n(\xi)|_{r=1}\, d\omega(\xi) = H^\wedge(\deg^{(i,k)}(n)) \mathbf{k}_n^{(i,k)} r^n Y_n(\eta)|_{r=1}. \tag{14.7.6}$$

It has been shown in Section 14.3 that every tensor spherical harmonic $\mathbf{y}_n \in \mathbf{harm}_n$ of order n can be expressed as linear combination of restrictions of homogeneous harmonic tensor polynomials of degrees $n-2, \ldots, n+2$ (cf. (14.3.1) and (14.3.2)). Therefore, with Theorem 3.6.1, the transformation matrices (14.3.3), (14.3.4), and their inverses (14.3.5), (14.3.6), we arrive at the following result which is the *second tensorial version of the Funk–Hecke formulae.*

Theorem 14.7.5. *Let $Y_n \in Harm_n$ be a spherical harmonic of order n, $H \in \mathcal{L}^1[-1,1]$, and $\eta \in \Omega$ fixed.*

If $n = 0$, then we have

$$\int_\Omega H(\xi \cdot \eta) \begin{pmatrix} \mathbf{o}_\xi^{(1,1)} Y_0(\xi) \\ \mathbf{o}_\xi^{(2,2)} Y_0(\xi) \\ \mathbf{o}_\xi^{(3,3)} Y_0(\xi) \end{pmatrix} d\omega(\xi)$$

$$= \begin{pmatrix} \frac{1}{3}\left(H^\wedge(0) + 2H^\wedge(2)\right) \mathbf{o}_\eta^{(1,1)} Y_0(\eta) + \frac{1}{3}\left(H^\wedge(0) - H^\wedge(2)\right) \mathbf{o}_\eta^{(2,2)} Y_0(\eta) \\ \frac{2}{3}\left(H^\wedge(0) - H^\wedge(2)\right) \mathbf{o}_\eta^{(1,1)} Y_0(\eta) + \frac{1}{3}\left(2H^\wedge(0) + H^\wedge(2)\right) \mathbf{o}_\eta^{(2,2)} Y_0(\eta) \\ H^\wedge(1) \mathbf{o}_\eta^{(3,3)} Y_0(\eta) \end{pmatrix}$$

If $n \geq 1$, then

$$\int_\Omega H(\xi\cdot\eta)\begin{pmatrix} \mathbf{o}_\xi^{(1,1)}Y_n(\xi) \\ \mathbf{o}_\xi^{(1,2)}Y_n(\xi) \\ \mathbf{o}_\xi^{(2,1)}Y_n(\xi) \\ \mathbf{o}_\xi^{(2,2)}Y_n(\xi) \\ \mathbf{o}_\xi^{(2,3)}Y_n(\xi) \end{pmatrix} d\omega(\xi)$$

$$= \quad (H^\wedge(n-2)M_{n-2} + H^\wedge(n)M_n + H^\wedge(n+2)M_{n+2}) \begin{pmatrix} \mathbf{o}_\eta^{(1,1)}Y_n(\eta) \\ \mathbf{o}_\eta^{(1,2)}Y_n(\eta) \\ \mathbf{o}_\eta^{(2,1)}Y_n(\eta) \\ \mathbf{o}_\eta^{(2,2)}Y_n(\eta) \\ \mathbf{o}_\eta^{(2,3)}Y_n(\eta) \end{pmatrix},$$

$$\int_\Omega H(\xi\cdot\eta)\begin{pmatrix} \mathbf{o}_\xi^{(1,3)}Y_n(\xi) \\ \mathbf{o}_\xi^{(3,1)}Y_n(\xi) \\ \mathbf{o}_\xi^{(3,2)}Y_n(\xi) \\ \mathbf{o}_\xi^{(3,3)}Y_n(\xi) \end{pmatrix} d\omega(\xi)$$

$$= \quad (H^\wedge(n-1)M_{n-1} + H^\wedge(n+1)M_{n+1}) \begin{pmatrix} \mathbf{o}_\eta^{(1,3)}Y_n(\eta) \\ \mathbf{o}_\eta^{(3,1)}Y_n(\eta) \\ \mathbf{o}_\eta^{(3,2)}Y_n(\eta) \\ \mathbf{o}_\eta^{(3,3)}Y_n(\eta) \end{pmatrix},$$

where the matrices $M_{n-2}, \ldots M_{n+2}$ *are given by*

$$(2n+1)M_{n-2} =$$

$$\begin{pmatrix} \frac{(2+n)(n+1)}{4n+6} & \frac{(2+n)(n+1)(n-1)}{(4n-2)n} & \frac{-(2+n)^2}{4n+6} & \frac{(2+n)(n+1)}{4n-2} & \frac{(2+n)(2n+1)}{2n} \\ \frac{(1-n)n}{4n+6} & \frac{(n-1)^2}{4n-2} & \frac{(n-1)n(2+n)}{(4n+6)(n+1)} & \frac{(1-n)n}{4n-2} & \frac{(1-n)(2n+1)}{2n+2} \\ \frac{(2+n)(n+1)}{-(4n+6)} & \frac{(2+n)(n+1)(n-1)}{(4n-2)n} & \frac{(2+n)^2}{4n+6} & \frac{(2+n)(n+1)}{2-4n} & \frac{(n+2)(2n+1)}{-2n} \\ \frac{(n-1)n}{4n+6} & \frac{(n-1)^2}{2-4n} & \frac{(1-n)n(2+n)}{(4n+6)(n+1)} & \frac{(n-1)n}{4n-2} & \frac{(n-1)(2n+1)}{2n+2} \\ \frac{(n+1)n}{4n+6} & \frac{(1-n)(n+1)}{4n-2} & \frac{-(2+n)n}{4n+6} & \frac{(n+1)n}{(4n-2)} & \frac{2n+1}{2} \end{pmatrix},$$

$$(2n+1)M_n =$$

$$\begin{pmatrix} \frac{n(7+5n)}{4n+6} & \frac{(2+n)(n+1)^2}{(2-4n)n} & \frac{(2+n)n}{-(4n+6)} & \frac{(6+3n)(n+1)}{2-4n} & \frac{(2+n)(2n+1)}{-2n} \\ \frac{(n+1)n}{-(4n+6)} & \frac{(n+1)(5n-3)}{4n-2} & \frac{\left(2-3n^2-3n\right)n}{(4n+6)(n+1)} & \frac{(n+1)n}{2-4n} & \frac{(n-1)2n+1)}{2n+2} \\ \frac{(n+1)n}{-(4n+6)} & \frac{(n+1)\left(2-3n^2-3n\right)}{(4n-2)n} & \frac{n(5n+8)}{4n+6} & \frac{(n+1)n}{2-4n} & \frac{(2+n)(2n+1)}{2n} \\ \frac{(3-3n)n}{4n+6} & \frac{(1-n)(n+1)}{4n-2} & \frac{(1-n)n^2}{(4n+6)(n+1)} & \frac{(n+1)(5n-2)}{4n-2} & \frac{(1-n)(2n+1)}{2n+2} \\ \frac{(n+1)n}{-(4n+6)} & \frac{(n-1)(n+1)}{4n-2} & \frac{(2+n)n}{4n+6} & \frac{(n+1)n}{2-4n} & \frac{2n+1}{2} \end{pmatrix},$$

$$(2n+1)M_{n+2} =$$
$$\begin{pmatrix} \frac{(2+n)(n+1)}{2n+3} & \frac{(2+n)(n+1)}{2n-1} & \frac{(2+n)(n+1)}{2n+3} & \frac{(2+n)(n+1)}{2n-1} & 0 \\ \frac{n^2}{2n+3} & \frac{n^2}{2n-1} & \frac{n^2}{2n+3} & \frac{n^2}{2n-1} & 0 \\ \frac{(n+1)^2}{2n+3} & \frac{(n+1)^2}{2n-1} & \frac{(n+1)^2}{2n+3} & \frac{(n+1)^2}{2n-1} & 0 \\ \frac{(n-1)n}{2n+3} & \frac{(n-1)n}{2n-1} & \frac{(n-1)n}{2n+3} & \frac{(n-1)n}{2n-1} & 0 \\ 0 & 0 & 0 & 0 & 0 \end{pmatrix},$$

$$(2n+1)M_{n-1} =$$
$$\begin{pmatrix} \frac{2+n^2+n}{2n+2} & -\frac{2+n^2+n}{2n} & \frac{(1-n)n}{2n+2} & \frac{(2+n)(n+1)}{2n} \\ -\frac{2+n^2+n}{2n+2} & \frac{2+n^2+n}{2n} & \frac{(n-1)n}{2n+2} & -\frac{(2+n)(n+1)}{2n} \\ \frac{(1-n)(2+n)}{2n+2} & \frac{(n-1)(2+n)}{2n} & \frac{(2+n)(3n+1)}{2n+2} & \frac{(2+n)(n+1)}{2n} \\ \frac{(n-1)(2+n)}{2n+2} & \frac{(1-n)(2+n)}{2n} & \frac{(n-1)n}{2n+2} & \frac{(n-1)(3n+2)}{2n} \end{pmatrix},$$

$$(2n+1)M_{n+1} =$$
$$\begin{pmatrix} \frac{(3n+5)n}{2n+2} & \frac{2+n^2+n}{2n} & \frac{(n-1)n}{2n+2} & -\frac{(2+n)(n+1)}{2n} \\ \frac{2+n^2+n}{2n+2} & \frac{(n+1)(3n-2)}{2n} & \frac{(1-n)n}{2n+2} & \frac{(2+n)(n+1)}{2n} \\ \frac{(n-1)(2+n)}{2n+2} & \frac{(1-n)(2+n)}{2n} & \frac{(n-1)n}{2n+2} & -\frac{(2+n)(n+1)}{2n} \\ \frac{(1-n)(2+n)}{2n+2} & \frac{(n-1)(2+n)}{2n} & \frac{(1-n)n}{2n+2} & \frac{(2+n)(n+1)}{2n} \end{pmatrix}.$$

Bibliographical Notes

Tensor spherical harmonics are used in many fields of application, and therefore many different approaches can be found in the literature: see for example G.E. BACKUS (1966, 1967), F.A. DAHLEN, M.L. SMITH (1975), W. FREEDEN *et al.* (1994), R.W. JAMES (1976), M.N. JONES (1980), and F.J. ZERILLI (1970). K.S. THORNE (1980) has collected many of the former studies in a review paper. Our approach is a straightforward generalization of the vector spherical harmonic theory. The tensor spherical harmonics of pure normal and mixed type are also considered by F.J. ZERILLI (1970), while the tangential ones have also been considered in the work of G.E. BACKUS (1966, 1967). The decomposition theorem has been proved by W. FREEDEN *et al.* (1994), where also a detailed view on tensor spherical splines is worked out. An intrinsic approach which emphasizes the tensor spherical harmonics to be eigenfunctions of a tensorial Beltrami operator is due to M. SCHREINER (1994).

15

TENSORIAL APPROXIMATION METHODS

Following the formalism established for tensor spherical harmonics we are now able to generalize our approximation methods directly to the tensorial framework. This can be done quite similarly to the method indicated in Chapter 13 for vector fields, which is why we only state the keypoints here.

Taking the orthonormal basis $\{\mathbf{y}_{n,j}^{(i,k)}\}_{i,k=1,2,3,n=0_{i,k},\ldots,j=1,\ldots,2n+1}$ defined in (14.2.3) we are able to introduce Sobolev spaces $\mathbf{h}^{(i,k)}(\{A_n\};\Omega)$ associated to a suitably given sequence $\{A_n\} \in \mathcal{A}$. Invariant pseudodifferential operators mapping scalar functions to tensor fields, tensor fields to scalar functions, and tensor fields to tensor fields can then be introduced in a straightforward way. In particular, the operators $\mathbf{o}^{(i,i)}$ and $O^{(i,i)}$, $i = 1,2,3$, can be interpreted to be pseudodifferential operators of order 0, while the operators $\mathbf{o}^{(i,k)}$ and $O^{(i,k)}$ are of order 1 for $(i,k) \in \{(1,2),(1,3),(2,1),(3,1)\}$, or of order 2 if $(i,k) \in \{(2,3),(3,2)\}$, respectively. Their symbols are given by

$$
\begin{aligned}
(\mathbf{o}^{(i,k)})^\wedge(n) &= (O^{(i,k)})^\wedge(n) \\
&= \begin{cases} 1 & \text{if } (i,k) = (1,1) \\ \sqrt{2} & \text{if } (i,k) \in \{(2,2),(3,3)\} \\ \sqrt{n(n+1)} & \text{if } (i,k) \in \{(1,2),(1,3),(2,1),(3,1)\} \\ \sqrt{n(n+1)(n(n+1)-2)} & \text{if } (i,k) \in \{(2,3),(3,2)\}. \end{cases}
\end{aligned}
$$

In order to develop a tensorial spline theory we have to start from a $\{(n+1/2)^{0_{i,k}}\}$-summable sequence $\{A_n\}$, which in turn constitutes nine Hilbert spaces

$$\mathbf{h}^{(i,k)} = \mathbf{h}^{(i,k)}(\{A_n\};\Omega)$$

each of them being equipped with a reproducing kernel given by

$$\mathbf{K}_{\mathbf{h}^{(i,k)}}(\xi,\eta) = \sum_{n=0_{i,k}}^{\infty} \sum_{j=1}^{2n+1} |A_n|^{-2} \mathbf{y}_{n,j}^{(i,k)}(\xi) \otimes \mathbf{y}_{n,j}^{(i,k)}(\eta), \quad (\xi,\eta) \in \Omega^2,$$

respectively. If the sequence $\{A_n\}$ is chosen in such a way that the scalar kernel

$$\sum_{n=0}^{\infty} \sum_{j=1}^{2n+1} |A_n|^{-2} Y_{n,j}(\xi) Y_{n,j}(\eta), \quad \xi, \eta \in \Omega,$$

is expressible in terms of elementary functions, then it may happen that also the tensorial kernel $\mathbf{K}_{\mathbf{h}^{(i,k)}}(\xi, \eta)$ is available in elementary form (cf. W. FREEDEN *et al.* (1994)). For the explicit calculation the tensorial addition theorem (Theorem 14.6.1) has to be applied as essential tool.

The space $\mathbf{h}^{(i,k)}$ can be split in canonical way into the orthogonal sum $\mathbf{h}^{(i,k)} = \mathbf{h}_{0,\ldots,m}^{(i,k)} \oplus \mathbf{h}_{0,\ldots,m}^{(i,k)\perp}$, where $\mathbf{h}_{0,\ldots,m}^{(i,k)}$ consists of all tensor spherical harmonics of order less or equal m and type (i,k). By use of the reproducing kernel in $\mathbf{h}_{0,\ldots,m}^{(i,k)\perp}$ given by

$$\mathbf{K}_{\mathbf{h}_{0,\ldots,m}^{(i,k)\perp}}(\xi, \eta) = \sum_{n=m+1}^{\infty} \sum_{j=1}^{2n+1} |A_n|^{-2} \mathbf{y}_{n,j}^{(i,k)}(\xi) \otimes \mathbf{y}_{n,j}^{(i,k)}(\eta), \quad (\xi, \eta) \in \Omega^2,$$

tensor spherical splines are derivable as follows: let $X_N = \{\eta_1, \ldots, \eta_N\} \subset \Omega$, $N \geq \sum_{n=0_{i,k}}^{m} (2n+1)$, be a system of points such that the matrix

$$\left(\sum_{n=0_{i,k}}^{m} \frac{2n+1}{4\pi} P_n(\eta_l \cdot \eta_m) \right)_{l,m}$$

is regular, where the indices l, m satisfy $1 \leq l, m \leq \sum_{n=0_{i,k}}^{m} (2n+1)$. Then a *tensor spherical spline of type* (i,k) is understood to be a tensor field of the form

$$\mathbf{s}^{(i,k)} = \mathbf{p}^{(i,k)} + \sum_{l=1}^{N} a_l^{(i,k)} O_{\eta_l}^{(i,k)} \mathbf{K}_{\mathbf{h}_{0,\ldots,m}^{(i,k)\perp}}(\eta_l, \cdot),$$

where $\mathbf{p}^{(i,k)}$ is a member of the space $\mathbf{h}_{0,\ldots,m}^{(i,k)}$ of tensor spherical harmonics of order less or equal m and type (i,k), and the coefficients $a_l^{(i,k)}$ have to satisfy the linear system

$$\sum_{l=1}^{N} a_l^{(i,k)} O_{\eta_l}^{(i,k)} \mathbf{y}_{n,j}^{(i,k)}(\eta_l) = 0,$$

$n = 0_{i,k}, \ldots, m$, $j = 1, \ldots, 2n+1$. Now, interpolation problems with tensorial data can be solved similarly as described in the vectorial case (cf. Chapter 13).

Starting from an approximate identity $\{I_h\}$ in $\mathcal{L}^2(\Omega)$ corresponding to the kernel $\{K_h\} \in \mathcal{L}^2[-1,1]$, we are able to construct approximate identities in $\mathbf{l}^2(\Omega)$ as follows:

$$\mathbf{K}_h^{(1,1,1,1)}(\xi,\eta) = \mathbf{o}_\xi^{(1,1)}\mathbf{o}_\eta^{(1,1)} K_h(\xi\cdot\eta).$$

For $(i,i) \in \{(2,2),(3,3)\}$ we set

$$\mathbf{K}_h^{(i,i,i,i)}(\xi,\eta) = \frac{1}{2}\mathbf{o}_\xi^{(i,i)}\mathbf{o}_\eta^{(i,i)} K_h(\xi\cdot\eta),$$

while for $(i,k) \in \{(1,2),(1,3),(2,1),(3,1)\}$ we let

$$\mathbf{K}_h^{(i,k,i,k)}(\xi,\eta) = \mathbf{o}_\xi^{(i,k)}\mathbf{o}_\eta^{(i,k)} \int_\Omega G(\Delta^*;\xi,\zeta)K_h(\zeta\cdot\eta)\,d\omega(\zeta).$$

Finally, for $(i,k) \in \{(2,3),(3,2)\}$, we put

$$\mathbf{K}_h^{(i,k,i,k)}(\xi,\eta) = \frac{1}{2}\mathbf{o}_\xi^{(i,k)}\mathbf{o}_\eta^{(i,k)} \int_\Omega G(\Delta^*(\Delta^*-2);\xi,\zeta)K_h(\zeta\cdot\eta)\,d\omega(\zeta).$$

Summing up all kernels we get

$$\mathbf{K}_h(\xi,\eta) = \sum_{i,k=1}^{3} \mathbf{K}_h^{(i,k,i,k)}(\xi,\eta), \quad (\xi,\eta)\in\Omega^2.$$

The kernel $\mathbf{K}_h(\cdot,\cdot)$ allows the definition of the operator $I_h : \mathbf{l}^2(\Omega) \to \mathbf{l}^2(\Omega)$ as follows:

$$I_h(\mathbf{f}) = \mathbf{K}_h * \mathbf{f} = \int_\Omega \mathbf{K}_h(\cdot,\eta)\ \mathbf{f}(\eta)\,d\omega(\eta), \quad \mathbf{f}\in\mathbf{l}^2(\Omega).$$

As a consequence, we obtain for all $\mathbf{f} \in \mathbf{l}^2(\Omega)$

$$\lim_{\substack{h\to 1\\ h<1}} \|I_h(\mathbf{f}) - \mathbf{f}\|_{\mathbf{l}^2(\Omega)} = 0.$$

Thus, the tensorial analogue of an approximate identity is developed. In this fashion, tensorial Toeplitz and Gabor transforms can be obtained. Moreover, combined approximation methods become available. The details are omitted.

If $\{\Psi_\rho\}$, $\rho\in(0,\infty)$, is a spherical wavelet of order 0, we define tensorial kernels $\mathbf{\Psi}_\rho^{(i,k,i,k)}(\cdot,\cdot)$ in analogy to the definition of $\mathbf{K}_h(\cdot,\cdot)$ with Ψ_ρ in the role of K_h. This enables us to introduce the vectorial wavelet transform of order 0 as follows:

$$(WT^{(i,k)})(\mathbf{f})(\rho;\eta) = \int_\Omega \mathbf{\Psi}_\rho^{(i,k,i,k)}(\xi,\eta)\cdot\mathbf{f}(\xi)\,d\omega(\xi).$$

As in the vectorial theory, an inverse of this transform can easily be obtained on the space $\mathbf{l}^2(\Omega)$.

Scale discrete tensorial wavelets can be derived similarly to the vectorial situation. If φ_0 is a generator of a scaling function, and $\psi_0, \tilde{\psi}_0$ are admissible generators satisfying

$$\tilde{\psi}_0(x)\psi_0(x) = (\varphi_0(x/2))^2 - (\varphi_0(x))^2, \quad x\in[0,\infty)$$

we introduce the scale discrete tensorial scaling function of type (i,k), $i,k = 1,2,3$, to be

$$\mathbf{\Phi}_j^{(i,k,i,k)}(\xi,\eta) = \sum_{n=0}^{\infty}\frac{2n+1}{4\pi}\varphi_j(n)\mathbf{P}_n^{(i,k,i,k)}(\xi,\eta), \quad \xi,\eta\in\Omega.$$

Consequently, the tensorial wavelets are given by

$$\mathbf{\Psi}_j^{(i,k,i,k)}(\xi,\eta) = \sum_{n=0}^{\infty}\frac{2n+1}{4\pi}\psi_j(n)\mathbf{P}_n^{(i,k,i,k)}(\xi,\eta),$$

$$\tilde{\mathbf{\Psi}}_j^{(i,k,i,k)}(\xi,\eta) = \sum_{n=0}^{\infty}\frac{2n+1}{4\pi}\tilde{\psi}_j(n)\mathbf{P}_n^{(i,k,i,k)}(\xi,\eta), \quad \xi,\eta\in\Omega.$$

The wavelet transform and the reconstruction formula of tensorial wavelets can be easily derived as in the vectorial situation. Concrete examples are obtainable in an obvious way.

REFERENCES

AGTERBERG, F.P. (1974) *Geomathematics.* Elsevier Scientific Publishing Company.

ANDREWS, L.C. (1985) *Special Functions for Engineers and Applied Mathematicians.* Macmillian.

ARENSTORF, R. JOHNSON, D. (1979) Uniform Distribution of Integral Points on 3-Dimensional Spheres via Modular Forms. *J. Number Theory,* **11**, 218–238.

ARONSZAJN, N. (1950) Theory of Reproducing Kernels. *Trans. Am. Math. Soc.*, **68**, 337–404.

ASKEY, R. (1966) Norm Inequalities for Some Orthogonal Series. *Bull. Amer. Math. Soc.*, **72**, 808-823.

BACKUS, G.E. (1966) Potentials for Tangent Tensor Fields on Spheroids. *Archs. Ration. Mech. Anal.*, **22**, 210-252.

BACKUS, G.E. (1967) Converting Vector and Tensor Equations to Scalar Equations in Spherical Coordinates. *Geophys. J.R. Astron. Soc.*, **13**, 61-101.

BÄCKER, M. (1995) *Sphärische Wavelets and ihre numerische Realisierung.* Diploma Thesis, University of Kaiserslautern, Geomathematics Group.

BARNHILL, R., PIPER, B., RESCORLA, K. (1987) Interpolation to Arbitrary Data on a Surface. In: *Geometric Modeling* (G. Farin, ed.), SIAM, Philadelphia, PA, 281-289.

BANK, R.E., WELFERT, B.D., YSERENTANT, H. (1990) A Class of Iterative Methods for Solving Saddle Point Problems. *Numer. Math.*, **56**, 645-666.

BAYER, M. (1996) *Spherical Wavelets and Their Application to Linear Inverse Satellite Problems.* Diploma Thesis, University of Kaiserslautern, Geomathematics Group.

BEN-MENAHEM, A., SINGH, S.J. (1981) *Seismic Waves and Sources.* Springer.

Berens, H., Butzer, P.L., Pawelke S. (1968) Limitierungsverfahren mehrdimensionaler Kugelfunktionen und deren Saturationsverhalten. *Publ. Res. Inst. Math. Sci. Ser.* A **4**, 201-268.

Blatt, J., Weisskopf, V. (1952) *Theoretical Nuclear Physics.* Appendix B, Wiley.

Bochner, S. (1950) Quasi-analytic Functions, Lapalce Operator, Positive Kernels. *Ann. of Math.*, **51**, 68-91.

Bochner, S. (1954) Positive Zonale Functions on Spheres. *Proc. Nat. Acad. Sci. USA*, **40**, 1141-1147.

Brand, R. (1994) *Approximation Using Spherical Singular Integrals and Its Application to Digital Terrain Modelling.* Diploma Thesis, University of Kaiserslautern, Geomathematics Group.

Brand, R., Freeden, W., Fröhlich, J. (1996) An Adaptive Hierarchical Approximation Method on the Sphere Using Axisymmetric Locally Supported Basis Functions. *Computing*, **57**, 187-212.

Brudowsky, J.M. (1992) *Über die elementaren Darstellungen Cauchy-Navierscher Splines.* Diploma Thesis, University of Kaiserslautern, Geomathematics Group.

Brudowsky, J.M., Freeden, W., Tücks, M. (1992) Deformation Analysis of the Earth by a Constructive Interpolation Model. In: *Proceedings of the 6th European Conference on Mathematics in Industry* (F. Hodnett, ed.), 87-90.

Burridge, R. (1969) Spherically Symmetric Differential Equations, the Rotation Group, and Tensor Spherical Functions. *Proc. Camb. Phil. Soc.*, **65**, 157-175.

Calderon A.P. (1964) Intermediate Spaces and Interpolation, the Complex Method. *Studia Math.*, **24**, 113-190.

Calderon A.P., Zygmund A. (1955) On a Problem of Mihlin. *Trans. Amer. Math. Soc.*, **78**, 205-224.

Chen, K.K. (1928) On the Cesàro-summability of the Laplace's Series of Hyperspherical Functions. *Sci. Rep. Tôhoku Imp. Univ. Ser.* 1, **17**, 1073-1089.

Chui, C.K. (1991) On Overview of Wavelets. In: *Approximation Theory and Functional Analysis*, Academic Press, 47-71.

Chui, C.K. (1992) *An Introduction to Wavelets.* Academic Press, San Diego, London.

CLEMENS, A. (1997) *Geopotential Determination by Spherical Wavelets.* Diploma Thesis, University of Kaiserslautern, Geomathematics Group.

COIFMAN, R.R., MEYER, Y., QUAKE, S., WICKERHAUSER, M.V. (1990) *Signal Processing and Compresion with Wave Packets.* Yale University, New Haven. Preprint.

CORPUT, J.G. VAN DER (1935) Verteilungsfunktionen I, II. *Nederl. Akad. Wetensch. Proc. Ser. B*, **38**, 813-321, 1058-1066.

COURANT, R., HILBERT, D. (1924) *Methoden der Mathematischen Physik I,II.* Springer.

CRAVEN, P., WAHBA, G. (1979) Smoothing Noisy Data with Spline Functions. *Numer. Math.*, **31**, 377-403.

CUI, J. (1995) *Finite Pointset Methods on the Sphere and Their Application in Physical Geodesy.* Doctoral Thesis, University of Kaiserslautern, Geomathematics Group.

CUI, J., FREEDEN, W., WITTE, B. (1992) Gleichmäßige Approximation mittels sphärischer Finite-Elemente und ihre Anwendung auf die Geodäsie. *Z. f. Vermessungswes.*, **117**, 266-278.

CUI, J., FREEDEN, W. (1997) Equidistribution on the Sphere. *SIAM J. Sci. Stat. Comput.*, **18**, 595-609.

DAHLEN, F.A., SMITH, M.L. (1975) The Influence of Rotation on the Free Oscillations of the Earth. *Trans. R. Soc. A*, **279**, 583-629.

DAHLKE, S., MAASS, P. (1996) A Continuous Wavelet Transform on Tangent Bundles of Spheres. *J. Fourier Anal.*, **2**, 379-396.

DAHLKE, S., DAHMEN, W., SCHMITT, W., WEINREICH, I. (1995) Multiresolution Analysis and Wavelets on S^2 and S^3. *Numer. Funct. Anal. and Optimiz.*, **16**, 19-41.

DAUBECHIES, I. (1990) The Wavelet Transform, Time-Frequency Localization and Signal Analysis. *IEEE Trans. Inform. Theory.*, **36**, 961-1005.

DAUBECHIES, I. (1992) *Ten Lectures on Wavelets.* CBMS-NSF Series in Applied Mathematics, **61**, SIAM Publications, Philadelphia.

DAUBECHIES, I., GROSSMANN, A., MEYER, Y. (1985) Painless Nonorthogonal Expansions. *J. Math. Phys.*, **27**, 1271-1283.

DAVIS, P.J. (1963) *Interpolation and Approximation.* Blaisdell Publishing Company.

DAVIS, P.J., RABINOWITZ, P. (1975) *Methods of Numerical Integration, Computerscience and Applied Mathematics.* Academic Press.

DONGARRA, J.J., BUNCH, J.R., MOLER, C.B., STEWART, G.W. (1979) *LINPACK Users' Guide.* SIAM, Philadelphia.

DRISCOLL, J.R., HEALY, R.M. (1994) Computing Fourier Transforms and Convolutions on the 2-Sphere. *Adv. Appl. Math.*, **15**, 202-250.

DUVAL-DESTIN, M., MUSCHIETTI, M.A., TORRESANI, B. (1993) Continuous Wavelet Decompositions, Multiresolution, and Contrast Analysis. *SIAM J. Math. Anal.*, **24**, 739-755.

DYN, N. (1989) Interpolation and Approximation by Radial and Related Functions. In: *Approximation Theory VI*, Vol. 1 (C.K. Chui, L.L. Schumaker, J.D. Ward, eds.), Academic Press, 211-234.

DYN, N., LEVIN, D., RIPPA, S. (1986) Numerical Procedures for Global Surface Fitting of Scattered Data by Radial Functions. *SIAM J. Sci. Stat. Comput.*, **7**, 639-659.

DYN, N., NARCOWICH, F.J., WARD, J.D. (1996) Variational Principles and Sobolev-Type Estimates for Generalized Interpolation on a Riemannian Manifold, Center of Approximation Theory, Texas A & M University, Preprint.

EDMONDS, A.R. (1957) *Angular Momentum in Quantum Mechanics.* Princeton University Press.

ELDEN, L. (1984) A Note on the Computation of the Generalized Cross-Validation Function for Ill-Conditioned Least Squares Problems. *BIT*, **24**, 467-472.

ERDELY, A. (1938) Die Funksche Integralgleichung der Kugelflächenfunktion und ihre Übertragung auf die Überkugel. *Math. Ann.*, **115**, 456-465.

ESKIN, G.I. (1981) *Boundary Value Problems for Elliptic Pseudodifferential Equations.* Translations of Math. Monographs, Vol. 52, American Mathematical Society, Providence, RI.

EULER, H.J., GROTEN, E., HAUSCH, W., STOCK, B. (1985) Precise Geoid Computation of Mountaineous Coastlines in View of Vertical Datum Determinations. *Veröffentlichungen des Zentralinstituts für Physik der Erde*, **81** (1), 109-139.

EULER, H.J., GROTEN, E., HAUSCH, W., KLING, T. (1986) New Results Obtained for Detailed Geoid Approximations. *Boll. di Geodesia e Scienze Affini*, **4**, 429-452.

FICHTENHOLZ, G.M. (1973) *Differential und Integralrechnung I.* VEB Deutscher Verlag der Wissenschaft.

FOLEY, T. (1990) Interpolation of Scattered Data on a Spherical Domain. In: *Algorithms for Approximation II* (J.C. Mason, M.G. Cox, eds.) Chapman & Hall, 303-310.

FOLEY, T., HAGEN, H. (1994) Advances in Scattered Data Interpolation. *Surv. Math. Ind.*, **4**, 71-84.

FRANKE, R. (1983) Lecture Notes on Global Basis Function Methods for Scattered Data. *International Symposium on Surface Approximation*, University of Milano, Italy.

FRANKE, R. (1985) Thin Plate Splines With Tension. *Comp. Aided Geom. Design*, **2**, 87-95.

FRANKE, R. (1987) Recent Advances in the Approximation of Surfaces from Scattered Data. In: *Topics in Multivariate Approximation* (L.K. Chui, L.L. Schumaker, F. Utreras, eds.) Academic Press, 79-98.

FRANKE, R., BAKER, E., GOERSS, J. (1988) The Use of Observed Data for the Initial-value Problem in Numerical Weather Prediction. *Comput. Maths. Applics.*, **16**, 169-184.

FREEDEN, W. (1978) An Application of a Summation Formula to Numerical Computation of Integrals over a Sphere. *Bull. Géod.*, **52**, 165-175.

FREEDEN, W. (1979) *Über eine Klasse von Integralformeln der Mathematischen Geodäsie.* Veröff. Geod. Inst. RWTH Aachen, Heft 27.

FREEDEN, W. (1980A) On the Approximation of External Gravitational Potential with Closed Systems of (Trial) Functions. *Bull. Geod.*, **54**, 1-20.

FREEDEN, W. (1980B) On Integral Formulas of the (Unit) Sphere and Their Application to Numerical Computation of Integrals. *Computing*, **25**, 131-146.

FREEDEN, W. (1981A) On Spherical Spline Interpolation and Approximation. *Math. Meth. in the Appl. Sci.*, **3**, 551-575.

FREEDEN, W. (1981B) On Approximation by Harmonic Splines. *Manuscr. Geod.*, **6**, 193-244.

FREEDEN, W. (1982A) Multidimensional Euler Summation Formulas and Numerical Cubature. *Internat. Series Numer. Math.*, **57**, 77-88.

FREEDEN, W. (1982B) On Spline Methods in Geodetic Approximation Problems. *Math. Meth. in the Appl. Sci.*, **4**, 382-396.

FREEDEN, W. (1982C) *On the Permanence Property in Spherical Spline Interpolation.* The Ohio State University, Department of Geodetic Science and Surveying, Columbus, Ohio, OSU Report No. 341.

FREEDEN, W. (1983) *Least Squares Approximation by Linear Combinations of (Multi-)Poles.* The Ohio State University, Department of Geodetic Science and Surveying, Columbus, Ohio, OSU Report No. 344.

FREEDEN, W. (1984A) Spherical Spline Interpolation: Basic Theory and Computational Aspects. *J. Comput. Appl. Math.*, **11**, 367-375.

FREEDEN, W. (1984B) Ein Konvergenzsatz in sphärischer Spline-Interpolation. *Z. f. Vermessungswes.*, **109**, 569-576.

FREEDEN, W. (1985A) Basis Systems and Their Role in the Approximation of the Earth's Gravitational Field. *Gerlands Beitr. Geophysik*, **94**, 19-34.

FREEDEN, W. (1985B) *Computation of Spherical Harmonics and Approximation by Spherical Harmonic Expansions.* The Ohio State University, Department of Geodetic Science and Surveying, Columbus, Ohio, OSU Report No. 362.

FREEDEN, W. (1985C) Ein Beitrag zur numerischen Berechnung des Dirichletschen Außenraumproblems der Potentialtheorie mittels harmonischer Splines. *Z. Angew. Math. Mech.*, **65**, 260-262.

FREEDEN, W. (1987A) Harmonic Splines for Solving Boundary Value Problems of Potential Theory. In: *Algorithms for Approximation* (J.C. Mason, M.G. Cox, eds.), The Institute of Mathematics and its Applications, Conference Series, vol. 10, Clarendon Press, Oxford, 507-529.

FREEDEN, W. (1987B) A Spline Interpolation Method for Solving Boundary Value Problems of Potential Theory from Discretely Given Data. *Numer. Meth. of Partial Diff. Equations*, **3**, 375-398.

FREEDEN, W. (1987C) Metaharmonic Splines for Solving the Exterior Dirichlet Problem of the Helmholtz Equation. In: *Topics in Multivariate Approximation Theory* (C.K. Chui, L.L. Schumaker, F. Utreras, eds.), *Academic Press*, 99-110.

FREEDEN, W. (1990A) Some Applications of Approximation Theory to the First Boundary Value Problem of Elastostatics. *Internat. Series Numer. Math.*, **94**, 121-129.

FREEDEN, W. (1990B) Spherical Spline Approximation and its Application in Physical Geodesy. In: *Geophysical Data Inversion Methods and Applications* (A. Vogel, C.O. Ofeagbu, R. Gorenflo, B. Ursin, eds.), *Vieweg Publication*, 79-104.

FREEDEN, W., FLECK, J. (1987) Numerical Integration by Means of Adapted Euler Summation Formulas. *Numer. Math.*, **51**, 37-64.

FREEDEN, W., GERVENS, T. (1988) Vector Spherical Spline Interpolation. *Internat. Series Numeric. Math.*, **90**, 157-171.

FREEDEN, W., GERVENS, T., MASON, J.C. (1990) A Minimum Norm Interpolation Method for Determining the Displacement Field of a Homogeneous Isotropic Elastic Body from Discrete Data. *IMA J. Appl. Math.*, **44**, 55-76.

FREEDEN, W., GERVENS, T. (1991) Vector Spherical Spline Interpolation (Basic Theory and Computational Aspects). *Math. Meth. in the Appl. Sci.*, **16**, 151-183.

FREEDEN, W., GERVENS, T., SCHREINER, M. (1994) Tensor Spherical Harmonics and Tensor Spherical Splines. *Manuscr. Geod.*, **19**., 70-100.

FREEDEN, W., GROTEN, E., SCHREINER, M., SÖHNE, W., TÜCKS, M. (1996A) Deformation Analysis Using Navier Spline Interpolation (with an Application to the Lake Blåsjø Area), *Allg. Vermessungs-Nachrichten*, **96**, 3, 120-146.

FREEDEN, W., HERMANN, P. (1986) Uniform Approximation by Spherical Spline Interpolation. *Math. Z.*, **193**, 265-275.

FREEDEN, W., MASON, J.C. (1990) *Uniform Piecewise Approximation on the Sphere.* In: *Algorithms for Approximation II* (J.C. Mason, M.G. Cox, eds.), Chapman & Hall, 320-333.

FREEDEN, W., REUTER, R. (1982) Remainder Terms in Numerical Integration Formulas of the Sphere. *Internat. Series Numeric. Math.*, **61**, 151-170.

FREEDEN, W., REUTER, R. (1984) Exact Computation of Spherical Harmonics. *Computing*, **32**, 365-378.

FREEDEN, W., REUTER, R. (1990) An Efficient Algorithm for the Generation of Homogeneous Harmonic Polynomials. In: *Scientific Software Systems* (J.C. Mason, M.G. Cox, eds.), Chapman & Hall, 166-180.

FREEDEN, W., RUMMEL, R., SANSÒ, F., SÜNKEL, H. (1996B) Gravitational Field Determination. In: *Progress in Industrial Mathematics At the European Consortium of Mathematics in Industry '94* (H. Neunzert, ed.), Wiley-Teubner, 413-453.

FREEDEN, W., SCHNEIDER, F. (1997) Wavelet Approximation on Closed Surfaces and Their Application to Boundary-value Problems of Potential Theory. *Math. Meth. in the Appl. Sci.* (accepted for publication).

FREEDEN, W., SCHNEIDER, F., SCHREINER, M. (1997) Gradiometry - An Inverse Problem in Modern Satellite Geodesy. *GAMM–SIAM Symposium on Inverse Problems: Geophysical Applications*, 179-239.

FREEDEN, W., SCHREINER, M. (1994) Satellite Gradiometry–from Mathematical and Numerical Point of View. *A Major STEP for Geodesy*, Report 1994 of the STEP Geodesy Working Group, 35-44.

FREEDEN, W., SCHREINER, M. (1995A) Non-orthogonal Expansions on the Sphere. *Math. Meth. in the Appl. Sci.*, **18**, 83-120.

FREEDEN, W., SCHREINER, M. (1995B) New Wavelet Methods for Approximating Harmonic Functions. In: *Geodetic Theory Today* (F. Sansò, ed.), International Association of Geodesy Symposia, **114**, Springer, 112-121.

FREEDEN, W., SCHREINER, M. (1995C) Satellite Gradiometry - A New Approach In: *Geodetic Theory Today* (F. Sansò, ed.), International Association of Geodesy Symposia, **114**, Springer, 122-139.

FREEDEN, W., SCHREINER, M. (1997) Orthogonal and Non-orthogonal Multiresolution Analysis, Scale Discrete and Exact Fully Discrete Wavelet Transform on the Sphere. *Constr. Approx.* (accepted for publication).

FREEDEN, W., SCHREINER, M., FRANKE, R. (1996C) A Survey on Spherical Spline Approximation. *Surv. Math. Ind.*, Springer (accepted for publication).

FREEDEN, W., WINDHEUSER, U. (1995) Earth's Gravitational Potential and Its MRA Approximation by Harmonic Singular Integrals. *Z. Angew. Math. Mech.*, **75**, SII, 633-634.

FREEDEN, W., WINDHEUSER, U. (1996) Spherical Wavelet Transform and Its Discretization. *Adv. Comput. Math.*, **5**, 51-94.

FREEDEN, W., WINDHEUSER, U. (1997) Combined Spherical Harmonic and Wavelet Expansion — A Future Concept in Earth's Gravitational Determination. *Appl. Comput. Harm. Anal.*, **4**, 1-37.

FREEDEN, W., WITTE, B. (1982) A Combined (Spline-) Interpolation and Smoothing Method for the Determination of the External Gravitational Potential from Heterogeneous Data. *Bull. Géod.*, **56**, 53-62.

FUNK, H. (1916) Beiträge zur Theorie der Kugelfunktion. *Math. Ann.*, **77**, 136-152.

GABOR D. (1946) Theory of Communications. *J. Inst. Elec. Eng.* (London), **93**, 429-457.

GAPOSCHKIN, E.M. (1980) Averaging on the Surface of a Sphere. *J. Geophysical Research*, Vol. 85, No. B6.

GAUSS C.F. (1839) Allgemeine Theorie des Erdmagnetismus. *Werke*, **5**, 121-193.

GEL'FAND, I.M., SHAPIRO, Z.Y. (1956) Representations of the Group of Rotation in Three-dimensional Space and their Applications. *Am. Math. Soc. Transl.*, **2**, 207.

GERVENS, T. (1989) Vektorkugelfunktionen mit Anwendungen in der Theorie der elastischen Verformungen für die Kugel. Doctoral Thesis, RWTH Aachen.

GIRARD, D. (1989) A Fast 'Monte-Carlo Cross-Validation' Procedure for Large Least Squares Problems With Noisy Data. *Numer. Math.*, **56**, 1-23.

GLOCKNER, O. (1997) *Spherical Panel Clustering and Its Numerical Aspects in Spline and Wavelet Computation.* Diploma Thesis, University of Kaiserslautern, Geomathematics Group.

GÖTTELMANN, J. (1996) Locally Supported Wavelets on the Sphere. *Preprint*, No. 14, Dept. of Math., University of Mainz.

GRADSHTEYN, I.S., RYZHIK, I.W. (1980) *Table of Integrals, Series and Products.* Academic Press.

GRAFAREND, E.W. (1982) Six Lectures on Geodesy and Global Geodynamics. *Mitt. Geod. Inst. TU Graz*, Folge 41 (H. Moritz, H. Sünkel, eds.), 531-685.

GRAFAREND, E.W. (1986) Three-Dimensional Deformation Analysis: Global Vector Spherical Harmonic and Local Finite Element Representation. *Tectonophysics*, **130**, 337-359

GRAFAREND, E.W., KRARUP, T., SYFFUS, R. (1996) An Algorithm for the Inverse of a Multivariate Homogeneous Polynomial of Degree n. *J. Geodesy*, **70**, 276-286.

GRAVEN, P., WAHBA, G. (1979) Smoothing Noisy Data with Spline Functions Estimating the Correct Degree of Smoothing by the Method of Generalized Cross Validation. *Numer. Math.*, **31**, 377-403.

GREVILLE, T.N.E. (1969) Introduction to Spline Functions. In: *Theory and Applications of Spline Functions* (T.N.E. Greville, ed.), Academic Press, 1-35

GRONWALL, T. (1914) On the Degree of Convergence of Laplace Series. *Trans. Am. Math. Soc.*, **15**, 1-30.

GROSSMANN, A., MORLET J. (1984) Decomposition of Hardy Functions into Square Integrable Wavelets of Constant Shape. *SIAM J. Math. Anal.*, **15**, (4), 723-736.

GROTEN, E. (1979) *Geodesy and the Earth's Gravity Field* I, II. Dümmler.

GU, C. (1989) RKPACK and its Applications: Fitting Smoothing Spline Models, *TR* 857, Dept. of Statistics, University of Wisconsin, Madison.

GU, C. WAHBA, G. (1991) Minimizing GCV/GML Scores with Multiple Smoothing Parameters via the Newton Method, *SIAM J. Sci. Statist. Comput.*, **12**, 383-398.

GURTIN, M.E. (1971) *Theory of Elasticity.* Handbuch der Physik, Vol. VI, 2nd ed. Springer.

GYSEN, H. v., MERRY, L.C. (1990) Towards a Cross-Validated Spherical Spline Geoid for the South Western Cape, South Africa. In: *Sea Surface Topography and the Geoid* (I.I. Mueller, ed.), Internat. Association of Geodesy Symposia, **104**, 53-60, Springer.

HARDY, R.L. (1971) Multiquadric Equations of Topography and Other Irregular Surfaces. *J. Geophysical Res.*, **76**, 1905-1915.

HECKE, E. (1918) Über orthogonal-invariante Integralgleichungen. *Math. Ann.*, **78**, 398-404.

HEIL, C.E., WALNUT D.F. (1989) Continuous and Discrete Wavelet Transform. *SIAM Review*, **31**, 628-666.

HEINE, E. (1878) *Handbuch der Kugelfunktionen*, G. Reimer.

HEISKANEN, W.A., MORITZ, H. (1967) *Physical Geodesy.* W.H. Freeman and Company.

HILBERT, D. (1912) *Grundzüge einer allgemeinen Theorie der linearen Integralgleichung.* Teubner.

HLAWKA, E. (1981) Gleichverteilung auf Produkten von Sphären. *J. Reine Angew. Math.*, **330**, 1-45.

HOBSON, E.W. (1955) *The Theory of Spherical and Ellipsoidal Harmonics.* (Reprint) Chelsea Publishing Company.

HOCHSTADT, H. (1971) *The Functions of Mathematical Physics.* Wiley-Intersciences.

HOLSCHNEIDER, M. (1996) Continuous Wavelet Transforms on the Sphere. *J. Math. Phys.*, **37** (8), 4156-4165.

HÖRMANDER, L. (1976) The Boundary Problems of Physical Geodesy. *Arch. Rational Mech. Anal.*, **62**, 1-52.

HOSCHEK, J., LASSER, D. (1992) *Grundlagen der geometrischen Datenverarbeitung*, Teubner.

HOSCHEK, J., SEEMANN, G. (1992) Spherical Splines. *Math. Modelling Numer. Anal.*, **26**, 1-22.

HUTCHINSON, M.F., DE HOOG, F.R. (1987) Smoothing Noisy Data With Spline Functions. *Numer. Math.*, **50**, 311-319.

JAMES, R.W. (1976) New Tensor Spherical Harmonics for Application to the Partial Differential Equations of Mathematical Physics. *Phil. Trans. R. Soc. A*, **281**, 197-221.

JONES, M.N. (1980) Tensor Spherical Harmonics and Their Application to the Navier Equation of Dynamic Elasticity. *Gerlands Beitr. Geophysik*, **89**, 135-148.

KATSAMBALOS, K. (1979) *The Effect of the Smoothing Operator on Potential Coefficients Determination.* The Ohio State University, Department of Geodetic Science and Surveying, Columbus, Ohio, OSU Report No. 287.

KELLOGG, O.D. (1929) *Foundations of Potential Theory.* Frederick Ungar Publishing Company.

KLING, T., BECKER, M. EULER, H.-J., GROTEN, E. (1987) *Studien zur detaillierten Geoidberechnung.* Deutsche Geodätische Kommission bei der Bayrischen Akademie der Wissenschaften, Series B, Angewandte Geodäsie, No. 285.

KOGBETLIANTZ, E., (1924) Recherches sur la sommabilité des séries ultrasphériques par la méthode des moyennes arithmétiques. *J. Math. Pures Appl.*, **9** (3), 107-187.

KUIPERS, L., NIEDERREITER H. (1974) *Uniform Distribution of Sequences.* John Wiley & Sons.

LAGALLY, M., FRANZ, W. (1964) *Vorlesung über Vektorrechnung.* Akad. Verlagsgesellschaft.

LANSER, M. (1997) *Band–limited Wavelet Approximation on the Sphere — Theoretical and Numerical Aspects.* Diploma Thesis, University of Kaiserslautern, Geomathematics Group.

LAPLACE, P.S. DE (1785) Théorie des attractions des sphéroides et de la figure des planètes. *Mèm de l'Acad.*, Paris.

LAWSON, C. (1984) C^1-Surface Interpolation for Data on a Sphere. *Rocky Mountain J. Math.*, **14**, 177-202.

LEBEDEV, N.N. (1973) *Spezielle Funktionen und ihre Anwendungen.* Bibliographisches Institut.

LEGENDRE, A.M. (1785) Recherches sur l'attraction des sphèroides homogènes. *Mèm. math. phys. près. à l'Acad. Aci. par. divers savantes*, **10**, 411-434.

LENSE, J. (1954) *Kugelfunktionen.* Leipzig Akad. Verlagsgesellschaft. Mathematik und ihre Anwendungen in Physik und Technik, Reihe A, Bd. 23.

LORENTZ, G.G. (1966) *Approximation of Function.* Holt, Rinehart and Winston.

LOUIS, A.K., MAAS, P., RIEDER, A. (1994) *Wavelets*, Teubner Studienbücher.

MADYCH, W., NELSON, S. (1988) Multivariate Interpolation and Conditionally Positive Definite Functions. *Approx. Theory Appl.*, **4**, 77-89.

MAGNUS, W., OBERHETTINGER, F., SONI, R.P. (1966) *Formulas and Theorems for the Special Functions of Mathematical Physics.* Die Grundlehren der mathematischen Wissenschaften in Einzeldarstellungen, Bd. 52, Springer.

MALYSEV, A.V. (1957) The Distribution of Integer Points in a Four-Dimensional Sphere. *Doklady Akad. Nauk* SSSR, **114**, 25-28.

MARR, D. (1982) *Vision.* Freeman, San Francisco.

MARTIN, A. (1991) Elastic Scattering and Orthogonal Polynomials. In: *Orthogonal Polynomials and Their Applications* (C. Brezinski, L. Gori, A. Ronveaux, eds.), Baltzer Scientific Publishing, 131-138.

MEISSL, P.A. (1971) *A Study of Covariance Functions Related to the Earth's Disturbing Potential.* The Ohio State University, Department of Geodetic Science and Surveying, Columbus, OSU Report No. 152.

MEISSL, P.A. (1981) The Use of Finite Elements in Physical Geodesy. The Ohio State University, Department of Geodetic Science and Surveying, Columbus, OSU Report No. 313.

MEYER, Y. (1993) *Wavelets. Algorithms & Applications.* SIAM Philadelphia.

MICCHELLI, C. (1986) Interpolation of Scattered Data: Distance Matrices and Conditionally Positive Definite Functions. *Constr. Approx.*, **2**, 11-22.

MORSE, P.M., FESHBACH, H. (1953) *Methods of Theoretical Physics.* McGraw-Hill, New York.

MOSES, H.E. (1974) The Use of Vector Spherical Harmonics in Global Meteorology and Aeronomy. *J. Atmospheric Sci.*, **31**, 1490-1500.

MÜLLER, C. (1952) Über die ganzen Lösungen der Wellengleichung. *Math. Ann.*, **124**, 235-264.

MÜLLER, C. (1966) *Spherical Harmonics.* Lecture Notes in Mathematics, **17**, Springer.

MÜLLER, C. (1969) *Foundation of the Mathematical Theory of Electromagnetic Waves.* Die Grundlehren der mathematischen Wissenschaften in Einzeldarstellungen, Bd. 155, Springer.

NARCOWICH, F.J., WARD, J.D. (1996) Nonstationary Wavelets on the m-Sphere for Scattered Data. *Appl. Comp. Harm. Anal.*, **3**, 324-336.

NEUMANN, F. (1887) *Vorlesungen über die Theorie des Potentials und der Kugelfunktionen*, Teubner.

NIEDERREITER, H. (1992) *Random Number Generation and Quasi-Monte Carlo Method*, SIAM, Philadelphia.

NIELSON, G., RAMARAJ, R. (1987) Interpolation Over a Sphere. *Computer Aided Geometric Design*, **4**, 41-57.

PEANO, G. (1913) Resto nelle formule di quadratura expresso con un integrale definito. *Atti. Accad. Naz. Lincei. Rend.*, **22**, 562-569.

PELLINEN, L.P. (1966) A Method for Expanding the Gravity Potential of the Earth in Spherical Harmonics. Translation ACIC TC 1282, NTIS: AD-661819, Moscow.

PERRIN, F., PERNIER, J., BERTRAND, O., ECHALLIER, J.F. (1989) Spherical Spline for Scalp Potential and Current Density Mapping. *Electroencephalogr. Clin. Neurophysiol.*, **72**, 184-187.

POLLACK, H.N. (1973) Spherical Harmonic Representation of the Gravitational Potential of a Point Mass, a Spherical Cap, and a Spherical Rectangle. *J. Geophys. Res.*, **78**, 1760-1768.

POLLARD, H. (1948) The Mean Convergence of Orthogonal Series, II. *Trans. Amer. Math. Soc.*, **63**, 355-367.

POTTMANN, H. ECK, M.: (1990) Modified Multiquadratic Methods for Scattered Data Interpolation over a Sphere. *Computer Aided Geometric Design*, **7**, 313-321.

POTTS, D., TASCHE, M. (1995) *Interpolatory Wavelets on the Sphere.* In: *Approximation Theory VIII* (C.K. Chui, L.L. Schumaker, eds.), 335-342.

POWELL, M.J.D. (1992) Wavelets Subdivision Algorithms and Radial Basis Functions. In: *Adv. in Numer. Anal., Vol. II*, (W. Light, ed.), Clarendon Press, 105-210 .

PRUDNIKOV, A.P., BRYCHKOV, Y.A., MARICHEV, O.I. (1986) *Integrals and Series.* Vol. 1,2,3; Gordon and Breach.

RAGOZIN, D.L. (1972) Uniform Convergence of Spherical Harmonic Expansion. *Math. Ann.*, **195**, 87-94.

RAGOZIN, D.L. (1983) Error Bounds for Erivative Estimates Based on Spline Smoothing of Exact or Noisy Data. *J. Approx. Theory*, **37**, 335-355.

RAPP, R.H., WANG, M., PAVLIS, N. (1991) *The Ohio State 1991 Geopotential and Sea Surface Topography Harmonic Coefficient Model.* The Ohio State University, Department of Geodetic Science and Surveying, Columbus, OSU Report No. 410.

REIMER, M. (1990) *Constructive Theory of Multivariate Functions.* BI-Wissenschaftsverlag.

REINSCH, C.H. (1975) Smoothing by Spline Functions. *Numer. Math.*, **10**, 177-183.

RENKA, R. (1984) Interpolation of Data on the Surface of a Sphere. *ACM Trans. Math. Soft.*, **11**, 417-482.

REUTER, R. (1982) *Über Integralformeln der Einheitsphäre und harmonische Splinefunktionen.* Veröff. Geod. Inst. RWTH Aachen, Heft 33.

RUBIN, B. (1994) Continuous Wavelet Transform on a Sphere. Hebrew University, Jerusalem, Preprint.

RUDIN, W. (1950) Uniqueness Theory for Laplace Series. *Trans. Amer. Math. Soc.*, **68**, 287-303.

RUMMEL, R. (1979) Determination of the Short-wavelength Components of the Gravity Field from Satellite-to-Satellite Tracking or Satellite Gradiometry. *Manuscr. Geod.*, **4**, 107-148.

RUMMEL, R. (1986) Satellite Gradiometry. In: *Lecture Notes in Earth Sciences 7, Mathematical and Numerical Techniques in Physical Geodesy* (H. Sünkel, ed.), Springer, 318-363.

RUMMEL, R., VAN GELDEREN, M. (1992) Spectral Analysis of the Full Gravity Tensor. *Geophys. J. Int.*, **111**, 159-169.

RUMMEL, R., VAN GELDEREN, M. (1995) Meissl Scheme–Spectral Characteristics of Physical Geodesy. *Manuscr. Geod.*, **20**, 379-385.

RVACHEV, V.A. (1990) Compactly Supported Solutions of Functional-Differential Equations and Their Applications. *Russian Math. Surveys*, **45**, No. 1, 87-120.

SAFF, E.B., KUIJLAARS, A. (1997) Distributing Many Points on a Sphere. *Math. Intelligencer*, **19**, 5-11.

SARD, A. (1963) *Linear Approximation.* Meth. Surveys, No. 9, American Mathematical Society Providence, RI.

SCHAFFELD, H.J. (1988) *Eine Finite-Elemente-Methode und ihre Anwendung zur Erstellung von Digitalen Geländemodellen.* Veröff. Geod. Inst. RWTH Aachen, Heft 42.

SCHNEIDER, F. (1994) *Spherical Splines With Local Supports and Their Application to Scattered Data Distribution.* Diploma Thesis, University of Kaiserslautern, Geomathematics Group.

SCHNEIDER, F. (1996A) A Hierarchical Spherical Spline Interpolation Method and Its Application to a Given Set of Air Pressure Data, *Student Proccedings of the 8th Conference of the European Consortium for Mathematics in Industry, ECMI 94*, (H. Neunzert, ed.), 219-233.

SCHNEIDER, F. (1996B) The Solution of Linear Inverse Problems in Satellite Geodesy by Means of Spherical Spline Approximation, *J. Geodesy*, **71**, 2-15.

SCHNEIDER, F. (1997) *Inverse Problems in Satellite Geodesy and Their Approximate Solution by Splines and Wavelets.* Doctoral Thesis, University of Kaiserslautern, Geomathematics Group, Shaker.

SCHOENBERG, I.J. (1938) Metric Spaces and Completely Monotonic Functions. *Ann. Math.*, **39**, 811-841.

SCHOENBERG, I.J. (1942) Positive Definite Functions on Spheres. *Duke Math. J.*, **9**, 98-108.

SCHOENBERG, I.J. (1964A) On Trigonometric Spline Interpolation. *J. Math. Mech.*, **13**, 795-825.

SCHOENBERG, I.J. (1964B) Spline Functions and the Problem of Graduation. *Proc. Nat. Acad. Sci.* USA, **52**, 942-950.

SCHREINER, M. (1994) *Tensor Spherical Harmonics and Their Application in Satellite Gradiometry.* Doctoral Thesis, University of Kaiserslautern, Geomathematics Group.

SCHREINER, M. (1995) The Combination of Spline Approximation and Numerical Filtering in Satellite Gradiometry. *Z. Angew. Math. Mech.*, **75**, 665-666.

SCHREINER, M. (1996A) Uniqueness Problems in Satellite Gradiometry. In: *Progress in Industrial Mathematics At the European Consortium of Mathematics in Industry '94* (H. Neunzert, ed.), Wiley-Teubner, 480-486.

SCHREINER, M. (1996B) A Pyramid Scheme for Spherical Wavelets. *AGTM Report*, No. 170, Geomathematics Group, Kaiserslautern.

SCHREINER, M. (1997A) Locally Supported Kernels for Spherical Spline Interpolation. *J. Approx. Theory*, **89**, 172-194.

SCHREINER, M. (1997B) On a New Condition for Strictly Positive Definite Functions on Spheres. *Proc. Am. Math. Soc.*, **125**, 531-539.

SCHRÖDER, P., SWELDENS, W. (1995) Spherical Wavelets. Preprint.

SEELEY, R.T., (1966) Spherical Harmonics. *Amer. Math. Monthly*, **73**, 115-121.

SHAPIRO, V.L. (1961) Topics in Fourier and Geometric Analysis. *Mem. Amer. Math., Soc.*, **39**.

SHURE, L., PARKER, R.L., BACKUS, G.E. (1982) Harmonic Splines for Geomagnetic Modelling. *Phys. Earth Planet. Inter.*, **28**, 215-229.

SJÖBERG, L. (1980) A Recurrence Relation for the β_n-Function. *Bull. Géod.*, **54**, 69-72.

SNEDDON, I.N. (1956) *Spezielle Funktionen der Mathematischen Physik und Chemie.* Bibliographisches Institut.

STEIN, E.M., WEISS, G. (1971) *Introduction to Fourier Analysis on Euclidean Spaces.* Princeton University Press.

STOER, J., BULIRSCH, R. (1980) *Introduction to Numerical Analysis.* Springer, New York.

STOKES, G.G. (1849) On the Variation of Gravity on the Surface of the Earth. *Trans. Cambridge Phil. Soc.*, **8**, 672-695.

SVENSSON, S.L. (1983) Pseudodifferential Operators—a New Approach to the Boundary Value Problems of Physical Geodesy. *Manuscr. Geod.*, **8**, 1-40.

SVENSSON, S.L. (1984) Finite Elements on the Sphere. *J. Approx. Theory*, **40**, 246-260.

SZEGÖ, G. (1959) *Orthogonal Polynomials.* Amer. Math. Soc. Colloq. Publ., vol. 23, American Mathematical Society, Providence, RI.

TAIJERON, H.J., GIBSON, A.G., CHANDLER, C. (1994) Spline Interpolation and Smoothing on Hyperspheres. *SIAM J. Sci. Comput.*, **15** (5), 1111-1125.

TERRAS, A.A. (1985) *Harmonic Analysis on Symmetric Spaces I*, Springer.

THORNE, K.S. (1980) Multipole Expansions of Gravitational Radiation. *Rev. Mod. Phys.*, **52** (2), (I), 299-339.

TORRESANI B. (1993) Phase Space Decomposition: Local Fourier Analysis on Spheres. Preprint.

TRICOMI, F.G. (1970) *Vorlesungen über Orthogonalreihen.* Die Grundlehren der mathematischen Wissenschaften in Einzeldarstellungen, Bd. 76, Springer

TÜCKS, M. (1996) *Navier-Splines und Ihre Anwendung in der Deformationsanalyse.* Doctoral Thesis, University of Kaiserslautern, Geomathematics Group, Shaker.

UTRERAS, F.I. (1981) Optimal Smoothing of Noisy Data Using Spline Functions. *SIAM J. Sci. Stat. Comput.*, **2**, 349-362.

UTRERAS, F.I. (1990) Recent Results on Multivariate Smoothing Splines. *Internat. Series Numer. Math.*, **94**, 299-312.

VILENKIN N.J. (1968) *Special Functions and the Theory of Group Representations.* Am. Math. Soc. Transl., **22**.

WAHBA, G. (1975) Smoothing Noisy Data by Spline Functions. *Numer. Math.* **24**, 383-393.

WAHBA, G. (1981) Spline Interpolation and Smoothing on the Sphere. *SIAM J. Sci. Stat. Comput.* **2**, 5-16. Also errata: *SIAM J. Sci. Stat. Comput.*, **3**, 385-386.

WAHBA, G. (1982) Vector Splines on the Sphere with Application to the Estimation of Vorticity and Divergence from Discrete Noisy Data. *Internat. Series Numer. Math.*, **61**, 407-429.

WAHBA, G. (1984) Surface Fitting With Scattered Noisy Data on Euclidean d-Space and on the Sphere. *Rocky Montain J. Math.*, **14**, 281-299.

WAHBA, G. (1990) *Spline Models for Observational Data.* CBMS-NSF Regional Conference Series in Applied Mathematics, **59**, SIAM, Philadelphia.

WAHBA, G., WENDELBERGER, J. (1980) Some New Mathematical Methods for Variational Objective Analysis Using Splines and Cross Validation. *Monthly Weather Rev.*, **108**, 1122-1143.

WEYL, H. (1916) Über die Gleichverteilung von Zahlen mod. Eins. *Math. Ann.*, **77**, 313-352.

WHITTAKER, E. (1923) On a New Method of Graduation. *Proc. Edin. Math. Soc.*, **41**, 63-75.

WINDHEUSER, U. (1995) *Sphärische Wavelets: Theorie und Anwendung in der Physikalischen Geodäsie.* Doctoral Thesis, University of Kaiserslautern, Geomathematics Group.

WINDHEUSER, U. (1996) Spherical Wavelet Packets and Its Application in Physical Geodesy. In: *Progress in Industrial Mathematics At the European Consortium of Mathematics in Industry '94* (H. Neunzert, ed.), Wiley-Teubner, 487-493.

XU, Y., CHENEY, E.W. (1992) Strictly Positive Definite Functions on Sphere. *Proc. Am. Math. Soc.*, **116**, 997-981.

YAMABE, H. (1950) On an Extension of Helly's Theorem. *Osaka Math. J.*, **2**, 15-17.

YOSIDA, K. (1980) *Functional Analysis.* 6th edition, Springer.

ZERILLI, F.J. (1970) Tensor Harmonics in Canonical Form for Gravitational Radiation and Other Applications. *J. Math. Phys.*, **11**, 2203-2208.

INDEX